Toyota Camry and Lexus ES 300 Automotive Repair Manual

by Robert Maddox, Jay Storer
and John H Haynes
Member of the Guild of Motoring Writers

Models covered:
All Toyota Camry, Avalon and Camry Solara
and Lexus ES 300 models
1997 through 2001

(92007-8U11)   ABCDE FG

Haynes Publishing Group
Sparkford Nr Yeovil
Somerset BA22 7JJ England

Haynes North America, Inc
859 Lawrence Drive
Newbury Park
California 91320 USA
www.haynes.com

Acknowledgements

We are grateful for the help and cooperation of the Toyota Motor Corporation for their assistance with technical information and certain illustrations.

© Haynes North America, Inc. 1999, 2000

With permission from J.H. Haynes & Co. Ltd.

A book in the Haynes Automotive Repair Manual Series

Printed in Malaysia

All rights reserved. No part of this book may be reproduced or transmitted in any form or by any means, electronic or mechanical, including photocopying, recording or by any information storage or retrieval system, without permission in writing from the copyright holder.

ISBN-10: 1-56392-404-8

ISBN-13: 978-1-56392-404-0

Library of Congress Card Number 00-110005

While every attempt is made to ensure that the information in this manual is correct, no liability can be accepted by the authors or publishers for loss, damage or injury caused by any errors in, or omissions from, the information given.

Contents

Introductory pages
About this manual	0-5
Introduction to the Toyota Camry, Avalon, Camry Solara and Lexus ES 300	0-5
Vehicle identification numbers	0-6
Stereo anti-theft system precaution	0-6
Buying parts	0-7
Maintenance techniques, tools and working facilities	0-7
Booster battery (jump) starting	0-15
Jacking and towing	0-15
Automotive chemicals and lubricants	0-16
Conversion factors	0-17
Fraction/decimal/millimeter equivalents	0-18
Safety first!	0-19
Troubleshooting	0-20

Chapter 1
Tune-up and routine maintenance — 1-1

Chapter 2 Part A
Four-cylinder engine — 2A-1

Chapter 2 Part B
V6 engine — 2B-1

Chapter 2 Part C
General engine overhaul procedures — 2C-1

Chapter 3
Cooling, heating and air conditioning systems — 3-1

Chapter 4
Fuel and exhaust systems — 4-1

Chapter 5
Engine electrical systems — 5-1

Chapter 6
Emissions and engine control systems — 6-1

Chapter 7 Part A
Manual transaxle — 7A-1

Chapter 7 Part B
Automatic transaxle — 7B-1

Chapter 8
Clutch and driveaxles — 8-1

Chapter 9
Brakes — 9-1

Chapter 10
Suspension and steering systems — 10-1

Chapter 11
Body — 11-1

Chapter 12
Chassis electrical system — 12-1

Wiring diagrams — 12-24

Index — IND-1

Haynes mechanic, author and photographer with 1997 Avalon

About this manual

Its purpose

The purpose of this manual is to help you get the best value from your vehicle. It can do so in several ways. It can help you decide what work must be done, even if you choose to have it done by a dealer service department or a repair shop; it provides information and procedures for routine maintenance and servicing; and it offers diagnostic and repair procedures to follow when trouble occurs.

We hope you use the manual to tackle the work yourself. For many simpler jobs, doing it yourself may be quicker than arranging an appointment to get the vehicle into a shop and making the trips to leave it and pick it up. More importantly, a lot of money can be saved by avoiding the expense the shop must pass on to you to cover its labor and overhead costs. An added benefit is the sense of satisfaction and accomplishment that you feel after doing the job yourself.

Using the manual

The manual is divided into Chapters. Each Chapter is divided into numbered Sections, which are headed in bold type between horizontal lines. Each Section consists of consecutively numbered paragraphs.

At the beginning of each numbered Section you will be referred to any illustrations which apply to the procedures in that Section. The reference numbers used in illustration captions pinpoint the pertinent Section and the Step within that Section. That is, illustration 3.2 means the illustration refers to Section 3 and Step (or paragraph) 2 within that Section.

Procedures, once described in the text, are not normally repeated. When it's necessary to refer to another Chapter, the reference will be given as Chapter and Section number. Cross references given without use of the word "Chapter" apply to Sections and/or paragraphs in the same Chapter. For example, "see Section 8" means in the same Chapter.

References to the left or right side of the vehicle assume you are sitting in the driver's seat, facing forward.

Even though we have prepared this manual with extreme care, neither the publisher nor the author can accept responsibility for any errors in, or omissions from, the information given.

NOTE

A **Note** provides information necessary to properly complete a procedure or information which will make the procedure easier to understand.

CAUTION

A **Caution** provides a special procedure or special steps which must be taken while completing the procedure where the Caution is found. Not heeding a Caution can result in damage to the assembly being worked on.

WARNING

A **Warning** provides a special procedure or special steps which must be taken while completing the procedure where the Warning is found. Not heeding a Warning can result in personal injury.

Introduction to the Toyota Camry, Avalon, Camry Solara and Lexus ES 300

This manual covers the Toyota Camry and Lexus ES 300, Camry Solara and Avalon models. The Camry, Lexus ES 300 and Avalon are four-door sedans, while the Camry Solara model is available in 2-door coupe or convertible body styles.

The transversely mounted inline four-cylinder and V6 engines used in these models are equipped with electronic port fuel injection.

The engine drives the front wheels through either a five-speed manual or a four-speed automatic transaxle via independent driveaxles.

Independent suspension, featuring coil spring/strut damper units, is used on all four wheels. The power-assisted rack and pinion steering unit is mounted behind the engine.

The brakes are disc-type at the front with either drum or discs at the rear, depending on model, with power assist standard. Anti-lock brakes (ABS) are available on all models.

Vehicle identification numbers

Modifications are a continuing and unpublicized process in vehicle manufacturing. Since spare parts manuals and lists are compiled on a numerical basis, the individual vehicle numbers are essential to correctly identify the component required.

Vehicle Identification Number (VIN)

This very important identification number is stamped on a plate attached to the dashboard inside the windshield on the driver's side of the vehicle (see illustration). It can also be found on the certification label located on the driver's side door post. The VIN also appears on the Vehicle Certificate of Title and Registration. It contains information such as where and when the vehicle was manufactured, the model year and the body style.

Certification label

The certification label is attached to the end of the driver's door post (see illustration). The plate contains the name of the manufacturer, the month and year of production, the Gross Vehicle Weight Rating (GVWR), the Gross Axle Weight Rating (GAWR) and the certification statement.

Engine identification numbers

The engine serial number can be found in a variety of locations, depending on engine type (see illustrations). The Camry and Camry Solara models can be equipped with either the 5S-FE four-cylinder or the 1MZ-FE V6. The Avalon and Lexus ES 300 models are available in the covered years only with the 1MZ-FE V6 engine.

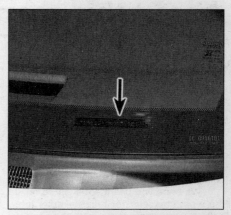

The Vehicle Identification Number (VIN) is located on a plate (arrow) on top of the dash (visible through the windshield)

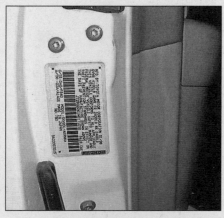

The vehicle certification label is located at the rear of the driver's door

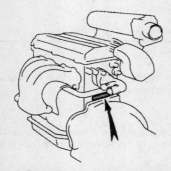

The engine serial number on the four-cylinder engine is located on the rear of the block below the cylinder head

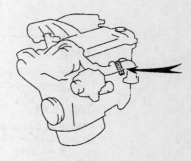

The V6 engine serial number is located on the front side of the block under the exhaust manifold

Stereo anti-theft system precaution

Stereo systems displaying ANTI-THEFT SYSTEM on the cassette tape slot cover have a built-in theft deterrent system designed to render the stereo inoperative should the stereo be stolen. If the power source to the stereo is cut, the anti-theft system will activate, so even if the power source is immediately reconnected the stereo will not function. If your vehicle is equipped with this anti-theft system, do not disconnect the cable from the negative terminal of the battery, remove the stereo or disconnect related components unless you have the individual ID (code) number for the stereo.

Your owner's manual will give specific information on selecting your three-digit audio security code. Keep a copy of the code at home, not in the vehicle.

If you discover that the system is inoperative after disconnecting and reconnecting the power source, enter the ID number. If the wrong number is entered, Err will appear on the display. You may make up to nine errors - a tenth error will activate the system and HELP will appear on the display. If this occurs, contact your local Toyota dealer service department.

Buying parts

Replacement parts are available from many sources, which generally fall into one of two categories - authorized dealer parts departments and independent retail auto parts stores. Our advice concerning these parts is as follows:

Retail auto parts stores: Good auto parts stores will stock frequently needed components which wear out relatively fast, such as clutch components, exhaust systems, brake parts, tune-up parts, etc. These stores often supply new or reconditioned parts on an exchange basis, which can save a considerable amount of money. Discount auto parts stores are often very good places to buy materials and parts needed for general vehicle maintenance such as oil, grease, filters, spark plugs, belts, touch-up paint, bulbs, etc. They also usually sell tools and general accessories, have convenient hours, charge lower prices and can often be found not far from home.

Authorized dealer parts department: This is the best source for parts which are unique to the vehicle and not generally available elsewhere (such as major engine parts, transmission parts, trim pieces, etc.).

Warranty information: If the vehicle is still covered under warranty, be sure that any replacement parts purchased - regardless of the source - do not invalidate the warranty!

To be sure of obtaining the correct parts, have engine and chassis numbers available and, if possible, take the old parts along for positive identification.

Maintenance techniques, tools and working facilities

Maintenance techniques

There are a number of techniques involved in maintenance and repair that will be referred to throughout this manual. Application of these techniques will enable the home mechanic to be more efficient, better organized and capable of performing the various tasks properly, which will ensure that the repair job is thorough and complete.

Fasteners

Fasteners are nuts, bolts, studs and screws used to hold two or more parts together. There are a few things to keep in mind when working with fasteners. Almost all of them use a locking device of some type, either a lockwasher, locknut, locking tab or thread adhesive. All threaded fasteners should be clean and straight, with undamaged threads and undamaged corners on the hex head where the wrench fits. Develop the habit of replacing all damaged nuts and bolts with new ones. Special locknuts with nylon or fiber inserts can only be used once. If they are removed, they lose their locking ability and must be replaced with new ones.

Rusted nuts and bolts should be treated with a penetrating fluid to ease removal and prevent breakage. Some mechanics use turpentine in a spout-type oil can, which works quite well. After applying the rust penetrant, let it work for a few minutes before trying to loosen the nut or bolt. Badly rusted fasteners may have to be chiseled or sawed off or removed with a special nut breaker, available at tool stores.

If a bolt or stud breaks off in an assembly, it can be drilled and removed with a special tool commonly available for this purpose. Most automotive machine shops can perform this task, as well as other repair procedures, such as the repair of threaded holes that have been stripped out.

Flat washers and lockwashers, when removed from an assembly, should always be replaced exactly as removed. Replace any damaged washers with new ones. Never use a lockwasher on any soft metal surface (such as aluminum), thin sheet metal or plastic.

Fastener sizes

For a number of reasons, automobile manufacturers are making wider and wider use of metric fasteners. Therefore, it is important to be able to tell the difference between standard (sometimes called U.S. or SAE) and metric hardware, since they cannot be interchanged.

All bolts, whether standard or metric, are sized according to diameter, thread pitch and length. For example, a standard 1/2 - 13 x 1 bolt is 1/2 inch in diameter, has 13 threads per inch and is 1 inch long. An M12 - 1.75 x 25 metric bolt is 12 mm in diameter, has a thread pitch of 1.75 mm (the distance between threads) and is 25 mm long. The two bolts are nearly identical, and easily confused, but they are not interchangeable.

In addition to the differences in diameter, thread pitch and length, metric and standard bolts can also be distinguished by examining the bolt heads. To begin with, the distance across the flats on a standard bolt head is measured in inches, while the same dimension on a metric bolt is sized in millimeters (the same is true for nuts). As a result, a standard wrench should not be used on a metric bolt and a metric wrench should not be used on a standard bolt. Also, most standard bolts have slashes radiating out from the center of the head to denote the grade or strength of the bolt, which is an indication of the amount of torque that can be applied to it. The greater the number of slashes, the greater the strength of the bolt. Grades 0 through 5 are commonly used on automobiles. Metric bolts have a property class (grade) number, rather than a slash, molded into their heads to indicate bolt strength. In this case, the higher the number, the stronger the bolt. Property class numbers 8.8, 9.8 and 10.9 are commonly used on automobiles.

Strength markings can also be used to distinguish standard hex nuts from metric hex nuts. Many standard nuts have dots stamped into one side, while metric nuts are marked with a number. The greater the number of dots, or the higher the number, the greater the strength of the nut.

Metric studs are also marked on their ends according to property class (grade). Larger studs are numbered (the same as metric bolts), while smaller studs carry a geometric code to denote grade.

It should be noted that many fasteners, especially Grades 0 through 2, have no distinguishing marks on them. When such is the case, the only way to determine whether it is standard or metric is to measure the thread pitch or compare it to a known fastener of the same size.

Standard fasteners are often referred to as SAE, as opposed to metric. However, it should be noted that SAE technically refers to a non-metric fine thread fastener only. Coarse thread non-metric fasteners are referred to as USS sizes.

Bolt strength marking (standard/SAE/USS; bottom - metric)

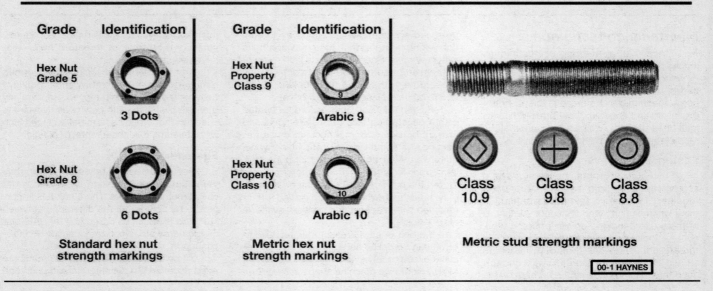

Standard hex nut strength markings

Metric hex nut strength markings

Metric stud strength markings

Maintenance techniques, tools and working facilities 0-9

Since fasteners of the same size (both standard and metric) may have different strength ratings, be sure to reinstall any bolts, studs or nuts removed from your vehicle in their original locations. Also, when replacing a fastener with a new one, make sure that the new one has a strength rating equal to or greater than the original.

Tightening sequences and procedures

Most threaded fasteners should be tightened to a specific torque value (torque is the twisting force applied to a threaded component such as a nut or bolt). Overtightening the fastener can weaken it and cause it to break, while undertightening can cause it to eventually come loose. Bolts, screws and studs, depending on the material they are made of and their thread diameters, have specific torque values, many of which are noted in the Specifications at the beginning of each Chapter. Be sure to follow the torque recommendations closely. For fasteners not assigned a specific torque, a general torque value chart is presented here as a guide. These torque values are for dry (unlubricated) fasteners threaded into steel or cast iron (not aluminum). As was previously mentioned, the size and grade of a fastener determine the amount of torque that can safely be applied to it. The figures listed here are approximate for Grade 2 and Grade 3 fasteners. Higher grades can tolerate higher torque values.

Fasteners laid out in a pattern, such as cylinder head bolts, oil pan bolts, differential cover bolts, etc., must be loosened or tightened in sequence to avoid warping the component. This sequence will normally be shown in the appropriate Chapter. If a specific pattern is not given, the following procedures can be used to prevent warping.

	Ft-lbs	Nm
Metric thread sizes		
M-6	6 to 9	9 to 12
M-8	14 to 21	19 to 28
M-10	28 to 40	38 to 54
M-12	50 to 71	68 to 96
M-14	80 to 140	109 to 154
Pipe thread sizes		
1/8	5 to 8	7 to 10
1/4	12 to 18	17 to 24
3/8	22 to 33	30 to 44
1/2	25 to 35	34 to 47
U.S. thread sizes		
1/4 - 20	6 to 9	9 to 12
5/16 - 18	12 to 18	17 to 24
5/16 - 24	14 to 20	19 to 27
3/8 - 16	22 to 32	30 to 43
3/8 - 24	27 to 38	37 to 51
7/16 - 14	40 to 55	55 to 74
7/16 - 20	40 to 60	55 to 81
1/2 - 13	55 to 80	75 to 108

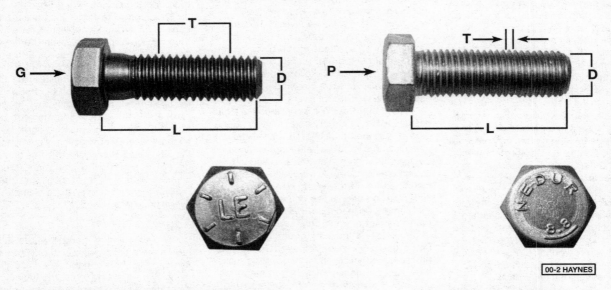

Standard (SAE and USS) bolt dimensions/grade marks

- G Grade marks (bolt strength)
- L Length (in inches)
- T Thread pitch (number of threads per inch)
- D Nominal diameter (in inches)

Metric bolt dimensions/grade marks

- P Property class (bolt strength)
- L Length (in millimeters)
- T Thread pitch (distance between threads in millimeters)
- D Diameter

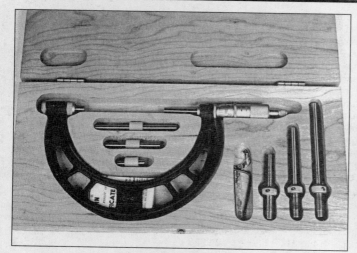

Micrometer set

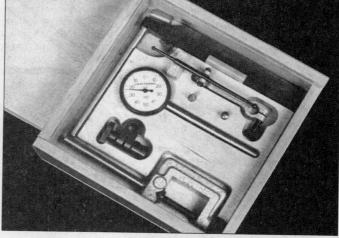

Dial indicator set

Initially, the bolts or nuts should be assembled finger-tight only. Next, they should be tightened one full turn each, in a criss-cross or diagonal pattern. After each one has been tightened one full turn, return to the first one and tighten them all one-half turn, following the same pattern. Finally, tighten each of them one-quarter turn at a time until each fastener has been tightened to the proper torque. To loosen and remove the fasteners, the procedure would be reversed.

Component disassembly

Component disassembly should be done with care and purpose to help ensure that the parts go back together properly. Always keep track of the sequence in which parts are removed. Make note of special characteristics or marks on parts that can be installed more than one way, such as a grooved thrust washer on a shaft. It is a good idea to lay the disassembled parts out on a clean surface in the order that they were removed. It may also be helpful to make sketches or take instant photos of components before removal.

When removing fasteners from a component, keep track of their locations. Sometimes threading a bolt back in a part, or putting the washers and nut back on a stud, can prevent mix-ups later. If nuts and bolts cannot be returned to their original locations, they should be kept in a compartmented box or a series of small boxes. A cupcake or muffin tin is ideal for this purpose, since each cavity can hold the bolts and nuts from a particular area (i.e. oil pan bolts, valve cover bolts, engine mount bolts, etc.). A pan of this type is especially helpful when working on assemblies with very small parts, such as the carburetor, alternator, valve train or interior dash and trim pieces. The cavities can be marked with paint or tape to identify the contents.

Whenever wiring looms, harnesses or connectors are separated, it is a good idea to identify the two halves with numbered pieces of masking tape so they can be easily reconnected.

Gasket sealing surfaces

Throughout any vehicle, gaskets are used to seal the mating surfaces between two parts and keep lubricants, fluids, vacuum or pressure contained in an assembly.

Many times these gaskets are coated with a liquid or paste-type gasket sealing compound before assembly. Age, heat and pressure can sometimes cause the two parts to stick together so tightly that they are very difficult to separate. Often, the assembly can be loosened by striking it with a soft-face hammer near the mating surfaces. A regular hammer can be used if a block of wood is placed between the hammer and the part. Do not hammer on cast parts or parts that could be easily damaged. With any particularly stubborn part, always recheck to make sure that every fastener has been removed.

Avoid using a screwdriver or bar to pry apart an assembly, as they can easily mar the gasket sealing surfaces of the parts, which must remain smooth. If prying is absolutely necessary, use an old broom handle, but keep in mind that extra clean up will be necessary if the wood splinters.

After the parts are separated, the old gasket must be carefully scraped off and the gasket surfaces cleaned. Stubborn gasket material can be soaked with rust penetrant or treated with a special chemical to soften it so it can be easily scraped off. A scraper can be fashioned from a piece of copper tubing by flattening and sharpening one end. Copper is recommended because it is usually softer than the surfaces to be scraped, which reduces the chance of gouging the part. Some gaskets can be removed with a wire brush, but regardless of the method used, the mating surfaces must be left clean and smooth. If for some reason the gasket surface is gouged, then a gasket sealer thick enough to fill scratches will have to be used during reassembly of the components. For most applications, a non-drying (or semi-drying) gasket sealer should be used.

Hose removal tips

Warning: *If the vehicle is equipped with air conditioning, do not disconnect any of the A/C hoses without first having the system depressurized by a dealer service department or a service station.*

Hose removal precautions closely parallel gasket removal precautions. Avoid scratching or gouging the surface that the hose mates against or the connection may leak. This is especially true for radiator hoses. Because of various chemical reactions, the rubber in hoses can bond itself to the metal spigot that the hose fits over. To remove a hose, first loosen the hose clamps that secure it to the spigot. Then, with slip-joint pliers, grab the hose at the clamp and rotate it around the spigot. Work it back and forth until it is completely free, then pull it off. Silicone or other lubricants will ease removal if they can be applied between the hose and the outside of the spigot. Apply the same lubricant to the inside of the hose and the outside of the spigot to simplify installation.

As a last resort (and if the hose is to be replaced with a new one anyway), the rubber can be slit with a knife and the hose peeled from the spigot. If this must be done, be careful that the metal connection is not damaged.

If a hose clamp is broken or damaged, do not reuse it. Wire-type clamps usually weaken with age, so it is a good idea to replace them with screw-type clamps whenever a hose is removed.

Tools

A selection of good tools is a basic requirement for anyone who plans to maintain and repair his or her own vehicle. For the owner who has few tools, the initial investment might seem high, but when compared to the spiraling costs of professional auto maintenance and repair, it is a wise one.

To help the owner decide which tools are needed to perform the tasks detailed in this manual, the following tool lists are offered: *Maintenance and minor repair,*

Maintenance techniques, tools and working facilities 0-11

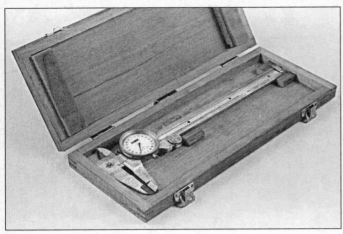
Dial caliper

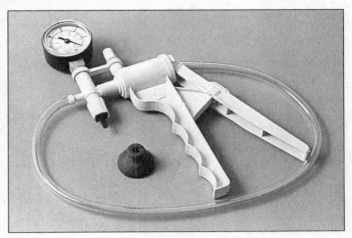
Hand-operated vacuum pump

Timing light

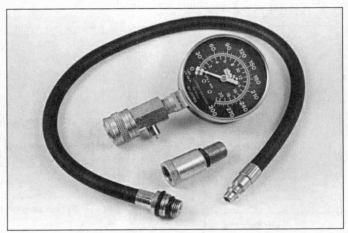
Compression gauge with spark plug hole adapter

Damper/steering wheel puller

General purpose puller

Hydraulic lifter removal tool

Repair/overhaul and *Special*.

The newcomer to practical mechanics should start off with the *maintenance and minor repair* tool kit, which is adequate for the simpler jobs performed on a vehicle. Then, as confidence and experience grow, the owner can tackle more difficult tasks, buying additional tools as they are needed.

Eventually the basic kit will be expanded into the *repair and overhaul* tool set. Over a period of time, the experienced do-it-yourselfer will assemble a tool set complete enough for most repair and overhaul procedures and will add tools from the special category when it is felt that the expense is justified by the frequency of use.

Maintenance and minor repair tool kit

The tools in this list should be considered the minimum required for performance of routine maintenance, servicing and minor repair work. We recommend the purchase of combination wrenches (box-end and open-

Maintenance techniques, tools and working facilities

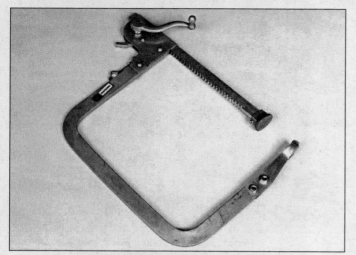

Valve spring compressor

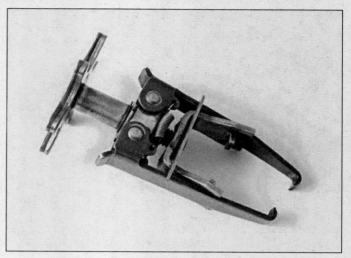

Valve spring compressor

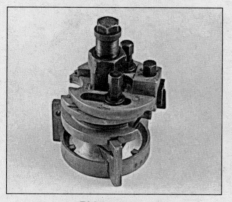

Ridge reamer

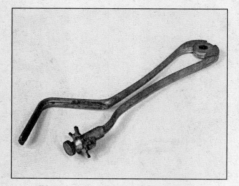

Piston ring groove cleaning tool

Ring removal/installation tool

Ring compressor

end combined in one wrench). While more expensive than open end wrenches, they offer the advantages of both types of wrench.

Combination wrench set (1/4-inch to 1 inch or 6 mm to 19 mm)
Adjustable wrench, 8 inch
Spark plug wrench with rubber insert
Spark plug gap adjusting tool
Feeler gauge set
Brake bleeder wrench
Standard screwdriver (5/16-inch x 6 inch)
Phillips screwdriver (No. 2 x 6 inch)
Combination pliers - 6 inch
Hacksaw and assortment of blades
Tire pressure gauge
Grease gun
Oil can
Fine emery cloth
Wire brush
Battery post and cable cleaning tool
Oil filter wrench
Funnel (medium size)
Safety goggles
Jackstands (2)
Drain pan

Note: *If basic tune-ups are going to be part of routine maintenance, it will be necessary to purchase a good quality stroboscopic timing light and combination tachometer/dwell meter. Although they are included in the list of special tools, it is mentioned here because they are absolutely necessary for tuning most vehicles properly.*

Repair and overhaul tool set

These tools are essential for anyone who plans to perform major repairs and are in addition to those in the maintenance and minor repair tool kit. Included is a comprehensive set of sockets which, though expensive, are invaluable because of their versatility, especially when various extensions and drives are available. We recommend the 1/2-inch drive over the 3/8-inch drive. Although the larger drive is bulky and more expensive, it has the capacity of accepting a very wide range of large sockets. Ideally, however, the mechanic should have a 3/8-inch drive set and a 1/2-inch drive set.

Socket set(s)
Reversible ratchet
Extension - 10 inch
Universal joint
Torque wrench (same size drive as sockets)
Ball peen hammer - 8 ounce
Soft-face hammer (plastic/rubber)
Standard screwdriver (1/4-inch x 6 inch)
Standard screwdriver (stubby - 5/16-inch)
Phillips screwdriver (No. 3 x 8 inch)
Phillips screwdriver (stubby - No. 2)
Pliers - vise grip
Pliers - lineman's
Pliers - needle nose
Pliers - snap-ring (internal and external)
Cold chisel - 1/2-inch

Maintenance techniques, tools and working facilities

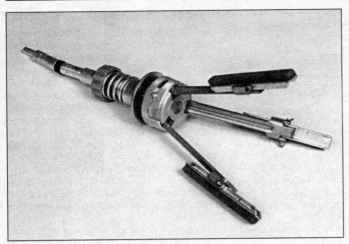

Cylinder hone

Brake hold-down spring tool

Scribe
Scraper (made from flattened copper tubing)
Centerpunch
Pin punches (1/16, 1/8, 3/16-inch)
Steel rule/straightedge - 12 inch
Allen wrench set (1/8 to 3/8-inch or 4 mm to 10 mm)
A selection of files
Wire brush (large)
Jackstands (second set)
Jack (scissor or hydraulic type)

Note: *Another tool which is often useful is an electric drill with a chuck capacity of 3/8-inch and a set of good quality drill bits.*

Special tools

The tools in this list include those which are not used regularly, are expensive to buy, or which need to be used in accordance with their manufacturer's instructions. Unless these tools will be used frequently, it is not very economical to purchase many of them. A consideration would be to split the cost and use between yourself and a friend or friends. In addition, most of these tools can be obtained from a tool rental shop on a temporary basis.

This list primarily contains only those tools and instruments widely available to the public, and not those special tools produced by the vehicle manufacturer for distribution to dealer service departments. Occasionally, references to the manufacturer's special tools are included in the text of this manual. Generally, an alternative method of doing the job without the special tool is offered. However, sometimes there is no alternative to their use. Where this is the case, and the tool cannot be purchased or borrowed, the work should be turned over to the dealer service department or an automotive repair shop.

Valve spring compressor
Piston ring groove cleaning tool
Piston ring compressor
Piston ring installation tool
Cylinder compression gauge
Cylinder ridge reamer
Cylinder surfacing hone
Cylinder bore gauge
Micrometers and/or dial calipers
Hydraulic lifter removal tool
Balljoint separator
Universal-type puller
Impact screwdriver
Dial indicator set
Stroboscopic timing light (inductive pick-up)
Hand operated vacuum/pressure pump
Tachometer/dwell meter
Universal electrical multimeter
Cable hoist
Brake spring removal and installation tools
Floor jack

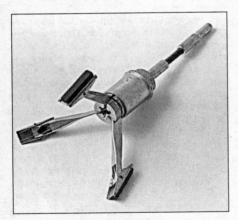

Brake cylinder hone

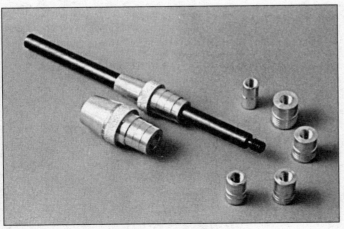

Clutch plate alignment tool

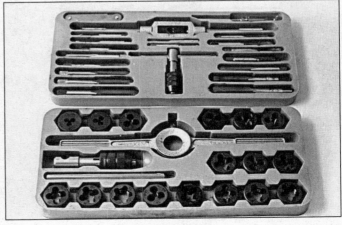

Tap and die set

Buying tools

For the do-it-yourselfer who is just starting to get involved in vehicle maintenance and repair, there are a number of options available when purchasing tools. If maintenance and minor repair is the extent of the work to be done, the purchase of individual tools is satisfactory. If, on the other hand, extensive work is planned, it would be a good idea to purchase a modest tool set from one of the large retail chain stores. A set can usually be bought at a substantial savings over the individual tool prices, and they often come with a tool box. As additional tools are needed, add-on sets, individual tools and a larger tool box can be purchased to expand the tool selection. Building a tool set gradually allows the cost of the tools to be spread over a longer period of time and gives the mechanic the freedom to choose only those tools that will actually be used.

Tool stores will often be the only source of some of the special tools that are needed, but regardless of where tools are bought, try to avoid cheap ones, especially when buying screwdrivers and sockets, because they won't last very long. The expense involved in replacing cheap tools will eventually be greater than the initial cost of quality tools.

Care and maintenance of tools

Good tools are expensive, so it makes sense to treat them with respect. Keep them clean and in usable condition and store them properly when not in use. Always wipe off any dirt, grease or metal chips before putting them away. Never leave tools lying around in the work area. Upon completion of a job, always check closely under the hood for tools that may have been left there so they won't get lost during a test drive.

Some tools, such as screwdrivers, pliers, wrenches and sockets, can be hung on a panel mounted on the garage or workshop wall, while others should be kept in a tool box or tray. Measuring instruments, gauges, meters, etc. must be carefully stored where they cannot be damaged by weather or impact from other tools.

When tools are used with care and stored properly, they will last a very long time. Even with the best of care, though, tools will wear out if used frequently. When a tool is damaged or worn out, replace it. Subsequent jobs will be safer and more enjoyable if you do.

How to repair damaged threads

Sometimes, the internal threads of a nut or bolt hole can become stripped, usually from overtightening. Stripping threads is an all-too-common occurrence, especially when working with aluminum parts, because aluminum is so soft that it easily strips out.

Usually, external or internal threads are only partially stripped. After they've been cleaned up with a tap or die, they'll still work. Sometimes, however, threads are badly damaged. When this happens, you've got three choices:

1) Drill and tap the hole to the next suitable oversize and install a larger diameter bolt, screw or stud.
2) Drill and tap the hole to accept a threaded plug, then drill and tap the plug to the original screw size. You can also buy a plug already threaded to the original size. Then you simply drill a hole to the specified size, then run the threaded plug into the hole with a bolt and jam nut. Once the plug is fully seated, remove the jam nut and bolt.
3) The third method uses a patented thread repair kit like Heli-Coil or Slimsert. These easy-to-use kits are designed to repair damaged threads in straight-through holes and blind holes. Both are available as kits which can handle a variety of sizes and thread patterns. Drill the hole, then tap it with the special included tap. Install the Heli-Coil and the hole is back to its original diameter and thread pitch.

Regardless of which method you use, be sure to proceed calmly and carefully. A little impatience or carelessness during one of these relatively simple procedures can ruin your whole day's work and cost you a bundle if you wreck an expensive part.

Working facilities

Not to be overlooked when discussing tools is the workshop. If anything more than routine maintenance is to be carried out, some sort of suitable work area is essential.

It is understood, and appreciated, that many home mechanics do not have a good workshop or garage available, and end up removing an engine or doing major repairs outside. It is recommended, however, that the overhaul or repair be completed under the cover of a roof.

A clean, flat workbench or table of comfortable working height is an absolute necessity. The workbench should be equipped with a vise that has a jaw opening of at least four inches.

As mentioned previously, some clean, dry storage space is also required for tools, as well as the lubricants, fluids, cleaning solvents, etc. which soon become necessary.

Sometimes waste oil and fluids, drained from the engine or cooling system during normal maintenance or repairs, present a disposal problem. To avoid pouring them on the ground or into a sewage system, pour the used fluids into large containers, seal them with caps and take them to an authorized disposal site or recycling center. Plastic jugs, such as old antifreeze containers, are ideal for this purpose.

Always keep a supply of old newspapers and clean rags available. Old towels are excellent for mopping up spills. Many mechanics use rolls of paper towels for most work because they are readily available and disposable. To help keep the area under the vehicle clean, a large cardboard box can be cut open and flattened to protect the garage or shop floor.

Whenever working over a painted surface, such as when leaning over a fender to service something under the hood, always cover it with an old blanket or bedspread to protect the finish. Vinyl covered pads, made especially for this purpose, are available at auto parts stores.

Booster battery (jump) starting

Observe the following precautions when using a booster battery to start a vehicle:
a) *Before connecting the booster battery, make sure the ignition switch is in the Off position.*
b) *Turn off the lights, heater and other electrical loads.*
c) *Your eyes should be shielded. Safety goggles are a good idea.*
d) *Make sure the booster battery is the same voltage as the dead one in the vehicle.*
e) *The two vehicles MUST NOT TOUCH each other.*
f) *Make sure the transmission is in Neutral (manual transaxle) or Park (automatic transaxle).*
g) *If the booster battery is not a maintenance-free type, remove the vent caps and lay a cloth over the vent holes.*

Connect the red jumper cable to the positive (+) terminals of each battery.

Connect one end of the black cable to the negative (-) terminal of the booster battery. The other end of this cable should be connected to a good ground on the engine block **(see illustration)**. Make sure the cable will not come into contact with the fan, drivebelts or other moving parts of the engine.

Start the engine using the booster battery, then, with the engine running at idle speed, disconnect the jumper cables in the reverse order of connection

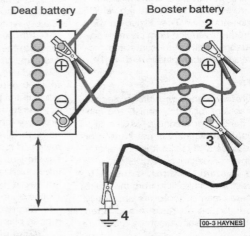

Make the booster battery cable connections in the numerical order shown (note that the negative cable of the booster battery is NOT attached to the negative terminal of the dead battery)

Jacking and towing

Jacking

The jack supplied with the vehicle should only be used for raising the vehicle for changing a tire or placing jackstands under the frame. **Warning:** *Never crawl under the vehicle or start the engine when the jack is being used as the only means of support.*

All vehicles are supplied with a scissors-type jack. When jacking the vehicle, it should be engaged with the seam notch, between the two dimples **(see illustration)**.

The vehicle should be on level ground with the wheels blocked and the transmission in Park (automatic) or Reverse (manual). Pry off the hub cap (if equipped) using the tapered end of the lug wrench. Loosen the lug nuts one-half turn and leave them in place until the wheel is raised off the ground.

Place the jack under the side of the vehicle in the indicated position. Use the supplied wrench to turn the jackscrew clockwise until the wheel is raised off the ground. Remove the lug nuts, pull off the wheel and replace it with the spare.

With the beveled side in, replace the lug nuts and tighten them until snug. Lower the vehicle by turning the jackscrew counter-clockwise. Remove the jack and tighten the nuts in a diagonal pattern to the torque listed in the Chapter 1 Specifications. If a torque wrench is not available, have the torque checked by a service station as soon as possible. Replace the hubcap by placing it in position and using the heel of your hand or a rubber mallet to seat it.

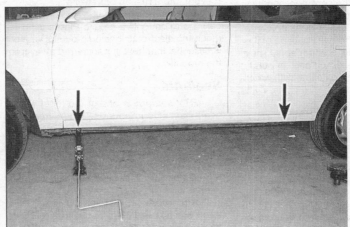

The jack fits over the rocker panel flange, between the two notches (there are two jacking points on each side of the vehicle)

Towing

Manual transmission-equipped vehicles can be towed with all four wheels on the ground. Automatic transmission-equipped models should only be towed with all four wheels on the ground if speeds do not exceed 35 mph and the distance is not over 50 miles, otherwise transmission damage can result.

Towing equipment specifically designed for this purpose should be used and should be attached to the main structural members of the vehicle, not the bumper or brackets.

Safety is a major consideration when towing and all applicable state and local laws must be obeyed. A safety chain system must be used for all towing.

While towing, the parking brake should be released and the transmission should be in Neutral. The steering must be unlocked (ignition switch in the Off position). Remember that power steering and power brakes will not work with the engine off.

Traction control

On models equipped with Traction-Control system, push in the TRAC switch (on the dashboard or floor console, depending on model) anytime the vehicle is on a "rolling road" tester such as a speedometer test machine or chassis dynamometer. The TRAC OFF indicator light should illuminate when the system is turned off.

Automotive chemicals and lubricants

A number of automotive chemicals and lubricants are available for use during vehicle maintenance and repair. They include a wide variety of products ranging from cleaning solvents and degreasers to lubricants and protective sprays for rubber, plastic and vinyl.

Cleaners

Carburetor cleaner and choke cleaner is a strong solvent for gum, varnish and carbon. Most carburetor cleaners leave a dry-type lubricant film which will not harden or gum up. Because of this film it is not recommended for use on electrical components.

Brake system cleaner is used to remove grease and brake fluid from the brake system, where clean surfaces are absolutely necessary. It leaves no residue and often eliminates brake squeal caused by contaminants.

Electrical cleaner removes oxidation, corrosion and carbon deposits from electrical contacts, restoring full current flow. It can also be used to clean spark plugs, carburetor jets, voltage regulators and other parts where an oil-free surface is desired.

Demoisturants remove water and moisture from electrical components such as alternators, voltage regulators, electrical connectors and fuse blocks. They are non-conductive, non-corrosive and non-flammable.

Degreasers are heavy-duty solvents used to remove grease from the outside of the engine and from chassis components. They can be sprayed or brushed on and, depending on the type, are rinsed off either with water or solvent.

Lubricants

Motor oil is the lubricant formulated for use in engines. It normally contains a wide variety of additives to prevent corrosion and reduce foaming and wear. Motor oil comes in various weights (viscosity ratings) from 0 to 50. The recommended weight of the oil depends on the season, temperature and the demands on the engine. Light oil is used in cold climates and under light load conditions. Heavy oil is used in hot climates and where high loads are encountered. Multi-viscosity oils are designed to have characteristics of both light and heavy oils and are available in a number of weights from 5W-20 to 20W-50.

Gear oil is designed to be used in differentials, manual transmissions and other areas where high-temperature lubrication is required.

Chassis and wheel bearing grease is a heavy grease used where increased loads and friction are encountered, such as for wheel bearings, balljoints, tie-rod ends and universal joints.

High-temperature wheel bearing grease is designed to withstand the extreme temperatures encountered by wheel bearings in disc brake equipped vehicles. It usually contains molybdenum disulfide (moly), which is a dry-type lubricant.

White grease is a heavy grease for metal-to-metal applications where water is a problem. White grease stays soft under both low and high temperatures (usually from -100 to +190-degrees F), and will not wash off or dilute in the presence of water.

Assembly lube is a special extreme pressure lubricant, usually containing moly, used to lubricate high-load parts (such as main and rod bearings and cam lobes) for initial start-up of a new engine. The assembly lube lubricates the parts without being squeezed out or washed away until the engine oiling system begins to function.

Silicone lubricants are used to protect rubber, plastic, vinyl and nylon parts.

Graphite lubricants are used where oils cannot be used due to contamination problems, such as in locks. The dry graphite will lubricate metal parts while remaining uncontaminated by dirt, water, oil or acids. It is electrically conductive and will not foul electrical contacts in locks such as the ignition switch.

Moly penetrants loosen and lubricate frozen, rusted and corroded fasteners and prevent future rusting or freezing.

Heat-sink grease is a special electrically non-conductive grease that is used for mounting electronic ignition modules where it is essential that heat is transferred away from the module.

Sealants

RTV sealant is one of the most widely used gasket compounds. Made from silicone, RTV is air curing, it seals, bonds, waterproofs, fills surface irregularities, remains flexible, doesn't shrink, is relatively easy to remove, and is used as a supplementary sealer with almost all low and medium temperature gaskets.

Anaerobic sealant is much like RTV in that it can be used either to seal gaskets or to form gaskets by itself. It remains flexible, is solvent resistant and fills surface imperfections. The difference between an anaerobic sealant and an RTV-type sealant is in the curing. RTV cures when exposed to air, while an anaerobic sealant cures only in the absence of air. This means that an anaerobic sealant cures only after the assembly of parts, sealing them together.

Thread and pipe sealant is used for sealing hydraulic and pneumatic fittings and vacuum lines. It is usually made from a Teflon compound, and comes in a spray, a paint-on liquid and as a wrap-around tape.

Chemicals

Anti-seize compound prevents seizing, galling, cold welding, rust and corrosion in fasteners. High-temperature anti-seize, usually made with copper and graphite lubricants, is used for exhaust system and exhaust manifold bolts.

Anaerobic locking compounds are used to keep fasteners from vibrating or working loose and cure only after installation, in the absence of air. Medium strength locking compound is used for small nuts, bolts and screws that may be removed later. High-strength locking compound is for large nuts, bolts and studs which aren't removed on a regular basis.

Oil additives range from viscosity index improvers to chemical treatments that claim to reduce internal engine friction. It should be noted that most oil manufacturers caution against using additives with their oils.

Gas additives perform several functions, depending on their chemical makeup. They usually contain solvents that help dissolve gum and varnish that build up on carburetor, fuel injection and intake parts. They also serve to break down carbon deposits that form on the inside surfaces of the combustion chambers. Some additives contain upper cylinder lubricants for valves and piston rings, and others contain chemicals to remove condensation from the gas tank.

Miscellaneous

Brake fluid is specially formulated hydraulic fluid that can withstand the heat and pressure encountered in brake systems. Care must be taken so this fluid does not come in contact with painted surfaces or plastics. An opened container should always be resealed to prevent contamination by water or dirt.

Weatherstrip adhesive is used to bond weatherstripping around doors, windows and trunk lids. It is sometimes used to attach trim pieces.

Undercoating is a petroleum-based, tar-like substance that is designed to protect metal surfaces on the underside of the vehicle from corrosion. It also acts as a sound-deadening agent by insulating the bottom of the vehicle.

Waxes and polishes are used to help protect painted and plated surfaces from the weather. Different types of paint may require the use of different types of wax and polish. Some polishes utilize a chemical or abrasive cleaner to help remove the top layer of oxidized (dull) paint on older vehicles. In recent years many non-wax polishes that contain a wide variety of chemicals such as polymers and silicones have been introduced. These non-wax polishes are usually easier to apply and last longer than conventional waxes and polishes.

Conversion factors

Length (distance)
Inches (in)	X	25.4	= Millimetres (mm)	X 0.0394	= Inches (in)
Feet (ft)	X	0.305	= Metres (m)	X 3.281	= Feet (ft)
Miles	X	1.609	= Kilometres (km)	X 0.621	= Miles

Volume (capacity)
Cubic inches (cu in; in^3)	X	16.387	= Cubic centimetres (cc; cm^3)	X 0.061	= Cubic inches (cu in; in^3)
Imperial pints (Imp pt)	X	0.568	= Litres (l)	X 1.76	= Imperial pints (Imp pt)
Imperial quarts (Imp qt)	X	1.137	= Litres (l)	X 0.88	= Imperial quarts (Imp qt)
Imperial quarts (Imp qt)	X	1.201	= US quarts (US qt)	X 0.833	= Imperial quarts (Imp qt)
US quarts (US qt)	X	0.946	= Litres (l)	X 1.057	= US quarts (US qt)
Imperial gallons (Imp gal)	X	4.546	= Litres (l)	X 0.22	= Imperial gallons (Imp gal)
Imperial gallons (Imp gal)	X	1.201	= US gallons (US gal)	X 0.833	= Imperial gallons (Imp gal)
US gallons (US gal)	X	3.785	= Litres (l)	X 0.264	= US gallons (US gal)

Mass (weight)
Ounces (oz)	X	28.35	= Grams (g)	X 0.035	= Ounces (oz)
Pounds (lb)	X	0.454	= Kilograms (kg)	X 2.205	= Pounds (lb)

Force
Ounces-force (ozf; oz)	X	0.278	= Newtons (N)	X 3.6	= Ounces-force (ozf; oz)
Pounds-force (lbf; lb)	X	4.448	= Newtons (N)	X 0.225	= Pounds-force (lbf; lb)
Newtons (N)	X	0.1	= Kilograms-force (kgf; kg)	X 9.81	= Newtons (N)

Pressure
Pounds-force per square inch (psi; lbf/in^2; lb/in^2)	X	0.070	= Kilograms-force per square centimetre (kgf/cm^2; kg/cm^2)	X 14.223	= Pounds-force per square inch (psi; lbf/in^2; lb/in^2)
Pounds-force per square inch (psi; lbf/in^2; lb/in^2)	X	0.068	= Atmospheres (atm)	X 14.696	= Pounds-force per square inch (psi; lbf/in^2; lb/in^2)
Pounds-force per square inch (psi; lbf/in^2; lb/in^2)	X	0.069	= Bars	X 14.5	= Pounds-force per square inch (psi; lbf/in^2; lb/in^2)
Pounds-force per square inch (psi; lbf/in^2; lb/in^2)	X	6.895	= Kilopascals (kPa)	X 0.145	= Pounds-force per square inch (psi; lbf/in^2; lb/in^2)
Kilopascals (kPa)	X	0.01	= Kilograms-force per square centimetre (kgf/cm^2; kg/cm^2)	X 98.1	= Kilopascals (kPa)

Torque (moment of force)
Pounds-force inches (lbf in; lb in)	X	1.152	= Kilograms-force centimetre (kgf cm; kg cm)	X 0.868	= Pounds-force inches (lbf in; lb in)
Pounds-force inches (lbf in; lb in)	X	0.113	= Newton metres (Nm)	X 8.85	= Pounds-force inches (lbf in; lb in)
Pounds-force inches (lbf in; lb in)	X	0.083	= Pounds-force feet (lbf ft; lb ft)	X 12	= Pounds-force inches (lbf in; lb in)
Pounds-force feet (lbf ft; lb ft)	X	0.138	= Kilograms-force metres (kgf m; kg m)	X 7.233	= Pounds-force feet (lbf ft; lb ft)
Pounds-force feet (lbf ft; lb ft)	X	1.356	= Newton metres (Nm)	X 0.738	= Pounds-force feet (lbf ft; lb ft)
Newton metres (Nm)	X	0.102	= Kilograms-force metres (kgf m; kg m)	X 9.804	= Newton metres (Nm)

Vacuum
Inches mercury (in. Hg)	X	3.377	= Kilopascals (kPa)	X 0.2961	= Inches mercury
Inches mercury (in. Hg)	X	25.4	= Millimeters mercury (mm Hg)	X 0.0394	= Inches mercury

Power
Horsepower (hp)	X	745.7	= Watts (W)	X 0.0013	= Horsepower (hp)

Velocity (speed)
Miles per hour (miles/hr; mph)	X	1.609	= Kilometres per hour (km/hr; kph)	X 0.621	= Miles per hour (miles/hr; mph)

Fuel consumption*
Miles per gallon, Imperial (mpg)	X	0.354	= Kilometres per litre (km/l)	X 2.825	= Miles per gallon, Imperial (mpg)
Miles per gallon, US (mpg)	X	0.425	= Kilometres per litre (km/l)	X 2.352	= Miles per gallon, US (mpg)

Temperature
Degrees Fahrenheit = (°C x 1.8) + 32   Degrees Celsius (Degrees Centigrade; °C) = (°F - 32) x 0.56

*It is common practice to convert from miles per gallon (mpg) to litres/100 kilometres (l/100km), where mpg (Imperial) x l/100 km = 282 and mpg (US) x l/100 km = 235

Fraction/Decimal/Millimeter Equivalents

DECIMALS TO MILLIMETERS

Decimal	mm	Decimal	mm
0.001	0.0254	0.500	12.7000
0.002	0.0508	0.510	12.9540
0.003	0.0762	0.520	13.2080
0.004	0.1016	0.530	13.4620
0.005	0.1270	0.540	13.7160
0.006	0.1524	0.550	13.9700
0.007	0.1778	0.560	14.2240
0.008	0.2032	0.570	14.4780
0.009	0.2286	0.580	14.7320
		0.590	14.9860
0.010	0.2540		
0.020	0.5080		
0.030	0.7620		
0.040	1.0160	0.600	15.2400
0.050	1.2700	0.610	15.4940
0.060	1.5240	0.620	15.7480
0.070	1.7780	0.630	16.0020
0.080	2.0320	0.640	16.2560
0.090	2.2860	0.650	16.5100
		0.660	16.7640
0.100	2.5400	0.670	17.0180
0.110	2.7940	0.680	17.2720
0.120	3.0480	0.690	17.5260
0.130	3.3020		
0.140	3.5560		
0.150	3.8100		
0.160	4.0640	0.700	17.7800
0.170	4.3180	0.710	18.0340
0.180	4.5720	0.720	18.2880
0.190	4.8260	0.730	18.5420
		0.740	18.7960
0.200	5.0800	0.750	19.0500
0.210	5.3340	0.760	19.3040
0.220	5.5880	0.770	19.5580
0.230	5.8420	0.780	19.8120
0.240	6.0960	0.790	20.0660
0.250	6.3500		
0.260	6.6040		
0.270	6.8580	0.800	20.3200
0.280	7.1120	0.810	20.5740
0.290	7.3660	0.820	21.8280
		0.830	21.0820
0.300	7.6200	0.840	21.3360
0.310	7.8740	0.850	21.5900
0.320	8.1280	0.860	21.8440
0.330	8.3820	0.870	22.0980
0.340	8.6360	0.880	22.3520
0.350	8.8900	0.890	22.6060
0.360	9.1440		
0.370	9.3980		
0.380	9.6520		
0.390	9.9060		
		0.900	22.8600
0.400	10.1600	0.910	23.1140
0.410	10.4140	0.920	23.3680
0.420	10.6680	0.930	23.6220
0.430	10.9220	0.940	23.8760
0.440	11.1760	0.950	24.1300
0.450	11.4300	0.960	24.3840
0.460	11.6840	0.970	24.6380
0.470	11.9380	0.980	24.8920
0.480	12.1920	0.990	25.1460
0.490	12.4460	1.000	25.4000

FRACTIONS TO DECIMALS TO MILLIMETERS

Fraction	Decimal	mm	Fraction	Decimal	mm
1/64	0.0156	0.3969	33/64	0.5156	13.0969
1/32	0.0312	0.7938	17/32	0.5312	13.4938
3/64	0.0469	1.1906	35/64	0.5469	13.8906
1/16	0.0625	1.5875	9/16	0.5625	14.2875
5/64	0.0781	1.9844	37/64	0.5781	14.6844
3/32	0.0938	2.3812	19/32	0.5938	15.0812
7/64	0.1094	2.7781	39/64	0.6094	15.4781
1/8	0.1250	3.1750	5/8	0.6250	15.8750
9/64	0.1406	3.5719	41/64	0.6406	16.2719
5/32	0.1562	3.9688	21/32	0.6562	16.6688
11/64	0.1719	4.3656	43/64	0.6719	17.0656
3/16	0.1875	4.7625	11/16	0.6875	17.4625
13/64	0.2031	5.1594	45/64	0.7031	17.8594
7/32	0.2188	5.5562	23/32	0.7188	18.2562
15/64	0.2344	5.9531	47/64	0.7344	18.6531
1/4	0.2500	6.3500	3/4	0.7500	19.0500
17/64	0.2656	6.7469	49/64	0.7656	19.4469
9/32	0.2812	7.1438	25/32	0.7812	19.8438
19/64	0.2969	7.5406	51/64	0.7969	20.2406
5/16	0.3125	7.9375	13/16	0.8125	20.6375
21/64	0.3281	8.3344	53/64	0.8281	21.0344
11/32	0.3438	8.7312	27/32	0.8438	21.4312
23/64	0.3594	9.1281	55/64	0.8594	21.8281
3/8	0.3750	9.5250	7/8	0.8750	22.2250
25/64	0.3906	9.9219	57/64	0.8906	22.6219
13/32	0.4062	10.3188	29/32	0.9062	23.0188
27/64	0.4219	10.7156	59/64	0.9219	23.4156
7/16	0.4375	11.1125	15/16	0.9375	23.8125
29/64	0.4531	11.5094	61/64	0.9531	24.2094
15/32	0.4688	11.9062	31/32	0.9688	24.6062
31/64	0.4844	12.3031	63/64	0.9844	25.0031
1/2	0.5000	12.7000	1	1.0000	25.4000

Safety first!

Regardless of how enthusiastic you may be about getting on with the job at hand, take the time to ensure that your safety is not jeopardized. A moment's lack of attention can result in an accident, as can failure to observe certain simple safety precautions. The possibility of an accident will always exist, and the following points should not be considered a comprehensive list of all dangers. Rather, they are intended to make you aware of the risks and to encourage a safety conscious approach to all work you carry out on your vehicle.

Essential DOs and DON'Ts

DON'T rely on a jack when working under the vehicle. Always use approved jackstands to support the weight of the vehicle and place them under the recommended lift or support points.

DON'T attempt to loosen extremely tight fasteners (i.e. wheel lug nuts) while the vehicle is on a jack - it may fall.

DON'T start the engine without first making sure that the transmission is in Neutral (or Park where applicable) and the parking brake is set.

DON'T remove the radiator cap from a hot cooling system - let it cool or cover it with a cloth and release the pressure gradually.

DON'T attempt to drain the engine oil until you are sure it has cooled to the point that it will not burn you.

DON'T touch any part of the engine or exhaust system until it has cooled sufficiently to avoid burns.

DON'T siphon toxic liquids such as gasoline, antifreeze and brake fluid by mouth, or allow them to remain on your skin.

DON'T inhale brake lining dust - it is potentially hazardous (see *Asbestos* below).

DON'T allow spilled oil or grease to remain on the floor - wipe it up before someone slips on it.

DON'T use loose fitting wrenches or other tools which may slip and cause injury.

DON'T push on wrenches when loosening or tightening nuts or bolts. Always try to pull the wrench toward you. If the situation calls for pushing the wrench away, push with an open hand to avoid scraped knuckles if the wrench should slip.

DON'T attempt to lift a heavy component alone - get someone to help you.

DON'T rush or take unsafe shortcuts to finish a job.

DON'T allow children or animals in or around the vehicle while you are working on it.

DO wear eye protection when using power tools such as a drill, sander, bench grinder, etc. and when working under a vehicle.

DO keep loose clothing and long hair well out of the way of moving parts.

DO make sure that any hoist used has a safe working load rating adequate for the job.

DO get someone to check on you periodically when working alone on a vehicle.

DO carry out work in a logical sequence and make sure that everything is correctly assembled and tightened.

DO keep chemicals and fluids tightly capped and out of the reach of children and pets.

DO remember that your vehicle's safety affects that of yourself and others. If in doubt on any point, get professional advice.

Asbestos

Certain friction, insulating, sealing, and other products - such as brake linings, brake bands, clutch linings, torque converters, gaskets, etc. - may contain asbestos. Extreme care must be taken to avoid inhalation of dust from such products, since it is hazardous to health. If in doubt, assume that they do contain asbestos.

Fire

Remember at all times that gasoline is highly flammable. Never smoke or have any kind of open flame around when working on a vehicle. But the risk does not end there. A spark caused by an electrical short circuit, by two metal surfaces contacting each other, or even by static electricity built up in your body under certain conditions, can ignite gasoline vapors, which in a confined space are highly explosive. Do not, under any circumstances, use gasoline for cleaning parts. Use an approved safety solvent.

Always disconnect the battery ground (-) cable at the battery before working on any part of the fuel system or electrical system. Never risk spilling fuel on a hot engine or exhaust component. It is strongly recommended that a fire extinguisher suitable for use on fuel and electrical fires be kept handy in the garage or workshop at all times. Never try to extinguish a fuel or electrical fire with water.

Fumes

Certain fumes are highly toxic and can quickly cause unconsciousness and even death if inhaled to any extent. Gasoline vapor falls into this category, as do the vapors from some cleaning solvents. Any draining or pouring of such volatile fluids should be done in a well ventilated area.

When using cleaning fluids and solvents, read the instructions on the container carefully. Never use materials from unmarked containers.

Never run the engine in an enclosed space, such as a garage. Exhaust fumes contain carbon monoxide, which is extremely poisonous. If you need to run the engine, always do so in the open air, or at least have the rear of the vehicle outside the work area.

If you are fortunate enough to have the use of an inspection pit, never drain or pour gasoline and never run the engine while the vehicle is over the pit. The fumes, being heavier than air, will concentrate in the pit with possibly lethal results.

The battery

Never create a spark or allow a bare light bulb near a battery. They normally give off a certain amount of hydrogen gas, which is highly explosive.

Always disconnect the battery ground (-) cable at the battery before working on the fuel or electrical systems.

If possible, loosen the filler caps or cover when charging the battery from an external source (this does not apply to sealed or maintenance-free batteries). Do not charge at an excessive rate or the battery may burst.

Take care when adding water to a non maintenance-free battery and when carrying a battery. The electrolyte, even when diluted, is very corrosive and should not be allowed to contact clothing or skin.

Always wear eye protection when cleaning the battery to prevent the caustic deposits from entering your eyes.

Household current

When using an electric power tool, inspection light, etc., which operates on household current, always make sure that the tool is correctly connected to its plug and that, where necessary, it is properly grounded. Do not use such items in damp conditions and, again, do not create a spark or apply excessive heat in the vicinity of fuel or fuel vapor.

Secondary ignition system voltage

A severe electric shock can result from touching certain parts of the ignition system (such as the spark plug wires) when the engine is running or being cranked, particularly if components are damp or the insulation is defective. In the case of an electronic ignition system, the secondary system voltage is much higher and could prove fatal.

Troubleshooting

Contents

Symptom	Section
Engine	
CHECK ENGINE light	See Chapter 6
Engine backfires	15
Engine diesels (continues to run) after switching off	18
Engine hard to start when cold	3
Engine hard to start when hot	4
Engine lacks power	14
Engine lopes while idling or idles erratically	8
Engine misses at idle speed	9
Engine misses throughout driving speed range	10
Engine rotates but will not start	2
Engine runs with oil pressure light on	17
Engine stalls	13
Engine starts but stops immediately	6
Engine stumbles on acceleration	11
Engine surges while holding accelerator steady	12
Engine will not rotate when attempting to start	1
Oil puddle under engine	7
Pinging or knocking engine sounds during acceleration or uphill	16
Starter motor noisy or excessively rough in engagement	5

Engine electrical system
Alternator light fails to go out	20
Battery will not hold a charge	19
Alternator light fails to come on when key is turned on	21

Fuel system
Excessive fuel consumption	22
Fuel leakage and/or fuel odor	23

Cooling system
Coolant loss	28
External coolant leakage	26
Internal coolant leakage	27
Overcooling	25
Overheating	24
Poor coolant circulation	29

Clutch
Clutch pedal stays on floor	39
Clutch slips (engine speed increases with no increase in vehicle speed)	35
Fluid in area of master cylinder dust cover and on pedal	31
Fluid on release cylinder	32
Grabbing (chattering) as clutch is engaged	36
High pedal effort	40
Noise in clutch area	38
Pedal feels spongy when depressed	33
Pedal travels to floor - no pressure or very little resistance	30
Transaxle rattling (clicking)	37
Unable to select gears	34

Manual transaxle
Clicking noise in turns	44
Clunk on acceleration or deceleration	43
Knocking noise at low speeds	41
Leaks lubricant	50
Locked in gear	51
Noise most pronounced when turning	42
Noisy in all gears	48
Noisy in neutral with engine running	46
Noisy in one particular gear	47
Slips out of gear	49
Vibration	45

Automatic transaxle
Engine will start in gears other than Park or Neutral	56
Fluid leakage	52
General shift mechanism problems	54
Transaxle fluid brown or has burned smell	53
Transaxle slips, shifts roughly, is noisy or has no drive in forward or reverse gears	57
Transaxle will not downshift with accelerator pedal pressed to the floor	55

Driveaxles
Clicking noise in turns	58
Shudder or vibration during acceleration	59
Vibration at highway speeds	60

Brakes
Brake pedal feels spongy when depressed	68
Brake pedal travels to the floor with little resistance	69
Brake roughness or chatter (pedal pulsates)	63
Dragging brakes	66
Excessive brake pedal travel	65
Excessive pedal effort required to stop vehicle	64
Grabbing or uneven braking action	67
Noise (high-pitched squeal when the brakes are applied)	62
Parking brake does not hold	70
Vehicle pulls to one side during braking	61

Suspension and steering systems
Abnormal or excessive tire wear	72
Abnormal noise at the front end	77
Cupped tires	82
Erratic steering when braking	79
Excessive pitching and/or rolling around corners or during braking	80
Excessive play or looseness in steering system	86
Excessive tire wear on inside edge	84
Excessive tire wear on outside edge	83
Hard steering	75
Poor returnability of steering to center	76
Rattling or clicking noise in steering gear	87
Shimmy, shake or vibration	74
Suspension bottoms	81
Tire tread worn in one place	85
Vehicle pulls to one side	71
Wander or poor steering stability	78
Wheel makes a thumping noise	73

Troubleshooting 0-21

This section provides an easy reference guide to the more common problems which may occur during the operation of your vehicle. These problems and their possible causes are grouped under headings denoting various components or systems, such as Engine, Cooling system, etc. They also refer you to the chapter and/or section which deals with the problem.

Remember that successful troubleshooting is not a mysterious art practiced only by professional mechanics. It is simply the result of the right knowledge combined with an intelligent, systematic approach to the problem. Always work by a process of elimination, starting with the simplest solution and working through to the most complex - and never overlook the obvious. Anyone can run the gas tank dry or leave the lights on overnight, so don't assume that you are exempt from such oversights.

Finally, always establish a clear idea of why a problem has occurred and take steps to ensure that it doesn't happen again. If the electrical system fails because of a poor connection, check the other connections in the system to make sure that they don't fail as well. If a particular fuse continues to blow, find out why - don't just replace one fuse after another. Remember, failure of a small component can often be indicative of potential failure or incorrect functioning of a more important component or system.

Engine

1 Engine will not rotate when attempting to start

1 Battery terminal connections loose or corroded (Chapter 1).
2 Battery discharged or faulty (Chapter 1).
3 Automatic transaxle not completely engaged in Park (Chapter 7) or clutch not completely depressed (Chapter 8).
4 Broken, loose or disconnected wiring in the starting circuit (Chapters 5 and 12).
5 Starter motor pinion jammed in flywheel ring gear (Chapter 5).
6 Starter solenoid faulty (Chapter 5).
7 Starter motor faulty (Chapter 5).
8 Ignition switch faulty (Chapter 12).
9 Starter pinion or flywheel teeth worn or broken (Chapter 5).

2 Engine rotates but will not start

1 Fuel tank empty.
2 Battery discharged (engine rotates slowly) (Chapter 5).
3 Battery terminal connections loose or corroded (Chapter 1).
4 Leaking fuel injector(s), faulty fuel pump, pressure regulator, etc. (Chapter 4).
5 Fuel not reaching fuel rail (Chapter 4).

6 Ignition components damp or damaged (Chapter 5).
7 Worn, faulty or incorrectly gapped spark plugs (Chapter 1).
8 Broken, loose or disconnected wiring in the starting circuit (Chapter 5).
9 Broken, loose or disconnected wires at the ignition coil(s) or faulty coil(s) (Chapter 5).
10 Faulty camshaft or crankshaft position sensor (Chapter 6).

3 Engine hard to start when cold

1 Battery discharged or low (Chapter 1).
2 Malfunctioning fuel system (Chapter 4).
3 Injector(s) leaking (Chapter 4).
4 Faulty coolant temperature sensor (Chapter 6).

4 Engine hard to start when hot

1 Air filter clogged (Chapter 1).
2 Fuel not reaching the fuel injection system (Chapter 4).
3 Corroded battery connections, especially ground (Chapter 1).

5 Starter motor noisy or excessively rough in engagement

1 Pinion or flywheel gear teeth worn or broken (Chapter 5).
2 Starter motor mounting bolts loose or missing (Chapter 5).

6 Engine starts but stops immediately

1 Loose or faulty electrical connections at coil(s) or alternator (Chapter 5).
2 Insufficient fuel reaching the fuel injector(s) (Chapters 1 and 4).
3 Vacuum leak at the gasket between the intake manifold/plenum and throttle body (Chapters 1 and 4).

7 Oil puddle under engine

1 Oil pan gasket and/or oil pan drain bolt washer leaking (Chapter 2).
2 Oil pressure sending unit leaking (Chapter 2).
3 Valve covers leaking (Chapter 2).
4 Engine oil seals leaking (Chapter 2).
5 Oil pump housing leaking (Chapter 2).

8 Engine lopes while idling or idles erratically

1 Vacuum leakage (Chapters 2 and 4).
2 Leaking EGR valve (Chapter 6).

3 Air filter clogged (Chapter 1).
4 Fuel pump not delivering sufficient fuel to the fuel injection system (Chapter 4).
5 Leaking head gasket (Chapter 2).
6 Timing belt and/or pulleys worn (Chapter 2).
7 Camshaft lobes worn (Chapter 2).

9 Engine misses at idle speed

1 Spark plugs worn or not gapped properly (Chapter 1).
2 Faulty spark plug wires (Chapter 1).
3 Vacuum leaks (Chapter 1).
4 Fault in engine management system (Chapter 6).
5 Uneven or low compression (Chapter 2).

10 Engine misses throughout driving speed range

1 Fuel filter clogged and/or impurities in the fuel system (Chapter 1).
2 Low fuel output at the injector(s) (Chapter 4).
3 Faulty or incorrectly gapped spark plugs (Chapter 1).
4 Fault in engine management system (Chapter 6).
5 Leaking spark plug wires (Chapters 1 or 5).
6 Faulty emission system components (Chapter 6).
7 Low or uneven cylinder compression pressures (Chapter 2).
8 Weak or faulty ignition system (Chapter 5).
9 Vacuum leak in fuel injection system, intake manifold/plenum, air control valve or vacuum hoses (Chapter 4).

11 Engine stumbles on acceleration

1 Spark plugs fouled (Chapter 1).
2 Fuel injection system faulty (Chapter 4).
3 Fuel filter clogged (Chapters 1 and 4).
4 Fault in engine management system (Chapter 6).
5 Intake manifold or plenum air leak (Chapters 2 and 4).

12 Engine surges while holding accelerator steady

1 Intake air leak (Chapter 4).
2 Fuel pump faulty (Chapter 4).
3 Loose fuel injector wire harness connectors (Chapter 4).
4 Defective ECM or information sensor (Chapter 6).

13 Engine stalls

1 Idle speed incorrect (Chapter 1).
2 Fuel filter clogged and/or water and impurities in the fuel system (Chapters 1 and 4).
3 Ignition components damp or damaged (Chapter 5).
4 Faulty emissions system components (Chapter 6).
5 Faulty or incorrectly gapped spark plugs (Chapter 1).
6 Faulty spark plug wires (Chapter 1).
7 Vacuum leak in the fuel injection system, intake manifold or vacuum hoses (Chapters 2 and 4).
8 Valve clearances incorrectly set (Chapter 1).

14 Engine lacks power

1 Fault in engine management system (Chapter 6).
2 Faulty spark plug wires (Chapters 1 and 5).
3 Faulty or incorrectly gapped spark plugs (Chapter 1).
4 Fuel injection system malfunction (Chapter 4).
5 Faulty coil(s) (Chapter 5).
6 Brakes binding (Chapter 9).
7 Automatic transaxle fluid level incorrect (Chapter 1).
8 Clutch slipping (Chapter 8).
9 Fuel filter clogged and/or impurities in the fuel system (Chapters 1 and 4).
10 Emissions control systems not functioning properly (Chapter 6).
11 Low or uneven cylinder compression pressures (Chapter 2).
12 Obstructed exhaust system (Chapter 4).

15 Engine backfires

1 Emission control system not functioning properly (Chapter 6).
2 Fault in engine management system (Chapter 6).
3 Faulty secondary ignition system (cracked spark plug insulator, faulty plug wires) (Chapters 1 and 5).
4 Fuel injection system malfunction (Chapter 4).
5 Vacuum leak at fuel injector(s), intake manifold, air control valve or vacuum hoses (Chapters 2 and 4).
6 Valve clearances incorrectly set and/or valves sticking (Chapter 1).

16 Pinging or knocking engine sounds during acceleration or uphill

1 Incorrect grade of fuel.

2 Fault in engine management system (Chapter 6).
3 Fuel injection system faulty (Chapter 4).
4 Improper or damaged spark plugs or wires (Chapter 1).
5 EGR valve not functioning (Chapter 6).
6 Vacuum leak (Chapters 2 and 4).
7 Defective knock sensor (Chapter 6).

17 Engine runs with oil pressure light on

1 Low oil level (Chapter 1).
2 Idle rpm below specification (Chapter 4).
3 Short in wiring circuit (Chapter 12).
4 Faulty oil pressure sender (Chapter 2).
5 Worn engine bearings and/or oil pump (Chapter 2).

18 Engine diesels (continues to run) after switching off

1 Idle speed too high (Chapter 1).
2 Excessive engine operating temperature (Chapter 3).
3 Fault in engine management system (Chapter 6).

Engine electrical system

19 Battery will not hold a charge

1 Alternator drivebelt defective or not adjusted properly (Chapter 1).
2 Battery electrolyte level low (Chapter 1).
3 Battery terminals loose or corroded (Chapter 1).
4 Alternator not charging properly (Chapter 5).
5 Loose, broken or faulty wiring in the charging circuit (Chapter 5).
6 Short in vehicle wiring (Chapter 12).
7 Internally defective battery (Chapters 1 and 5).

20 Alternator light fails to go out

1 Faulty alternator or charging circuit (Chapter 5).
2 Alternator drivebelt defective or out of adjustment (Chapter 1).
3 Alternator voltage regulator inoperative (Chapter 5).

21 Alternator light fails to come on when key is turned on

1 Warning light bulb defective (Chapter 12).
2 Fault in the printed circuit, dash wiring or bulb holder (Chapter 12).

Fuel system

22 Excessive fuel consumption

1 Dirty or clogged air filter element (Chapter 1).
2 Fault in engine management system (Chapter 6).
3 Emissions systems not functioning properly (Chapter 6).
4 Fuel injection system not functioning properly (Chapter 4).
5 Low tire pressure or incorrect tire size (Chapter 1).

23 Fuel leakage and/or fuel odor

1 Leaking fuel feed or return line (Chapters 1 and 4).
2 Tank overfilled.
3 Evaporative canister filter clogged (Chapters 1 and 6).
4 Fuel injection system not functioning properly (Chapter 4).

Cooling system

24 Overheating

1 Insufficient coolant in system (Chapter 1).
2 Water pump defective (Chapter 3).
3 Radiator core blocked or grille restricted (Chapter 3).
4 Thermostat faulty (Chapter 3).
5 Electric coolant fan blades broken or cracked (Chapter 3).
6 Radiator cap not maintaining proper pressure (Chapter 3).
7 Fault in engine management system (Chapter 6).

25 Overcooling

1 Faulty thermostat (Chapter 3).
2 Inaccurate temperature gauge sending unit (Chapter 3)

26 External coolant leakage

1 Deteriorated/damaged hoses; loose clamps (Chapters 1 and 3).
2 Water pump defective (Chapter 3).
3 Leakage from radiator core or coolant reservoir bottle (Chapter 3).
4 Engine drain or water jacket core plugs leaking (Chapter 2).

27 Internal coolant leakage

1 Leaking cylinder head gasket (Chapter 2).

Troubleshooting 0-23

2 Cracked cylinder bore or cylinder head (Chapter 2).

28 Coolant loss

1 Too much coolant in system (Chapter 1).
2 Coolant boiling away because of overheating (Chapter 3).
3 Internal or external leakage (Chapter 3).
4 Faulty radiator cap (Chapter 3).

29 Poor coolant circulation

1 Inoperative water pump (Chapter 3).
2 Restriction in cooling system (Chapters 1 and 3).
3 Water pump drivebelt defective/out of adjustment (Chapter 1).
4 Thermostat sticking (Chapter 3).

Clutch

30 Pedal travels to floor - no pressure or very little resistance

1 Master or release cylinder faulty (Chapter 8).
2 Hose/pipe burst or leaking (Chapter 8).
3 Connections leaking (Chapter 8).
4 No fluid in reservoir (Chapter 8).
5 If fluid level in reservoir rises as pedal is depressed, master cylinder center valve seal is faulty (Chapter 8).
6 If there is fluid on dust seal at master cylinder, piston primary seal is leaking (Chapter 8).
7 Broken release bearing or fork (Chapter 8).

31 Fluid in area of master cylinder dust cover and on pedal

Rear seal failure in master cylinder (Chapter 8).

32 Fluid on release cylinder

Release cylinder plunger seal faulty (Chapter 8).

33 Pedal feels spongy when depressed

Air in system (Chapter 8).

34 Unable to select gears

1 Faulty transaxle (Chapter 7).
2 Faulty clutch disc (Chapter 8).
3 Release lever and bearing not assembled properly (Chapter 8).
4 Faulty pressure plate (Chapter 8).
5 Pressure plate-to-flywheel bolts loose (Chapter 8).

35 Clutch slips (engine speed increases with no increase in vehicle speed)

1 Clutch plate worn (Chapter 8).
2 Clutch plate is oil soaked by leaking rear main seal (Chapter 8).
3 Clutch plate not seated. It may take 30 or 40 normal starts for a new one to seat.
4 Warped pressure plate or flywheel (Chapter 8).
5 Weak diaphragm spring (Chapter 8).
6 Clutch plate overheated. Allow to cool.

36 Grabbing (chattering) as clutch is engaged

1 Oil on clutch plate lining, burned or glazed facings (Chapter 8).
2 Worn or loose engine or transaxle mounts (Chapters 2 and 7).
3 Worn splines on clutch plate hub (Chapter 8).
4 Warped pressure plate or flywheel (Chapter 8).
5 Burned or smeared resin on flywheel or pressure plate (Chapter 8).

37 Transaxle rattling (clicking)

1 Release lever loose (Chapter 8).
2 Clutch plate damper spring failure (Chapter 8).
3 Low engine idle speed (Chapter 1).

38 Noise in clutch area

1 Fork shaft improperly installed (Chapter 8).
2 Faulty bearing (Chapter 8).

39 Clutch pedal stays on floor

1 Clutch master cylinder piston binding in bore (Chapter 8).
2 Broken release bearing or fork (Chapter 8).

40 High pedal effort

1 Piston binding in bore (Chapter 8).
2 Pressure plate faulty (Chapter 8).
3 Incorrect size master or release cylinder (Chapter 8).

Manual transaxle

41 Knocking noise at low speeds

1 Worn driveaxle constant velocity (CV) joints (Chapter 8).
2 Worn side gear shaft counterbore in differential case (Chapter 7A).*

42 Noise most pronounced when turning

Differential gear noise (Chapter 7A).*

43 Clunk on acceleration or deceleration

1 Loose engine or transaxle mounts (Chapters 2 and 7A).
2 Worn differential pinion shaft in case.*
3 Worn side gear shaft counterbore in differential case (Chapter 7A).*
4 Worn or damaged driveaxle inner CV joints (Chapter 8).

44 Clicking noise in turns

Worn or damaged outer CV joint (Chapter 8).

45 Vibration

1 Rough wheel bearing (Chapters 1 and 10).
2 Damaged driveaxle (Chapter 8).
3 Out of round tires (Chapter 1).
4 Tire out of balance (Chapters 1 and 10).
5 Worn CV joint (Chapter 8).

46 Noisy in neutral with engine running

1 Damaged input gear bearing (Chapter 7A).*
2 Damaged clutch release bearing (Chapter 8).

47 Noisy in one particular gear

1 Damaged or worn constant mesh gears (Chapter 7A).*
2 Damaged or worn synchronizers (Chapter 7A).*
3 Bent reverse fork (Chapter 7A).*
4 Damaged fourth speed gear or output gear (Chapter 7A).*
5 Worn or damaged reverse idler gear or idler bushing (Chapter 7A).*

48 Noisy in all gears

1 Insufficient lubricant (Chapters 1 and 7A).
2 Damaged or worn bearings (Chapter 7A).*
3 Worn or damaged input gear shaft and/or output gear shaft (Chapter 7A).*

49 Slips out of gear

1 Worn or improperly adjusted linkage (Chapter 7A).
2 Transaxle loose on engine (Chapter 7A).
3 Shift linkage does not work freely, binds (Chapter 7A).
4 Input gear bearing retainer broken or loose (Chapter 7A).*
5 Dirt between clutch cover and engine housing (Chapter 7A).
6 Worn shift fork (Chapter 7A).*

50 Leaks lubricant

1 Side gear shaft seals worn (Chapter 7).
2 Excessive amount of lubricant in transaxle (Chapters 1 and 7A).
3 Loose or broken input gear shaft bearing retainer (Chapter 7A).*
4 Input gear bearing retainer O-ring and/or lip seal damaged (Chapter 7A).*

51 Locked in gear

Lock pin or interlock pin missing (Chapter 7A).*

Although the corrective action necessary to remedy the symptoms described is beyond the scope of this manual, the above information should be helpful in isolating the cause of the condition so that the owner can communicate clearly with a professional mechanic.

Automatic transaxle

Note: *Due to the complexity of the automatic transaxle, it is difficult for the home mechanic to properly diagnose and service this component. For problems other than the following, the vehicle should be taken to a dealer or transaxle shop.*

52 Fluid leakage

1 Automatic transaxle fluid is a deep red color. Fluid leaks should not be confused with engine oil, which can easily be blown onto the transaxle by air flow.
2 To pinpoint a leak, first remove all built-up dirt and grime from the transaxle housing with degreasing agents and/or steam cleaning. Then drive the vehicle at low speeds so air flow will not blow the leak far from its source. Raise the vehicle and determine where the leak is coming from. Common areas of leakage are:
a) Pan (Chapters 1 and 7)
b) Dipstick tube (Chapters 1 and 7)
c) Transaxle oil lines (Chapter 7)
d) Speed sensor (Chapter 7)
e) Differential drain plug (Chapters 1 and 7B)

53 Transaxle fluid brown or has a burned smell

Transaxle fluid overheated (Chapter 1).

54 General shift mechanism problems

1 Chapter 7, Part B, deals with checking and adjusting the shift linkage on automatic transaxles. Common problems which may be attributed to poorly adjusted linkage are:
a) Engine starting in gears other than Park or Neutral.
b) Indicator on shifter pointing to a gear other than the one actually being used.
c) Vehicle moves when in Park.
2 Refer to Chapter 7B for the shift linkage adjustment procedure.

55 Transaxle will not downshift with accelerator pedal pressed to the floor

Throttle valve cable out of adjustment (Chapter 7B).

56 Engine will start in gears other than Park or Neutral

Neutral start switch malfunctioning (Chapter 7B).

57 Transaxle slips, shifts roughly, is noisy or has no drive in forward or reverse gears

There are many probable causes for the above problems, but the home mechanic should be concerned with only one possibility - fluid level. Before taking the vehicle to a repair shop, check the level and condition of the fluid as described in Chapter 1. Correct the fluid level as necessary or change the fluid and filter if needed. If the problem persists, have a professional diagnose the cause.

Driveaxles

58 Clicking noise in turns

Worn or damaged outboard CV joint (Chapter 8).

59 Shudder or vibration during acceleration

1 Excessive toe-in (Chapter 10).
2 Incorrect spring heights (Chapter 10).
3 Worn or damaged inboard or outboard CV joints (Chapter 8).
4 Sticking inboard CV joint assembly (Chapter 8).

60 Vibration at highway speeds

1 Out of balance front wheels and/or tires (Chapters 1 and 10).
2 Out of round front tires (Chapters 1 and 10).
3 Worn CV joint(s) (Chapter 8).

Brakes

Note: *Before assuming that a brake problem exists, make sure that:*
a) *The tires are in good condition and properly inflated (Chapter 1).*
b) *The front end alignment is correct (Chapter 10).*
c) *The vehicle is not loaded with weight in an unequal manner.*

61 Vehicle pulls to one side during braking

1 Incorrect tire pressures (Chapter 1).
2 Front end out of alignment (have the front end aligned).
3 Front, or rear, tires not matched to one another.
4 Restricted brake lines or hoses (Chapter 9).
5 Malfunctioning drum brake or caliper assembly (Chapter 9).
6 Loose suspension parts (Chapter 10).
7 Loose calipers (Chapter 9).
8 Excessive wear of brake shoe or pad material or disc/drum on one side.

62 Noise (high-pitched squeal when the brakes are applied)

Front and/or rear disc brake pads worn out. The noise comes from the wear sensor rubbing against the disc (does not apply to all vehicles). Replace pads with new ones immediately (Chapter 9).

Troubleshooting

63 Brake roughness or chatter (pedal pulsates)

1. Excessive lateral runout (Chapter 9).
2. Uneven pad wear (Chapter 9).
3. Defective disc (Chapter 9).

64 Excessive brake pedal effort required to stop vehicle

1. Malfunctioning power brake booster (Chapter 9).
2. Partial system failure (Chapter 9).
3. Excessively worn pads or shoes (Chapter 9).
4. Piston in caliper or wheel cylinder stuck or sluggish (Chapter 9).
5. Brake pads or shoes contaminated with oil or grease (Chapter 9).
6. New pads or shoes installed and not yet seated. It will take a while for the new material to seat against the disc or drum.

65 Excessive brake pedal travel

1. Partial brake system failure (Chapter 9).
2. Insufficient fluid in master cylinder (Chapters 1 and 9).
3. Air trapped in system (Chapters 1 and 9).

66 Dragging brakes

1. Incorrect adjustment of brake light switch (Chapter 9).
2. Master cylinder pistons not returning correctly (Chapter 9).
3. Restricted brakes lines or hoses (Chapters 1 and 9).
4. Incorrect parking brake adjustment (Chapter 9).

67 Grabbing or uneven braking action

1. Malfunction of proportioning valve (Chapter 9).
2. Malfunction of power brake booster unit (Chapter 9).
3. Binding brake pedal mechanism (Chapter 9).

68 Brake pedal feels spongy when depressed

1. Air in hydraulic lines (Chapter 9).
2. Master cylinder mounting bolts loose (Chapter 9).
3. Master cylinder defective (Chapter 9).

69 Brake pedal travels to the floor with little resistance

1. Little or no fluid in the master cylinder reservoir caused by leaking caliper piston(s) (Chapter 9).
2. Loose, damaged or disconnected brake lines (Chapter 9).

70 Parking brake does not hold

Parking brake linkage improperly adjusted (Chapters 1 and 9).

Suspension and steering systems

Note: Before attempting to diagnose the suspension and steering systems, perform the following preliminary checks:

a) *Tires for wrong pressure and uneven wear.*
b) *Steering universal joints from the column to the rack and pinion for loose connectors or wear.*
c) *Front and rear suspension and the rack and pinion assembly for loose or damaged parts.*
d) *Out-of-round or out-of-balance tires, bent rims and loose and/or rough wheel bearings.*

71 Vehicle pulls to one side

1. Mismatched or uneven tires (Chapter 10).
2. Broken or sagging springs (Chapter 10).
3. Wheel alignment (Chapter 10).
4. Front brake dragging (Chapter 9).

72 Abnormal or excessive tire wear

1. Wheel alignment (Chapter 10).
2. Sagging or broken springs (Chapter 10).
3. Tire out of balance (Chapter 10).
4. Worn strut damper (Chapter 10).
5. Overloaded vehicle.
6. Tires not rotated regularly.

73 Wheel makes a thumping noise

1. Blister or bump on tire (Chapter 10).
2. Improper strut damper action (Chapter 10).

74 Shimmy, shake or vibration

1. Tire or wheel out-of-balance or out-of-round (Chapter 10).
2. Loose or worn wheel bearings (Chapters 1, 8 and 10).
3. Worn tie-rod ends (Chapter 10).
4. Worn lower balljoints (Chapters 1 and 10).
5. Excessive wheel runout (Chapter 10).
6. Blister or bump on tire (Chapter 10).

75 Hard steering

1. Lack of lubrication at balljoints, tie-rod ends and rack and pinion assembly (Chapter 10).
2. Front wheel alignment (Chapter 10).
3. Low tire pressure(s) (Chapters 1 and 10).

76 Poor returnability of steering to center

1. Lack of lubrication at balljoints and tie-rod ends (Chapter 10).
2. Binding in balljoints (Chapter 10).
3. Binding in steering column (Chapter 10).
4. Lack of lubricant in steering gear assembly (Chapter 10).
5. Front wheel alignment (Chapter 10).

77 Abnormal noise at the front end

1. Lack of lubrication at balljoints and tie-rod ends (Chapters 1 and 10).
2. Damaged strut mounting (Chapter 10).
3. Worn control arm bushings or tie-rod ends (Chapter 10).
4. Loose stabilizer bar (Chapter 10).
5. Loose wheel nuts (Chapters 1 and 10).
6. Loose suspension bolts (Chapter 10)

78 Wander or poor steering stability

1. Mismatched or uneven tires (Chapter 10).
2. Lack of lubrication at balljoints and tie-rod ends (Chapters 1 and 10).
3. Worn strut assemblies (Chapter 10).
4. Loose stabilizer bar (Chapter 10).
5. Broken or sagging springs (Chapter 10).
6. Wheels out of alignment (Chapter 10).

79 Erratic steering when braking

1. Wheel bearings worn (Chapter 10).
2. Broken or sagging springs (Chapter 10).
3. Leaking wheel cylinder or caliper (Chapter 10).
4. Warped rotors or drums (Chapter 10).

80 Excessive pitching and/or rolling around corners or during braking

1. Loose stabilizer bar (Chapter 10).

Troubleshooting

2 Worn strut dampers or mountings (Chapter 10).
3 Broken or sagging springs (Chapter 10).
4 Overloaded vehicle.

81 Suspension bottoms

1 Overloaded vehicle.
2 Worn strut dampers (Chapter 10).
3 Incorrect, broken or sagging springs (Chapter 10).

82 Cupped tires

1 Front wheel or rear wheel alignment (Chapter 10).
2 Worn strut dampers (Chapter 10).
3 Wheel bearings worn (Chapter 10).
4 Excessive tire or wheel runout (Chapter 10).
5 Worn balljoints (Chapter 10).

83 Excessive tire wear on outside edge

1 Inflation pressures incorrect (Chapter 1).
2 Excessive speed in turns.
3 Front end alignment incorrect (excessive toe-in). Have professionally aligned.
4 Suspension arm bent or twisted (Chapter 10).

84 Excessive tire wear on inside edge

1 Inflation pressures incorrect (Chapter 1).
2 Front end alignment incorrect (toe-out). Have professionally aligned.
3 Loose or damaged steering components (Chapter 10).

85 Tire tread worn in one place

1 Tires out of balance.
2 Damaged or buckled wheel. Inspect and replace if necessary.
3 Defective tire (Chapter 1).

86 Excessive play or looseness in steering system

1 Wheel bearing(s) worn (Chapter 10).
2 Tie-rod end loose (Chapter 10).
3 Steering gear loose (Chapter 10).
4 Worn or loose steering intermediate shaft (Chapter 10).

87 Rattling or clicking noise in steering gear

1 Steering gear loose (Chapter 10).
2 Steering gear defective.

Chapter 1
Tune-up and routine maintenance

Contents

	Section		Section
Air filter replacement	16	Fuel tank cap gasket replacement	33
Automatic transaxle differential lubricant level check	18	Interior ventilation filter replacement	21
Automatic transaxle/differential fluid change	27	Introduction	2
Automatic transaxle fluid level check	7	Maintenance schedule	1
Battery check, maintenance and charging	10	Manual transaxle lubricant change	28
Brake check	15	Manual transaxle lubricant level check	19
Chassis and body fastener check	29	Positive Crankcase Ventilation (PCV) valve check and replacement	34
CHECK ENGINE light on	See Chapter 6	Power steering fluid level check	6
Clutch/brake pedal freeplay, height check and adjustment	23	Spark plug check and replacement	30
Cooling system check	13	Spark plug wire check and replacement	31
Cooling system servicing (draining, flushing and refilling)	25	Steering, suspension and driveaxle boot check	20
Drivebelt check, adjustment and replacement	11	Tire and tire pressure checks	5
Engine oil and oil filter change	8	Tire rotation	14
Evaporative emissions control system check	26	Tune-up general information	3
Exhaust system check	22	Underhood hose check and replacement	12
Fluid level checks	4	Valve clearance check and adjustment	32
Fuel filter replacement	24	Windshield wiper blade inspection and replacement	9
Fuel system check	17		

Specifications

Recommended lubricants and fluids

Note: *Listed here are manufacturer recommendations at the time this manual was written. Manufacturers occasionally upgrade their fluid and lubricant specifications, so check with your auto parts store for current recommendations.*

Engine oil type	API grade SJ or ILSAC GF-2 multigrade and fuel-efficient oil
Viscosity	See accompanying chart
Fuel	Unleaded fuel, 87 octane or higher
Coolant	50 to 70-percent ethylene-glycol based antifreeze and water
Automatic transaxle fluid type	
Camry/Avalon/Camry Solara	DEXRON III automatic transmission fluid
Lexus ES 300	
1999 and earlier	DEXRON III automatic transmission fluid
2000 and later	Toyota ATF Type T-IV automatic transmission fluid
Automatic transaxle differential fluid type	DEXRON III automatic transmission fluid
Manual transaxle lubricant type	API GL-4 or GL-5 75W-90 gear oil
Brake fluid type	DOT 3 brake fluid
Clutch fluid type	DOT 3 brake fluid
Power steering system fluid	Dexron II or III automatic transmission fluid

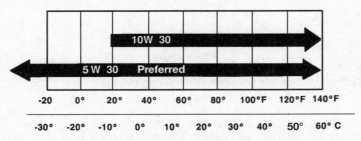

Engine oil viscosity chart

Capacities

Engine oil (including filter)
    Four-cylinder engine .... 3.8 qts
    V6 engine .... 5.0 qts
Coolant
    Four-cylinder engine .... 7.3 qts
    V6 engine .... 9.6 qts
Transaxle
    Automatic
        Four-cylinder engine
            Drain and refill .... 2.6 qts
            Differential .... 1.7 qts
        V6 engine
            Camry/Avalon/Solara
                Drain and refill .... 5.0 qts
                Differential .... 1.7 qts
            Lexus ES 300
                1997
                      Drain and refill .... 3.7 qts
                      Differential .... 1 qt
                1998 (drain and refill) .... 5.0 qts
                1999 and later (drain and refill) .... 3.7 qts
    Manual
        Four-cylinder engine .... 2.3 qts
        V6 engine .... 4.4 qts

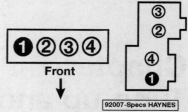

Cylinder numbering and coil terminal location - four-cylinder engine

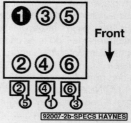

Cylinder numbering and coil terminal location - V6 engine

Ignition system

Spark plug
    Camry and Solara
        Type .... NGK BKR6EKPB-11 or equivalent
        Gap .... 0.043 inch
    Avalon
        1999 and earlier
            Type .... NGK BKR6EKPB-11 or equivalent
            Gap .... 0.043 inch
        2000 and later
            Type .... NGK IFR6A-11 or equivalent
            Gap .... 0.043 inch
    Lexus ES 300
        1998 and earlier
            Type .... NGK BKR6EKPB-11 or equivalent
            Gap .... 0.043 inch
        1999 and later
            Type .... NGK IFR6A-11 or equivalent
            Gap .... 0.043 inch
Spark plug wire resistance .... Less than 25,000 ohms
Engine firing order
    Four-cylinder engine 1-3-4-2
    V6 engine 1-2-3-4-5-6

Valve clearance (engine cold)

Four-cylinder engine
    Intake .... 0.007 to 0.011 inch
    Exhaust .... 0.011 to 0.015 inch
V6 engine
    Intake .... 0.006 to 0.010 inch
    Exhaust .... 0.010 to 0.014 inch

Cooling system

Thermostat rating
    Starts to open .... 176 to 183-degrees F
    Fully open .... 203-degrees F

Accessory drivebelt tension (with Burroughs or Nippondenso tension gauge)

Four-cylinder engine
    Alternator (with air conditioning)
        New belt .... 140 to 190 lbs
        Used belt .... 100 to 120 lbs

Chapter 1 Tune-up and routine maintenance

Alternator (without air conditioning)	
New belt	100 to 150 lbs
Used belt	75 to 115 lbs
Power steering pump	
New belt	95 to 145 lbs
Used belt	60 to 100 lbs
V6 engine	
Alternator	
New belt	170 to 180 lbs
Used belt	95 to 135 lbs
Power steering pump	
New belt	150 to 185 lbs
Used belt	95 to 135 lbs

Clutch

Clutch pedal freeplay	7/32 to 5/8 inch
Pedal height	
Four-cylinder engine	6-1/8 to 6-5/8 inches
V6 engine	6-3/8 to 6-3/4 inches
Pushrod play at pedal top	3/64 to 13/64 inch

Brakes

Disc brake pad lining thickness (minimum)	1/16 inch
Drum brake shoe lining thickness (minimum)	1/16 inch
Brake pedal	
Freeplay	1/32 to 1/4 inch
Free height	6 to 6-3/8 inches
Brake light switch clearance	1/32 to 3/32 inch
Pedal reserve height	2-3/4 inches
Parking brake adjustment	back off 8 notches from contact

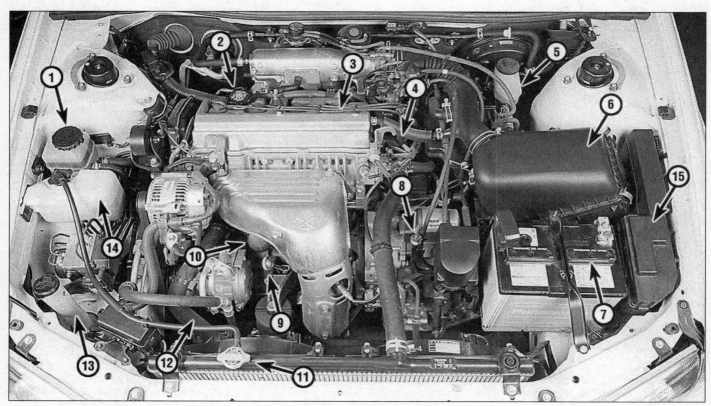

Typical four-cylinder engine compartment layout

1	Power steering fluid reservoir	6	Air filter housing	11	Radiator cap
2	Engine oil filler cap	7	Battery	12	Radiator hose
3	Spark plug wires	8	Automatic transaxle fluid dipstick	13	Windshield washer fluid reservoir
4	Ignition coil pack	9	Engine oil dipstick	14	Engine coolant reservoir
5	Brake master cylinder reservoir	10	Engine oil filter	15	Main fuse/relay box

Chapter 1 Tune-up and routine maintenance

Suspension and steering
Steering wheel freeplay limit.. 1-3/16 inches
Balljoint allowable movement .... 0 inch

Torque specifications — Ft-lbs (unless otherwise indicated)
Automatic transaxle
    Pan bolts .... 43 in-lbs
    Strainer bolts .... 96 in-lbs
    Drain plug 36
Manual transaxle drain and filler plugs 36
Wheel lug nuts 76
Chassis and body
    Front seat mounting bolts 32
    Front suspension subframe (see Chapter 7A)
        Bolt A 134
        Bolt B 24
        Nut C 27
    Rear suspension
        Crossmember-to-body nuts/bolts 38
        Stopper plate nuts 28
Spark plugs 13

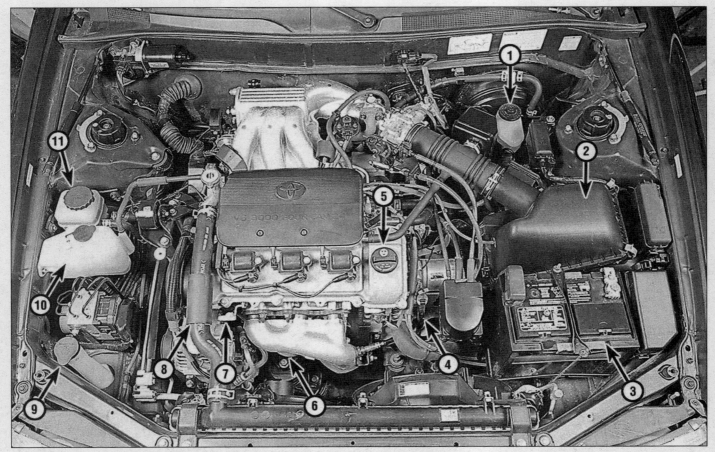

Typical V6 engine compartment layout

1   Brake master cylinder reservoir
2   Air filter housing
3   Battery
4   Automatic transaxle fluid dipstick
5   Engine oil filler cap
6   Engine oil filter
7   Engine oil dipstick
8   Radiator hose
9   Windshield washer fluid reservoir
10 Engine coolant reservoir
11 Power steering fluid reservoir

Chapter 1 Tune-up and routine maintenance 1-5

Typical front underside components

1. Engine oil drain plug
2. Front brake caliper
3. Outer driveaxle boot
4. Steering gear boot
5. Automatic transaxle drain plug
6. Exhaust pipe
7. Inner driveaxle boot
8. Brake hose

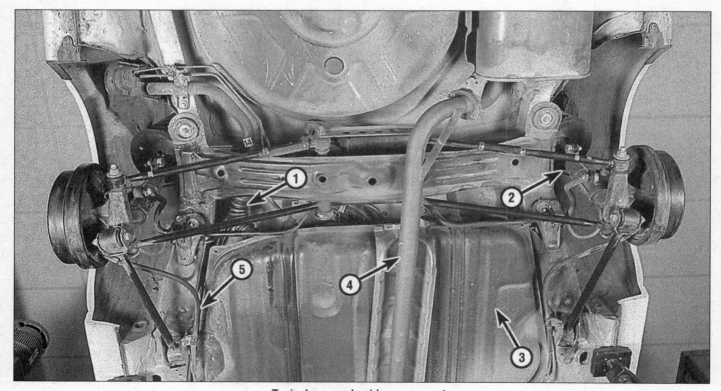

Typical rear underside components

1. Fuel pipe-to-tank hose
2. Shock and spring assembly
3. Fuel tank
4. Exhaust pipe
5. Parking brake cable

Toyota Camry and Lexus 300, Avalon and Solara Maintenance Schedule

The maintenance intervals in this manual are provided with the assumption that you, not the dealer, will be doing the work. These are the minimum maintenance intervals recommended by the factory for vehicles that are driven daily. If you wish to keep your vehicle in peak condition at all times, you may wish to perform some of these procedures even more often. Because frequent maintenance enhances the efficiency, performance and resale value of your car, we encourage you to do so. If you drive in dusty areas, tow a trailer, idle or drive at low speeds for extended periods or drive for short distances (less than four miles) in below freezing temperatures, shorter intervals are also recommended.

When your vehicle is new, it should be serviced by a factory authorized dealer service department to protect the factory warranty. In many cases, the initial maintenance check is done at no cost to the owner.

Every 250 miles or weekly, whichever comes first

Check the engine oil level (Section 4)
Check the engine coolant level (Section 4)
Check the windshield washer fluid level (Section 4)
Check the brake and clutch fluid levels (Section 4)
Check the tires and tire pressures (Section 5)

Every 3000 miles or 3 months, whichever comes first

All items listed above plus:
Check the power steering fluid level (Section 6)
Check the automatic transaxle fluid level (Section 7)
Change the engine oil and oil filter (Section 8)

Every 6000 miles or 6 months, whichever comes first

Inspect and replace if necessary the windshield wiper blades (Section 9)
Check and service the battery (Section 10)
Check and adjust if necessary the engine drivebelts (Section 11)
Inspect and replace if necessary all underhood hoses (Section 12)
Check the cooling system (Section 13)
Rotate the tires (Section 14)

Every 15,000 miles or 12 months, whichever comes first

All items listed above plus:
Inspect the brake system (Section 15)*
Replace the air filter (Section 16)
Inspect the fuel system (Section 17)
Check the automatic transaxle differential lubricant level (Section 18)*
Check the manual transaxle lubricant level (Section 19)*
Inspect the suspension, steering components and driveaxle boots (Section 20)
Inspect the exhaust system (Section 22)
Replace interior ventilation filter (Section 21)*

Every 30,000 miles or 24 months, whichever comes first

All items listed above plus:
Check the clutch pedal for proper freeplay and height (Section 23)
Replace the fuel filter (Section 24)
Service the cooling system (drain, flush and refill) (Section 25)
Inspect the evaporative emissions control system (Section 26)
Change the automatic transaxle and on some models, differential fluid (Section 27)**
Change the manual transaxle lubricant (Section 28)
Check and tighten critical chassis and body fasteners (Section 29)
Replace the spark plugs (Section 30)
Inspect and replace if necessary the spark plug wires (Section 31)

Every 60,000 miles or 48 months, whichever comes first

Inspect and if necessary adjust the valve clearance (Section 32)
Replace the fuel tank cap gasket (Section 33)
Check and replace if necessary the PCV valve (Section 34)

Every 90,000 miles

Replace the timing belt (Chapter 2A or 2B) ***

* *This item is affected by "severe" operating conditions as described below. If your vehicle is operated under "severe" conditions, perform all maintenance indicated with an asterisk (*) at 3000 mile/3 month intervals. Severe conditions are indicated if you mainly operate your vehicle under one or more of the following conditions:*

Operating in dusty areas
Towing a trailer
Idling for extended periods and/or low speed operation
Operating when outside temperatures remain below freezing and when most trips are less than 4 miles

** *If operated under one or more of the following conditions, change the automatic transaxle fluid and differential lubricant every 15,000 miles:*

In heavy city traffic where the outside temperature regularly reaches 90-degrees F (32-degrees C) or higher In hilly or mountainous terrain
Frequent trailer pulling

*** *Replace the timing belt at 60,000 miles only if operated under severe conditions.*

Chapter 1 Tune-up and routine maintenance

2 Introduction

This chapter is designed to help the home mechanic maintain his vehicle for peak performance, economy, safety and long life.

On the following pages is a master maintenance schedule, followed by sections dealing specifically with each item on the schedule. Visual checks, adjustments, component replacement and other helpful items are included. Refer to the accompanying illustrations of the engine compartment and the underside of the vehicle for the location of various components.

Servicing your vehicle in accordance with the mileage/time maintenance schedule and the following Sections will provide it with a planned maintenance program that should result in a long and reliable service life. This is a comprehensive plan, so maintaining some items but not others at the specified service intervals won't produce the same results.

As you service your vehicle, you will discover that many of the procedures can - and should - be grouped together because of the nature of the particular procedure you're performing or because of the close proximity of two otherwise unrelated components to one another.

For example, if the vehicle is raised for any reason, you should inspect the exhaust, suspension, steering and fuel systems while you're under the vehicle. When you're rotating the tires, it makes good sense to check the brakes and wheel bearings since the wheels are already removed.

Finally, let's suppose you have to borrow or rent a torque wrench. Even if you only need to tighten the spark plugs, you might as well check the torque of as many critical fasteners as time allows.

The first step of this maintenance program is to prepare yourself before the actual work begins. Read through all sections pertinent to the procedures you're planning to do, then make a list of and gather together all the parts and tools you will need to do the job. If it looks as if you might run into problems during a particular segment of some procedure, seek advice from your local parts man or dealer service department.

3 Tune-up general information

The term tune-up is used in this manual to represent a combination of individual operations rather than one specific procedure.

If, from the time the vehicle is new, the routine maintenance schedule is followed closely and frequent checks are made of fluid levels and high wear items, as suggested throughout this manual, the engine will be kept in relatively good running condition and the need for additional work will be minimized.

More likely than not, however, there will be times when the engine is running poorly due to lack of regular maintenance. This is even more likely if a used vehicle, which has not received regular and frequent maintenance checks, is purchased. In such cases, an engine tune-up will be needed outside of the regular routine maintenance intervals.

The first step in any tune-up or engine diagnosis to help correct a poor running engine would be a cylinder compression check. A check of the engine compression (Chapter 2 Part C) will give valuable information regarding the overall performance of many internal components and should be used as a basis for tune-up and repair procedures. If, for instance, a compression check indicates serious internal engine wear, a conventional tune-up will not help the running condition of the engine and would be a waste of time and money. Also in Chapter 2, Part C is information on checking engine vacuum, which also gives information on the engine's state-of-tune and condition.

The following series of operations are those most often needed to bring a generally poor-running engine back into a proper state of tune.

Minor tune-up

Clean, inspect and test the battery (Section 10)
Check all engine related fluids (Section 4)
Check and adjust the drivebelts (Section 11)
Inspect the spark plug wires (Section 31)
Check the air filter (Section 16)
Check the cooling system (Section 13)
Check all underhood hoses (Section 12)

Major tune-up

All items listed under Minor tune-up, plus . . .
Replace the spark plugs (Section 30)
Check the charging system (Chapter 5)
Check the fuel system (Section 17)
Replace the air filter (Section 16)
Replace the spark plug wires (Section 31)

4 Fluid level checks (every 250 miles or weekly)

1 Fluids are an essential part of the lubrication, cooling, brake, clutch and other systems. Because these fluids gradually become depleted and/or contaminated during normal operation of the vehicle, they must be periodically replenished. See *Recommended lubricants and fluids* and *capacities* at the beginning of this Chapter before adding fluid to any of the following components. **Note:** *The vehicle must be on level ground before fluid levels can be checked.*

Engine oil

Refer to illustrations 4.2, 4.4 and 4.6

2 The engine oil level is checked with a dipstick located at the front side of the engine **(see illustration)**. The dipstick extends through a metal tube from which it

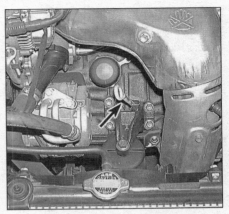

4.2 The engine oil dipstick is mounted on the front (radiator) side of the engine

protrudes down into the engine oil pan.

3 The oil level should be checked before the vehicle has been driven, or about 15 minutes after the engine has been shut off. If the oil is checked immediately after driving the vehicle, some of the oil will remain in the upper engine components, producing an inaccurate reading on the dipstick.

4 Pull the dipstick from the tube and wipe all the oil from the end with a clean rag or paper towel. Insert the clean dipstick all the way back into its metal tube and pull it out again. Observe the oil at the end of the dipstick. At its highest point, the level should be between the L and F marks **(see illustration)**.

5 It takes one quart of oil to raise the level from the L mark to the F mark on the dipstick. Do not allow the level to drop below the L mark or oil starvation may cause engine damage. Conversely, overfilling the engine (adding oil above the F mark) may cause oil-fouled spark plugs, oil leaks or oil seal failures.

6 Remove the threaded cap from the valve cover to add oil **(see illustration)**. Use a funnel to prevent spills. After adding the oil, install the filler cap hand tight. Start the

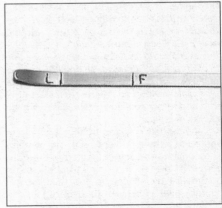

4.4 The oil level should be at or near the F mark on the dipstick - if it isn't, add enough oil to bring the level to or near the F mark (it takes one quart to raise the level from the L to F mark)

4.6 The threaded oil filler cap is located on the valve cover - to prevent dirt from contaminating the engine, always make sure the area around this opening is clean before removing the cap

4.8 Make sure the coolant level is between the Full and Low lines - if it's below the Low line, add a sufficient quantity of the specified mixture of antifreeze and water

4.14 The windshield washer fluid reservoir is located at the right front corner of the engine compartment - fluid can be added after flipping up the cap

engine and look carefully for any small leaks around the oil filter or drain plug. Stop the engine and check the oil level again after it has had sufficient time to drain from the upper block and cylinder head galleys.

7   Checking the oil level is an important preventive maintenance step. A continually dropping oil level indicates oil leakage through damaged seals, from loose connections, or past worn rings or valve guides. If the oil looks milky in color or has water droplets in it, a cylinder head gasket may be blown. The engine should be checked immediately. The condition of the oil should also be checked. Each time you check the oil level, slide your thumb and index finger up the dipstick before wiping off the oil. If you see small dirt or metal particles clinging to the dipstick, the oil should be changed (see Section 8).

Engine coolant

Refer to illustration 4.8

Warning: *Do not allow antifreeze to come in contact with your skin or painted surfaces of the vehicle. Flush contaminated areas immediately with plenty of water. Don't store new coolant or leave old coolant lying around where it's accessible to children or pets – they're attracted by its sweet smell. Ingestion of even a small amount of coolant can be fatal! Wipe up garage floor and drip pan spills immediately. Keep antifreeze containers covered and repair cooling system leaks as soon as they're noticed.*

8   All models covered by this manual are equipped with a coolant recovery system. The coolant reservoir is located in the front corner of the engine compartment and is connected by a hose to the base of the coolant filler cap **(see illustration)**. If the coolant heats up during engine operation, coolant can escape through the pressurized filler cap and connecting hose into the reservoir. As the engine cools, the coolant is automatically drawn back into the cooling system to maintain the correct level.

9   The coolant level should be checked regularly. It must be between the Full and Low lines on the tank. The level will vary with the temperature of the engine. When the engine is cold, the coolant level should be at or slightly above the Low mark on the tank. Once the engine has warmed up, the level should be at or near the Full mark. If it isn't, allow the fluid in the tank to cool, then remove the cap from the reservoir and add coolant to bring the level up to the Full line. Use only ethylene/glycol type coolant and water in the mixture ratio recommended by your owner's manual. Do not use supplemental inhibitors or additives. If only a small amount of coolant is required to bring the system up to the proper level, water can be used. However, repeated additions of water will dilute the recommended antifreeze and water solution. In order to maintain the proper ratio of antifreeze and water, it is advisable to top up the coolant level with the correct mixture. Refer to your owner's manual for the recommended ratio.

10   If the coolant level drops within a short time after replenishment, there may be a leak in the system. Inspect the radiator, hoses, engine coolant filler cap, drain plugs, air bleeder plugs and water pump. If no leak is evident, have the radiator cap pressure tested. **Warning:** *Never remove the radiator pressure cap when the engine is running or has just been shut down, because the cooling system is hot. Escaping steam and scalding liquid could cause serious injury.*

11   If it is necessary to open the radiator cap, wait until the system has cooled completely, then wrap a thick cloth around the cap and turn it to the first stop. If any steam escapes, wait until the system has cooled further, then remove the cap.

12   When checking the coolant level, always note its condition. It should be relatively clear. If it is brown or rust colored, the system should be drained, flushed and refilled. Even if the coolant appears to be normal, the corrosion inhibitors wear out with use, so it must be replaced at the specified intervals.

13   Do not allow antifreeze to come in contact with your skin or painted surfaces of the vehicle. Flush contacted areas immediately with plenty of water.

Windshield washer fluid

Refer to illustration 4.14

14   Fluid for the windshield washer system is stored in a plastic reservoir which is located at the right front corner of the engine compartment **(see illustration)**. In milder climates, plain water can be used to top up the reservoir, but the reservoir should be kept no more than two-thirds full to allow for expansion should the water freeze. In colder climates, the use of a specially designed windshield washer fluid, available at your dealer and any auto parts store, will help lower the freezing point of the fluid. Mix the solution with water in accordance with the manufacturer's directions on the container. Do not use regular antifreeze. It will damage the vehicle's paint.

Brake and clutch fluid

Refer to illustration 4.16

15   The brake master cylinder is mounted on the front of the power booster unit in the engine compartment. The clutch cylinder, used with a manual transaxle, is located next to the master cylinder.

16   To check the fluid level of the brake master cylinder reservoir, simply look at the MAX and MIN marks on the reservoir **(see illustration)**. To check the fluid level of the clutch master cylinder reservoir, note whether the fluid level is even with the maximum level line. The level should be within the specified distance from the maximum fill line for both reservoirs.

17   If the level is low for either reservoir, wipe the top of the reservoir cover with a clean rag to prevent contamination of the brake or clutch system before lifting the cover.

Chapter 1 Tune-up and routine maintenance   1-9

4.16 The brake fluid should be kept between the Min and Max marks on the reservoir - turn and lift up the cap to add fluid

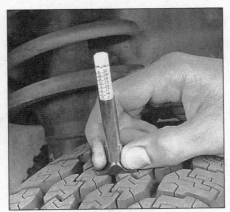

5.2 Use a tire tread depth gauge to monitor tire wear - they are available at auto parts stores and service stations and cost very little

18 Add only the specified brake fluid to the brake or clutch reservoir (refer to *Recommended lubricants and fluids* at the front of this chapter or to your owner's manual). Mixing different types of brake fluid can damage the system. Fill the brake master cylinder reservoir only to the dotted line - this brings the fluid to the correct level when you put the cover back on. **Warning:** *Use caution when filling either reservoir - brake fluid can harm your eyes and damage painted surfaces. Do not use brake fluid that has been opened for more than one year or has been left open. Brake fluid absorbs moisture from the air.*

Excess moisture can cause a dangerous loss of braking.

19 While the reservoir cap is removed, inspect the master cylinder reservoir for contamination. If deposits, dirt particles or water droplets are present, the system should be drained and refilled (see Chapter 8 for clutch reservoir or Chapter 9 for brake reservoir).

20 After filling the reservoir to the proper level, make sure the lid is properly seated to prevent fluid leakage and/or system pressure loss.

21 The brake fluid in the master cylinder will drop slightly as the brake pads at each wheel wear down during normal operation. If the master cylinder requires repeated replenishing to keep it at the proper level, this is an indication of leakage in the brake system, which should be corrected immediately. Check all brake lines and connections, along with the wheel cylinders and booster (see Section 16 for more information).

22 If, upon checking the master cylinder fluid level, you discover one or both reservoirs empty or nearly empty, the brake system should be bled (see Chapter 9).

5 Tire and tire pressure checks (every 250 miles or weekly)

Refer to illustrations 5.2, 5.3, 5.4a, 5.4b and 5.8

1 Periodic inspection of the tires may spare you from the inconvenience of being stranded with a flat tire. It can also provide you with vital information regarding possible problems in the steering and suspension systems before major damage occurs.

2 Normal tread wear can be monitored with a simple, inexpensive device known as a tread depth indicator **(see illustration)**. When the tread depth reaches the specified minimum, replace the tire(s).

3 Note any abnormal tread wear **(see illustration)**. Tread pattern irregularities such as cupping, flat spots and more wear on one side than the other are indications of front end alignment and/or balance problems. If

UNDERINFLATION

CUPPING

Cupping may be caused by:
- Underinflation and/or mechanical irregularities such as out-of-balance condition of wheel and/or tire, and bent or damaged wheel.
- Loose or worn steering tie-rod or steering idler arm.
- Loose, damaged or worn front suspension parts.

OVERINFLATION

INCORRECT TOE-IN OR EXTREME CAMBER

FEATHERING DUE TO MISALIGNMENT

5.3 This chart will help you determine the condition of the tires, the probable cause(s) of abnormal wear and the corrective action necessary

Chapter 1 Tune-up and routine maintenance

5.4a If a tire looses air on a steady basis, check the valve core first to make sure it's snug (a special inexpensive wrench is commonly available at auto parts stores)

5.4b If the valve core is tight, raise the corner of the vehicle with the low tire and spray a soapy water solution onto the tread as the tire is turned slowly - leaks will cause small bubbles to appear

5.8 To extend the life of the tires, check the air pressure at least once a week with an accurate gauge (don't forget the spare)

any of these conditions are noted, take the vehicle to a tire shop or service station to correct the problem.

4 Look closely for cuts, punctures and embedded nails or tacks. Sometimes a tire will hold its air pressure for a short time or leak down very slowly even after a nail has embedded itself into the tread. If a slow leak persists, check the valve stem core to make sure it is tight **(see illustration)**. Examine the tread for an object that may have embedded itself into the tire or for a "plug" that may have begun to leak (radial tire punctures are repaired with a plug that is fitted in a puncture). If a puncture is suspected, it can be easily verified by spraying a solution of soapy water onto the puncture area **(see illustration)**. The soapy solution will bubble if there is a leak. Unless the puncture is inordinately large, a tire shop or gas station can usually repair the punctured tire.

5 Carefully inspect the inner sidewall of each tire for evidence of brake fluid leakage. If you see any, inspect the brakes immediately.

6 Correct tire air pressure adds miles to the lifespan of the tires, improves mileage and enhances overall ride quality. Tire pressure cannot be accurately estimated by looking at a tire, particularly if it is a radial. A tire pressure gauge is therefore essential. Keep an accurate gauge in the glovebox. The pressure gauges fitted to the nozzles of air hoses at gas stations are often inaccurate.

7 Always check tire pressure when the tires are cold. "Cold," in this case, means the vehicle has not been driven over a mile in the three hours preceding a tire pressure check. A pressure rise of four to eight pounds is not uncommon once the tires are warm.

8 Unscrew the valve cap protruding from the wheel or hubcap and push the gauge firmly onto the valve **(see illustration)**. Note the reading on the gauge and compare this figure to the recommended tire pressure shown on the tire placard in the glovebox. Be sure to reinstall the valve cap to keep dirt and moisture out of the valve stem mechanism.

Check all four tires and, if necessary, add enough air to bring them up to the recommended pressure levels.

9 Don't forget to keep the spare tire inflated to the specified pressure (consult your owner's manual). Note that the air pressure specified for a compact spare is significantly higher than the pressure of the regular tires.

6 Power steering fluid level check (every 3000 miles or 3 months)

Refer to illustrations 6.4a and 6.4b

1 Unlike manual steering, the power steering system relies on fluid which may, over a period of time, require replenishing.
2 The fluid reservoir for the power steering pump is located on the right (passenger side) inner fender panel near the front of the engine.
3 For the check, the front wheels should be pointed straight ahead and the engine should be off.
4 The reservoir is translucent plastic and the fluid level can be checked visually **(see illustrations)**.

6.4a The power steering fluid reservoir (arrow) is located on the right side of the engine compartment

5 If additional fluid is required, pour the specified type directly into the reservoir, using a funnel to prevent spills.
6 If the reservoir requires frequent fluid additions, all power steering hoses, hose connections, the power steering pump and the rack and pinion assembly should be carefully checked for leaks.

7 Automatic transaxle fluid level check (every 3000 miles or 3 months)

Refer to illustrations 7.4a and 7.4b

1 The level of the automatic transaxle fluid should be carefully maintained. Low fluid level can lead to slipping or loss of drive, while overfilling can cause foaming, loss of fluid and transaxle damage.
2 The transaxle fluid level should only be checked when the transaxle is hot (at its normal operating temperature). If the vehicle has just been driven over 10 miles (15 miles in a frigid climate), and the fluid temperature is 160 to 175-degrees F, the transaxle is hot. **Caution:** *If the vehicle has just been driven for a long time at high speed or in city traffic*

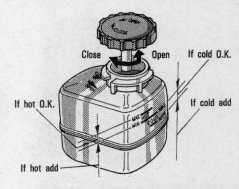

6.4b The reservoir is translucent so the fluid level can be checked either hot or cold without removing the cap

Chapter 1 Tune-up and routine maintenance 1-11

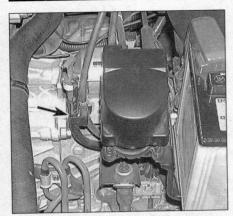

7.4a The automatic transaxle dipstick is located next to the battery

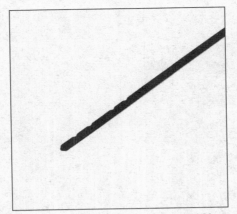

7.4b If the automatic transaxle fluid is cold, the level should be between the lower two notches; if it's at normal operating temperature, the level should be between the upper notches on the dipstick

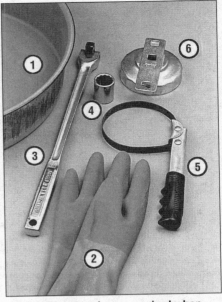

8.2 These tools are required when changing the engine oil and filter

1. **Drain pan** - It should be fairly shallow in depth, but wide to prevent spills
2. **Rubber gloves** - When removing the drain plug and filter, you will get oil on your hands (the gloves will prevent burns)
3. **Breaker bar** - Sometimes the oil drain plug is tight, and a long breaker bar is needed to loosen it
4. **Socket** - To be used with the breaker bar or a ratchet (must be the correct size to fit the drain plug)
5. **Filter wrench** - This is a metal band-type wrench, which requires clearance around the filter to be effective
6. **Filter wrench** - This type fits on the bottom of the filter and can be turned with a ratchet or breaker bar (different-size wrenches are available for different types of filters)

in hot weather, or if it has been pulling a trailer, an accurate fluid level reading cannot be obtained. Allow the fluid to cool down for about 30 minutes.

3 If the vehicle has not just been driven, park the vehicle on level ground, set the parking brake and start the engine. While the engine is idling, depress the brake pedal and move the selector lever through all the gear ranges, beginning and ending in Park.

4 With the engine still idling, remove the dipstick from its tube **(see illustration)**. Check the level of the fluid on the dipstick **(see illustration)** and note its condition.

5 Wipe the fluid from the dipstick with a clean rag and reinsert it back into the filler tube until the cap seats.

6 Pull the dipstick out again and note the fluid level. If the transaxle is cold, the level should be in the COLD or COOL range on the dipstick. If it is hot, the fluid level should be in the HOT range. If the level is at the low side of either range, add the specified automatic transaxle fluid through the dipstick tube with a funnel.

7 Add just enough of the recommended fluid to fill the transaxle to the proper level. It takes about one pint to raise the level from the low mark to the high mark when the fluid is hot, so add the fluid a little at a time and keep checking the level until it is correct.

8 The condition of the fluid should also be checked along with the level. If the fluid at the end of the dipstick is black or a dark reddish brown color, or if it emits a burned smell, the fluid should be changed (see Section 27). If you are in doubt about the condition of the fluid, purchase some new fluid and compare the two for color and smell.

8 Engine oil and oil filter change (every 3000 miles or 3 months)

Refer to illustrations 8.2, 8.7, 8.13, and 8.15

1 Frequent oil changes are the best preventive maintenance the home mechanic can give the engine, because aging oil becomes diluted and contaminated, which leads to premature engine wear.

2 Make sure that you have all the necessary tools before you begin this procedure **(see illustration)**. You should also have plenty of rags or newspapers handy for mopping up any spills.

3 Access to the underside of the vehicle is greatly improved if the vehicle can be lifted on a hoist, driven onto ramps or supported by jack stands. **Warning:** *Do not work under a vehicle which is supported only by a bumper, hydraulic or scissors-type jack.*

4 If this is your first oil change, get under the vehicle and familiarize yourself with the location of the oil drain plug. The engine and exhaust components will be warm during the actual work, so try to anticipate any potential problems before the engine and accessories are hot.

5 Park the vehicle on a level spot. Start the engine and allow it to reach its normal operating temperature (the needle on the temperature gauge should be at least above the bottom mark). Warm oil and sludge will flow out more easily. Turn off the engine when it's warmed up. Remove the filler cap.

6 Raise the vehicle and support it securely on jack stands. **Warning:** *To avoid personal injury, never get beneath a vehicle when it is supported by only by a jack. The jack provided with your vehicle is designed solely for raising the vehicle to remove and replace the wheels. Always use jack stands to support the vehicle when it becomes necessary to place your body underneath the vehicle.*

7 Being careful not to touch the hot exhaust components, place the drain pan under the drain plug in the bottom of the pan and remove the plug **(see illustration)**. You may want to wear gloves while unscrewing the plug the final few turns if the engine is really hot.

8 Allow the old oil to drain into the pan. It may be necessary to move the pan farther under the engine as the oil flow slows to a trickle. Inspect the old oil for the presence of metal shavings and chips.

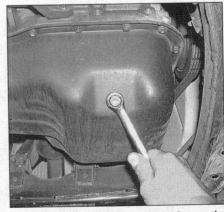

8.7 Use the proper size box-end wrench or socket to remove the oil drain plug without rounding off the corners

Chapter 1 Tune-up and routine maintenance

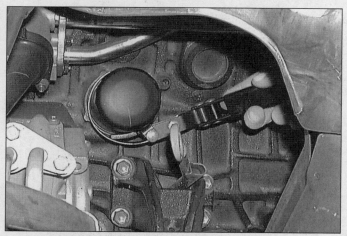

8.13 On four-cylinder models, the oil filter is mounted on the side of the engine block so it's a good idea to pack rags around it before removal to minimize the mess - since it is usually on very tight, you'll need a special wrench for removal - DO NOT use the wrench to tighten the new filter

8.15 Lubricate the oil filter gasket with clean engine oil before installing the filter on the engine

9   After all the oil has drained, wipe off the drain plug with a clean rag. Even minute metal particles clinging to the plug would immediately contaminate the new oil.

10 Clean the area around the drain plug opening, reinstall the plug and tighten it securely, but do not strip the threads.

11 Move the drain pan into position under the oil filter.

12 Loosen the oil filter **(see illustration)** by turning it counterclockwise with an oil filter wrench. Once the filter is loose, use your hands to unscrew it from the block. Just as the filter is detached from the block, immediately tilt the open end up to prevent the oil inside the filter from spilling out. **Warning:** *The engine exhaust manifold may still be hot, so be careful.*

13 With a clean rag, wipe off the mounting surface on the block. If a residue of old oil is allowed to remain, it will smoke when the block is heated up. It will also prevent the new filter from seating properly. Also make sure that the none of the old gasket remains stuck to the mounting surface. It can be removed with a scraper if necessary.

14 Compare the old filter with the new one to make sure they are the same type. Smear some engine oil on the rubber gasket of the new filter and screw it into place **(see illustration).** Because overtightening the filter will damage the gasket, do not use a filter wrench to tighten the filter. Tighten it by hand until the gasket contacts the seating surface. Then seat the filter by giving it an additional 3/4-turn.

15 Remove all tools, rags, etc. from under the vehicle, being careful not to spill the oil in the drain pan, then lower the vehicle.

16 Add new oil to the engine through the oil filler cap in the valve cover. Use a spout or funnel to prevent oil from spilling onto the top of the engine. Pour three quarts of fresh oil into the engine. Wait a few minutes to allow the oil to drain into the pan, then check the

level on the oil dipstick (see Section 4 if necessary). If the oil level is at or near the F mark, install the filler cap hand tight, start the engine and allow the new oil to circulate.

17 Allow the engine to run for about a minute. While the engine is running, look under the vehicle and check for leaks at the oil pan drain plug and around the oil filter. If either is leaking, stop the engine and tighten the plug or filter slightly.

18 Wait a few minutes to allow the oil to trickle down into the pan, then recheck the level on the dipstick and, if necessary, add enough oil to bring the level to the F mark.

19 During the first few trips after an oil change, make it a point to check frequently for leaks and proper oil level.

20 The old oil drained from the engine cannot be reused in its present state and should be discarded. Oil reclamation centers, auto garages and gas stations will normally accept the oil, which can be refined and used again. After the oil has cooled, it can be drained into a suitable container (capped plastic jugs, topped bottles, milk cartons, etc.) for transport to one of these disposal sites.

9 Windshield wiper blade inspection and replacement (every 6000 miles or 6 months)

Refer to illustrations 9.6, 9.7, 9.8, 9.9 and 9.10

1   The windshield wiper and blade assembly should be inspected periodically for damage, loose components and cracked or worn blade elements.

2   Road film can build up on the wiper blades and affect their efficiency, so they should be washed regularly with a mild detergent solution.

3   The action of the wiping mechanism can loosen bolts, nuts and fasteners, so they

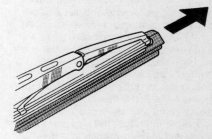

9.6 Pull the type A wiper element tabs out of the frame

should be checked and tightened, as necessary, at the same time the wiper blades are checked.

4   If the wiper blade elements are cracked, worn or warped, or no longer clean adequately, they should be replaced with new ones.

5   Lift the arm assembly away from the glass for clearance.

Type A

6   Working from the end of the wiper closest to the arm, pull the rubber blade element from the frame and discard it **(see illustration).**

7   To install a new rubber wiper element, insert the end with the small protrusions into the replacement hole and work the rubber along the slot in the blade frame **see illustration).**

Type B

8   Squeeze the end of the wiper element and pull it out of the frame **(see illustration).**

9   Insert the end of the new element without cutouts into the end of the frame **(see illustration).**

10 Work the rubber along the slot in the blade frame until the cutouts are locked into place by wiper frame claw **(see illustration).**

Chapter 1 Tune-up and routine maintenance

1-13

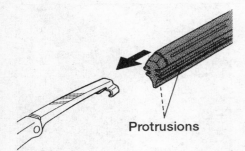

9.7 To install a new element, insert the end of the blade with the small protrusions into the frame

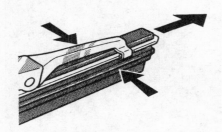

9.8 Squeeze the end of the type B wiper element and pull it straight out of the frame

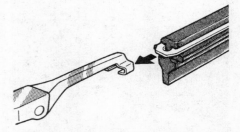

9.9 Insert the end without the cutout into the frame

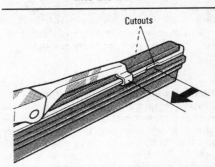

9.10 Slide the element into place until the cutouts seat into the locking claw

10 Battery check, maintenance and charging (every 6000 miles or 6 months)

Refer to illustrations 10.1, 10.7a, 10.7b, 10.8a, 10.8b and 10.13

Warning: *Certain precautions must be followed when checking and servicing the battery. Hydrogen gas, which is highly flammable, is always present in the battery cells, so keep lighted tobacco and all other open flames and sparks away from the battery. The electrolyte inside the battery is actually dilute sulfuric acid, which will cause injury if splashed on your skin or in your eyes. It will also ruin clothes and painted surfaces. When removing the battery cables, always detach the negative cable first and hook it up last!*

Check

1 A routine preventive maintenance program for the battery in your vehicle is the only way to ensure quick and reliable starts. But before performing any battery maintenance, make sure that you have the proper equipment necessary to work safely around the battery **(see illustration)**.

2 There are also several precautions that should be taken whenever battery maintenance is performed. Before servicing the battery, always turn the engine and all accessories off and disconnect the cable from the negative terminal of the battery.

3 The battery produces hydrogen gas, which is both flammable and explosive. Never create a spark, smoke or light a match around the battery. Always charge the battery in a ventilated area.

4 Electrolyte contains poisonous and corrosive sulfuric acid. Do not allow it to get in your eyes, on your skin or your clothes. Never ingest it. Wear protective safety glasses when working near the battery. Keep children away from the battery.

5 Note the external condition of the battery. If the positive terminal and cable clamp on your vehicle's battery is equipped with a rubber protector, make sure that it's not torn or damaged. It should completely cover the terminal. Look for any corroded or loose connections, cracks in the case or cover, or loose hold-down clamps. Also check the entire length of each cable for cracks and frayed conductors.

10.1 Tools and materials required for battery maintenance

1 **Face shield/safety goggles** - When removing corrosion with a brush, the acidic particles can easily fly up into your eyes
2 **Baking soda** - A solution of baking soda and water can be used to neutralize corrosion
3 **Petroleum jelly** - A layer of this on the battery posts will help prevent corrosion
4 **Battery post/cable cleaner** - This wire brush cleaning tool will remove all traces of corrosion from the battery posts and cable clamps
5 **Treated felt washers** - Placing one of these on each post, directly under the cable clamps, will help prevent corrosion
6 **Puller** - Sometimes the cable clamps are very difficult to pull off the posts, even after the nut/bolt has been completely loosened. This tool pulls the clamp straight up and off the post without damage
7 **Battery post/cable cleaner** - Here is another cleaning tool which is a slightly different version of Number 4 above, but it does the same thing
8 **Rubber gloves** - Another safety item to consider when servicing the battery; remember that's acid inside the battery!

10.7a Battery terminal corrosion usually appears as light, fluffy powder

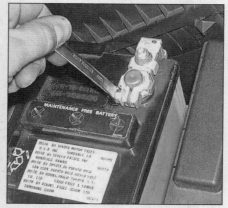

10.7b Loosen the battery cable clamp bolt with a wrench - sometimes special battery pliers are required for this procedure if corrosion has deteriorated the hex nut - always remove the negative cable first and reconnect it last!

10.8a When cleaning the cable clamps, all corrosion must be removed (the inside of the clamp is tapered to match the taper on the post, so don't remove too much material)

6   Some models with sealed batteries have a battery condition indicator on top of the battery. Compare the color showing in the window to the condition color chart on the battery. You may catch a low-charge battery condition before it strands you on the roadside. If the color indicate a low state of charge, charge the battery and examine the charging system (see Chapter 5 and this Section).

Maintenance

7   If corrosion, which looks like white, fluffy deposits (see illustration) is evident, particularly around the terminals, the battery should be removed for cleaning. Loosen the cable clamp bolts with a wrench, being careful to remove the negative cable first, and slide them off the terminals (see illustration). Then disconnect the hold-down clamp bolt and nut, remove the clamp and lift the battery from the engine compartment.

8   Clean the cable clamps thoroughly with a battery brush or a terminal cleaner and a solution of warm water and baking soda. Wash the terminals and the top of the battery case with the same solution but make sure that the solution doesn't get into the battery. When cleaning the cables, terminals and battery top, wear safety goggles and rubber gloves to prevent any solution from coming in contact with your eyes or hands. Wear old clothes too - even diluted, sulfuric acid splashed onto clothes will burn holes in them. If the terminals have been extensively corroded, clean them up with a terminal cleaner (see illustrations). Thoroughly wash all cleaned areas with plain water.

9   Whenever the battery is removed for cleaning or charging, inspect the battery carrier before reinstalling the battery in the engine compartment. If the carrier is dirty or covered with corrosion, clean it in the same solution of warm water and baking soda. Inspect the metal brackets which support the carrier to make sure that they are not covered with corrosion. If they are, wash them off. If corrosion is extensive, sand the brackets down to bare metal and spray them with a zinc-based primer (available in spray cans at auto paint and body supply stores).

10   Reinstall the battery back into the engine compartment. Make sure that no parts or wires are laying on the carrier during installation of the battery. Information on removing and installing the battery can be found in Chapter 5. Information on jump starting can be found at the front of this manual. For more detailed battery checking procedures, refer to the Haynes Automotive Electrical Manual.

11   Install a pair of specially-treated felt washers around the terminals (available at auto parts stores), then coat the terminals and the cable clamps with petroleum jelly or grease to prevent further corrosion. Install the cable clamps and tighten the nuts, being careful to install the negative cable last.

12   Install the hold-down clamp and nuts. Tighten the nuts only enough to hold the battery firmly in place. Overtightening these nuts can crack the battery case.

13   Make sure that the battery tray is in good condition and the hold-down clamp bolts are tight (see illustration). If the battery is removed from the tray, make sure no parts remain in the bottom of the tray when the battery is reinstalled. When reinstalling the hold-down clamp bolts, do not overtighten them.

Charging

Warning: *When batteries are being charged, hydrogen gas, which is very explosive and flammable, is produced. Do not smoke or allow open flames near a charging or a recently charged battery. Wear eye protection when near the battery during charging. Also, make sure the charger is unplugged before connecting or disconnecting the battery from the charger.*

14   Slow-rate charging is the best way to restore a battery that's discharged to the point where it will not start the engine. It's also a good way to maintain the battery charge in a vehicle that's only driven a few miles between starts. Maintaining the battery charge is particularly important in the winter when the battery must work harder to start the engine and electrical accessories that drain the battery are in greater use.

15   It's best to use a one or two-amp bat-

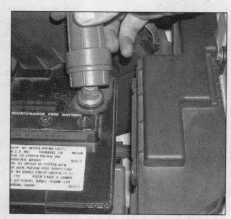

10.8b Regardless of the type of tool used on the battery posts, a clean, shiny surface should be the result

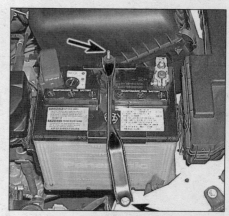

10.13 Make sure the battery hold-down nuts (arrows) are tight

Chapter 1 Tune-up and routine maintenance

tery charger (sometimes called a "trickle" charger). They are the safest and put the least strain on the battery. They are also the least expensive. For a faster charge, you can use a higher amperage charger, but don't use one rated more than 1/10th the amp/hour rating of the battery. Rapid boost charges that claim to restore the power of the battery in one to two hours are hardest on the battery and can damage batteries not in good condition. This type of charging should only be used in emergency situations.

16 The average time necessary to charge a battery should be listed in the instructions that come with the charger. As a general rule, a trickle charger will charge a battery in 12 to 16 hours.

11 Drivebelt check, adjustment and replacement (every 6000 miles or 6 months)

Refer to illustrations 11.3a, 11.3b, 11.5 and 11.6

Check

1 The alternator and air conditioning compressor drivebelts, also referred to as V-ribbed belts or simply "fan" belts, are located at the right end of the engine. The good condition and proper adjustment of the alternator belt is critical to the operation of the engine. Because of their composition and the high stresses to which they are subjected, drivebelts stretch and deteriorate as they get older. They must therefore be periodically inspected.

2 The number of belts used on a particular vehicle depends on the accessories fitted. One belt transmits power from the crankshaft to the alternator and air conditioning. If the vehicle is equipped with power steering, the pump is driven by it's own belt.

3 With the engine off, open the hood and locate the drivebelts at the left end of the engine. With a flashlight, check each belt for separation of the adhesive rubber on both sides of the core, core separation from the belt side, a severed core, separation of the ribs from the adhesive rubber, cracking or separation of the ribs, and torn or worn ribs or cracks in the inner ridges of the ribs **(see illustrations)**. Also check for fraying and glazing, which gives the belt a shiny appearance. Both sides of the belt should be inspected, which means you will have to twist the belt to check the underside. Use your fingers to feel the belt where you can't see it. If any of the above conditions are evident, replace the belt (go to Step 8).

4 To check the tension of each belt in accordance with factory specifications, install either a Nippondenso or Burroughs belt tension gauge on the belt. Measure the tension in accordance with the tool manufacturer's instructions and compare your measurement to the specified drivebelt tension for either a used or new belt. **Note:** *A "used" belt is defined as any belt which has been operated more than five minutes on the engine; a "new" belt is one that has been used for less than five minutes.*

5 If you don't have either of the above tools, and cannot borrow one, the following rule of thumb method is recommended: Push firmly on the belt with your thumb at a distance halfway between the pulleys and note how far the belt can be pushed (deflected). Measure this deflection with a ruler **(see illustration)**. The belt should deflect 1/4-inch if the distance from pulley center to pulley center is between 7 and 11 inches; the belt should deflect 1/2-inch if the distance from pulley center to pulley center is between 12 and 16 inches.

Adjustment

6 If the alternator/air conditioner compressor belt must be adjusted, loosen the alternator pivot bolt located on the front left corner of the block. Loosen the locking bolt and turn the adjusting bolt **(see illustration)**. Measure the belt tension in accordance with one of the above methods. Repeat this step until the air conditioning compressor drivebelt is adjusted.

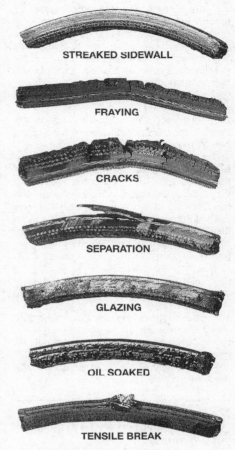

11.3a Check a V-belt for signs of wear like these - if the belt looks worn - replace it

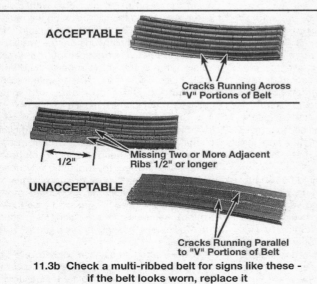

11.3b Check a multi-ribbed belt for signs like these - if the belt looks worn, replace it

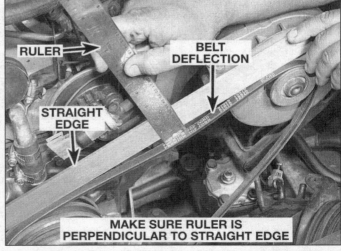

11.5 Measuring drivebelt deflection with a straightedge and ruler

Chapter 1 Tune-up and routine maintenance

7 Adjust the power steering pump belt by loosening the adjustment bolt that secures the pump to the slotted bracket and pivot the pump (away from the engine to tighten the belt, toward it to loosen it). Repeat the procedure until the drivebelt tension is correct and tighten the bolt.

Replacement

8 To replace a belt, follow the above procedures for drivebelt adjustment but slip the belt off the crankshaft pulley and remove it. If you are replacing the power steering pump belt, you will have to remove the air conditioning compressor belt first because of the way they are arranged on the crankshaft pulley. Because of this and because belts tend to wear out more or less together, it is a good idea to replace both belts at the same time. Mark each belt and its appropriate pulley groove so the replacement belts can be fitted in their proper positions.

9 Take the old belts to the parts store in order to make a direct comparison for length, width and design.

10 After replacing the drivebelt, make sure that it fits properly in the ribbed grooves in the pulleys. It is essential that the belt be properly centered.

11 Adjust the belt(s) in accordance with the procedure outlined above.

12 Underhood hose check and replacement (every 6000 miles or 6 months)

Caution: *Replacement of air conditioning hoses must be left to a dealer service department or air conditioning shop that has the equipment to depressurize the system safely. Never remove air conditioning components or hoses until the system has been evacuated and the refrigerant recovered by a dealer service department or air-conditioning shop.*

General

1 High temperatures in the engine compartment can cause the deterioration of the rubber and plastic hoses used for engine, accessory and emission systems operation. Periodic inspection should be made for cracks, loose clamps, material hardening and leaks.

2 Information specific to the cooling system hoses can be found in Section 13.

3 Some, but not all, hoses are secured to the fittings with clamps. Where clamps are used, check to be sure they haven't lost their tension, allowing the hose to leak. If clamps aren't used, make sure the hose has not expanded and/or hardened where it slips over the fitting, allowing it to leak.

Vacuum hoses

4 It's quite common for vacuum hoses, especially those in the emissions system, to be color coded or identified by colored stripes molded into them. Various systems require hoses with different wall thickness,

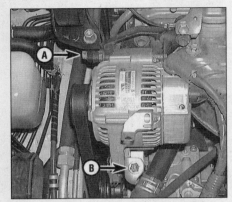

11.6 After loosening the pivot bolt (A) and the adjustment pinch bolt, turn the adjustment bolt (B) clockwise to tighten the belt, or counterclockwise to loosen the belt

collapse resistance and temperature resistance. When replacing hoses, be sure the new ones are made of the same material.

5 Often the only effective way to check a hose is to remove it completely from the vehicle. If more than one hose is removed, be sure to label the hoses and fittings to ensure correct installation.

6 When checking vacuum hoses, be sure to include any plastic T-fittings in the check. Inspect the fittings for cracks and the hose where it fits over the fitting for distortion, which could cause leakage.

7 A small piece of vacuum hose (1/4-inch inside diameter) can be used as a stethoscope to detect vacuum leaks. Hold one end of the hose to your ear and probe around vacuum hoses and fittings, listening for the "hissing" sound characteristic of a vacuum leak. **Warning:** *When probing with the vacuum hose stethoscope, be very careful not to come into contact with moving engine components such as the drivebelts, cooling fan, etc.*

Fuel hose

Warning: *There are certain precautions which must be taken when inspecting or servicing fuel system components. Work in a well ventilated area and do not allow open flames (cigarettes, appliance pilot lights, etc.) or bare light bulbs near the work area. Mop up any spills immediately and do not store fuel soaked rags where they could ignite.*

8 Check all rubber fuel lines for deterioration and chafing. Check especially for cracks in areas where the hose bends and just before fittings, such as where a hose attaches to the fuel filter.

9 High quality fuel line should be used for fuel line replacement. Never, under any circumstances, use unreinforced vacuum line, clear plastic tubing or water hose for fuel lines.

10 Spring-type clamps are commonly used on fuel lines. These clamps often lose their tension over a period of time, and can be "sprung" during removal. Replace all spring-type clamps with screw clamps whenever a hose is replaced.

Metal lines

11 Sections of metal line are often used for fuel line between the fuel pump and fuel injection unit. Check carefully to be sure the line has not been bent or crimped and that cracks have not started in the line.

12 If a section of metal fuel line must be replaced, only seamless steel tubing should be used, since copper and aluminum tubing don't have the strength necessary to withstand normal engine vibration.

13 Check the metal brake lines where they enter the master cylinder and brake proportioning unit (if used) for cracks in the lines or loose fittings. Any sign of brake fluid leakage calls for an immediate thorough inspection of the brake system.

13 Cooling system check (every 6000 miles or 6 months)

Refer to illustration 13.4

1 Many major engine failures can be attributed to a faulty cooling system. If the vehicle is equipped with an automatic transaxle, the cooling system also cools the transaxle fluid and thus plays an important role in prolonging transaxle life.

2 The cooling system should be checked with the engine cold. Do this before the vehicle is driven for the day or after the engine has been shut off for at least three hours. **Warning:** *Never remove the radiator pressure cap when the engine is running or has just been shut down, because the cooling system is hot. Escaping steam and scalding liquid could cause serious injury.*

3 Remove the radiator pressure cap by turning it to the left until it reaches a stop. If you hear a hissing sound (indicating there is still pressure in the system), wait until it stops. Now press down on the cap with the palm of your hand and continue turning to the left until the cap can be removed. Thoroughly clean the cap, inside and out, with clean water. Also clean the filler neck on the radiator. All traces of corrosion should be removed. The coolant inside the radiator should be relatively transparent. If it's rust colored, the system should be drained and refilled (see Section 24). If the coolant level isn't up to the top, add additional antifreeze/coolant mixture (see Section 4).

4 Carefully check the large upper and lower radiator hoses along with the smaller diameter heater hoses which run from the engine to the bulkhead. Inspect each hose along its entire length, replacing any hose which is cracked, swollen or shows signs of deterioration. Cracks may become more apparent if the hose is squeezed **(see illustration)**. Regardless of condition, it's a good idea to replace hoses with new ones every two years.

5 Make sure that all hose connections are tight. A leak in the cooling system will usually show up as white or rust colored deposits on the areas adjoining the leak. If wire-type

Chapter 1 Tune-up and routine maintenance   1-17

Check for a chafed area that could fail prematurely.

Check for a soft area indicating the hose has deteriorated inside.

Overtightening the clamp on a hardened hose will damage the hose and cause a leak.

Check each hose for swelling and oil-soaked ends. Cracks and breaks can be located by squeezing the hose.

13.4 Hoses, like drivebelts, have a habit of failing at the worst possible time - to prevent the inconvenience of a blown radiator or heater hose, inspect them carefully as shown here

clamps are used at the ends of the hoses, it may be a good idea to replace them with more secure screw-type clamps.
6 Use compressed air or a soft brush to remove bugs, leaves, etc. from the front of the radiator or air conditioning condenser. Be careful not to damage the delicate cooling fins or cut yourself on them.
7 Every other inspection, or at the first indication of cooling system problems, have the cap and system pressure tested. If you don't have a pressure tester, most gas stations and garages will do this for a minimal charge.

14 Tire rotation (every 6000 miles or 6 months)

Refer to illustration 14.2

1 The tires should be rotated at the specified intervals and whenever uneven wear is noticed. Since the vehicle will be raised and the tires removed anyway, check the brakes (see Section 15) at this time.
2 Radial tires must be rotated in a specific pattern **(see illustration)**. Most models are equipped with non-directional tires, but some sport models may be equipped with directional tires, which have a specific rotational pattern. When choosing replacement tires, examine the sidewalls. Directional tires have arrows on the sidewall that indicate the direction they must turn, and a set of these tires includes two left-side tires and two right-side tires. The left and right side tires must not be rotated to the other side.
3 Refer to the information in *Jacking and towing* at the front of this manual for the proper procedures to follow when raising the vehicle and changing a tire. If the brakes are to be checked, do not apply the parking brake as stated. Make sure the tires are blocked to prevent the vehicle from rolling.
4 Preferably, the entire vehicle should be raised at the same time. This can be done on a hoist or by jacking up each corner and then

lowering the vehicle onto jack stands placed under the frame rails. Always use four jack stands and make sure the vehicle is firmly supported.
5 After rotation, check and adjust the tire pressures as necessary and be sure to check the wheel lug nut torque.
6 For further information on the wheels and tires, refer to Chapter 10.

15 Brake check (every 15,000 miles or 12 months)

Note: For detailed photographs of the brake system, refer to Chapter 9.

1 In addition to the specified intervals, the brakes should be inspected every time the wheels are removed or whenever a defect is suspected. Any of the following symptoms could indicate a potential brake system defect: The vehicle pulls to one side when the brake pedal is depressed; the brakes make squealing or dragging noises when applied; brake travel is excessive; the pedal pulsates; brake fluid leaks, usually onto the inside of the tire or wheel.
2 The disc brake pads have built-in wear indicators which should make a high pitched squealing or scraping noise when they are worn to the replacement point. When you hear this noise, replace the pads immediately or expensive damage to the discs can result.
3 Loosen the wheel nuts.
4 Raise the vehicle and place it securely on jack stands.
5 Remove the wheels (see *Jacking and towing* at the front of this book, or your owner's manual, if necessary).

Disc brakes

Refer to illustration 15.6

6 There are two pads - an outer and an inner - in each caliper. The pads are visible through inspection holes in each caliper **(see illustration)**.
7 Check the pad thickness by looking at each end of the caliper and through the inspection hole in the caliper body. If the lin-

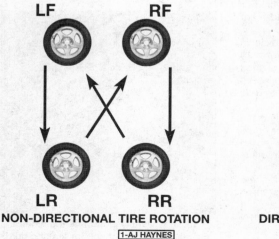
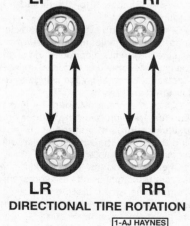

14.2 The recommended tire rotation pattern for these models

15.6 You'll find an inspection hole like this in each caliper through which you can view the inner and outer brake pad lining

15.13 A quick check of the remaining drum brake shoe lining material can be made by removing the rubber plug in the backing plate and looking through the inspection hole

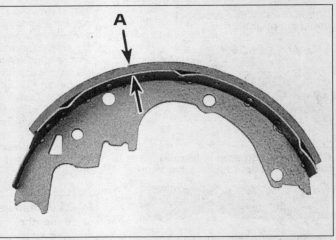

15.15 If the lining is bonded to the brake shoe, measure the lining thickness from the outer surface to the metal shoe, as shown here; if the lining is riveted to the shoe, measure from the lining outer surface to the rivet head

ing material is less than the thickness listed in this Chapter's Specifications, replace the pads. **Note:** *Keep in mind that the lining material is riveted or bonded to a metal backing plate and the metal portion is not included in this measurement.*

8   If it is difficult to determine the exact thickness of the remaining pad material by the above method, or if you are at all concerned about the condition of the pads, remove the caliper(s), then remove the pads from the calipers for further inspection (refer to Chapter 9).

9   Once the pads are removed from the calipers, clean them with brake cleaner and remeasure them with a small steel pocket ruler or a venire caliper.

10   Measure the disc thickness with a micrometer to make sure that it still has service life remaining. If any disc is thinner than the specified minimum thickness, replace it (refer to Chapter 9). Even if the disc has service life remaining, check its condition. Look for scoring, gouging and burned spots. If these conditions exist, remove the disc and have it resurfaced (see Chapter 9).

11   Before installing the wheels, check all brake lines and hoses for damage, wear, deformation, cracks, corrosion, leakage, bends and twists, particularly in the vicinity of the rubber hoses at the calipers. Check the clamps for tightness and the connections for leakage. Make sure that all hoses and lines are clear of sharp edges, moving parts and the exhaust system. If any of the above conditions are noted, repair, reroute or replace the lines and/or fittings as necessary (see Chapter 9).

12   Some models are equipped with disc brakes on the rear wheels which incorporate a drum-type parking brake into the rear discs. The inspection procedure for the parking brake lining is the same as for the rear drum brake lining described below.

Rear drum brakes

Refer to illustrations 15.13, 15.15 and 15.17

13   To check the brake shoe lining thickness without removing the brake drums, remove the rubber plug from the backing plate and use a flashlight to inspect the linings **(see illustration)**. For a more thorough brake inspection, follow the procedure below.

14   Refer to Chapter 9 and remove the rear brake drums. **Warning:** *Brake dust produced by lining wear and deposited on brake components may contain asbestos, which is hazardous to your health. DO NOT blow it out with compressed air and DO NOT inhale it!*

15   Note the thickness of the lining material on the rear brake shoes **(see illustration)** and look for signs of contamination by brake fluid and grease. If the lining material is within 1/16-inch of the recessed rivets or metal shoes, replace the brake shoes with new ones. The shoes should also be replaced if they are cracked, glazed (shiny lining surfaces) or contaminated with brake fluid or grease. See Chapter 9 for the replacement procedure.

16   Check the shoe return and hold-down springs and the adjusting mechanism to make sure they're fitted correctly and in good condition. Deteriorated or distorted springs, if not replaced, could allow the linings to drag and wear prematurely.

17   Check the wheel cylinders for leakage by carefully peeling back the rubber boots **(see illustration)**. If brake fluid is noted behind the boots, the wheel cylinders must be replaced (see Chapter 9).

18   Check the drums for cracks, score marks, deep scratches and hard spots, which will appear as small discolored areas. If imperfections cannot be removed with emery cloth, the drums must be resurfaced by an automotive machine shop (see Chapter 9 for more detailed information).

19   Refer to Chapter 9 and install the brake drums.

20   Install the wheels and snug the wheel nuts finger tight.

21   Remove the jack stands and lower the vehicle.

22   Tighten the wheel nuts to the torque listed in this Chapter's Specifications.

Brake booster check

23   Sit in the driver's seat and perform the following sequence of tests.

24   With the brake fully depressed, start the engine - the pedal should move down a little when the engine starts.

25   With the engine running, depress the brake pedal several times - the travel distance should not change.

26   Depress the brake, stop the engine and hold the pedal in for about 30 seconds - the pedal should neither sink nor rise.

27   Restart the engine, run it for about a minute and turn it off. Then firmly depress the

15.17 Carefully peel back the wheel cylinder boot and check for leaking fluid indicating that the cylinder must be replaced or rebuilt

Chapter 1 Tune-up and routine maintenance 1-19

16.1 Detach the four air cleaner housing clips (arrows indicate two)

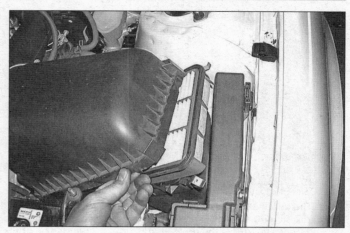

16.2 Lift the cover up and remove the filter

brake several times - the pedal travel should decrease with each application.

28 If your brakes do not operate as described above when the preceding tests are performed, the brake booster is either in need of repair or has failed. Refer to Chapter 9 for the removal procedure.

Parking brake

29 Slowly pull up on the parking brake handle or push down on the parking brake pedal and count the number of clicks you hear until the handle is up (or the pedal down) as far as it will go. The adjustment is correct if you hear the specified number of clicks. If you hear more or fewer clicks, it's time to adjust the parking brake (refer to Chapter 9).

30 An alternative method of checking the parking brake is to park the vehicle on a steep hill with the parking brake set and the transaxle in Neutral. If the parking brake cannot prevent the vehicle from rolling, it is in need of adjustment (see Chapter 9).

16 Air filter replacement (every 15,000 miles or 12 months)

Refer to illustrations 16.1 and 16.2

1 The air filter is located inside a housing at the left (drivers) side of the engine compartment. To remove the air filter, release the four spring clips retaining the two halves of the air cleaner housing **(see illustration)**.

2 Detach the air intake hose from the cover, then lift the cover up and remove the air filter element **(see illustration)**.

3 Inspect the outer surface of the filter element. If it is dirty, replace it. If it is only moderately dusty, it can be reused by blowing it clean from the back to the front surface with compressed air. **Warning:** *Always wear eye protection when using compressed air!* Because it is a pleated paper type filter, it cannot be washed or oiled. If it cannot be cleaned satisfactorily with compressed air, discard and replace it. **Caution:** *Never drive the vehicle with the air cleaner removed.*

Excessive engine wear could result and backfiring could even cause a fire under the hood.
4 Installation is the reverse of removal.

17 Fuel system check (every 15,000 miles or 12 months)

Refer to illustration 17.5

Warning: *Certain precautions should be observed when inspecting or servicing the fuel system components. Work in a well ventilated area and do not allow open flames (cigarettes, appliance pilot lights, etc.) near the work area. Mop up spills immediately and do not store fuel-soaked rags where they could ignite. It is a good idea to keep a dry chemical (Class B) fire extinguisher near the work area any time the fuel system is being serviced.*

1 If you smell fuel while driving or after the vehicle has been sitting in the sun, inspect the fuel system immediately.

2 Remove the fuel filler cap and inspect if for damage and corrosion. The gasket should have an unbroken sealing imprint. If the gasket is damaged or corroded, remove it and install a new one (Section 33).

3 Inspect the fuel feed and return lines for

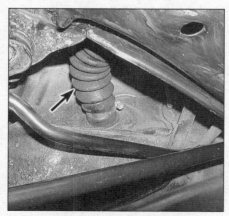

17.5 Check the fuel tank filler hose connection (arrow) for leaks and damage

cracks. Make sure that the threaded flare nut type connectors which secure the metal fuel lines to the fuel injection system and the banjo bolts which secure the banjo fittings to the in-line fuel filter are tight.

4 Since some components of the fuel system - the fuel tank and part of the fuel feed and return lines, for example - are underneath the vehicle, they can be inspected more easily with the vehicle raised on a hoist. If that's not possible, raise the vehicle and support it securely on jack stands.

5 With the vehicle raised and safely supported, inspect the fuel tank and filler neck for punctures, cracks and other damage. The connection between the filler neck and the tank is particularly critical. Sometimes a rubber filler neck will leak because of loose clamps or deteriorated rubber **(see illustration)**. These are problems a home mechanic can usually rectify. **Warning:** *Do not, under any circumstances, try to repair a fuel tank (except rubber components). A welding torch or any open flame can easily cause fuel vapors inside the tank to explode.*

6 Carefully check all rubber hoses and metal lines leading away from the fuel tank. Check for loose connections, deteriorated hoses, crimped lines and other damage. Carefully inspect the lines from the tank to the fuel injection system. Repair or replace damaged sections as necessary (see Chapter 4).

18 Automatic transaxle differential lubricant level check (not all models) (every 15,000 miles or 12 months)

Refer to illustration 18.2

1 The automatic transaxle differential on most models has a separate lubricant supply with a check/fill plug which must be removed to check the level. If the vehicle is raised to gain access to the plug, be sure to support it safely on jack stands - DO NOT crawl under the vehicle when it's supported only by the jack.

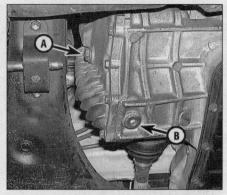

18.2 The automatic transaxle differential check/fill plug (A) is located on the rear of the transaxle - (B) is the drain plug (4-cylinder model)

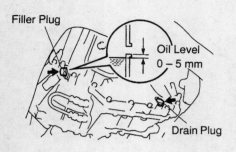

19.1 Remove the filler plug to check the manual transaxle lubricant level - the level should be at the bottom of the hole

20.7 To check the balljoints attempt to move the lower arm up and down with a prybar to make sure here is no play in the balljoint (if there is, replace it)

20.8 Push on the balljoint boot to check for tears and grease leaks

2 Remove the check/fill plug from the front of the transaxle **(see illustration)**.
3 Use your little finger as a dipstick to make sure the lubricant level is even with the bottom of the plug hole. If not, use a syringe or a gear oil pump to add the recommended lubricant (see this Chapter's Specifications) until it just starts to run out of the opening.
4 Install the plug and tighten it securely.

19 Manual transaxle lubricant level check (every 15,000 miles or 12 months)

Refer to illustration 19.1
1 The manual transaxle does not have a dipstick. To check the fluid level, raise the vehicle and support it securely on jack stands. On the lower front side of the transaxle housing, you will see a plug **(see illustration)**. Remove it. If the lubricant level is correct, it should be up to the lower edge of the hole.
2 If the transaxle needs more lubricant (if the level is not up to the hole), use a syringe or a gear oil pump to add more. Stop filling the transaxle when the lubricant begins to run out the hole.
3 Install the plug and tighten it securely. Drive the vehicle a short distance, then check for leaks.

20 Steering, suspension and driveaxle boot check (every 15,000 miles or 12 months)

Steering check

Note: *For detailed illustrations of the steering and suspension components, refer to Chapter 10.*
1 With the vehicle on the ground and the front wheels pointed straight ahead, rock the steering wheel gently back and forth. If freeplay is excessive, a front wheel bearing, main shaft yoke, intermediate shaft yoke, lower arm balljoint or steering system joint is worn or the steering gear is out of adjustment or broken. Steering wheel freeplay is the amount of travel (measured at the rim of the steering wheel) between the initial steering input and the point at which the front wheels begin to turn (indicated by slight resistance). Refer to Chapter 10 for the appropriate repair procedure.
2 Other symptoms, such as excessive vehicle body movement over rough roads, swaying (leaning) around corners and binding as the steering wheel is turned, may indicate faulty steering and/or suspension components.

Suspension check

Refer to illustrations 20.7 and 20.8
3 Check the shock absorbers by pushing down and releasing the vehicle several times at each corner. If the vehicle does not come back to a level position within one or two bounces, the shocks/struts are worn and must be replaced. When bouncing the vehicle up and down, listen for squeaks and noises from the suspension components. Additional information on suspension components can be found in Chapter 10.
4 Raise the vehicle with a floor jack and support it securely on jack stands. See *Jacking and towing* at the front of this book for the proper jacking points.
5 Check the tires for irregular wear patterns and proper inflation. See Section 5 in this Chapter for information regarding tire wear and Chapter 10 for the wheel bearing replacement procedures.
6 Inspect the universal joint between the steering shaft and the steering gear housing. Check the steering gear housing for lubricant leakage or oozing. Make sure that the dust seals and boots are not damaged and that the boot clamps are not loose. Check the steering linkage for looseness or damage. Check the track rod ends for excessive play. Look for loose bolts, broken or disconnected parts and deteriorated rubber bushings on all suspension and steering components. While an assistant turns the steering wheel from side to side, check the steering components for free movement, chafing and binding. If the steering components do not seem to be reacting with the movement of the steering wheel, try to determine where the slack is located.
7 Check the balljoints for wear by placing a wood block under each tire. Lower the jack until there is about half a load on the coil spring and place jack stands under the subframe. Make sure that the front wheels are in a straight forward position and block the wheel with chocks. Try to move each lower arm up and down with a prybar **(see illustration)** to ensure that its balljoint has no play. If any balljoint does have play, replace it. See Chapter 10 for the front balljoint replacement procedure.
8 Inspect the balljoint boots for damage and leaking grease **(see illustration)**. Replace the balljoints with new ones if they are damaged (see Chapter 10).

Driveaxle boot check

Refer to illustration 20.10
9 The driveaxle boots are very important because they prevent dirt, water and foreign material from entering and damaging the constant velocity (CV) joints. Oil and grease can cause the boot material to deteriorate prematurely, so it's a good idea to wash the boots with soap and water.
10 Inspect the boots for tears and cracks as well as loose clamps **(see illustration)**. If there is any evidence of cracks or leaking lubricant, they must be replaced as described in Chapter 8.

Chapter 1 Tune-up and routine maintenance   1-21

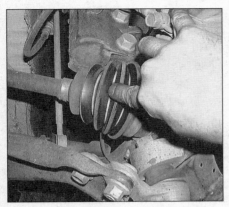

20.10 Flex the driveaxle boots by hand to check for tears, cracks and leaking grease

21 Interior ventilation filter replacement (every 15,000 miles or 12 months)

Refer to illustration 21.4

1 1999 and later models are equipped with an air filter in the blower housing that cleans the air before it enters the passenger's compartment.
2 To remove the air filter, remove the glove box (see Chapter 11).
3 Remove the filter cover mounting screws.
4 Lift the cover up and remove the air filter element **(see illustration)**.
5 Installation is the reverse of removal.

22 Exhaust system check (every 15,000 miles or 12 months)

Refer to illustration 22.2

1 With the engine cold (at least three hours after the vehicle has been driven), check the complete exhaust system from its starting point at the engine to the end of the tailpipe. This should be done on a hoist where unrestricted access is available.
2 Check the pipes and connections for evidence of leaks, severe corrosion or damage. Make sure that all brackets and hangers are in good condition and tight **(see illustration)**.
3 At the same time, inspect the underside of the body for holes, corrosion, open seams, etc. which may allow exhaust gases to enter the passenger compartment. Seal all body openings with silicone or body putty.
4 Rattles and other noises can often be traced to the exhaust system, especially the mounts and hangers. Try to move the pipes, silencer and catalytic converter. If the components can come in contact with the body or suspension parts, secure the exhaust system with new mounts.
5 Check the running condition of the engine by inspecting inside the end of the tailpipe. The exhaust deposits here are an indication of engine state-of-tune. If the pipe

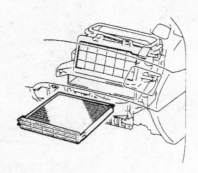

21.4 Typical ventilation filter details

is black and sooty or coated with white deposits, the engine is in need of a tune-up, including a thorough fuel system inspection.

23 Clutch/brake pedal height and freeplay - check and adjustment (every 30,000 miles or 24 months)

Pedal height

Refer to illustrations 23.1 and 23.2

1 The height of the clutch and brake pedal is the distance the pedal sits off the floor **(see illustration)**. If the pedal height is not within Specifications, it must be adjusted.
2 To adjust the clutch pedal, loosen the locknut and back the pushrod out for clearance. Turn the pushrod to adjust the pedal height in the middle of the specified range, then retighten the locknut **(see illustration)**.
3 At the brake pedal, loosen the locknut on the brake switch and retract the switch. Before measuring the brake pedal height, make sure the pedal is in the fully-returned position. Measure the pedal height and adjust if necessary (see Step 2).
4 Adjust the brake pedal switch by turning it clockwise until the switch body just contacts the pedal arm, then rotate it counter-

23.2 Loosen the pedal pushrod locknut, then adjust the pushrod to achieve proper pedal height

22.2 Check the exhaust system for damage, or worn rubber hangers (arrow)

23.1 To check the pedal height, measure the distance between the natural resting place of the pedal and the floor

clockwise to gain the specified clearance at the beginning of this Chapter and tighten the switch locknut.

Pedal freeplay

Refer to illustration 23.5

5 The freeplay is the pedal slack, or the distance the pedal can be depressed before it begins to have any effect on the clutch or brake system **(see illustration)**. If the pedal freeplay is not within the specified range, it must be adjusted.

23.5 Pedal freeplay is the distance between the natural resting point of the pedal to the point at which resistance is felt

1-22 Chapter 1 Tune-up and routine maintenance

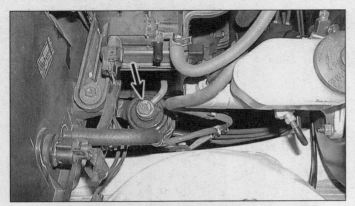

24.4 Disconnect the fuel line fitting at the top and bottom of the fuel filter

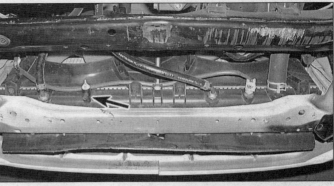

25.4 On most models you will have to remove a splash panel for access to the drain fitting located at the bottom of the radiator (arrow)

6   To adjust the pedal freeplay, loosen the locknut on the clutch pushrod. Then back out the pushrod to adjust the freeplay to the specified range, then retighten the locknut.
7   Before adjusting brake pedal freeplay, depress the brake pedal several times (with the engine off). Measure the freeplay and adjust if necessary. Loosen the locknut on the pushrod, then back off the pushrod to adjust the pedal freeplay to the specified range and retighten the locknut.

Brake pedal reserve distance

8   With the parking brake released and the engine running, depress the pedal with normal braking effort and have an assistant measure the distance from the center of the pedal pad to the floor. If the distance is less than specified, refer to Chapter 9 and troubleshoot the brake system.

24 Fuel filter replacement (every 30,000 miles or 24 months)

Refer to illustration 24.4

1   Refer to Chapter 4 and relieve the fuel system pressure.
2   Disconnect the negative battery cable. **Caution:** *If the stereo in your vehicle is equipped with an anti-theft system, make sure you have the correct activation code before disconnecting the battery.*
3   If necessary for access, remove the air cleaner assembly (see Chapter 4) and evaporative canister (see Chapter 6).
4   If equipped with a union bolt at the top of the filter, use a backup wrench to hold the filter, loosen the bolt with another wrench and disconnect the fitting from the filter **(see illustration)**. On models equipped with a quick-connect fitting at the top of the filter, remove the fitting protector and disconnect the fitting from the filter (see Chapter 4). Loosen the fitting at the bottom of the fuel filter with a flare nut wrench.
5   Remove the bracket bolts from the bulkhead and remove the filter and bracket assembly.
6   Note that the inlet and outlet pipes are clearly labeled on their respective ends of the filter and that the flanged end of the filter faces down. Make sure the new filter is fitted so that it's facing the proper direction as noted above. When correctly fitted, the filter should be fitted so that the outlet faces up and the inlet faces down.
7   Using new sealing washers provided by the filter manufacturer, install the inlet and fittings and tighten them securely. If equipped with a quick-connect fitting, push the fuel line on the filter until it snaps in place and install the fitting protector.
8   The remainder of installation is the reverse of the removal procedure.

25 Cooling system servicing (draining, flushing and refilling) (every 30,000 miles or 24 months)

Warning: *Do not allow engine coolant (antifreeze) to come in contact with your skin or painted surfaces of the vehicle. Rinse off spills immediately with plenty of water. Antifreeze is highly toxic if ingested. Never leave antifreeze laying around in an open container or in puddles on the floor; children and pets are attracted by it's sweet smell and may drink it. Check with local authorities about disposing of used antifreeze. Many communities have collection centers which will see that antifreeze is disposed of safely.*

1   Periodically, the cooling system should be drained, flushed and refilled to replenish the antifreeze mixture and prevent formation of rust and corrosion, which can impair the performance of the cooling system and cause engine damage. When the cooling system is serviced, all hoses and the radiator cap should be checked and replaced if necessary.

Draining

Refer to illustrations 25.4, 25.5a and 25.5b

2   Apply the parking brake and block the wheels. If the vehicle has just been driven, wait several hours to allow the engine to cool down before beginning this procedure.
3   Once the engine is completely cool, remove the radiator cap.
4   Move a large container under the radiator drain to catch the coolant. Attach a 3/8-inch inner diameter hose to the drain fitting to direct the coolant into the container (some models are already equipped with a hose), then open the drain fitting (a pair of pliers may be required to turn it) **(see illustration)**.
5   After the coolant stops flowing out of the radiator, move the container under the engine block drain plug **(see illustrations)**. Loosen the plug and allow the coolant in the block to drain.
6   While the coolant is draining, check the condition of the radiator hoses, heater hoses and clamps (refer to Section 13 if necessary).

25.5a On four-cylinder engines, the coolant drain plug is located on the back side of the engine block (arrow)

25.5b The V6 engine has a coolant drain like this (arrow) located on both sides of the block

Chapter 1 Tune-up and routine maintenance 1-23

26.2a Check the charcoal canister for damage and the hose connections (arrows) for cracks and damage

7 Replace any damaged clamps or hoses (see Chapter 3).

Flushing

8 Once the system is completely drained, remove the thermostat from the engine (see Chapter 3). Then reinstall the thermostat housing without the thermostat. This will allow the system to be flushed.
9 Reinstall the engine block drain plugs and tighten the radiator drain plug. Turn your heating system controls to Hot, so that the heater core will be flushed at the same time as the rest of the cooling system.
10 Disconnect the upper radiator hose from the radiator. Place a garden hose in the upper radiator inlet, turn the water on and flush the system until the water runs clear out of the upper radiator hose.
11 In severe cases of contamination or clogging of the radiator, remove the radiator (see Chapter 3) and have a radiator repair facility clean and repair it if necessary. Many deposits can be removed by the chemical action of a cleaner available at auto parts stores. Follow the procedure outlined in the manufacturer's instructions. **Note:** *When the coolant is regularly drained and the system refilled with the correct antifreeze/water mixture, there should be no need to use chemical cleaners or descalers.*
12 After flushing, drain the radiator and remove the block drain plugs once again to drain the water from the system.

Refilling

13 Close and tighten the radiator drain. Install and tighten the block drain plug.
14 Place the heater temperature control in the maximum heat position.
15 Slowly add new coolant (a 50/50 mixture of water and antifreeze) to the radiator until it's full. Add coolant to the reservoir up to the lower mark.
16 Leave the radiator cap off and run the engine in a well-ventilated area until the thermostat opens (coolant will begin flowing through the radiator and the upper radiator hose will become hot).
17 Turn the engine off and let it cool. Add more coolant mixture to bring the level back up to the lip on the radiator filler neck.
18 Squeeze the upper radiator hose to expel air, then add more coolant mixture if necessary. Replace the radiator cap.
19 Start the engine, allow it to reach normal operating temperature and check for leaks.

26 Evaporative emissions control system check (every 30,000 miles or 24 months)

Refer to illustrations 26.2a and 26.2b

1 The function of the evaporative emissions control system is to draw fuel vapors from the fuel tank and fuel system, store them in a charcoal canister and then burn them during normal engine operation.
2 The most common symptom of a fault in the evaporative emissions system is a strong fuel odor in or around the vehicle. If a fuel odor is detected, inspect the charcoal canister. Check the canister and all hoses for damage and deterioration **(see illustrations)**.
3 The evaporative emissions control system is explained in more detail in Chapter 6.

27 Automatic transaxle/differential fluid change (not all models) (every 30,000 miles or 24 months)

Refer to illustrations 27.7, 27.8a, 27.8b, 27.9, 27.12, 27.14a and 27.14b

1 At the specified time intervals, the automatic transaxle and differential fluid should be drained and replaced. **Note:** *Although the manufacturer doesn't specify it, it is a good idea to clean the transaxle fluid strainer periodically to remove accumulated dirt and metal particles.*
2 Before beginning work, purchase the specified transaxle fluid (see Recommended fluids and lubricants at the front of this Chapter).
3 Other tools necessary for this job include jack stands to support the vehicle in a raised position, a 10 mm hex bit or Allen wrench, a drain pan capable of holding at least eight pints, newspapers and clean rags.
4 The fluid should be drained immediately after the vehicle has been driven. Hot fluid is more effective than cold fluid at removing built up sediment. **Warning:** *Fluid temperature can exceed 350-degrees F in a hot transaxle. Wear protective gloves.*
5 After the vehicle has been driven to warm up the fluid, raise it and place it on jack stands for access to the transaxle and differential drain plugs.
6 Move the necessary equipment under the vehicle, being careful not to touch any of the hot exhaust components.
7 Place the drain pan under the drain plug in the transaxle pan and remove the drain plug **(see illustration)**. Be sure the drain pan is in position, as fluid will come out with some force. Once the fluid is drained, reinstall the drain plug securely. If you aren't going to clean the strainer, proceed to Step 14.
8 To clean the strainer, remove the front transaxle pan bolts, then loosen the rear bolts and carefully pry the pan loose with a screwdriver and allow the remaining fluid to drain **(see illustrations)**. Once the fluid had drained, remove the bolts and lower the pan.

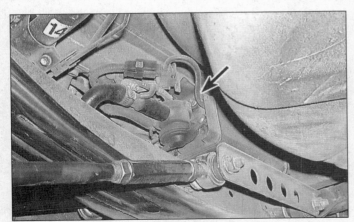

26.2b On later models, the EVAP canister (arrow) is located at the rear of the vehicle, ahead of the fuel tank

27.7 Use a 10 mm hex bit or Allen wrench to remove the automatic transaxle drain plug

1-24 Chapter 1 Tune-up and routine maintenance

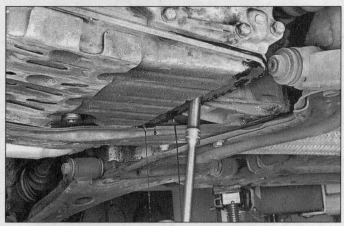

27.8a After loosening the front bolts, remove the rear pan bolts and . . .

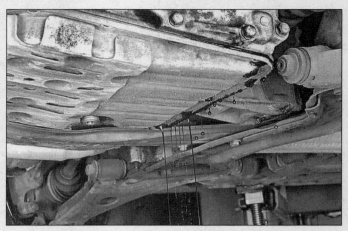

27.8b . . . allow the remaining fluid to drain out

27.9 Remove the strainer bolts and lower the strainer (be careful, there will be some residual fluid)

27.12 Noting their locations, remove the magnets and wash them and the pan in solvent before installing them

9 Remove the strainer retaining bolts, disconnect the clip (some models) and lower the strainer from the transaxle **(see illustration)**. Be careful when lowering the strainer as it contains residual fluid.
10 Wash the strainer thoroughly in clean transmission fluid.
11 Place the strainer in position, connect the clip (if equipped) and install the bolts. Tighten the bolts to the torque listed in this Chapter's Specifications.
12 Carefully clean the gasket surfaces of the fluid pan, removing all traces of old gasket material. Noting their location, remove the magnets, wash the pan in clean solvent and dry it with compressed air. **Warning:** Always wear eye protection when using compressed air! Be sure to clean and reinstall the magnets in the pan **(see illustration)**.
13 Install a new gasket, place the fluid pan in position and install the bolts in their original positions. Tighten the bolts to the torque listed in this Chapter's Specifications.
14 Locate the differential drain plug (this

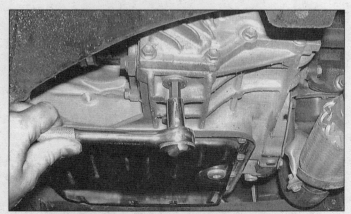

27.14a Remove the four-cylinder model automatic transaxle differential drain plug with a hex bit or Allen wrench

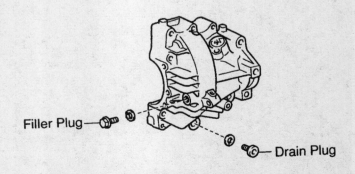

27.14b V6 model automatic transaxle differential drain and fill plug locations

Chapter 1 Tune-up and routine maintenance

29.1a Periodically check the tightness of the nuts and bolts on front suspension subframe (arrows, two front bolts are accessed after plastic splash shields are removed) . . .

won't be necessary on 1999 and later Lexus ES 300 transaxles as they don't have a separate differential housing). Place the drain pan underneath the plug, remove it with a hex bit or Allen wrench and drain the fluid (see illustrations). When the differential fluid has drained, reinstall the plug securely.

15 Referring to Section 18, add new fluid to the differential until it begins to run out of the filler hole (see *Recommended lubricants and fluids* at the beginning of this Chapter for the specified fluid type and capacity). **Caution:** *Do not overfill. The automatic transaxle and the differential are separate units.*

16 Lower the vehicle.

17 With the engine off, add new fluid to the transaxle through the dipstick tube (see *Recommended fluids and lubricants* for the recommended fluid type and capacity). Use a funnel to prevent spills. It is best to add a little fluid at a time, continually checking the level with the dipstick (see Section 7). Allow the fluid time to drain into the pan.

18 Start the engine and gearchange the selector into all positions from P through L, then gearchange into P and apply the parking brake.

19 With the engine idling, check the fluid level. Add fluid up to the Cool level on the dipstick.

28 Manual transaxle lubricant change (every 30,000 miles or 24 months)

1 The fluid should be drained immediately after the vehicle has been driven. Hot fluid is more effective than cold fluid at removing built up sediment.

2 After the vehicle has been driven to warm up the fluid, raise it and place it on jack stands for access to the drain plug.

3 Remove the drain plug(s) and drain the fluid (see Section 19).

4 Reinstall the drain plug securely.

5 Add new fluid until it begins to run out of the filler hole. See *Recommended lubricants and fluids* for the specified lubricant type.

29 Chassis and body fastener check (every 30,000 miles or 24 months)

Refer to illustrations 29.1a and 29.1b

1 Tighten the following fasteners to the torque values listed in this Chapter's Specifications **(see illustrations)**:

  a) Front seat mounting bolts.
  b) Both front and rear suspension member-to-body mounting bolts and nuts (left and right sides).

30 Spark plug check and replacement (every 30,000 miles or 24 months)

Refer to illustrations 30.1a, 30.1b, 30.1c, 30.4a, 30.4b, 30.7, 30.8, 30.10, 30.12a and 30.12b

1 Spark plug replacement requires a

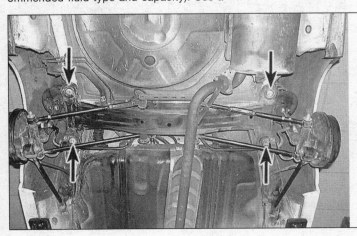

29.1b . . . and rear suspension mounts (arrows)

1-26 Chapter 1 Tune-up and routine maintenance

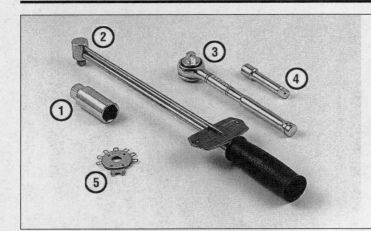

30.1a Tools required for changing spark plugs
1. **Spark plug socket** - This will have special padding inside to protect the spark plug's porcelain insulator
2. **Torque wrench** - Although not mandatory, using this tool is the best way to ensure the plugs are tightened properly
3. **Ratchet** - Standard hand tool to fit the spark plug socket
4. **Extension** - Depending on model and accessories, you may need special extensions and universal joints to reach one or more of the plugs
5. **Spark plug gap gauge** - This gauge for checking the gap comes in a variety of styles. Make sure the gap for your engine is included

spark plug socket that fits onto a ratchet. This socket is lined with a rubber grommet to protect the porcelain insulator of the spark plug and to hold the plug while you insert it into the spark plug hole. You will also need a wire-type feeler gauge to check and adjust the spark plug gap and a torque wrench to tighten the new plugs to the specified torque **(see illustration)**. On engines equipped with a V-bank cover, remove the three 5 mm cap nuts and detach the cover for access **(see illustrations)**.

2 If you are replacing the plugs, purchase the new plugs, adjust them to the proper gap and then replace each plug one at a time. **Note:** *The manufacturer specifies that only platinum or iridium-tipped spark plugs be used on these models. When buying new spark plugs, it's essential that you obtain the correct plugs for your specific vehicle. This information can be found in the Specifications Section at the beginning of this Chapter, on the Vehicle Emissions Control Information (VECI) label located on the underside of the hood or in the owner's manual. If these sources specify different plugs, purchase the spark plug type specified on the VECI label because that information is provided specifically for your engine.*

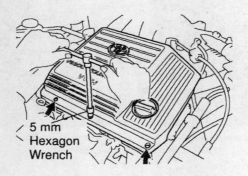

30.1b Remove the three 5 mm cap nuts

3 Inspect each of the new plugs for defects. If there are any signs of cracks in the porcelain insulator of a plug, don't use it.
4 Check the electrode gaps of the new plugs. **Note:** *Do not adjust the gap on the iridium spark plugs used on some later models because using a gapping tool on them could damage the iridium plating on the electrodes. These spark plugs are pre-gapped by the manufacturer.* On other types of spark

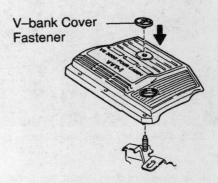

30.1c Rotate the fastener counterclockwise, remove it and detach the V-bank cover - when installing, press the fastener into place

plugs, check the gap by inserting the wire gauge of the proper thickness between the electrodes at the tip of the plug **(see illustration)**. The gap between the electrodes should be identical to that listed in this Chapter's Specifications or on the VECI label. If the gap is incorrect, use the notched adjuster on the feeler gauge body to bend the curved side

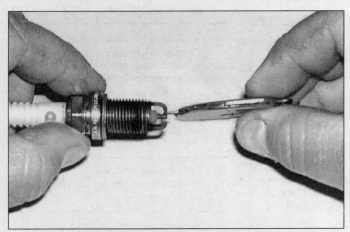

30.4a Spark plug manufacturers recommend using a wire-type gauge when checking the gap - if the wire does not slide between the electrodes with a slight drag, adjustment is required

30.4b To change the gap, bend the side electrodes only, be very careful not to crack or chip the porcelain insulator surrounding the center electrode

Chapter 1 Tune-up and routine maintenance   1-27

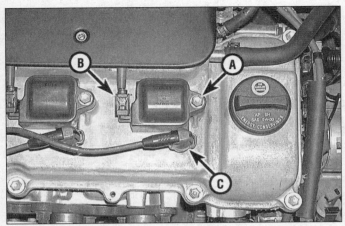

30.7 On V6 engines, remove the retaining bolt (A), disconnect the electrical connector (B) and detach the individual coils to reach the front spark plugs - the spark plug wires (C) lead to the rear bank spark plugs

30.8 When removing spark plug wires, pull only on the boot using a twisting/pulling motion

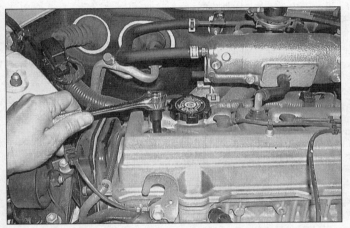

30.10 Because they are deeply recessed, the proper spark plug socket and an extension will be required when removing or installing the spark plugs

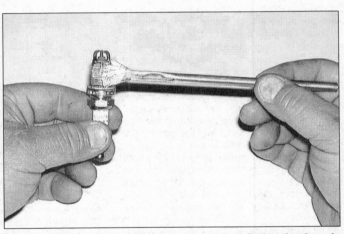

30.12a A light coat of anti-seize compound applied to the threads of the spark plugs will keep the threads in the cylinder head from being damaged the next time the plugs are removed

electrode slightly (see illustration).

5 If the side electrode is not exactly over the center electrode, use the notched adjuster to align them. **Caution:** *If the gap of a new plug must be adjusted, bend only the base of the negative electrode - do not touch the tip.*

6 For access to the rear bank plugs on V6 engines, disconnect any hoses or components that would interfere with access and move them out of the way.

7 On V6 models, remove the bolts and detach each ignition coil assembly from the spark plugs (front bank only on most models) **(see illustration)**.

8 To prevent the possibility of mixing up spark plug wires, work on one spark plug at a time. Remove the wire and boot from one spark plug. Grasp the boot - not the cable - as shown, give it a half twisting motion and pull straight up **(see illustration)**.

9 If compressed air is available, blow any dirt or foreign material away from the spark plug area before proceeding (a common bicycle pump will also work). **Warning:** *Always wear eye protection when using compressed air!*

10 Remove the spark plug **(see illustration)**.

11 Whether you are replacing the plugs at this time or intend to re-use the old plugs, compare each old spark plug with those shown on the inside of the back cover to determine the overall running condition of the engine.

12 Apply a small amount of anti-seize compound to the spark plug threads **(see illustration)**. It's often difficult to insert spark plugs into their holes without cross-threading them. To avoid this possibility, fit a short piece of rubber hose over the end of the spark plug **(see illustration)**. The flexible hose acts as a universal joint to help align the plug with the spark plug hole. Should the plug begin to cross-thread, the hose will slip on the spark plug, preventing thread damage. Tighten the plug to the torque listed in this Chapter's Specifications.

13 Attach the plug wire to the new spark plug, again using a twisting motion on the boot until it is firmly seated on the end of the spark plug.

14 Follow the above procedure for the remaining spark plugs, replacing them one at a time to prevent mixing up the spark plug wires.

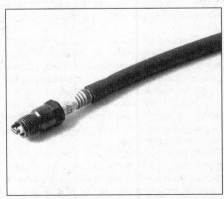

30.12b A section of rubber hose will aid in getting the spark plug threads started

31 Spark plug wire - check and replacement (every 30,000 miles or 24 months)

1 The spark plug wires should be checked whenever new spark plugs are fitted.

2 Begin this procedure by making a visual check of the spark plug wires while the engine is running. In a darkened garage (make sure there is ventilation) start the engine and observe each plug wire. Be careful not to come into contact with any moving engine parts. If there is a break in the wire, you will see arcing or a small spark at the damaged area. If arcing is noticed, obtain new wires.

3 The spark plug wires should be inspected one at a time to prevent mixing up the order, which is essential for proper engine operation. Each original plug wire should be numbered to help identify its location. If the number is illegible, a piece of tape can be marked with the correct number and wrapped around the plug wire. **Note:** *Four-cylinder engines have a spark plug wire for each spark plug, all routed to the coilpack. On V6 engines, there are three coilpacks, one mounted over each of the spark plugs on the front bank. From these coils, three spark plug wires are routed around the intake manifold to the rear bank spark plugs, which have a standard boot termination like four-cylinder plug wires.*

4 Disconnect the plug wire from the spark plug (refer to Section 30). A removal tool can be used for this purpose or you can grasp the rubber boot, twist the boot half a turn and pull the boot free. Do not pull on the wire itself.

5 Check inside the boot for corrosion, which will look like a white crusty powder.

6 Push the wire and boot back onto the end of the spark plug. It should fit tightly onto the end of the plug. If it doesn't, remove the wire and use pliers to carefully crimp the metal connector inside the wire boot until the fit is snug.

7 Using a clean rag, wipe the entire length of the wire to remove built-up dirt and grease. Once the wire is clean, check for burns, cracks and other damage. Do not bend the wire sharply, because the conductor might break.

8 Disconnect the wire from the coilpack. Use a small screwdriver to release the locking claw from the boot protector and pull only on the rubber boot. Check for corrosion and a tight fit. Replace the wire in the coilpack.

9 Inspect the remaining spark plug wires, making sure that each one is securely fastened at the coilpack and spark plug when the check is complete.

10 If new spark plug wires are required, purchase a set for your specific engine model. Remove and replace the wires one at a time to avoid mix-ups in the firing order.

11 For electrical checks of plug wires and coilpacks, refer to Chapter 5.

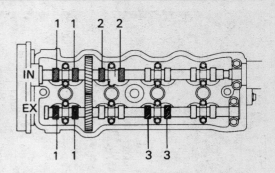

32.7a On four-cylinder engines, when the no. 1 piston is at TDC on the compression stroke, the valve clearance for the no. 1 and no. 3 cylinder exhaust valves and the no. 1 and no. 2 cylinder intake valves can be measured

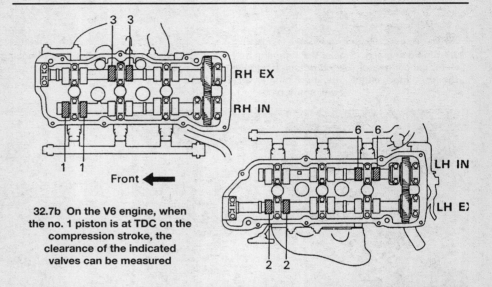

32.7b On the V6 engine, when the no. 1 piston is at TDC on the compression stroke, the clearance of the indicated valves can be measured

32 Valve clearance check and adjustment (every 60,000 miles or 48 months)

Refer to illustrations 32.7a, 32.7b, 32.7c, 32.8, 32.9a, 32.9b, 32.11a, 32.11b, 32.11c and 32.12

Note: *The following procedure requires the use of a special lifter tool. It is impossible to perform this task without it.*

1 Disconnect the negative cable from the battery. **Caution:** *If the stereo in your vehicle is equipped with an anti-theft system, make sure you have the correct activation code before disconnecting the battery.*

2 On four-cylinder models, disconnect the spark plug wires (Section 30) and remove any other components that will interfere with valve cover removal.

3 On V6 models, drain the coolant (Section 25), remove the air cleaner assembly and air intake plenum (Chapter 2B). Disconnect the spark plug wires from the rear bank spark plugs and remove the ignition coil/spark plug wire assembly from the front bank. Remove any other components that will interfere with valve cover removal.

4 Blow out the spark plug recess with compressed air, if available, to remove any debris that might fall into the cylinders, then remove the spark plugs (see Section 30). **Warning:** *Always wear eye protection when using compressed air!*

5 Remove the valve cover(s) (refer to Chapter 2A or 2B).

6 Refer to Chapter 2 and position the number 1 piston at TDC on the compression stroke.

7 Measure the clearance of the indicated valves with a feeler gauge of the specified thickness **(see illustrations)**. Record the clearance of each valve and note which are out of specification. This information will be used later to determine the required replacement shims.

8 On four-cylinder engines, turn the crankshaft one complete revolution and realign the timing marks. Measure the remaining valves **(see illustration)**.

9 On V6 engines, turn the crankshaft 2/3-turn (240-degrees) clockwise. Measure the valve clearance on the valves shown **(see illustration)**. Rotate the crankshaft a further 2/3-turn and measure the clearance on the remaining valves **(see illustration)**.

10 After the measuring and recording the clearance of each valve, turn the crankshaft pulley until the camshaft lobe above the first valve which you intend to adjust is pointing upward, away from the shim.

11 Position the notch in the lifter toward the

Chapter 1 Tune-up and routine maintenance

32.7c Measure the clearance for each valve with a feeler gauge of the specified thickness - if the clearance is correct, you should feel a slight drag on the gauge as you pull it out

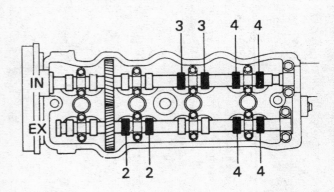

32.8 On four-cylinder engines, when the no. 4 piston is at TDC on the compression stroke, the valve clearance for the no. 2 and no. 4 exhaust valves and the no. 3 and no. 4 intake valves can be measured

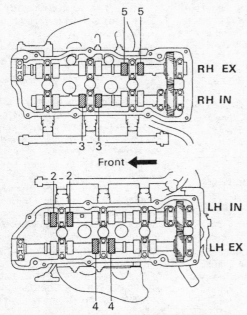

32.9a After the V6 engine has been rotated 240-degrees from TDC for the no. 1 piston, on the compression stroke, measure the clearance of the indicated valves

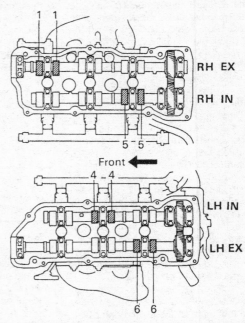

32.9b On the V6 engine, rotate the crankshaft an additional 2/3 of a revolution (240-degrees) and measure the clearance of the remaining valves

spark plug. Then depress the lifter with the special lifter tools (see illustrations). Place the special lifter tool in position as shown, with the longer jaw of the tool gripping the lower edge of the cast lifter boss and the upper, shorter jaw gripping the upper edge of the lifter itself. Depress the lifter by squeezing the handles of the lifter tool together, then hold the lifter down with the smaller tool and remove the larger one. Remove the adjusting shim with a small screwdriver or a pair of tweezers (see illustrations). Note that the wire hook on the end of some lifter tool handles can be used to clamp both handles together to keep the lifter depressed while the shim is removed.

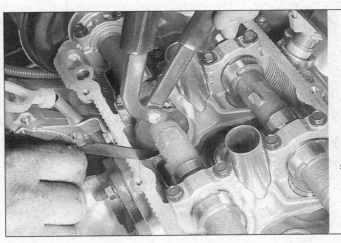

32.11a Install the lifter tool as shown and squeeze the handles together to depress the lifter, then hold the lifter down with the smaller tool so the shim can be removed

32.11b Keep pressure on the lifter with the smaller tool and remove the shim with a small screwdriver . . .

32.11c . . . a pair of tweezers or a magnet as shown here

12 Measure the thickness of the shim with a micrometer **(see illustration)**. To calculate the correct thickness of a replacement shim that will place the valve clearance within the specified value, use the following formula:

$N = T + (A - V)$
T = thickness of the old shim
A = valve clearance measured
N = thickness of the new shim
V = desired valve clearance (see this Chapter's Specifications)

13 Select a shim with a thickness as close as possible to the valve clearance calculated. Shims, which are available in 17 sizes in increments of 0.0020-inch (0.050 mm), range in size from 0.0984-inch (2.500 mm) to 0.1299-inch (3.300 mm). **Note:** *Through careful analysis of the shim sizes needed to bring the out-of-specification valve clearance within specification, it is often possible to simply move a shim that has to come out anyway to another lifter requiring a shim of that particular size, thereby reducing the number of new shims that must be purchased.*

14 Place the special lifter tool in position as shown in illustration 32.11a, with the longer jaw of the tool gripping the lower edge of the cast lifter boss and the upper, shorter jaw gripping the upper edge of the lifter itself, press down the lifter by squeezing the handles of the lifter tool together and install the new adjusting shim (note that the wire hook on the end of one lifter tool handle can be used to clamp the handles together to keep the lifter depressed while the shim is inserted). Measure the clearance with a feeler gauge to make sure that your calculations are correct.

15 Repeat this procedure until all the valves which are out of clearance have been corrected.

16 Installation of the spark plugs, valve cover, spark plug wires and boots, accelerator cable bracket, etc. is the reverse of removal.

33 Fuel tank cap gasket replacement (every 60,000 miles or 48 months)

Refer to illustration 33.2

1 Obtain a new gasket.
2 Remove the tank cap and carefully pry the old gasket out of the recess **(see illustration)**. Be very careful not to damage the sealing surface inside the cap.
3 Work the new gasket into the cap recess.
4 Install the cap, then remove it and make sure the gasket seals all the way around.

34 Positive Crankcase Ventilation (PCV) valve check and replacement (every 60,000 miles or 48 months)

Refer to illustration 34.4

1 The PCV valve and hose is located in the valve cover.
2 Disconnect the hose, pull the PCV valve from the cover, then reconnect the hose.
3 With the engine idling at normal operating temperature, place your finger over the valve opening. If there's no vacuum at the valve, check for a plugged hose or valve. Replace any plugged or deteriorated hoses.
4 Turn off the engine. Remove the PCV valve from the hose. Blow through the valve from the valve cover (cylinder head) end. If air will not pass through the valve in this direction, replace it with a new one **(see illustration)**.
5 When purchasing a replacement PCV valve, make sure it's for your particular vehicle and engine size. Compare the old valve with the new one to make sure they're the same.

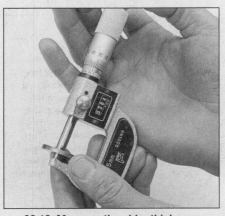
32.12 Measure the shim thickness with a micrometer

33.2 Use a small screwdriver to carefully pry out the old gasket - take care not to damage the cap

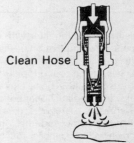
34.4 To check the PCV valve, first attach a clean section of hose to the cylinder head side of the valve and blow through it - air should pass through easily - then blow through the opposite side of the valve and verify that air doesn't pass through it

Chapter 2 Part A
Four-cylinder engine

Contents

	Section
Camshaft oil seal - replacement	9
Camshafts and valve lifters - removal, inspection and installation	10
CHECK ENGINE light on	See Chapter 6
Crankshaft front oil seal - replacement	8
Cylinder compression check	See Chapter 2C
Cylinder head - removal and installation	11
Drivebelt check, adjustment and replacement	See Chapter 1
Engine mounts - check and replacement	16
Engine oil and filter change	See Chapter 1
Engine overhaul - general information	See Chapter 2C
Engine - removal and installation	See Chapter 2C
Exhaust manifold - removal and installation	6
Flywheel/driveplate - removal and installation	14

	Section
General information	1
Intake manifold - removal and installation	5
Oil pan - removal and installation	12
Oil pump - removal, inspection and installation	13
Rear main oil seal - replacement	15
Repair operations possible with the engine in the vehicle	2
Spark plug replacement	See Chapter 1
Timing belt and sprockets - removal, inspection and installation	7
Top Dead Center (TDC) for number one piston - locating	3
Valve cover - removal and installation	4
Valves - servicing	See Chapter 2C
Water pump - removal and installation	See Chapter 3

Specifications

General
Engine designation	5S-FE
Displacement	134 cubic inches (2.2 liters)
Cylinder numbers (drivebelt end-to-transaxle end)	1-2-3-4
Firing order	1-3-4-2

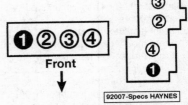

Cylinder numbering and coil terminal location

Cylinder head
Warpage limits	
Cylinder head-to-block surface	0.020 inch
Intake and exhaust manifolds	0.0118 inch

Timing belt
Idler pulley spring free length	1.811 inches

Camshaft
Journal diameter (all)	1.0614 to 1.0620 inches
Bearing oil clearance	
Standard	0.0010 to 0.0024 inch
Service limit	0.0039 inch
Runout limit	0.0016 inch
Lobe height	
Intake camshaft	
Standard	1.6539 to 1.6579 inches
Service limit	1.6496 inches
Exhaust camshaft	
Standard	1.5772 to 1.5811 inches
Service limit	1.5728 inches
Camshaft thrust clearance (endplay)	
Intake camshaft	
Standard	0.0018 to 0.0039 inch
Service limit	0.0047 inch
Exhaust camshaft	
Standard	0.0012 to 0.0033 inch
Service limit	0.0039 inch

Camshaft (continued)

Camshaft gear spring end gap	0.886 to 0.902 inch
Camshaft gear backlash	
Standard	0.0008 to 0.0079 inch
Service limit	0.0188 inch
Valve lifter	
Diameter	1.2191 to 1.2195 inches
Bore diameter	1.2205 to 1.2212 inches
Lifter oil clearance	
Standard	0.0009 to 0.0020 inch
Service limit	0.0028 inch

Oil pump

Driven rotor-to-case clearance	
Standard	0.0039 to 0.0063 inch
Service limit	0.0079 inch
Rotor tip clearance	
Standard	0.0016 to 0.0063 inch
Service limit	0.0079 inch

Torque specifications

	Ft-lbs (unless otherwise indicated)
Intake manifold nuts/bolts	168 in-lbs
Intake manifold brace bolts	29
Exhaust manifold nuts/bolts	36
Engine balancer assembly mounting bolts	36
Crankshaft pulley-to-crankshaft bolt	80
Flywheel/driveplate bolts	
Flywheel (manual transaxle)	65
Driveplate (automatic transaxle)	61
Idler pulley bolts	31
Cylinder head bolts	
Step 1	36
Step 2	Tighten an additional 90-degrees (1/4 turn)
Camshaft bearing cap bolts	168 in-lbs
Camshaft sprocket bolt	40
Oil pump bolts	78 in-lbs
Oil pump sprocket nut	18
Oil pick-up (strainer) nuts/bolts	48 in-lbs
Oil pan-to-block bolts	48 in-lbs
Spark plug tube nuts	33
Rear crankshaft oil seal retainer bolts	108 in-lbs
Engine mounts	
Front insulator top bolt	47
Front mount bracket to engine bolts	47
Front mount-to-chassis bolts	
Silver colored	32
Green colored	49
Rear mount upper insulator bolt	47
Rear mount bracket-to-engine	47
Rear mount insulator-to-chassis nuts	49
Right-hand engine movement control rod bolts	47
Right-hand mount bracket-to-engine bolts	38

1 General information

This Part of Chapter 2 is devoted to in-vehicle repair procedures for the four-cylinder engine. All information concerning engine removal and installation and engine block and cylinder head overhaul can be found in Part C of this Chapter.

The following repair procedures are based on the assumption that the engine is installed in the vehicle. If the engine has been removed from the vehicle and mounted on a stand, many of the steps outlined in this Part of Chapter 2 will not apply.

The Specifications included in this Part of Chapter 2 apply only to the procedures contained in this Part. Part C of Chapter 2 contains the Specifications necessary for cylinder head and engine block rebuilding.

During the years covered by this manual, the four-cylinder engine in the Camry is designated the 5S-FE. The engine design includes dual overhead camshafts (DOHC) and four valves per cylinder.

2 Repair operations possible with the engine in the vehicle

Many major repair operations can be accomplished without removing the engine from the vehicle.

Clean the engine compartment and the exterior of the engine with some type of degreaser before any work is done. It will make the job easier and help keep dirt out of the internal areas of the engine.

Depending on the components involved,

Chapter 2 Part A Four-cylinder engine

3.6 Align the crankshaft drivebelt pulley notch (arrow) with the 0 (zero) on the timing plate

4.5 The valve cover is held in place by the large spark plug tube nuts (arrows)

it may be helpful to remove the hood to improve access to the engine as repairs are performed (refer to Chapter 11 if necessary). Cover the fenders to prevent damage to the paint. Special pads are available, but an old bedspread or blanket will also work.

If vacuum, exhaust, oil or coolant leaks develop, indicating a need for gasket or seal replacement, the repairs can generally be made with the engine in the vehicle. The intake and exhaust manifold gaskets, oil pan gasket, crankshaft oil seals and cylinder head gasket are all accessible with the engine in place.

Exterior engine components, such as the intake and exhaust manifolds, the oil pan, the oil pump, the water pump, the starter motor, the alternator and the fuel system components can be removed for repair with the engine in place.

Since the cylinder head can be removed without pulling the engine, camshaft and valve component servicing can also be accomplished with the engine in the vehicle. Replacement of the timing belt and pulleys is also possible with the engine in the vehicle.

In extreme cases caused by a lack of necessary equipment, repair or replacement of piston rings, pistons, connecting rods and rod bearings is possible with the engine in the vehicle. However, this practice is not recommended because of the cleaning and preparation work that must be done to the components involved.

3 Top Dead Center (TDC) for number one piston - locating

Refer to illustration 3.6

1 Top Dead Center (TDC) is the highest point in the cylinder that each piston reaches as it travels up-and-down when the crankshaft turns. Each piston reaches TDC on the compression stroke and again on the exhaust stroke, but TDC generally refers to piston position on the compression stroke.

The timing marks on the vibration damper installed on the front of the crankshaft are referenced to the number one piston at TDC on the compression stroke.

2 Positioning the piston(s) at TDC is an essential part of procedures such as timing belt and sprocket replacement.

3 In order to bring any piston to TDC, the crankshaft must be turned using one of the methods outlined below. When looking at the timing belt end of the engine, normal crankshaft rotation is clockwise. **Warning:** *Before beginning this procedure, be sure to place the transmission in Neutral and remove the ignition key.*

   a) *The preferred method is to turn the crankshaft with a large socket and breaker bar attached to the large bolt threaded into the center of the crankshaft pulley.*
   b) *A remote starter switch, which may save some time, can also be used. Attach the switch leads to the S (switch) and B (battery) terminals on the starter motor. Once the piston is close to TDC, use a socket and breaker bar as described in the previous paragraph.*
   c) *If an assistant is available to turn the ignition switch to the Start position in short bursts, you can get the piston close to TDC without a remote starter switch. Use a socket and breaker bar as described in Paragraph a) to complete the procedure.*

4 Disable the ignition system by disconnecting the primary electrical connectors at the ignition coil pack/modules (see Chapter 5).

5 Remove the spark plugs and install a compression gauge in the number one cylinder. Turn the crankshaft clockwise with a socket and breaker bar as described above.

6 When the piston approaches TDC, compression will be noted on the compression gauge. Continue turning the crankshaft until the notch in the crankshaft damper is aligned with the TDC mark on the front cover **(see illustration)**. At this point number one cylinder is at TDC on the compression stroke. If the marks aligned but there was no compression, the piston was on the exhaust stroke. Continue rotating the crankshaft 360-degrees (1-turn).

7 After the number one piston has been positioned at TDC on the compression stroke, TDC for any of the remaining cylinders can be located by turning the crankshaft 180 degrees and following the firing order (refer to the Specifications). Rotating the engine 180 degrees past TDC #1 will put the engine at TDC compression for cylinder #3.

4 Valve cover - removal and installation

Refer to illustrations 4.5 and 4.7

Removal

1 Disconnect the negative cable from the battery. **Caution:** *If the stereo in your vehicle is equipped with an anti-theft system, make sure you have the correct activation code before disconnecting the battery.*

2 Detach the breather hose from the valve cover.

3 Remove the spark plug wires from the spark plugs, handling them by the boots, not pulling on the wires.

4 Remove the wiring harness bolt at the timing belt cover and move the harness aside.

5 Remove the spark plug tube nuts, then detach the cover and gasket from the head. The spark plug tube nuts are used to hold the cover in place **(see illustration)**. If the cover is stuck to the head, bump the end with a block of wood and a hammer to jar it loose. If that doesn't work, try to slip a flexible putty knife between the head and cover to break the seal. **Caution:** *Don't pry at the cover-to-head joint or damage to the sealing surfaces may occur, leading to oil leaks after the cover is reinstalled.*

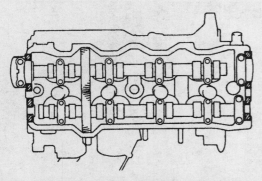

4.7 Apply sealant to the eight points indicated by the shaded areas before installing the valve cover

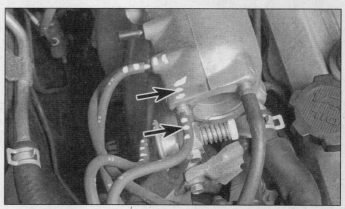

5.4a The various hoses should be marked to ensure correct reinstallation

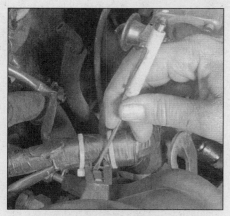

5.4b Press in on the clips to release the wiring harness retainers

5.4c Disconnect the two ground straps (arrows) from the firewall side of the intake manifold, then disconnect the knock sensor

5.6a Remove the bolts (arrows) from the intake manifold-to-block braces (seen from below)

Installation

6 The mating surfaces of the housing or cylinder head and cover must be clean when the cover is installed. Use a gasket scraper to remove all traces of sealant and old gasket material, then clean the mating surfaces with lacquer thinner or acetone. If there's residue or oil on the mating surfaces when the cover is installed, oil leaks may develop.

7 Apply RTV sealant to the gasket/seal joints **(see illustration)**. Install the spark plug tube grommets in the valve cover with the index marks facing the timing belt end of the engine.

8 Position a new valve cover gasket on the cylinder head, then install the valve cover and spark plug tube nuts.

9 Tighten the spark plug tube nuts to the torque listed in this Chapter's Specifications in three or four equal steps.

10 Reinstall the remaining parts, run the engine and check for oil leaks.

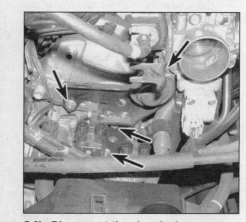

5.6b Disconnect the electrical connectors from the coil pack, then remove the mounting bolts (arrows) and remove the coil pack mount/intake brace

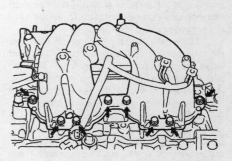

5.7 Locations of the intake manifold bolts/nuts

5 Intake manifold - removal and installation

Refer to illustrations 5.4a, 5.4b, 5.4c, 5.6a, 5.6b, 5.7 and 5.8

Warning: *Wait until the engine is completely cool before beginning this procedure.*

Removal

1 Disconnect the negative cable from the battery. **Caution:** *If the stereo in your vehicle is equipped with an anti-theft system, make sure you have the correct activation code before disconnecting the battery.*

2 Loosen the air cleaner hose clamp at the throttle body and remove the air cleaner top (four clamps), resonator and hose.

3 Disconnect the coolant hoses from the throttle body and plug them to prevent coolant leakage.

4 Label and detach any wire harness and control cables or hoses connected to the intake manifold **(see illustrations)**.

5 Carefully lift the wire harness over the manifold.

6 Unbolt the intake manifold-to-engine braces **(see illustrations)**.

7 Remove the mounting nuts/bolts **(see illustration)**, then detach the manifold from the engine.

Chapter 2 Part A Four-cylinder engine

5.8 Remove all traces of old gasket material and sealant with a scraper - be careful not to gouge the aluminum manifold

6.3 Remove the exhaust manifold heat insulator bolts (arrows)

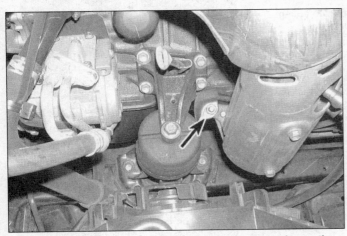

6.6 Unbolt the exhaust manifold flange brace (arrow) near the front engine mount

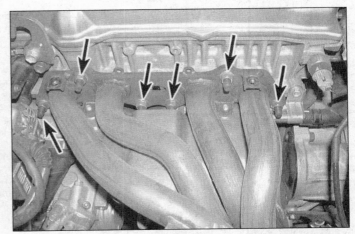

6.7 Remove the exhaust manifold fasteners (arrows) at the cylinder head

Installation

8 Use a scraper to remove all traces of old gasket material and sealant from the manifold and cylinder head **(see illustration)**, then clean the mating surfaces with lacquer thinner or acetone. If the gasket was leaking, have the manifold checked for warpage at an automotive machine shop and resurfaced if necessary.
9 Install a new gasket, then position the manifold on the head and install the nuts/bolts.
10 Tighten the nuts/bolts in three or four equal steps to the torque listed in this Chapter's Specifications. Work from the center out towards the ends to avoid warping the manifold.
11 Install the remaining parts in the reverse order of removal.
12 Before starting the engine, check the throttle linkage for smooth operation.
13 Check the coolant and add some, if necessary, to bring it to the appropriate level. Run the engine and check for coolant and vacuum leaks.
14 Road test the vehicle and check for proper operation of all accessories, including the cruise control system.

6 Exhaust manifold - removal and installation

Refer to illustrations 6.3, 6.6 and 6.7
Warning: *The engine must be completely cool before beginning this procedure.*

Removal

1 Disconnect the negative cable from the battery. **Caution:** *If the stereo in your vehicle is equipped with an anti-theft system, make sure you have the correct activation code before disconnecting the battery.*
2 Unplug the oxygen sensor wire harness and the clamp over the harness. If you're installing a new exhaust manifold, remove the sensor (see Chapter 6).
3 Remove the upper heat insulator from the manifold **(see illustration)**.
4 Apply penetrating oil to the exhaust manifold mounting nuts/bolts.
5 Disconnect the exhaust pipe from the exhaust manifold (see Chapter 4).
6 Remove the exhaust manifold brace (some models have two braces) and lower heat insulator **(see illustration)**.
7 Remove the nuts/bolts and detach the manifold and gasket **(see illustration)**.

Installation

8 Use a scraper to remove all traces of old gasket material and carbon deposits from the manifold and cylinder head mating surfaces. If the gasket was leaking, have the manifold checked for warpage at an automotive machine shop and resurfaced if necessary.
9 Position a new gasket over the cylinder head studs. **Note:** *The marks on the gasket should face out (away from the head) and the arrow should point toward the rear (transaxle end) of the engine.*
10 Install the manifold and thread the mounting nuts/bolts into place.
11 Working from the center out, tighten the nuts/bolts to the torque listed in this Chapter's Specifications in three or four equal steps.
12 Reinstall the remaining parts in the

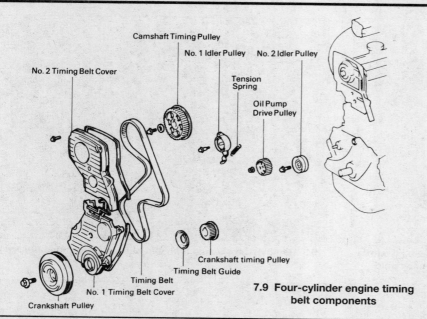

7.9 Four-cylinder engine timing belt components

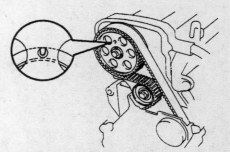

7.10 Align the upper camshaft sprocket timing marks with the number 1 cylinder at TDC

7.11 If you intend to re-use the timing belt, apply match marks on the sprocket and belt (arrow)

reverse order of removal.

13   Run the engine and check for exhaust leaks.

7 Timing belt and sprockets - removal, inspection and installation

Removal

** CAUTION **

The timing system is complex. Severe engine damage will occur if you make any mistakes. Do not attempt this procedure unless you are highly experienced with this type of repair. If you are at all unsure of your abilities, consult an expert. Double-check all your work and be sure everything is correct before you attempt to start the engine.

Refer to illustrations 7.9, 7.10, 7.11, 7.12, 7.13, 7.14a, 7.14b, 7.15, 7.16, 7.17a and 7.17b

1   Disconnect the negative cable from the battery. **Caution:** *If the stereo in your vehicle is equipped with an anti-theft system, make sure you have the correct activation code before disconnecting the battery.*

2   Block the rear wheels and set the parking brake.

3   Loosen the lug nuts on the right front wheel and raise the vehicle. Support the front of the vehicle securely on jackstands.

4   Remove the right front wheel and fender apron seal (see Chapter 11).

5   Remove the coolant overflow tank (see Chapter 3).

6   Remove the spark plugs and drivebelts (see Chapter 1).

7   Remove the alternator and bracket (see Chapter 5).

8   Support the engine from underneath with a jack (use a wood block on the jack and don't place the block under the oil pan drain plug) and remove the right engine mount and engine support rod (see Section 16)

9   Remove the upper timing belt cover screws, pull up the wiring harness (see Section 4) and remove the upper (no. 2) timing belt cover and gaskets **(see illustration)**.

10   Position the number one piston at TDC on the compression stroke (see Section 3). Make sure the small hole in the camshaft pulley is aligned with the TDC mark on the cam bearing cap **(see illustration)**.

11   If you plan to re-use the timing belt, apply match marks on the sprocket and belt and an arrow indicating direction of travel on the belt **(see illustration)**.

12   Loosen the upper (no. 1) idler pulley set bolt and unhook the spring **(see illustration)**. Slip the timing belt off the sprocket. If you're

7.12 Loosen the upper idler pulley set bolt (arrow) and unhook the spring

7.13 Remove the valve cover and hold the camshaft with a large wrench on the raised hex as the sprocket bolt is loosened - DO NOT use the timing belt tension to keep the sprocket from turning!

Chapter 2 Part A Four-cylinder engine

2A-7

7.14a Remove the flywheel/driveplate cover and use a prybar wedged against the ring gear teeth or a converter bolt to hold the crankshaft while loosening the pulley bolt with a breaker bar - the bolt is very tight, so use the appropriate tools

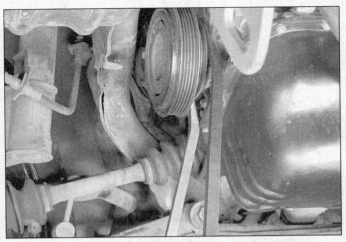

7.14b Often the crankshaft pulley can be removed with even applications of a pry bar - if you use a puller, it should be the type that attaches to the hub only (do not use a jaw-type puller)

7.15 Remove the lower timing belt cover bolts (arrows) and slip the cover and gaskets off the engine

7.16 If you plan to re-use the timing belt, apply match marks (arrow) on the belt and sprocket

removing the upper part of the belt only, for camshaft seal replacement or cylinder head removal, it isn't necessary to detach the belt from the crankshaft sprocket.

13 If the camshaft sprocket is worn or damaged, remove the valve cover, hold the rear (intake) camshaft with a large wrench and remove the bolt, then detach the sprocket **(see illustration)**.

14 Remove the crankshaft pulley bolt. With the flywheel inspection cover removed (see Chapter 7), wedge a large screwdriver into the flywheel/driveplate ring gear teeth or against a converter bolt to keep the engine from turning. Use a breaker bar and socket to loosen the pulley bolt **(see illustration)**. Remove the bolt and detach the pulley with prybars or a vibration damper puller **(see illustration)**. Do not use an outside-jaw gear puller! Before the pulley is removed completely, double-check that the crankshaft is still at TDC.

15 Remove the lower (no. 1) timing belt cover and gaskets **(see illustration)** and slip the belt guide off the crankshaft.

16 If you plan to re-use the timing belt,

7.17a The crankshaft sprocket should slide off the crankshaft easily

apply match marks on the sprocket and belt **(see illustration)**.

17 Slip the timing belt off the sprocket and remove it. If the sprocket is worn or damaged, or if you need to replace the crankshaft

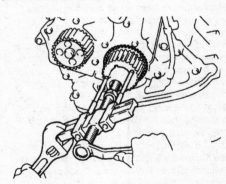

7.17b If the crankshaft sprocket is stuck, remove it with a puller

front oil seal, remove the sprocket from the crankshaft **(see illustrations)**.

Inspection

Refer to illustrations 7.18 and 7.20
Caution: Do not bend, twist or turn the timing belt inside out. Do not allow it to come in contact with oil, coolant or fuel. Do not utilize

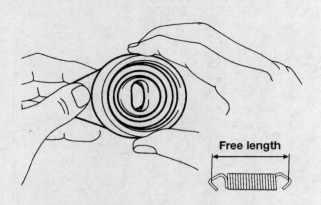

7.18 Check the idler pulley bearing for smooth operation and measure the free length of the tension spring for comparison to this Chapter's Specifications

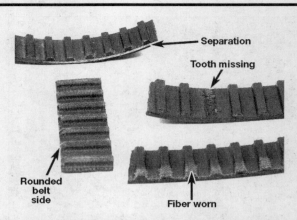

7.20 Check the timing belt for cracked and missing teeth - wear on one side of the belt indicates sprocket misalignment problems

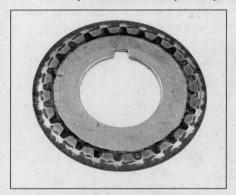

7.28 The belt guide should be installed with the tooth marks in contact with the timing belt and the cupped side facing out

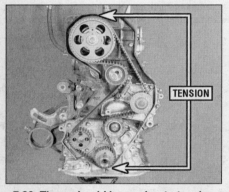

7.30 There should be moderate tension on the side of the belt facing the front of the vehicle

7.33 The marks should align as shown at Top Dead Center

timing belt tension to keep the camshaft or crankshaft from turning when installing the sprocket bolt(s). Do not turn the crankshaft or camshaft more than a few degrees (if necessary for tooth alignment) while the timing belt is removed.

18 Remove the idler pulleys and check the bearings for smooth operation and excessive play. Inspect the spring for damage and compare the free length to this Chapter's Specifications **(see illustration)**.

19 If the timing belt broke during engine operation, the belt may have been contaminated or overtightened.

20 If the belt teeth are cracked or missing **(see illustration)**, the water pump, oil pump or camshaft(s) may have seized. **Caution:** *If the timing belt broke during engine operation, the valves may have come in contact with the pistons, causing damage. Check the valve clearance (see Chapter 1) - bent valves usually will have excessive clearance, indicating damage that will require head removal to repair.*

21 If there is noticeable wear or cracks on the face of the belt, check to see if there are nicks or burrs on the idler pulleys.

22 If there is wear or damage on only one side of the belt, check the belt guide and the alignment of the sprockets.

23 Replace the timing belt with a new one if obvious wear or damage is noted or if it is the least bit questionable. Correct any problems which contributed to belt failure prior to belt installation. **Note:** *Professionals recommend replacing the belt whenever it is removed, since belt failure can lead to expensive engine damage. The manufacturer recommends changing the belt at 60,000-mile intervals.*

Installation

> ** CAUTION **
>
> Before starting the engine, carefully rotate the crankshaft by hand through at least two full revolutions (use a socket and breaker bar on the crankshaft pulley center bolt). If you feel any resistance, STOP! There is something wrong - most likely, valves are contacting the pistons. You must find the problem before proceeding. Check your work and see if any updated repair information is available.

Refer to illustrations 7.28, 7.30, 7.33 and 7.34

24 Remove all dirt, oil and grease from the timing belt area at the front of the engine.

25 If they were removed, install the idler pulleys and tension spring. The upper (no. 1) idler should be pulled back against spring tension as far as possible and the bolt temporarily tightened.

26 Recheck the camshaft and crankshaft timing marks to be sure they are properly aligned (see Step 11).

27 Install the timing belt on the crankshaft, oil pump, water pump and idler pulleys. If the original belt is being reinstalled, align the marks made during removal.

28 Slip the belt guide onto the crankshaft with the cupped side facing out **(see illustration)**.

29 Reinstall the lower timing belt cover and crankshaft pulley and recheck the TDC marks.

30 Slip the timing belt over the camshaft sprocket. Keep tension on the side nearest the front of the vehicle **(see illustration)**. If the original belt is being reinstalled, align the marks made during removal.

31 Loosen the upper (no. 1) idler pulley bolt 1/2-turn, allowing the spring to apply pressure to the idler pulley.

32 Slowly turn the crankshaft clockwise 1-7/8 revolutions until the crank pulley mark lines up with the 45-degree BTDC (before TDC for number one cylinder) mark on the lower belt cover, then tighten the idler pulley to Specifications.

33 Rotate the crankshaft two revolutions

Chapter 2 Part A Four-cylinder engine

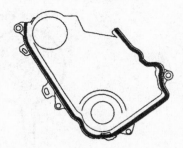

7.34 Install new adhesive gasket material to the timing belt covers in the areas indicated by black lines

8.2a Wrap tape around the screwdriver tip and carefully work the crankshaft front oil seal out of the bore - DO NOT nick or scratch the crankshaft in the process!

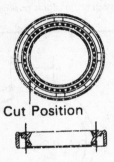

8.2b The seal may be removed more easily by carefully cutting the lip as indicated

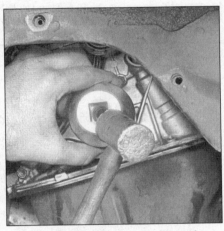

8.4 Gently drive the new seal into place with the spring side installed into the engine

and recheck the timing marks **(see illustration)**. With the crankshaft at TDC for number one cylinder, the camshaft sprocket hole must line up with the timing mark. If the marks are not aligned exactly as shown, repeat the belt installation procedure. **Caution:** *DO NOT start the engine until you're absolutely certain that the timing belt is installed correctly. Serious and costly engine damage could occur if the belt is installed wrong.*

34 Reinstall the remaining parts in the reverse order of removal. **Note:** *When reinstalling the covers, if the sealing material is cut or compressed, clean the covers and apply new self-stick gasket material* **(see illustration)**.

35 Run the engine and check for proper operation.

8 Crankshaft front oil seal - replacement

Refer to illustrations 8.2a, 8.2b and 8.4

1 Remove the timing belt and crankshaft sprocket (see Section 7).
2 Note how far the seal is recessed in the bore, then carefully pry it out of the oil pump housing with a screwdriver or seal removal tool **(see illustration)**. Don't scratch the housing bore or damage the crankshaft in the process (if the crankshaft is damaged, the new seal will end up leaking). **Note:** *The seal may be easier to remove if the old seal lip is cut with a sharp utility knife first* **(see illustration)**.
3 Clean the bore in the housing and coat the outer edge of the new seal with engine oil or multi-purpose grease. Apply multi-purpose grease to the seal lip.
4 Using a socket with an outside diameter slightly smaller than the outside diameter of the seal, carefully drive the new seal into place with a hammer **(see illustration)**. Make sure it's installed squarely and driven in to the same depth as the original. If a socket isn't available, a short section of large diameter pipe will also work. Check the seal after installation to make sure the spring didn't pop out of place.
5 Reinstall the crankshaft sprocket and timing belt (see Section 7).
6 Run the engine and check for oil leaks at the front seal.

9 Camshaft oil seal - replacement

Refer to illustration 9.3

1 Remove the timing belt, upper idler pulley, and camshaft sprocket (see Section 7).
2 Remove the upper timing belt rear cover.
3 Note how far the seal is seated in the bore, then carefully pry it out with a small screwdriver **(see illustration)**. Don't scratch the bore or damage the camshaft in the process (if the camshaft is damaged, the new seal will end up leaking).
4 Clean the bore and coat the outer edge of the new seal with engine oil or multi-purpose grease. Apply multi-purpose grease to the seal lip.
5 Using a socket with an outside diameter slightly smaller than the outside diameter of the seal, carefully drive the new seal into place with a hammer. Make sure it's installed squarely and driven in to the same depth as the original. If a socket isn't available, a short section of pipe will also work.
6 Reinstall the timing belt rear cover, camshaft sprocket, and timing belt (see Section 7).
7 Run the engine and check for oil leaks at the camshaft seal.

9.3 Carefully pry the camshaft seal out of the bore - DO NOT nick or scratch the camshaft journal

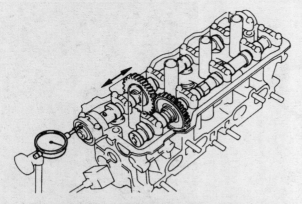

10.4 Mount a dial indicator as shown to measure camshaft endplay

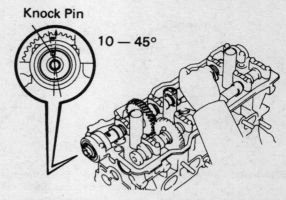

10.5 Turn the INTAKE camshaft until the knock pin is 10 to 45-degrees to the left of vertical (12 o'clock position)

10.6 Install a service bolt through the sub-gear, into the main gear (arrow)

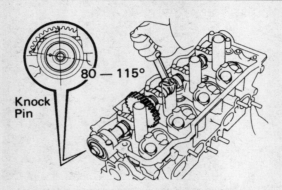

10.11 Turn the intake camshaft until the knock pin is 80 to 115-degrees to the left of vertical (12 o'clock position)

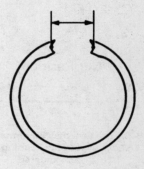

10.20 Measure the distance between the ends of the camshaft gear spring

10 Camshafts and valve lifters - removal, inspection and installation

Note: *Before beginning this procedure, obtain two 6 x 1.0 mm bolts 16 to 20 mm long. They will be referred to as service bolts in the text.*

Removal

Refer to illustrations 10.4, 10.5, 10.6 and 10.11

1 Remove the valve cover as described in Section 4.
2 Remove the distributor (see Chapter 5).
3 Remove the timing belt, camshaft sprocket and upper rear belt cover (see Sections 7 and 9).
4 Measure the camshaft thrust clearance (endplay) with a dial indicator (see illustration). If the clearance is greater than the service limit, replace the camshaft and/or the cylinder head.

Exhaust camshaft

5 Position the knock pin in the INTAKE camshaft at 10 to 45-degrees left of vertical (see illustration). This will position the exhaust camshaft lobes so the camshaft will be pushed out evenly by the valve spring pressure.
6 Secure the exhaust camshaft sub-gear to the main gear by installing one of the service bolts into a threaded hole (see illustration).
7 Remove the rear (transaxle end) exhaust camshaft bearing cap bolts and detach the bearing cap.
8 Loosen the number 1, 2 and 4 exhaust camshaft bearing cap bolts in 1/4-turn increments until the bolts can be removed by hand. Lift off the first, second and fourth bearing caps.
9 Finally, loosen the number 3 bearing cap bolts in 1/4-turn increments until they can be removed by hand, then detach the center (no. 3) cap. **Caution:** *As the center bearing cap bolts are being loosened, make sure the camshaft is moving up evenly. If one end or the other stops moving and the cam gets cocked, start over by reinstalling the bearing caps and resetting the knock pin. DO NOT try to pry or force the camshaft out.*
10 Lift the camshaft straight up and out of the head.

Intake camshaft

11 Position the knock pin in the intake camshaft at 80 to 115-degrees left of vertical (see illustration).
12 Remove the front (timing belt end) intake camshaft bearing cap bolts and detach the bearing cap and oil seal. **Caution:** *Do not pry the cap off. If it doesn't come loose easily, leave it in place without bolts.*
13 Loosen the number 1, 3 and 4 intake camshaft bearing cap bolts in 1/4-turn increments until the bolts can be removed by hand. Lift off the first, third and fourth bearing caps.
14 Finally, loosen the number 2 bearing cap bolts in 1/4-turn increments until they can be removed by hand, then detach the center (no. 2) bearing cap. **Caution:** *As the center bearing cap bolts are being loosened, make sure the camshaft is moving up evenly. If one end or the other stops moving and the cam gets cocked, start over by reinstalling the bearing caps and resetting the knock pin. DO NOT try to pry or force the camshaft out.*
15 Lift the camshaft straight up and out of the head.
16 To disassemble the exhaust camshaft gear, mount it in a vise with the jaws gripping

Chapter 2 Part A Four-cylinder engine

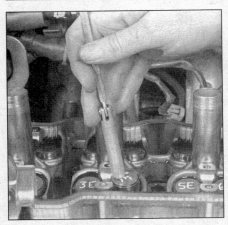

10.21 Wipe the oil off the valve shims and mark the intakes I and the exhausts E - a magnetic tool works well for removing lifters

10.22 Wipe off the oil and inspect each lifter for wear and scuffing

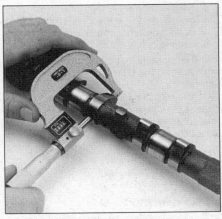

10.26 Measure the lobe heights on each camshaft - if any lobe height is less than the specified allowable minimum, replace that camshaft

10.27 Measure each journal diameter with a micrometer (if any journal measures less than the specified limit, replace the camshaft)

10.28a Lay a strip of Plastigage on each camshaft journal

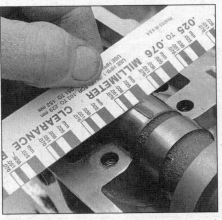

10.28b Compare the width of the crushed Plastigage to the scale on the envelope to determine the oil clearance

the large hex on the shaft.

17 Install a second service bolt in the unthreaded hole in the camshaft sub-gear. Using a screwdriver positioned against the service bolt just installed, rotate the sub-gear clockwise and remove the first service bolt from the threaded hole.

18 Remove the sub-gear snap-ring.

19 The wave washer, sub-gear and camshaft gear spring can now be removed from the camshaft.

Inspection

Refer to illustrations 10.20, 10.21, 10.22, 10.26, 10.27, 10.28a, 10.28b and 10.30

20 Measure the end gap (distance between the ends) of the camshaft gear spring **(see illustration)** and compare it to this Chapter's Specifications. If not as specified, replace the spring.

21 Carefully label, then remove the valve lifters and shims **(see illustration)**.

22 Inspect each lifter for scuffing and score marks **(see illustration)**.

23 Measure the outside diameter of each lifter and the corresponding lifter bore inside diameter. Subtract the lifter diameter from the lifter bore diameter to determine the oil clearance. Compare it to this Chapter's Specifications. If the oil clearance is excessive, a new head and/or new lifters will be required.

24 Store the lifters in a clean box, separated from each other, so they won't be damaged. Make sure the shims stay with the lifters (don't mix them up).

25 Visually examine the cam lobes and bearing journals for score marks, pitting, galling and evidence of overheating (blue, discolored areas). Look for flaking away of the hardened surface layer of each lobe.

26 Using a micrometer, measure the height of each camshaft lobe **(see illustration)**. Compare your measurements with this Chapter's Specifications. If the height for any one lobe is less than the specified minimum, replace the camshaft.

27 Using a micrometer, measure the diameter of each journal at several points **(see illustration)**. Compare your measurements with this Chapter's Specifications. If the diameter of any one journal is less than specified, replace the camshaft.

28 Check the oil clearance for each camshaft journal as follows:

a) Clean the bearing caps and the camshaft journals with lacquer thinner or acetone.

b) Carefully lay the camshaft(s) in place in the head. Don't install the lifters and don't use any lubrication.

c) Lay a strip of Plastigage on each journal **(see illustration)**.

d) Install the bearing caps with the arrows pointing toward the front (timing belt end) of the engine.

e) Tighten the bolts to the specified torque in 1/4-turn increments. **Note:** *Don't turn the camshaft while the Plastigage is in place.*

f) Remove the bolts and detach the caps.

g) Compare the width of the crushed Plastigage (at its widest point) to the scale on the Plastigage envelope **(see illustration)**. If the clearance is greater than specified, replace the camshaft and/or cylinder head.

i) Scrape off the Plastigage with your fingernail or the edge of a credit card - don't scratch or nick the journals or bearing caps.

29 Temporarily install the camshafts without installing the lifters or exhaust camshaft sub-gear.

30 Measure the gear backlash (the free play between the gear teeth) with a dial indicator **(see illustration)** and compare it to this Chapter's Specifications.

Installation

Refer to illustrations 10.34, 10.35, 10.36, 10.43, 10.45 and 10.46

Intake camshaft

31 Apply moly-base grease or engine assembly lube to the lifters, then install them in their original locations. Make sure the valve adjustment shims are in place in the lifters.

32 Apply moly-base grease or engine assembly lube to the camshaft lobes and bearing journals.

33 Position the intake camshaft in the cylinder head with the knock pin 80-degrees to the left of vertical **(see illustration 10.11)**.

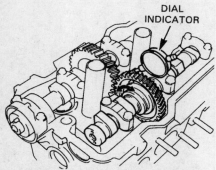

10.30 Position a dial indicator as shown here to measure gear backlash

34 Apply a thin coat of RTV sealant to the outer edge of the front bearing-cap-to-cylinder-head mating surface **(see illustration)**. **Note:** *The cap must be installed immediately or the sealer will dry prematurely.*

35 Install the bearing caps in numerical order with the arrows pointing toward the timing belt end of the engine **(see illustration)**.

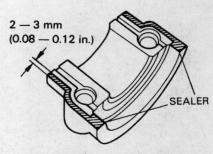

10.34 Apply sealant to the shaded areas of the front intake camshaft bearing cap

36 Following the recommended tightening sequence **(see illustration)**, tighten the bearing cap bolts in 1/4-turn increments to the torque listed in this Chapter's Specifications.

37 Refer to Section 9 and install a new camshaft oil seal.

Exhaust camshaft

38 Reassemble the exhaust camshaft gear. Install the cam gear spring, sub-gear and wave washer in the gear. Secure them with the snap-ring.

39 Reinstall the service bolt in the unthreaded hole, turn the sub-gear with a screwdriver and install the second service bolt in the threaded hole. Tighten it to clamp the sub-gear to the camshaft gear, then remove the first bolt.

40 Apply moly-base grease or engine assembly lube to the lifters, then install them in their original locations. Make sure the valve adjustment shims are in place in the lifters.

41 Apply moly-base grease or engine assembly lube to the camshaft lobes and bearing journals.

42 Rotate the INTAKE camshaft until the knock pin is positioned 10-degrees to the left of vertical **(see illustration 10.5)**.

43 Align the exhaust camshaft gear with the intake camshaft gear by matching up the timing marks on the gears **(see illustration)**. **Caution:** *There are also assembly reference marks on each gear, above the timing marks - do not mistake them for the timing marks.*

44 Roll the exhaust camshaft down into position. Turn the intake camshaft back and

10.35 Intake camshaft bearing cap arrangement - the arrows point toward the front or timing belt end of the engine

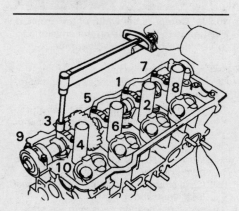

10.36 INTAKE camshaft bearing cap bolt tightening sequence

10.43 Align the camshaft timing gears as shown here

10.45 Exhaust camshaft bearing cap arrangement - the arrows point toward the front or timing belt end of the engine

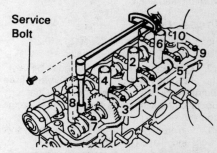

10.46 EXHAUST camshaft bearing cap bolt tightening sequence

Chapter 2 Part A Four-cylinder engine

2A-13

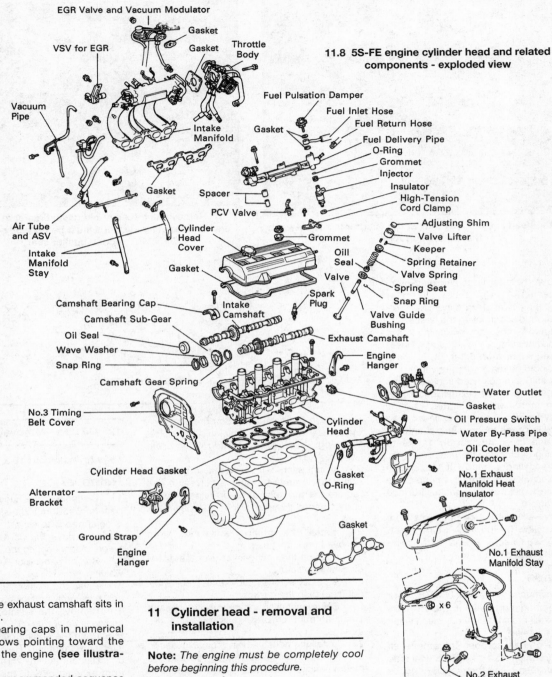

11.8 5S-FE engine cylinder head and related components - exploded view

forth a little until the exhaust camshaft sits in the bearings evenly.

45 Install the bearing caps in numerical order with the arrows pointing toward the timing belt end of the engine **(see illustration)**.

46 Following the recommended sequence **(see illustration)**, tighten the bearing cap bolts in 1/4-turn increments to the torque listed in this Chapter's Specifications.

47 Remove the service bolt from the camshaft gear.

48 Install the timing belt pulley on the intake camshaft and tighten the bolt to the torque listed in this Chapter's Specifications. Prevent the camshaft from turning by holding it with a wrench on the large hex **(see illustration 7.13)**.

49 Install the timing belt (see Section 7).

50 The remainder of installation is the reverse of the removal procedure.

11 Cylinder head - removal and installation

Note: *The engine must be completely cool before beginning this procedure.*

Removal

Refer to illustrations 11.8 and 11.14

1 Disconnect the negative cable from the battery. **Caution:** *If the stereo in your vehicle is equipped with an anti-theft system, make sure you have the correct activation code before disconnecting the battery.*

2 Drain the coolant from the engine block and radiator (see Chapter 1). On Solara models, remove the front suspension upper brace.

3 Drain the engine oil and remove the oil filter (see Chapter 1).

4 Remove the throttle body, fuel injectors and fuel rail (see Chapter 4).

5 Remove the intake manifold (see Section 5).

6 Remove the exhaust manifold (see Section 6).

7 Remove the timing belt, camshaft sprocket and upper idler pulley (see Section 7).

8 Remove the upper rear (no. 3) timing belt cover **(see illustration)**. **Note:** *Disconnect the camshaft position sensor at this time (see Chapter 6).*

9 Remove the camshafts and lifters (see

11.14 If the head is stuck, pry only at the overhang, not between the mating surfaces

11.17 Remove all traces of old gasket material - the cylinder head and block mating surfaces must be perfectly clean to ensure a good gasket seal

Section 10).
10 Remove the alternator and distributor (see Chapter 5).
11 Unbolt the power steering pump and set it aside without disconnecting the hoses.
12 Label and remove any remaining items, such as coolant fittings, tubes, cables, hoses or wires **(see illustration 5.4a)**. At this point the head should be ready for removal.
13 Using an 8 mm hex head socket bit and a breaker bar, loosen the cylinder head bolts in 1/4-turn increments until they can be removed by hand. Loosen the cylinder head bolts opposite of the recommended tightening sequence **(see illustration 11.25)** to avoid warping or cracking the head.
14 Lift the cylinder head off the engine block. If it's stuck, very carefully pry up at the transaxle end, beyond the gasket surface **(see illustration)**.
15 Remove all external components from the head to allow for thorough cleaning and inspection. See Chapter 2, Part C, for cylinder head servicing procedures.

Installation

Refer to illustrations 11.17 and 11.25

16 The mating surfaces of the cylinder head and block must be perfectly clean when the head is installed.
17 Use a gasket scraper to remove all traces of carbon and old gasket material **(see illustration)**, then clean the mating surfaces with lacquer thinner or acetone. If there's oil on the mating surfaces when the head is installed, the gasket may not seal correctly and leaks could develop. When working on the block, stuff the cylinders with clean shop rags to keep out debris. Use a vacuum cleaner to remove material that falls into the cylinders.
18 Check the block and head mating surfaces for nicks, deep scratches and other damage. If damage is slight, it can be removed with a file; if it's excessive, machining may be the only alternative.
19 Use a tap of the correct size to chase the threads in the cylinder head bolt holes,

then clean the holes with compressed air - make sure that nothing remains in the holes. **Warning:** *Wear eye protection when using compressed air!*
20 Mount each bolt in a vise and run a die down the threads to remove corrosion and restore the threads. Dirt, corrosion, sealant and damaged threads will affect torque readings.
21 Install the components that were removed from the head.
22 Position the new gasket over the dowel pins in the block.
23 Carefully set the head on the block without disturbing the gasket.
24 Before installing the head bolts, apply a small amount of clean engine oil to the threads.
25 Install the bolts in their original locations and tighten them finger tight. Following the recommended sequence, tighten the bolts in several steps to the torque listed in this Chapter's Specifications **(see illustration)**. **Note:** *The bolts should be lightly oiled on the threads and under the heads, and washers must be used.*
26 The remaining installation steps are the reverse of removal.
27 Check and adjust the valves as necessary (see Chapter 1).
28 Refill the cooling system, install a new oil filter and add oil to the engine (see Chapter 1).
29 Run the engine and check for leaks. Set the ignition timing (see Chapter 5) and road test the vehicle.

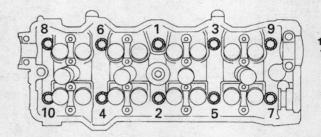

11.25 Cylinder head bolt TIGHTENING sequence

12 Oil pan - removal and installation

Refer to illustrations 12.8, 12.9 and 12.14

Removal

1 Disconnect the negative cable from the battery. **Caution:** *If the stereo in your vehicle is equipped with an anti-theft system, make sure you have the correct activation code before disconnecting the battery.*
2 Set the parking brake and block the rear wheels.
3 Raise the front of the vehicle and support it securely on jackstands.
4 Remove the splash shields under the engine, if equipped.
5 Drain the engine oil and remove the oil filter (see Chapter 1). Remove the oil dipstick.
6 Disconnect the three nuts holding the front exhaust pipe to the exhaust manifold, then the two bolts/nuts where the converter attaches to the rest of the exhaust system (see Chapter 4).
7 Unbolt the two bolts and remove the support bracket, then remove the front exhaust pipe. Remove the brace from the exhaust manifold (see Section 6).
8 Remove the front (radiator side) block-to-transaxle brace **(see illustration)**.
9 Remove the bolts and detach the oil pan. If it's stuck, pry it loose very carefully with a small screwdriver or putty knife **(see illustration)**. Don't damage the mating surfaces of the pan and block or oil leaks could develop.

Chapter 2 Part A Four-cylinder engine

12.8 Remove the bolts (arrows) on the front (radiator side) block-to-transaxle brace

12.9 Carefully pry the oil pan away from the block - if the mating surfaces are damaged, oil leaks could develop

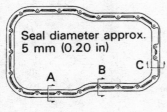

 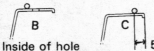

12.14 Apply a bead of RTV sealant to the oil pan flange

Installation

10 Use a scraper to remove all traces of old gasket material and sealant from the block and oil pan. Clean the mating surfaces with lacquer thinner or acetone.
11 Make sure the threaded bolt holes in the block are clean.
12 Check the oil pan flange for distortion, particularly around the bolt holes. If necessary, place the pan on a block of wood and use a hammer to flatten and restore the gasket surface.
13 Inspect the oil pump pick-up tube assembly for cracks and a blocked strainer. If the pick-up was removed, clean it thoroughly and install it now, using a new O-ring or gasket. Tighten the nuts/bolts to the torque listed in this Chapter's Specifications.
14 Apply a 5 mm wide bead of RTV sealant to the oil pan flange (see illustration). Note: *The oil pan must be installed within 5 minutes once the sealer has been applied.*
15 Carefully position the oil pan on the engine block and install the bolts. Working from the center out, tighten them to the torque listed in this Chapter's Specifications in three or four steps.
16 The remainder of installation is the reverse of removal. Be sure to add oil and install a new oil filter. Use new "doughnut" gaskets on each end of the front exhaust pipe.
17 Run the engine and check for oil pressure and leaks.

13 Oil pump - removal, inspection and installation

Removal

Refer to illustrations 13.2, 13.4, 13.5a, 13.5b, 13.6 and 13.8

1 Remove the oil pan (see Section 12).
2 Remove the nuts/bolts and detach the oil pick-up tube assembly (see illustration).
3 Support the engine securely from above and remove the lower idler pulley, crankshaft pulley, timing belt, and crankshaft sprocket (see Section 7).
4 Remove the 12 bolts and detach the oil pump case from the engine (see illustration). You may have to pry carefully between the front main bearing cap and the pump case with a screwdriver.
5 Remove the two remaining bolts (see illustration) and separate the pump body

13.2 The oil pick-up assembly and baffle plate are held in place with two nuts and two bolts (arrows)

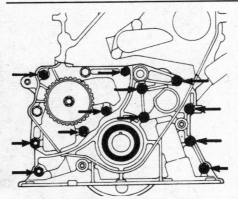

13.4 Remove the oil pump case-to-block bolts (arrows)

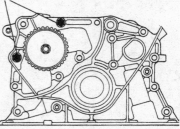

13.5a Remove the oil pump body-to-oil pump case bolts

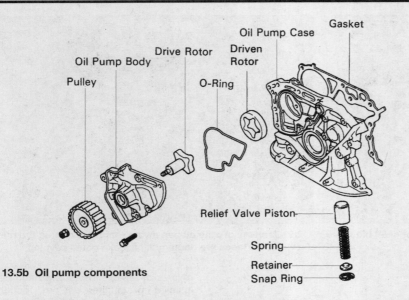

13.5b Oil pump components

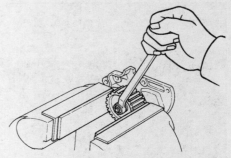

13.6 Hold the oil pump sprocket in a well-padded vise while the retaining nut is removed

13.8 Remove the snap-ring to disassemble the oil pressure relief valve

from the case. Lift out the driven rotor and remove the O-ring **(see illustration)**.

6 Clamp the pump sprocket in a well-padded vise **(see illustration)** and remove the sprocket nut and sprocket. Remove the drive rotor.

7 Use a scraper to remove all traces of sealant and old gasket material from the pump case and engine block, then clean the mating surfaces with lacquer thinner or acetone.

8 Remove the oil pressure relief valve snap-ring **(see illustration)**, retainer, spring and piston. **Warning:** *The spring is tightly compressed - be careful and wear eye protection.*

Inspection

Refer to illustrations 13.11a and 13.11b

9 Clean all components with solvent, then inspect them for wear and damage.

10 Check the oil pressure relief valve piston sliding surface and valve spring. If either the spring or the valve is damaged, they must be replaced as a set.

11 Check the driven rotor-to-case and drive rotor tip clearance with feeler gauges **(see illustrations)** and compare the results to this Chapter's Specifications. If the clearance is excessive, replace the rotors as a set. If necessary, replace the oil pump case and body.

Installation

Refer to illustration 13.14

12 Pry the old drive rotor shaft seal out with a screwdriver. Using a deep socket and a hammer, carefully drive a new seal into place. Apply multi-purpose grease to the seal lip.

13 Install a new crankshaft seal using the same procedure as outlined in the previous step. Apply multi-purpose grease to the seal lip.

14 Install a new O-ring, then lubricate the driven rotor with clean engine oil and place it in the pump case with the mark facing out **(see illustration)**.

15 Lubricate the shaft and install the drive rotor in the pump body, then reinstall the pulley and tighten the nut to the torque listed in this Chapter's Specifications.

16 Pack the pump cavity with petroleum jelly and attach the pump body to the case with the 16 mm-long bolts **(see illustration 13.5a)**.

17 Lubricate the oil pressure relief valve piston with clean engine oil and reinstall the valve components in the pump case.

18 Place a new gasket on the engine block

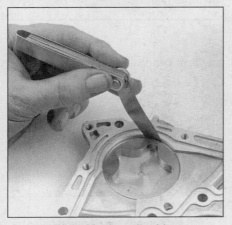

13.11a Measure the driven rotor-to-case clearance . . .

13.11b . . . and the rotor tip clearance with a feeler gauge

13.14 Oil pump case ready for pump body installation; note that the seal is in place, the O-ring is in place and the mark on the driven rotor is facing out (arrow)

Chapter 2 Part A Four-cylinder engine

14.3 Mark the flywheel/driveplate and the crankshaft so they can be reassembled in the same relative positions - most models will have eight flywheel bolts

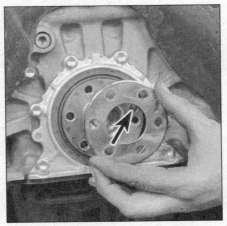

14.5 On vehicles with an automatic transaxle, there is a spacer plate on each side of the driveplate; when installing, line up the spacers with the locating pin (arrow)

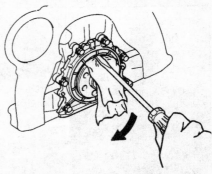

15.2 The quick way to replace the rear crankshaft oil seal is to simply pry the old one out with a screwdriver, lubricate the crankshaft journal and the lip of the new seal with multi-purpose grease and push the new seal into place - the seal lip is very stiff and can be easily damaged during installation if you're not careful

(the dowel pins should hold it in place).
19   Position the pump case against the block and install the mounting bolts.
20   Tighten the bolts to the torque listed in this Chapter's Specifications in three or four steps. Follow a criss-cross pattern to avoid warping the case.
21   Using a new gasket, install the oil pick-up tube assembly and baffle plate. Tighten the fasteners to the torque listed in this Chapter's Specifications.
22   Reinstall the remaining parts in the reverse order of removal.
23   Add oil, start the engine and check for oil pressure and leaks.
24   Recheck the engine oil level.

14 Flywheel/driveplate - removal and installation

Refer to illustrations 14.3 and 14.5

Removal

1   Raise the vehicle and support it securely on jackstands, then refer to Chapter 7 and remove the transaxle. If it's leaking, now would be a very good time to replace the front pump seal/O-ring (automatic transaxle only).
2   Remove the pressure plate and clutch disc (Chapter 8) (manual transaxle equipped vehicles). Now is a good time to check/replace the clutch components and pilot bearing.
3   Use a center punch or paint to make alignment marks on the flywheel/driveplate and crankshaft to ensure correct alignment during reinstallation **(see illustration)**.
4   Remove the bolts that secure the flywheel/driveplate to the crankshaft. If the crankshaft turns, wedge a screwdriver in the ring gear teeth to jam the flywheel.
5   Remove the flywheel/driveplate from the crankshaft. Since the flywheel is fairly heavy,

be sure to support it while removing the last bolt. Automatic transaxle equipped vehicles have spacers on both sides of the driveplate **(see illustration)**. Keep them with the driveplate.

Installation

6   Clean the flywheel to remove grease and oil. Inspect the surface for cracks, rivet grooves, burned areas and score marks. Light scoring can be removed with emery cloth. Check for cracked and broken ring gear teeth. Lay the flywheel on a flat surface and use a straightedge to check for warpage.
7   Clean and inspect the mating surfaces of the flywheel/driveplate and the crankshaft. If the crankshaft rear seal is leaking, replace it before reinstalling the flywheel/driveplate.
8   Position the flywheel/driveplate against the crankshaft. Be sure to align the marks made during removal. Note that some engines have an alignment dowel or staggered bolt holes to ensure correct installation. Before installing the bolts, apply thread locking compound to the threads.
9   Wedge a screwdriver in the ring gear teeth to keep the flywheel/driveplate from turning and tighten the bolts to the torque listed in this Chapter's Specifications. Follow a criss-cross pattern and work up to the final torque in three or four steps.
10   The remainder of installation is the reverse of the removal procedure.

15 Rear main oil seal - replacement

Refer to illustrations 15.2, 15.5 and 15.6

1   The transaxle must be removed from the vehicle for this procedure (see Chapter 7).
2   The seal can be replaced without removing the oil pan or removing the seal retainer. However, this method is not recommended because the lip of the seal is quite stiff and it's possible to cock the seal in the

retainer bore or damage it during installation. If you want to take the chance, pry out the old seal with a screwdriver **(see illustration)**. Apply multi-purpose grease to the crankshaft seal journal and the lip of the new seal and carefully push the new seal into place. The lip is stiff so carefully work it onto the seal journal of the crankshaft with a smooth object like the end of an extension as you tap the seal into place. Don't rush it or you may damage the seal.
3   The following method is recommended but requires removal of the oil pan (see Section 12) and the seal retainer.
4   After the oil pan has been removed, remove the bolts, detach the seal retainer and remove all the old gasket material.
5   Position the seal and retainer assembly between two wood blocks on a workbench and drive the old seal out from the back side with a screwdriver **(see illustration)**.
6   Drive the new seal into the retainer with

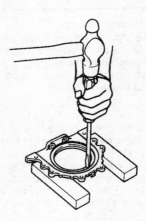

15.5 After removing the retainer assembly from the block, support it between two wood blocks and drive out the old seal with a screwdriver and hammer

15.6 Drive the new seal into the retainer with a block of wood or a section of pipe, if you have one large enough - make sure that you don't cock the seal in the retainer bore

a block of wood **(see illustration)** or a section of pipe slightly smaller in diameter than the outside diameter of the seal.
7   Lubricate the crankshaft seal journal and the lip of the new seal with multi-purpose grease. Position a new gasket on the engine block.
8   Slowly and carefully push the seal onto the crankshaft. The seal lip is stiff, so work it onto the crankshaft with a smooth object such as the end of an extension as you push the retainer against the block.
9   Install and tighten the retainer bolts to the torque listed in this Chapter's Specifications. The bottom sealing flange of the retainer must not extend below the bottom sealing flange (oil pan rail) of the block.
10   The remaining steps are the reverse of removal.

16 Engine mounts - check and replacement

Refer to illustration 16.4
1   Engine mounts seldom require attention, but broken or deteriorated mounts should be replaced immediately or the added strain placed on the driveline components may cause damage or wear.

Check
2   During the check, the engine must be raised slightly to remove the weight from the mounts.
3   Raise the vehicle and support it securely on jackstands, then position a jack under the engine oil pan. Place a large block of wood between the jack head and the oil pan, then

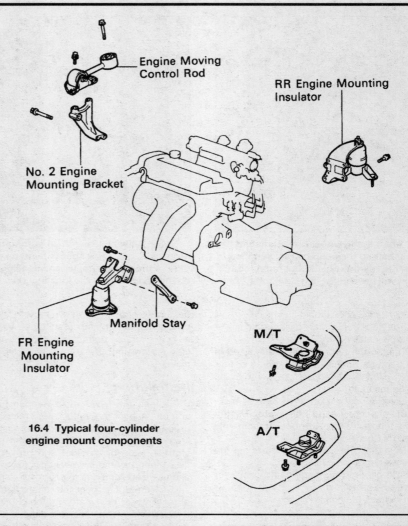

16.4 Typical four-cylinder engine mount components

carefully raise the engine just enough to take the weight off the mounts. Do not position the wood block under the drain plug. **Warning:** *DO NOT place any part of your body under the engine when it's supported only by a jack!*
4   Check the mounts **(see illustration)** to see if the rubber is cracked, hardened or separated from the metal plates. Sometimes the rubber will split right down the center.
5   Check for relative movement between the mount plates and the engine or frame (use a large screwdriver or pry bar to attempt to move the mounts). If movement is noted, lower the engine and tighten the mount fasteners.
6   Rubber preservative should be applied to the mounts to slow deterioration.

Replacement
7   Disconnect the negative battery cable from the battery, then raise the vehicle and support it securely on jackstands (if not already done). Support the engine as described in Step 3. **Caution:** *If the stereo in your vehicle is equipped with an anti-theft system, make sure you have the correct activation code before disconnecting the battery.*
8   To remove the right engine mount, remove the nut and withdraw the through-bolt from the frame bracket.
9   Remove the mount-to-bracket nuts and detach the mount.
10   To remove the rear engine mount, pull the rubber plugs from the chassis to access the three nuts.
11   To remove the front engine mount, remove the nut holding the insulator to the engine bracket, then the three bolts holding the insulator to the chassis.
12   Installation is the reverse of removal. Use thread locking compound on the mount bolts/nuts and be sure to tighten them securely.
13   See Chapter 7 for transaxle mount replacement.

Chapter 2 Part B
V6 engine

Contents

Section		Section
Camshafts and lifters - removal, inspection and installation 10		Oil pan - removal and installation.. 12
CHECK ENGINE light on .... See Chapter 6		Oil pump - removal, inspection and installation........................... 13
Cylinder compression check .... See Chapter 2C		Oil seals - replacement 9
Cylinder heads - removal and installation 11		Rear main oil seal - replacement... 15
Drivebelt check, adjustment and replacement .... See Chapter 1		Repair operations possible with the engine in the vehicle 2
Engine mounts - check and replacement 16		Spark plug replacement .... See Chapter 1
Engine oil and filter change .... See Chapter 1		Timing belt and sprockets - removal, inspection
Engine overhaul - general information .... See Chapter 2C		and installation 8
Engine - removal and installation................................. See Chapter 2C		Top Dead Center (TDC) for number one piston - locating 3
Exhaust manifolds - removal and installation 7		Valve covers - removal and installation...................................... 5
Flywheel/driveplate - removal and installation 14		Variable Valve Timing (VVT) system - description, check
General information 1		and component replacement... 4
Intake manifold - removal and installation 6		Water pump - removal and installation .... See Chapter 3

Specifications

General
Engine designation	1MZ-FE
Cylinder numbers (timing belt end-to-transaxle end)	
Right (firewall) side	1-3-5
Left (radiator) side	2-4-6
Firing order	1-2-3-4-5-6
Displacement...	183 cubic inches (3.0 liters)

Cylinder head
Warpage limits	
Cylinder head	0.0039 inch
Intake manifold..	0.0031 inch
Exhaust manifolds...	0.0196 inch

Camshaft and related components
Valve clearance (engine cold)	
Intake	0.005 to 0.009 inch
Exhaust...	0.011 to 0.015 inch
Bearing journal diameter	
Intake	1.0610 to 1.0616 inches
Exhaust..	1.0613 to 1.0620 inches
Bearing oil clearance	
Standard	
Intake	0.0014 to 0.0028 inch
Exhaust..	0.0010 to 0.0024 inch
Service limit	
Intake	0.0039 inch
Exhaust..	0.0035 inch
Lobe height	
Intake	
Standard	1.6579 to 1.6618 inches
Service Limit	1.6520 inches
Exhaust	
Standard	1.6520 to 1.6559 inches
Service limit...	1.6461 inches
Thrust clearance (endplay)	
Standard	0.0016 to 0.0035 inch
Service limit..	0.0047 inch

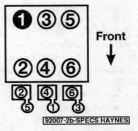

Cylinder numbering and coil terminal location

Camshaft and related components (continued)

Runout limit (total indicator reading)	0.0024 inch
Camshaft gear backlash	
Standard	0.0008 to 0.0079 inch
Service limit	0.0188 inch
Camshaft gear spring end gap	0.712 to 0.740 inch
Timing belt tensioner protrusion	0.394 to 0.425 inch
Lifters	
Outside diameter	1.2191 to 1.2195 inch
Bore diameter	1.2205 to 1.2211 inch
Lifter-to-bore (oil) clearance	
Standard	0.0009 to 0.0020 inch
Service limit	0.0028 inch

Oil pump

Driven rotor-to-pump body clearance	
Standard	0.0039 to 0.0069 inch
Service limit	0.0118 inch
Rotor tip clearance	
Standard	0.0043 to 0.0094 inch
Service limit	0.0138 inch
Rotor side clearance	
Standard	0.0012 to 0.0035 inch
Service limit	0.0059 inch

Torque specifications

Ft-lbs (unless otherwise indicated)

Air conditioning compressor bracket mounting bolts	19
Intake manifold bolts/nuts	132 in-lbs
Intake plenum bolts/nuts	32
Intake plenum brace bolts	168 in-lbs
Exhaust manifold nuts	36
Crankshaft pulley bolt	159
Timing belt cover bolts (no. 3)	74 in-lbs
Idler pulley bolts	
No. 1	25
No. 2	32
Timing belt tensioner bolts	20
Camshaft pulley bolts	94
Camshaft bearing cap bolts	144 in-lbs
Cylinder head bolts (12-point)	
Step 1	40
Step 2	Tighten an additional 90-degrees (1/4-turn)
Cylinder head bolt (recessed)	156 in-lbs
Oil pan bolts	
aluminum section	168 in-lbs
steel section	69 in-lbs
Oil pump mounting bolts	
10 mm bolt head	69 in-lbs
12 mm bolt head	168 in-lbs
14 mm bolt head	27
Oil pick-up tube mounting bolts	69 in-lbs
Flywheel/driveplate bolts*	61
Rear crankshaft oil seal retainer mounting bolts	69 in-lbs
Shock tower brace nuts (Solara models only)	59
Valve cover nuts	69 in-lbs
Variable Valve Timing (VVT) system (1999 through 2001 Lexus ES 300)	
Oil control valve filter plug	33
Oil control valve hold-down bolt	66 in-lbs
Camshaft sprocket/actuator assembly retaining nut**	110
Engine mounts	
Engine movement control rod mounting bolts	47
Front engine mount insulator-to-chassis bolts	
Green	48
Silver	32
Camry models only, all bolts (TMC made)	59
Front (shock absorber) mount bolts	35
Rear engine mount insulator nuts	48

* Apply thread locking compound to the threads prior to installation
** Reverse-threaded.

Chapter 2 Part B V6 engine

1 General information

The models covered by this manual are available with the 1MZ-FE V6 engine. The design is a 3.0-liter, DOHC (dual overhead cam) with four valves per cylinder (24 in all), an aluminum engine block, distributorless ignition, and two-piece oil pan.

This Part of Chapter 2 is devoted to in-vehicle repair procedures for the V6 engine. All information concerning engine removal and installation and engine block and cylinder head overhaul can be found in Part C of this Chapter.

The following repair procedures are based on the assumption that the engine is installed in the vehicle. If the engine has been removed from the vehicle and mounted on a stand, many of the steps outlined in this Part of Chapter 2 will not apply.

The Specifications included in this Part of Chapter 2 apply only to the procedures contained in this Part. Part C of Chapter 2 contains the Specifications necessary for cylinder head and engine block rebuilding.

2 Repair operations possible with the engine in the vehicle

Many major repair operations can be accomplished without removing the engine from the vehicle.

Clean the engine compartment and the exterior of the engine with some type of degreaser before any work is done. It will make the job easier and help keep dirt out of the internal areas of the engine.

Depending on the components involved, it may be helpful to remove the hood to improve access to the engine as repairs are performed (refer to Chapter 11 if necessary). Cover the fenders to prevent damage to the paint. Special pads are available, but an old bedspread or blanket will also work.

If vacuum, exhaust, oil or coolant leaks develop, indicating a need for gasket or seal replacement, the repairs can generally be made with the engine in the vehicle. The intake and exhaust manifold gaskets, oil pan gasket, crankshaft oil seals and cylinder head gaskets are all accessible with the engine in place.

Exterior engine components, such as the intake and exhaust manifolds, the oil pan, the oil pump, the water pump, the starter motor, the alternator, and the fuel system components can be removed for repair with the engine in place.

Since the cylinder heads can be removed without pulling the engine, valve component servicing can also be accomplished with the engine in the vehicle. Replacement of the camshafts, timing belt and pulleys is also possible with the engine in the vehicle.

In extreme cases caused by a lack of necessary equipment, repair or replacement of piston rings, pistons, connecting rods and rod bearings is possible with the engine in the vehicle. However, this practice is not recommended because of the cleaning and preparation work that must be done to the components involved.

3 Top Dead Center (TDC) for number one piston - locating

Refer to illustrations 3.6 and 3.7

1 Top Dead Center (TDC) is the highest point in the cylinder that each piston reaches as it travels up the cylinder bore. Each piston reaches TDC on the compression stroke and again on the exhaust stroke, but TDC generally refers to piston position on the compression stroke.

2 Positioning the piston(s) at TDC is an essential part of many procedures such as valve timing, camshaft and timing belt/pulley removal and distributor removal.

3 Before beginning this procedure, be sure to place the transaxle in Neutral and apply the parking brake or block the rear wheels. Also, disable the ignition system by disconnecting the primary electrical connectors at the ignition coil pack/modules (see Chapter 5). Remove the spark plugs (see Chapter 1).

4 In order to bring any piston to TDC, the crankshaft must be turned using one of the methods outlined below. When looking at the front of the engine, normal crankshaft rotation is clockwise.

   a) *The preferred method is to turn the crankshaft with a socket and ratchet attached to the bolt threaded into the front of the crankshaft. Apply pressure on the bolt in a clockwise direction only. Never turn the bolt counterclockwise.*

   b) *A remote starter switch, which may save some time, can also be used. Follow the instructions included with the switch. Once the piston is close to TDC, use a socket and ratchet as described in the previous paragraph.*

   c) *If an assistant is available to turn the ignition switch to the Start position in short bursts, you can get the piston close to TDC without a remote starter switch. Make sure your assistant is out of the vehicle, away from the ignition switch, then use a socket and ratchet as described in Paragraph (a) to complete the procedure.*

5 The V6 engine does not have a distributor, but rather three ignition coils, fired in sequence by signals from the computer. The three coils sit directly on top of cylinders 2, 4 and 6, with spark plug wires connecting these coils to the cylinders on the rear bank.

6 Remove the spark plug and install a compression pressure gauge in the number one spark plug hole (refer to Chapter 2C). It should be a gauge with a screw-in fitting and a hose at least six inches long **(see illustration)**.

7 Rotate the crankshaft using one of the methods described above while observing the compression gauge. When TDC for the compression stroke of number one cylinder is reached, compression pressure will show on the gauge as the marks are beginning to line up on the crankshaft pulley **(see illustration)**.

3.6 A compression gauge can be used in the number one plug hole (arrow) to assist in finding TDC on 1MZ-FE engines

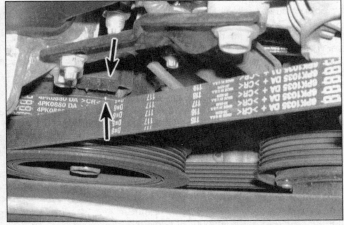

3.7 Turn the crankshaft until the notch in the pulley aligns with the zero (arrows) on the timing plate

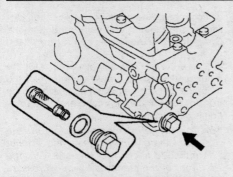

4.8a The oil control valve (OCV) filters are located at the rear of each cylinder head

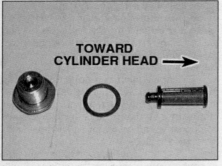

4.8b If you suspect a problem with the VVT system, inspect the OCV filter for obstructions; when you install the filter, use a new O-ring and make sure that the big end of the filter faces toward the head

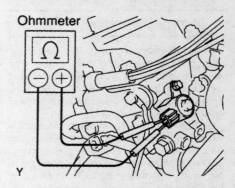

4.9 To measure the resistance of the OCV coil, unplug the electrical connector and hook up an ohmmeter across the two terminals; there should be 6.9 to 7.9 ohms resistance

If you go past the marks, release the gauge pressure and rotate the crankshaft around two more revolutions. *Note: The most positive method for finding TDC on the V6 engine is to examine the crankshaft timing marks and camshaft sprocket timing marks (see Section 8).*

4 Variable Valve Timing (VVT) system – description, check and component replacement

Description

Note: A Variable Valve Timing (VVT) system is used on 1999 through 2001 Lexus ES 300 models. The following procedures apply to the oil control valve, oil filter or intake camshaft sprocket/actuator on either cylinder head.

1 The VVT system varies intake camshaft timing by directing oil pressure to advance or retard the intake camshaft sprocket/actuator assembly. Changing the intake camshaft timing during certain engine conditions increases engine torque and fuel economy and reduces emissions.
2 System components include the Powertrain Control Module (PCM), an oil control valve (OCV) in each cylinder head, an oil filter for each OCV and an intake camshaft sprocket/actuator assembly.
3 The PCM uses inputs from the vehicle speed sensor (VSS), the throttle position sensor (TPS), the mass air flow (MAF) sensor and the engine coolant temperature (ECT) sensor to turn the oil control valve ON or OFF.
4 When the OCV is energized by the PCM, it directs a specified amount of oil pressure from the engine to advance or retard the intake camshaft sprocket/actuator assembly.
5 The intake camshaft sprocket/actuator assembly is equipped with an inner hub that is attached to the camshaft. The inner hub consists of a series of fixed vanes that use oil pressure as a wedge against the vanes to rotate the camshaft. The higher the oil pressure, the greater the amount of actuator rotation and, therefore, the amount of camshaft advance or retard. How does greater oil pressure advance *or* retard the cam? By directing the oil to the advance or retard side of the fixed vanes.
6 When oil is applied to the advance side of the vanes, the actuator can advance the camshaft up to 21 degrees in a clockwise direction. When oil is applied to the retard side of the vanes, the actuator will start to rotate the camshaft counterclockwise back to 0 degrees, which is the normal position of the actuator during engine operation under no load or at idle. The PCM can also send a signal to the oil control valve to stop oil flow to both (advance and retard) passages to hold camshaft advance in its current position.
7 Under light engine loads, the VVT system retards the camshaft timing to decrease valve overlap and stabilize engine output. Under medium engine loads, the VVT system advances the camshaft timing to increase valve overlap, thereby increasing fuel economy and decreasing exhaust emissions. Under heavy engine loads at low RPM, the VVT system advances the camshaft timing to help close the intake valve faster, which improves low to midrange torque. Under heavy engine loads at high RPM, the VVT system retards the camshaft timing to slow the closing of the intake valve to improve engine horsepower.

Check and component replacement

*Note 1: A problem in the VVT oil control valve circuit will set a diagnostic **trouble code** and turn on the Check Engine light on the dash. Refer to Chapter 6 for accessing trouble codes.*
Note 2: Most problems in the VVT system originate from the oil control valve(s) and filter(s). Regular engine oil and filter changes are necessary for trouble-free operation of the OCVs.
Note 3: Some checks and inspections of the VVT system require removal of the valve cover and the intake camshaft.

Oil control valve (OCV) filter

Refer to illustrations 4.8a and 4.8b
8 A clogged OCV filter screen is often the cause of VVT system problems. Remove the OCV filter from the rear of the cylinder head **(see illustration)** and then inspect the filter for clogging. Clean the filter if necessary and reinstall it using a new O-ring. Make sure that the big end of the filter faces toward the head **(see illustration)**. Be sure to tighten the filter plug to the torque listed in this Chapter's Specifications.

Oil control valve (OCV)

Refer to illustrations 4.9, 4.10, 4.11a and 4.11b
9 Disconnect the electrical connector from the oil control valve and measure the resistance between the terminals of the control valve **(see illustration)**. There should be 6.9 to 7.9 ohms resistance. If the indicated resistance is out of range, go to the next Step and replace the OCV. If the resistance is within range, go to Step 11.
10 To replace the OCV, unplug the electrical connector, remove the hold-down bolt **(see illustration)** and pull the OCV out of the cylinder head. Use a new O-ring when installing the new OCV. Be sure to tighten the OCV hold-down bolt to the torque listed in this Chapter's Specifications.

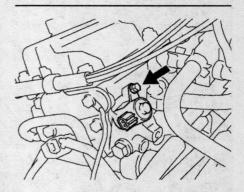

4.10 To remove the OCV, unplug the electrical connector, remove the hold-down bolt (arrow) and pull the OCV out of the cylinder head

Chapter 2 Part B V6 engine

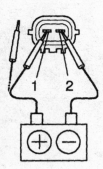

4.11a Using a pair of fused jumper wires, ground terminal No. 2 of the OCV connector, and then apply battery voltage to terminal No. 1

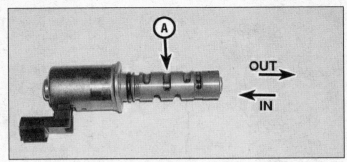

4.11b When voltage is applied, the OCV plunger (A), which you can see through the slots in the housing, should move out; when voltage is cut, the plunger should move in

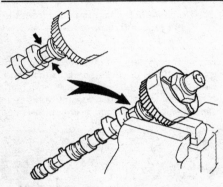

4.13 Secure the camshaft in a bench vise by clamping up the hexagonal "nut" part of the cam; be careful not to damage the camshaft or any of the cam lobes or journals

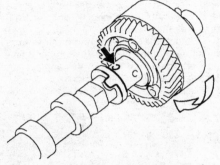

4.16 Apply 14 psi of air pressure to the advance side oil port (arrow) and try to rotate the actuator assembly by hand; the actuator should rotate freely from the locked position, with no obvious binding, for about 30 degrees in the advance angle direction (arrow)

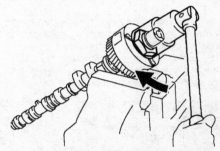

4.18 To replace the sprocket/actuator assembly, put the camshaft in a bench vise and then remove the 46mm nut by turning it *clockwise* (the nut is reverse-threaded)

11 With the OCV removed from the cylinder head (see previous Step), check the operation of the OCV plunger. Using a pair of fused jumper wires, ground terminal No. 2 of the OCV connector, and then apply battery voltage to terminal No. 1 **(see illustration)**. When voltage is applied, the plunger should move out; when voltage is cut, the plunger should move back inward **(see illustration)**. If the OCV doesn't operate as described, replace it (see Step 10). Always use a new O-ring when reinstalling the OCV.

Camshaft sprocket/actuator assembly

Refer to illustrations 4.13, 4.16, 4.18 and 4.20

12 Remove the valve cover (see Section 5), the timing belt (see Section 8) and the intake camshaft (see Section 10). **Note:** *Do NOT remove the exhaust camshaft or sprocket from the engine.*

13 Mount the camshaft in a bench vise. Secure the camshaft in the vise by clamping up the hexagonal "nut" part of the cam **(see illustration)**. **Caution:** *Be careful not to damage the camshaft or any of the cam lobes or journals.*

14 Verify that the sprocket/actuator will not rotate from the locked position. The locked position is a "neutral position" in which the actuator is placed during idle and no load conditions, and anytime that the VVT system is not activated by the PCM.

15 Using lacquer thinner or acetone, remove all traces of oil from the front cam journals and the VVT oil control orifices. Apply vinyl tape over all the oil control orifices except the advance side oil port.

16 Apply 14 psi of air pressure to the advance side oil port and try to rotate the actuator assembly by hand. The actuator should rotate freely, with no obvious binding, for about 30 degrees in the advance angle direction from the locked position **(see illustration)**. **Note:** *It is critical to have an air tight seal between the air gun nozzle and the advance oil port hole, because if air leaks out at the air nozzle, or at any of the other oil control orifices, the lock pin in the actuator won't be forced out of its locating hole. If leakage occurs, apply a little more air pressure to the advance side oil port to force the lock pin from the locating hole.*

17 If the actuator does not rotate freely as described, replace the Intake camshaft sprocket/actuator assembly.

18 To replace the sprocket/actuator assembly, put the camshaft in a bench vise (see Step 13), and then remove the 46mm nut by turning it *clockwise* (the nut is reverse-threaded) **(see illustration)**.

19 Remove the sprocket/actuator assembly. If it's hard to pull off the camshaft, tap it lightly with a plastic-tip hammer. **Caution:** *Do NOT try to disassemble the sprocket/actuator assembly - it's NOT rebuildable.*

20 Lubricate the sprocket/actuator seating surface on the camshaft with clean engine oil. Before installing the sprocket/actuator assembly, make sure that the lock pin (in the camshaft) and lock pin groove (inside the sprocket/actuator) are aligned **(see illustration)**. Install the sprocket/actuator assembly. Using a NEW sprocket/actuator assembly

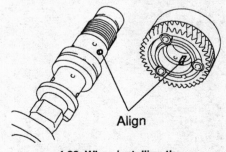

4.20 When installing the sprocket/actuator assembly, make sure that the lock pin (in the camshaft) and lock pin groove (inside the sprocket/actuator) are aligned

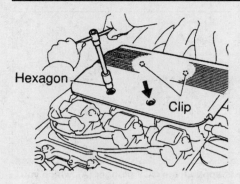

5.2 Using an Allen wrench, remove the two bolts (arrows) and pull up the V-bank cover, which has two clips under the "finned" area

5 Valve covers - removal and installation

Removal

Refer to illustrations 5.2, 5.3, 5.4a, 5.4b and 5.5

1 Disconnect the negative cable from the battery. **Caution:** *If the stereo in your vehicle is equipped with an anti-theft system, make sure you have the correct activation code before disconnecting the battery.*
2 Remove the V-bank cover **(see illustration)**. If you're removing the rear (firewall side) cover, remove the intake plenum assembly (see Section 6).
3 If you're working on the front (radiator side) or both covers, remove the coil connectors, coils and spark plugs from the front bank **(see illustration)**.
4 Detach the engine wiring harness from the left side of the engine, the number 3 timing belt cover, the rear of the engine and right-hand side **(see illustrations)**.
5 Remove the retaining nuts and sealing

5.3 Disconnect the ignition coils from the harness and remove the coils

retaining nut, tighten the retaining nut to the torque listed in this Chapter's Specifications (don't forget, it's reverse-threaded).
21 Install the intake camshaft (see Section 10), the timing belt (see Section 8) and the valve cover (see Section 5).

washers, then detach the cover(s) **(see illustration)**. If the cover is stuck to the head, bump the end with a wood block and a hammer to jar it loose. If that doesn't work, try to slip a flexible putty knife between the head

5.4a Remove two nuts and disconnect the left-hand engine harness . . .

5.4b . . . disconnect the wire clips at the timing belt cover and the five bolts retaining the right-hand harness, then move the harness away from the rear valve cover

5.5 Remove the bolts and sealing washers and remove the valve cover

5.8 Apply RTV sealant to the areas indicated (arrows) and install the cover with a new gasket

Chapter 2 Part B V6 engine

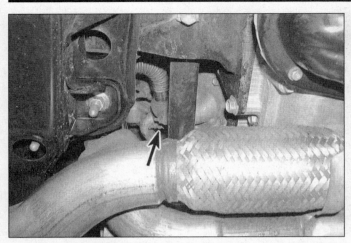

6.7a Disconnect the EGR pipe (arrow) from the exhaust manifold (shown from below)

6.7b Unbolt the two intake braces (arrows)

and cover to break the seal. **Caution:** *Don't pry at the cover-to-head joint or damage to the sealing surfaces may occur, leading to oil leaks after the cover is reinstalled.*

Installation

Refer to illustration 5.8

6 The mating surfaces of the cylinder head and cover must be clean when the cover is installed. Use a gasket scraper to remove all traces of sealant and old gasket material, then clean the mating surfaces with lacquer thinner or acetone. If there's residue or oil on the mating surfaces when the cover is installed, oil leaks may develop.

7 Install new spark plug tube seals.

8 Apply RTV sealant to the gasket/seal joints at the front and rear camshaft-to-head mounts and install the valve cover with a new gasket **(see illustration)**.

9 Tighten the nuts to the torque listed in this Chapter's Specifications in three or four equal steps.

10 Reinstall the remaining parts, run the engine and check for oil leaks.

6 Intake manifold - removal and installation

Warning: *Wait until the engine is completely cool before beginning this procedure.*

Removal

1 Relieve the fuel system pressure, then disconnect the negative cable from the battery. **Caution:** *If the stereo in your vehicle is equipped with an anti-theft system, make sure you have the correct activation code before disconnecting the battery.*

Upper intake plenum

Refer to illustrations 6.7a, 6.7b and 6.8

2 Drain the coolant into a clean container (see Chapter 1).

3 Remove the V-bank cover **(see illustration 5.2)**.

4 Remove the air cleaner assembly (see Chapter 4).

5 Disconnect the electrical connectors, vacuum and coolant bypass hoses from the throttle body, then disconnect the throttle linkage (see Chapter 4). The throttle body can remain attached to the intake plenum unless it is being removed for cleaning or gasket service.

6 On Solara models, remove the front suspension brace between the two shock towers.

7 Disconnect the EGR pipe at the intake plenum, then unbolt the two intake plenum braces at the firewall side of the rear cylinder head **(see illustrations)**.

8 Remove the bolts and nuts holding the plenum to the lower intake manifold and lift the plenum off **(see illustration)**.

Lower intake manifold

Refer to illustrations 6.10, 6.11 and 6.12

9 Disconnect the electrical connectors from the fuel injectors. **Note:** *The intake manifold can be removed with the injectors and fuel rails in place or removed, depending on the work to be done.*

10 Disconnect the air assist and heater hoses at the left end of the intake manifold **(see illustration)**. Also detach the fuel line

6.8 Remove the bolts (A) and nuts (B), then remove the intake plenum

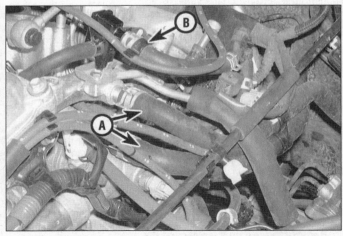

6.10 Detach the heater hoses (A) and the air assist hose (B)

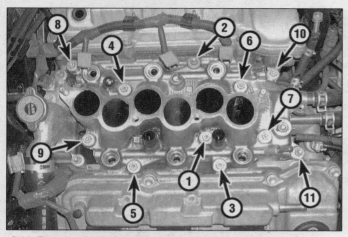

6.11 Remove the bolts and nuts (arrows), then remove the lower intake manifold - numbers indicate tightening sequence for reassembly

6.12 This water transfer hose (arrow) should be replaced whenever the intake manifold is off for other repairs

from the fuel rail (see Chapter 4).
11   Remove the eight mounting nuts/bolts, then detach the manifold from the engine **(see illustration)**. If it's stuck, don't pry between the gasket mating surfaces or damage may result.
12   There is a water transfer hose **(see illustration)** that is exposed only when the intake manifold is removed. Because of the difficulty in getting at this hose for replacement, we recommend that it be replaced with a new hose if the intake manifold is removed for other work.

Installation

13   Use a scraper to remove all traces of old gasket material and sealant from the manifold and cylinder heads, then clean the mating surfaces with lacquer thinner or acetone.
14   Install new gaskets, then position the manifold on the engine. Make sure the gaskets haven't shifted, then install the nuts/bolts.
15   Tighten the nuts/bolts, in three or four equal steps, to the torque listed in this Chapter's Specifications. Tighten the bolts in the sequence shown in **illustration 6.11**.
16   Install the remaining parts in the reverse order of removal. Use a new gasket between the intake manifold and the plenum.
17   Refill the cooling system. Run the engine and check for fuel, vacuum and coolant leaks.

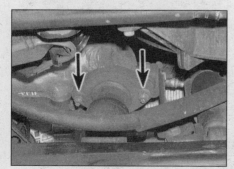

7.5 Remove the two nuts (arrows) at the exhaust pipe/manifold joint - front pipe shown, rear similar

7   Exhaust manifolds - removal and installation

Refer to illustration 7.5
Warning: *The engine must be completely cool before beginning this procedure.*
1   Disconnect the negative cable from the battery. **Caution:** *If the stereo in your vehicle is equipped with an anti-theft system, make sure you have the correct activation code before disconnecting the battery.*
2   Spray penetrating oil on the exhaust manifold fasteners and allow it to soak in.
3   Remove the heated oxygen sensors from the front and rear manifolds.
4   Remove the EGR pipe from the rear exhaust manifold and cylinder head (see Chapter 6).
5   Remove the nuts retaining the exhaust pipes to the front and rear exhaust manifolds **(see illustration)**.
6   Disconnect the electrical connector from the oxygen sensor at each manifold.
7   Unbolt the exhaust manifolds from the cylinder heads and slip them off the mounting studs.
8   Carefully inspect the manifolds and fasteners for cracks and damage.
9   Use a scraper to remove all traces of old gasket material and carbon deposits from the manifolds and cylinder head mating surfaces. If the gasket was leaking, check the manifolds for warpage on the cylinder head mounting surface by placing a straightedge over the surface and trying to insert a feeler gauge. If the clearance exceeds the limit listed in this Chapter's Specifications, have the manifold resurfaced at an automotive machine shop.
10   Position new gaskets over the cylinder head studs.
11   Install the manifolds and thread the mounting nuts into place.
12   Working from the center out, tighten the nuts to the torque listed in this Chapter's Specifications in three or four equal steps.

13   Reinstall the remaining parts in the reverse order of removal. Use new gaskets when connecting the exhaust pipes.
14   Run the engine and check for exhaust leaks.

8   Timing belt and sprockets - removal, inspection and installation

Removal

** CAUTION **
The timing system is complex. Severe engine damage will occur if you make any mistakes. Do not attempt this procedure unless you are highly experienced with this type of repair. If you are at all unsure of your abilities, consult an expert. Double-check all your work and be sure everything is correct before you attempt to start the engine.

Refer to illustrations 8.9, 8.11a, 8.11b, 8.12, 8.13, 8.14, 8.17, 8.21 and 8.24
1   Disconnect the negative cable from the battery. **Caution:** *If the stereo in your vehicle is equipped with an anti-theft system, make sure you have the correct activation code before disconnecting the battery.*
2   Remove the coolant overflow tank and windshield washer tank (see Chapter 3).
3   On Solara models, remove the four nuts and the front suspension upper brace from the strut towers.
4   Remove the drivebelts from the alternator and power steering pump (see Chapter 1).
5   Loosen the lug nuts on the right front wheel, but don't remove them yet.
6   Raise the front of the vehicle and support it securely on jackstands. Apply the parking brake and block the rear wheels. Remove the right front wheel.
7   Remove the right front inner fender apron (see Chapter 11).
8   Position the number one cylinder at TDC (see Section 3).

Chapter 2 Part B V6 engine 2B-9

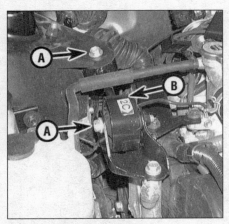

8.9 Remove the two through-bolts (A) and the engine movement control rod (B)

8.11b Remove the number 2 timing belt cover

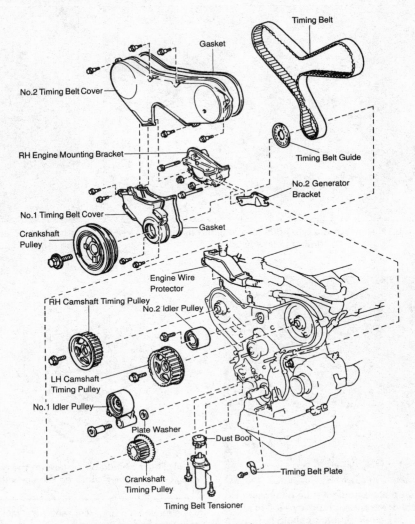

8.11a Timing belt and related components

9   Support the engine with a jack from below and remove the engine movement control rod and its bracket on the engine **(see illustration)**. Place a wood block on the jack head and do not place the jack directly under the oil pan drain plug.
10 Remove the spark plugs (see Chapter 1).
11 Remove the upper (no. 2) timing belt cover and gasket **(see illustrations)**. *Note: Unclip the wiring harness above the cover and push it back enough to remove the belt cover.*
12 Check to see if there are installation marks on the timing belt - if you intend to re-use the belt and the marks have been obscured, make new ones **(see illustration)**.
13 Make sure the camshaft pulley timing marks are properly aligned **(see illustration)**.

8.12 If you intend to re-use the belt and the original installation marks are obscured or missing, make new ones

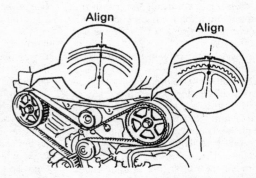

8.13 Both camshaft sprocket timing marks should align with the marks on the rear cover

8.14 Remove the two bolts (arrows) and detach the timing belt tensioner

8.17 If the camshaft sprockets are damaged, or the cams are to be removed, hold the hex portion of the camshaft (arrow) with a wrench while removing the sprocket bolt

8.21 A puller should not be necessary to remove the crankshaft pulley - if it is stuck use two prybars behind it

8.24 There should be a mark on the belt next to the drilled mark on the sprocket

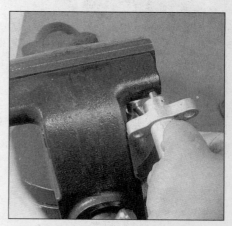

8.29 Check the tensioner for signs of leakage and test for leakdown by forcing it against an immovable object

14 Remove the timing belt tensioner **(see illustration)**. Be sure to remove the rubber boot as well; it may stick in the tensioner recess. **Caution:** *Loosen the bolts a little at a time, alternating from side-to-side until the tensioner pressure is off.*
15 Relieve the tension between the rear (right-hand) and front (left-hand) camshaft sprockets by turning the rear sprocket slightly clockwise. **Caution:** *Don't turn the sprocket any more than is necessary to provide enough slack to allow you to slip the belt off the sprockets.*
16 Remove the timing belt from the camshaft sprockets.
17 The camshaft sprockets can be removed at this point, if they are worn or damaged. Remove the valve cover(s) (see Section 5) and hold the camshaft with a wrench on the cast-in hex while loosening the sprocket bolt **(see illustration)**. Remove the bolt and detach the sprocket.
18 Remove the upper (no. 2) idler pulley **(see illustration 8.11a)**.
19 Remove the crankshaft (drivebelt) pulley bolt. Wedge a large screwdriver into the flywheel/driveplate ring gear teeth or against a converter bolt to keep the engine from turning. Use a breaker bar and socket to loosen the pulley bolt.
20 When the crankshaft pulley bolt is loosened, the TDC position of the crankshaft may be disturbed. Check and align again, if necessary. **Note:** *The crankshaft timing belt sprocket has a TDC alignment mark that lines up with a mark on the oil pump housing, making it easy to check the TDC alignment even after the crankshaft pulley and lower timing belt cover are removed.*
21 The crankshaft pulley should come off with strong hand pressure **(see illustration)**. If not, use two prybars behind it to lever it off. Do not use a jaw-type puller.
22 Remove the lower (no. 1) timing belt cover and gasket **(see illustration 8.11a)**.
23 Slip the timing belt guide off the crankshaft.
24 If you're re-using the belt, check for a mark on the belt adjacent to the drilled mark on the crankshaft sprocket **(see illustration)**. If the original mark is gone, make a new one, then slip the belt off the sprocket.
25 Using an Allen wrench, remove the no. 1 idler pulley and plate washer **(see illustration 8.11a)**.
26 If it's worn or damaged, or if you're replacing the crankshaft front oil seal, the crankshaft sprocket can now be removed. If it won't come off by hand, a steering wheel type puller may be needed to remove the sprocket. Be careful not to damage the crankshaft sensor portion of the sprocket during the removal process. If necessary, remove the lower sprocket retainer **(see illustration 8.33)**.

Inspection

Refer to illustrations 8.29 and 8.30

27 Refer to Chapter 2, Part A for the timing belt inspection procedures.
28 Check the belt tensioner for visible oil leakage. If there's only a faint trace of oil on the pushrod side, the tensioner seal is in satisfactory condition.
29 Hold the tensioner in both hands and push it forcefully against an immovable object **(see illustration)**. If the pushrod moves, replace the tensioner.
30 Measure the protrusion of the pushrod from the housing end **(see illustration)**.

Chapter 2 Part B V6 engine 2B-11

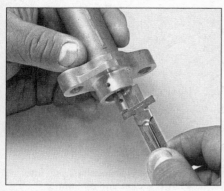

8.30 Measure the tensioner pushrod protrusion and compare it to the Specifications

8.33 Align the marks on the crankshaft timing sprocket with the marks on the oil pump case (arrow A), and install the sprocket retainer (arrow B) and bolt

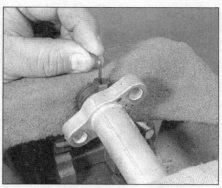

8.48 Restrain the tensioner pushrod by compressing the unit in a vise and inserting a pin approximately 0.050-inch (1.27 mm) in diameter - make sure the rubber boot is in place

Compare your measurement to this Chapter's Specifications. If the protrusion is not as specified, replace the tensioner.
31 Check that the idler pulleys turn smoothly.

Installation

> **** CAUTION ****
>
> Before starting the engine, carefully rotate the crankshaft by hand through at least two full revolutions (use a socket and breaker bar on the crankshaft pulley center bolt). If you feel any resistance, STOP! There is something wrong - most likely, valves are contacting the pistons. You must find the problem before proceeding. Check your work and see if any updated repair information is available.

Refer to illustrations 8.33 and 8.48

32 Remove all dirt, oil and grease from the timing belt area at the front of the engine.
33 Align the crankshaft timing sprocket keyway with the crankshaft key and install the sprocket with the flange side up against the engine **(see illustration 8.11a)**. Be careful not to damage the crankshaft sensor portion of the crankshaft sprocket. Check the alignment of the TDC marks on the sprocket and the oil pump housing, and reinstall the retainer and bolt **(see illustration)**.
34 Apply thread-locking compound to the first two or three threads on the lower (no. 1) idler pulley bolt, then position the idler pulley and washer and install the bolt. Tighten the bolt to the torque listed in this Chapter's Specifications.
35 Install the timing belt, starting at the crankshaft sprocket. If you're re-using the original belt, align the marks on the belt with the marks on the sprockets and covers. Install the belt over the lower (no. 1) idler and water pump pulleys.
36 Slip the belt guide over the end of the crankshaft with the cupped side facing out.
37 Install the lower (no. 1) timing belt cover and gasket **(see illustration 8.11a)**.
38 Slip the crankshaft (drivebelt) pulley onto the crankshaft, aligning the pulley keyway with the crankshaft key. Install the bolt and tighten it to the torque listed in this Chapter's Specifications. Prevent the crankshaft from turning by using the method described in Step 19.
39 Install the upper (no. 2) idler pulley. Tighten the bolt to the torque listed in this Chapter's Specifications. Make sure the pulley turns smoothly.
40 Install the front camshaft sprocket (if it was removed) on the camshaft with the flange side facing OUT. Align the pin hole in the sprocket with the pin in the end of the camshaft.
41 Install the sprocket retaining bolt and tighten it to the torque listed in this Chapter's Specifications. Use the method described in Step 17 to keep the camshaft from turning.
42 Recheck the timing marks to be sure the crankshaft hasn't turned **(see illustration 8.33)**. If you're re-using the original belt, the installation mark should line up as it did in Step 12. If not, change the position of the timing belt on the crankshaft sprocket. The mark on the front (left-hand) camshaft sprocket should be at the top (12 o'clock position), aligned with the mark on the rear (no. 3) timing cover.
43 The rear (right-hand) camshaft knock pin hole should be at the top (12 o'clock position). If necessary, remove the valve cover and turn the camshaft slightly with a wrench to align the sprocket with the mark on the rear (no. 3) cover **(see illustration 8.13)**.
44 Install the rear (right-hand) camshaft sprocket (if it was removed) on the camshaft with the flange side facing IN. Align the pin hole in the sprocket with the knock pin in the end of the camshaft.
45 Install the retaining bolt and tighten it to the torque listed in this Chapter's Specifications. Use the method described in Step 17 to keep the camshaft from turning. Be sure the timing mark is still aligned with the rear (no. 3) cover.
46 Turn the front (left-hand) camshaft sprocket clockwise slightly (about one tooth) with a pin spanner. If the special tool isn't available, grip the hex on the camshaft with a wrench and turn it **(see illustration 8.17)**. If you're re-using the original belt, align the installation mark with the camshaft timing mark. Slip the belt onto the sprocket, then turn the camshaft counterclockwise, back to its original position. There should now be slight tension on the belt.
47 Slip the belt onto the rear sprocket. If you're re-using the original belt, align the installation marks.
48 Using a press or vise, slowly compress the timing belt tensioner pushrod **(see illustration)**. Insert a metal pin, drill bit or Allen wrench through the holes in the pushrod and housing. Release the pressure from the press or vise.
49 Install the timing belt tensioner and tighten the bolts to the torque listed in this Chapter's Specifications. Remove the retaining pin.
50 Using a socket and breaker bar on the crankshaft pulley bolt, turn the crankshaft slowly (clockwise) through two complete revolutions (720-degrees). Recheck the timing marks **(see illustrations 3.7 and 8.13)**. *Caution: If the timing marks are not aligned exactly as shown, repeat the timing belt installation procedure. DO NOT start the engine until you're absolutely certain that the timing belt is installed correctly. Serious and costly engine damage could occur if the belt is installed wrong.*
51 Reinstall the engine movement control rod's mounting bracket.
52 Install the upper (no. 2) timing belt cover and gasket.
53 Install the engine movement control rod and braces and tighten the bolts securely.
54 Reinstall the remaining parts in the reverse order of removal.

9 Oil seals - replacement

Crankshaft oil seal

Refer to illustrations 9.3 and 9.5

1 Remove the timing belt and crankshaft timing belt sprocket (see Section 8).
2 Note how far the seal is recessed in the bore, then cut away the seal lip with a razor knife.

Chapter 2 Part B V6 engine

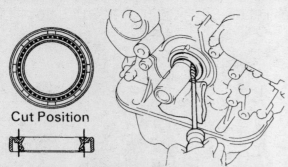

9.3 Cut away the crankshaft seal lip, wrap a screwdriver tip with tape and pry out the seal

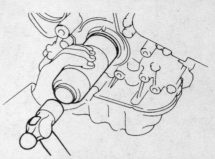

9.5 Lubricate the seal lip and drive the new crankshaft seal into place with a seal driver or a large socket and a hammer

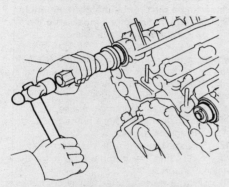

9.12 Lubricate the seal lip and tap the new camshaft seal into place with a seal driver or a large socket and a hammer

3   Carefully pry the seal out of the engine with a screwdriver or seal removal tool **(see illustration)**. If you use a screwdriver, wrap tape around the tip - don't scratch the housing bore or damage the crankshaft (if the crankshaft is damaged, the new seal will end up leaking).

4   Clean the bore in the engine and coat the outer edge of the new seal with engine oil or multi-purpose grease. Apply the same grease to the seal lip.

5   Using a seal driver or a socket with an outside diameter slightly smaller than the outside diameter of the seal, carefully drive the new seal into place with a hammer **(see illustration)**. Make sure it's installed squarely and driven in to the same depth as the original. Check the seal after installation to make sure the spring didn't pop out of place.

6   Reinstall the crankshaft timing sprocket and timing belt (see Section 8). Be careful not to scratch the crankshaft sensor portion of the sprocket.

7   Run the engine and check for oil leaks at the front seal.

Camshaft oil seals

Refer to illustration 9.12

8   Remove the timing belt and camshaft sprocket(s) (see Section 8).

9   Remove the bolts and detach the rear (no. 3) timing belt cover.

10   Note how far the seal is seated in the bore, then carefully pry it out with a screwdriver. Wrap the screwdriver tip with tape - don't scratch the bore or damage the camshaft (if the camshaft is damaged, the new seal will end up leaking).

11   Clean the bore and coat the outer edge of the new seal with engine oil or multi-purpose grease. Apply multi-purpose grease to the seal lip.

12   Using a seal driver or a socket with an outside diameter slightly smaller than the outside diameter of the seal, carefully drive the new seal into place with a hammer **(see illustration)**. Make sure it's installed squarely and driven in to the same depth as the original.

13   Reinstall the rear timing belt cover and tighten the bolts.

14   Reinstall the camshaft sprocket(s) and timing belt (see Section 8).

15   Run the engine and check for oil leaks at the camshaft seal.

10 Camshafts and lifters - removal, inspection and installation

Note: *Before beginning this procedure, obtain two 6 x 1.0 mm bolts 16 to 20 mm long. They will be referred to as service bolts in the text.*

Removal

Refer to illustrations 10.3, 10.4, 10.10, 10.12, 10.13 and 10.14

1   Position the engine at TDC (see Section 3), then remove the valve covers (see Section 5), the timing belt and the camshaft sprocket (see Section 8).

2   The following steps apply to the removal of each of the four camshafts. On each head,

10.3 Align the timing marks (arrow) on the backside of the camshaft gears

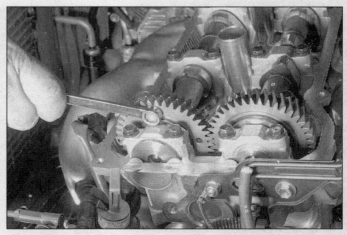

10.4 Install a service bolt through the sub-gear into the main gear

Chapter 2 Part B V6 engine

10.10 Mark up a cardboard box to store the lifters/shims and camshaft bearing caps - use a separate box for each set to avoid mix-ups and mark the FRONT, INTAKE and EXHAUST orientation

10.12 With the hex portion of the camshaft held in a vise, use a two-pin spanner to remove the tension from the subgear and remove the service bolt, then release the subgear

10.13 Remove the snap-ring with a pair of snap-ring pliers

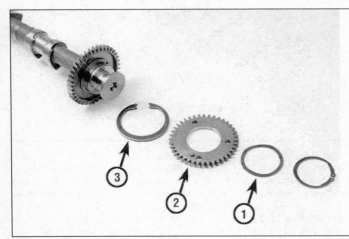

10.14 Remove the wave washer (1), the camshaft subgear (2) and the gear spring (3)

unthreaded hole in the camshaft sub-gear. Using a screwdriver positioned against the service bolt just installed, rotate the sub-gear clockwise and remove the first service bolt. The second bolt isn't needed if you have a two-pin spanner **(see illustration)**.

13 Remove the sub-gear snap-ring **(see illustration)**.

14 The wave washer, sub-gear and camshaft gear spring can now be removed from the camshaft **(see illustration)**. Be sure to keep the parts from the rear cylinder head cams separate from the front cylinder head parts.

Inspection

15 Refer to Chapter 2, Part A for camshaft, lifter and related component inspection procedures. Be sure to use the Specifications in this Part of Chapter 2 for the V6 engines.

Installation

Refer to illustrations 10.21a, 10.21b, 10.22, 10.23a, 10.23b, 10.24a, 10.24b, 10.27, 10.28a, 10.28b, 10.29a and 10.29b

16 Reassemble the exhaust camshaft gear(s) by installing the camshaft gear spring, sub-gear, wave washer and snap-ring.

17 Mount the camshaft in a padded vise.

18 Insert a service bolt into the unthreaded hole in the camshaft subgear. Using a screwdriver, align the holes of the camshaft driven gear and sub-gear by turning the camshaft sub-gear clockwise. Install a second service bolt in the threaded hole, tightening it to clamp the gears together. Remove the service bolt from the unthreaded hole. Repeat the procedure for the other exhaust camshaft.

19 Apply moly-base grease or engine assembly lube to the lifters, then install them in their original locations in the cylinder heads. Make sure the valve adjustment shims are in place in the lifters, and that all lifters are installed in their original bores.

20 Apply camshaft installation lubricant to the exhaust camshaft lobes, bearing journals and gear thrust faces.

the exhaust camshaft subgear is secured first, the intake cam removed, then the exhaust camshaft is removed.

3 Make sure the cam timing marks on the drive and driven gears are in alignment **(see illustration)**. **Note:** *On the right (rear) cylinder head, align the two dots on the intake camshaft gear with the two dots on the exhaust camshaft gear. On the left (front) cylinder head, align the one dot on the intake camshaft gear with the one dot on the exhaust camshaft gear.*

4 Secure the exhaust camshaft sub-gear to the driven gear with a service bolt installed in the threaded hole **(see illustration)**. **Caution:** *Since the camshaft thrust clearance is minimal, the camshafts must be held level as they are being removed. If they aren't, the portion of the cylinder head next to the cam gears may crack or be damaged by the gear leverage. Before lifting a camshaft out of the head, make certain that the torsional spring force of the sub-gear has been eliminated by the service bolt.*

5 Loosen the intake camshaft bearing cap bolts in 1/4-turn increments until they can be removed by hand. Follow the reverse of the recommended tightening sequence **(see illustrations 10.29a and 10.29b)**.

6 Remove the bearing caps and gently lift out the intake camshaft. Be sure to keep it level.

7 Loosen the exhaust camshaft bearing cap bolts in 1/4-turn increments until they can be removed by hand. Follow the reverse of the recommended tightening sequence **(see illustrations 10.24a and 10.24b)**.

8 Remove the exhaust camshaft bearing caps and oil seal and gently lift out the exhaust camshaft. Be sure to keep it level.

9 Repeat the steps for the other cylinder head.

10 Store the bearing caps in the correct order. **Note:** *If necessary, the valve lifters and shims can now be removed with a magnetic tool. Be sure to store them separately so they can be reinstalled in their original locations* **(see illustration)**.

11 To disassemble an exhaust camshaft gear, mount the cam in a vise with the jaws gripping the large hex on the shaft.

12 Install a second service bolt in the

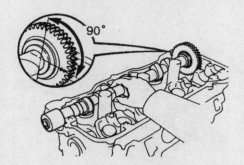 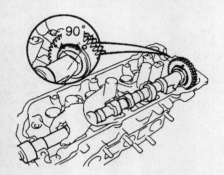 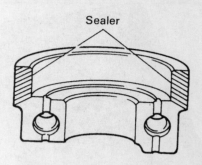

10.21a On the right (rear) cylinder, set the exhaust camshafts into place with the two dots facing the intake camshaft . . .

10.21b . . . on the left (front) cylinder head, set the exhaust camshaft into place with the one dot facing the intake camshaft

10.22 Apply RTV sealant to the shaded areas on the bearing cap

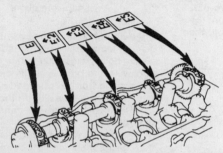 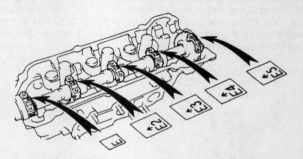

10.23a Install the exhaust camshaft bearing caps with the arrows pointing toward the timing belt end of the engine; this is the rear (right) cylinder head . . .

10.23b . . . and this is the front (left) cylinder head

21 Set the exhaust camshaft in place in the cylinder head with the timing mark on the subgear facing the center of the head **(see illustrations)**.
22 Apply a thin coat of RTV sealant to the outer edges of the front bearing cap cylinder head mating surfaces **(see illustration)**.
23 Install the bearing caps in numerical order with the arrows pointing toward the timing belt end of the engine **(see illustrations)**.
24 Tighten the bearing cap bolts in 1/4-turn increments to the torque listed in this Chapter's Specifications. Follow the recommended sequence **(see illustrations)**.

25 Refer to Section 9 and install a new camshaft oil seal.
26 Apply camshaft installation lubricant to the intake camshaft lobes, bearing journals and gear thrust faces.
27 Set the intake camshaft in place in the cylinder head, with the timing mark aligned with the exhaust camshaft timing mark **(see illustration)**. **Note:** *On the right (rear) cylinder head, align the two dots on the intake camshaft gear with the two dots on the exhaust camshaft gear. On the left (front) cylinder head, align the one dot on the intake camshaft gear with the one dot on the*

exhaust camshaft gear.
28 Install the bearing caps in numerical order with the arrows pointing toward the front (timing belt end) of the engine **(see illustrations)**.
29 Tighten the bearing cap bolts in 1/4-turn increments to the torque listed in this Chapter's Specifications. Follow the recommended sequence **(see illustrations)**.
30 Remove the service bolt from the exhaust camshaft subgear.
31 Reinstall the camshaft sprockets and the timing belt (see Section 8).
32 Reinstall the remaining components in

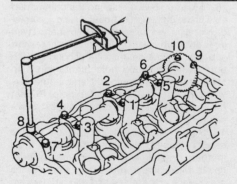 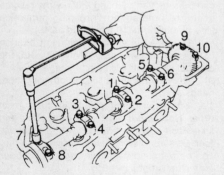 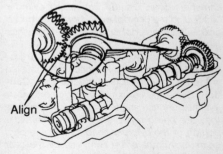

10.24a Exhaust camshaft bearing cap bolt tightening sequence - rear cylinder head

10.24b Exhaust camshaft bearing cap bolt tightening sequence - front cylinder head

10.27 Align the dots on the exhaust camshaft subgear with the dots on the intake gear (rear cylinder head shown, front head similar except align one dot with one dot)

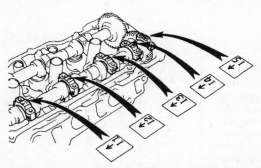

10.28a Install the intake camshaft bearing caps with the arrows pointing toward the timing belt end of the engine; this is the rear (right) cylinder head . . .

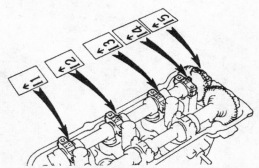

10.28b . . . and this is the front (left) cylinder head

the reverse order of removal.
33 Before reinstalling the valve covers, apply RTV sealant in the areas indicated **(see illustration 5.8)**. Clean the rubber half-circle plugs for the back of the heads and reinstall with new RTV sealant.
34 The remainder of the installation is the reverse of the disassembly sequence.
35 Run the engine, then check for leaks and proper operation.

11 Cylinder heads - removal and installation

Warning: *Wait until the engine is completely cool before beginning this procedure.*

Removal

Refer to illustrations 11.8 and 11.11

1 Disconnect the negative cable from the battery. **Caution:** *If the stereo in your vehicle is equipped with an anti-theft system, make sure you have the correct activation code before disconnecting the battery.*
2 Drain the cooling system, including the block (see Chapter 1).
3 Remove the air intake plenum and intake manifold (see Section 6).
4 Remove the exhaust manifold(s) (see Section 7).
5 Remove the alternator (see Chapter 5).
6 Remove the timing belt, camshaft sprockets and upper idler pulley (see Section 8). **Note:** *If the timing belt is not being replaced at the time the heads are removed, the timing belt and lower timing belt cover don't have to be completely removed. Follow the Steps in Section 8, marking the belt location at the lower cover at TDC, and removing the belt only from the camshaft sprockets and idler pulley.*
7 Remove the bolts holding the number 3

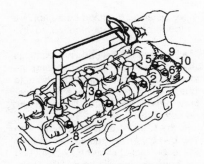

10.29a Intake camshaft bearing cap bolt tightening sequence - rear cylinder head

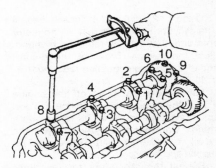

10.29b Intake camshaft bearing cap bolt tightening sequence - front cylinder head

timing belt cover **(see illustration 8.11a)**.
8 Disconnect the coolant sensor connectors and remove the water transfer casting **(see illustration)**.
9 Remove the camshaft(s) from the head(s) you intend to remove (see Section 10).
10 Remove the bolts at the rear of the cylinder heads and move the engine wiring harnesses away from the heads.
11 Using an 8 mm hex bit or Allen wrench, remove the recessed head bolts (one in each head) **(see illustration)**.

11.8 Remove the bolts and the coolant transfer casting (A) - the transfer hose (B) should be replaced whenever the intake manifold is removed

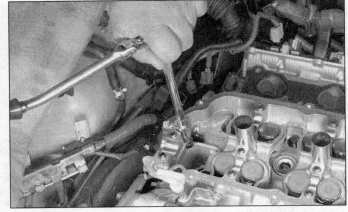

11.11 Remove the recessed bolt with an 8 mm hex bit socket

11.20 Be sure the new head gaskets are positioned right side up (check all holes and coolant passages for correct alignment) and over the block dowels

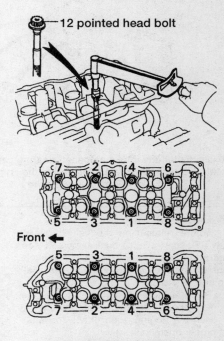

11.23 Cylinder head bolt TIGHTENING sequence

12 Using a 12-point socket, loosen the rest of the cylinder head bolts in 1/4-turn increments until they can be removed by hand. Follow the reverse order of the recommended tightening sequence **(see illustration 11.23)**.

13 Lift the cylinder head off the engine block. If the head is stuck, place a wood block against it and strike the wood with a hammer. **Caution:** *Don't pry between the head and block. The gasket surfaces may be damaged and leaks could result.*

14 Repeat the procedure for the other head.

Installation

Refer to illustrations 11.20 and 11.23

15 The mating surfaces of the cylinder heads and block must be perfectly clean when the heads are installed.

16 Use a gasket scraper to remove all traces of carbon and old gasket material, then clean the mating surfaces with lacquer thinner or acetone. If there's oil on the mating surfaces when the head is installed, the gasket may not seal correctly and leaks could develop. When working on the block, stuff the cylinders with clean shop rags to keep out debris. Use a vacuum cleaner to remove material that falls into the cylinders.

17 Check the block and head mating surfaces for nicks, deep scratches and other damage. If damage is slight, it can be removed with a file; if it's excessive, machining may be the only alternative.

18 Use a tap of the correct size to chase the threads in the cylinder head bolt holes, then clean the holes with compressed air - make sure that nothing remains in the holes. **Warning:** *Wear eye protection when using compressed air!*

19 Mount each bolt in a vise and run a die down the threads to remove corrosion and restore the threads. Dirt, corrosion, sealant and damaged threads will affect torque readings.

20 Position the new gaskets over the dowel pins in the block **(see illustration)**.

21 Carefully set the head on the block without disturbing the gasket.

22 Before installing the head bolts, apply a small amount of clean engine oil to the threads.

23 Install the bolts in their original locations and tighten them finger tight. Following the recommended sequence, tighten the bolts to the torque listed in this Chapter's Specifications **(see illustration)**. Don't tighten the recessed bolt at this time.

24 Mark the front of each bolt head with paint. You can also mark the socket you are using. Place the socket over the 12-point bolt so that you can observe the mark.

25 Following the same sequence, tighten each 12-point bolt an additional 1/4-turn (90-degrees) **(see illustration 11.23)**.

26 Tighten the recessed bolt to the torque listed in this Chapter's Specifications.

27 Repeat the entire procedure to install the other cylinder head.

28 The remaining installation steps are the reverse of removal.

29 Refill the cooling system, change the oil and filter (see Chapter 1), run the engine and check for leaks.

12 Oil pan - removal and installation

Removal

Refer to illustrations 12.7, 12.8, 12.9 and 12.11

1 Remove the hood (see Chapter 11).

2 Disconnect the negative cable from the battery. **Caution:** *If the stereo in your vehicle is equipped with an anti-theft system, make sure you have the correct activation code before disconnecting the battery.*

3 Raise the vehicle and support it securely on jackstands.

4 Remove the engine splash shields.

5 Drain the engine oil and remove the oil filter. The oil pan on the V6 engine is a two-part assembly, with an aluminum casting attached to the block and transaxle, and a lower stamped-steel pan section at the bottom.

6 Disconnect the front exhaust pipe from the catalytic converter, the two bolts and the support bracket, and the nuts holding the pipe to the front and rear exhaust manifolds (see Chapter 4 and Section 6).

7 Remove the flywheel housing cover **(see illustration)**.

8 Remove the two bolts holding the aluminum oil pan section to the transaxle **(see illustration)** and air conditioner compressor bracket (see Chapter 3).

9 Remove the 10 bolts and two nuts securing the steel oil pan section, and detach the steel pan **(see illustration)**. If it's stuck, pry it loose very carefully with a small screwdriver or putty knife. Don't damage the mating surfaces of the pan or oil leaks could develop.

12.7 Remove two bolts and the flywheel housing cover

Chapter 2 Part B V6 engine

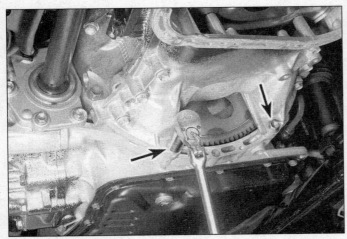

12.8 Remove the two bolts (arrows) securing the aluminum upper pan to the transaxle

12.9 Remove the steel section of the oil pan

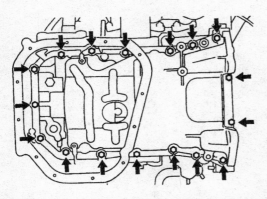

12.11 Remove the bolts retaining the aluminum portion of the pan (automatic transaxle models have 17 bolts, manual transaxle models have 19 bolts)

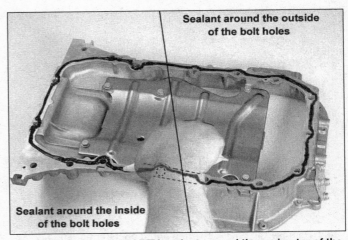

12.18 Apply the bead of RTV sealant around the perimeter of the aluminum pan section as shown

10 Remove the oil pump strainer/pickup **(see illustration 13.4)**.
11 Remove the bolts securing the aluminum oil pan section to the block **(see illustration)** (automatic transaxle models have 17 bolts, manual transaxle models have 19 bolts). **Note:** *Some bolts are within the area formerly covered by the steel pan.*
12 Remove the oil pan baffle plate, if necessary.

Installation

Refer to illustration 12.18

13 Use a scraper to remove all traces of old sealant from the block and oil pan. Clean the mating surfaces with lacquer thinner or acetone.
14 Make sure the threaded bolt holes in the block are clean.
15 Check the flange of the steel pan section for distortion, particularly around the bolt holes. If necessary, place the pan on a wood block and use a hammer to flatten and restore the gasket surface.

16 If the baffle had been removed, reinstall it now.
17 Clean the mating surfaces of the engine block and aluminum pan section, being careful not to gouge the soft metal, which could lead to leaks.
18 Apply a 4 mm wide bead of RTV sealant to the aluminum pan section **(see illustration)**.
19 Install the aluminum pan section within five minutes and uniformly tighten the bolts to the torque listed in this Chapter's Specifications in several passes. Work from the center out towards the ends of the pan.
20 Inspect the oil pump pick-up/strainer assembly for cracks and a blocked strainer. If the pick-up was removed, clean it with solvent or thinner and install it now, using a new gasket (see Section 13). Tighten the fasteners to the torque listed in this Chapter's Specifications.
21 Apply a 3 to 4 mm wide bead of RTV sealant to the flange of the steel oil pan section. **Note:** *The steel pan section must be installed within five minutes once the sealant has been applied.*

22 Carefully position the steel pan on the aluminum section and install the bolts. Working from the center out, tighten them to the torque listed in this Chapter's Specifications in three or four steps.
23 The remainder of installation is the reverse of removal. Be sure to add oil and install a new oil filter.
24 Run the engine and check for oil pressure and leaks.

13 Oil pump - removal, inspection and installation

Removal

Refer to illustrations 13.4, 13.7, 13.8, 13.9, 13.11 and 13.12

1 Remove the oil pan (see Section 12).
2 Remove the timing belt (see Section 8) and lower timing belt idler pulley.
3 Remove the crankshaft sprocket (see Section 8).

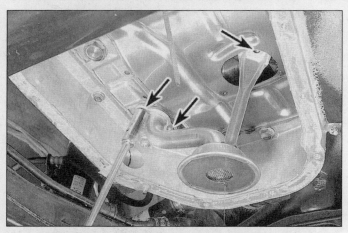

13.4 The oil pick-up tube is held in place with three fasteners (arrows)

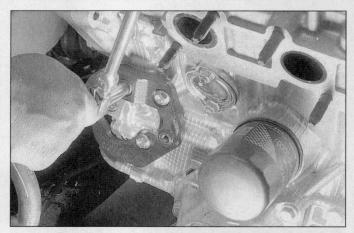

13.7 Unbolt the air conditioning compressor bracket from the block

13.8 Remove the power steering pump adjuster bar and pry the pump away from the oil pump body

13.9 Using a 10 mm hex bit or Allen wrench, remove the lower timing belt idler pulley

4 Remove the oil pick-up tube **(see illustration)**.
5 Remove the alternator (see Chapter 5).
6 Unbolt the air conditioning compressor and set it aside without disconnecting the refrigerant lines.
7 Remove the compressor bracket **(see illustration)**.
8 Remove the power steering pump adjusting bar and pry the pump away from the oil pump body **(see illustration)**.
9 Remove the lower timing belt idler pulley **(see illustration)**.
10 Remove the bolts and detach the oil pump from the engine. You may have to pry carefully between the front main bearing cap and the pump body with a screwdriver.
11 Remove the O-ring. Remove the oil

13.11 Remove the oil pressure relief plug, spring and valve

13.12 Use a large Phillips screwdriver or bit to remove the screws retaining the pump cover

Chapter 2 Part B V6 engine

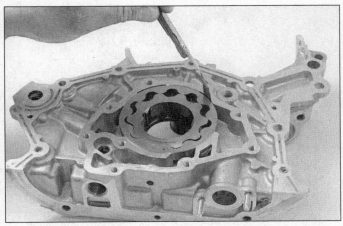

13.17a Measure the driven rotor-to-body clearance with a feeler gauge

13.17b Measure the rotor side clearance with a precision straightedge and feeler gauge

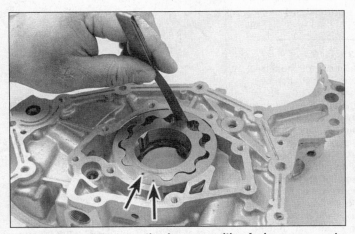

13.17c Measure the rotor tip clearance with a feeler gauge - note the rotor marks are facing out (when the pump body cover is installed, the marks will be against the cover)

13.26 Be sure to align the drive rotor and the crankshaft as the oil pump is installed, and install a new O-ring (arrow) on the block

pressure relief valve snap-ring, retainer, spring and valve (see illustration). **Warning:** *The spring is tightly compressed - be careful and wear eye protection.*

12 Use a large Phillips screwdriver to remove the screws retaining the body cover to the rear of the oil pump (see illustration).

13 Lift the cover off and remove the pump rotors.

14 Use a scraper to remove all traces of sealant and old gasket material from the pump body and engine block, then clean the mating surfaces with lacquer thinner or acetone.

Inspection

Refer to illustrations 13.17a, 13.17b and 13.17c

15 Clean all components with solvent, then inspect them for wear and damage.

16 Check the oil pressure relief valve sliding surface and valve spring. If either the spring or the valve is damaged, they must be replaced as a set.

17 Check the clearance of the following components with a feeler gauge and compare the measurements to this Chapter's Specifications (see illustrations):

a) Driven rotor-to-oil pump body clearance
b) Rotor side clearance
c) Rotor tip clearance

Installation

Refer to illustration 13.26

18 Pry the old crankshaft seal out with a screwdriver.

19 Apply multi-purpose grease or engine oil to the outer edge of the new seal and carefully drive it into place with a seal driver and a hammer. Also apply multi-purpose grease to the seal lip.

20 Place the drive and driven rotors into the pump body with the marks facing out (see illustration 13.17c).

21 Pack the pump cavity with petroleum jelly and install the cover. Tighten the screws securely following a criss-cross pattern.

22 Lubricate the oil pressure relief valve with engine oil and install the valve components in the pump body.

23 Use acetone or lacquer thinner and a clean rag to remove all traces of oil from the gasket surfaces.

24 Apply a 2 to 3 mm wide bead of anaerobic sealant to the oil pump. Avoid using an excessive amount of sealant, especially around oil passages and bolt holes. Assembly must be completed within five minutes of sealant application, otherwise the material must be removed and reapplied.

25 Position a new O-ring on the block.

26 Engage the spline teeth on the oil pump drive rotor with the large teeth on the crankshaft and slide the pump into place (see illustration).

27 Install the oil pump mounting bolts in their original locations and tighten them to the torque listed in this Chapter's Specifications in a criss-cross pattern.

28 Using a new gasket, install the oil pickup tube and tighten the fasteners to the torque listed in this Chapter's Specifications.

29 Reinstall the remaining parts in the reverse order of removal.

30 Add oil, start the engine and check for oil leaks.

31 Recheck the engine oil level.

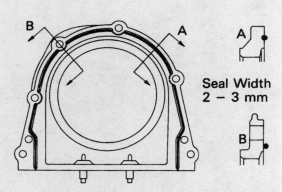

15.1 Apply sealant to the oil seal retainer-to-block mating surface

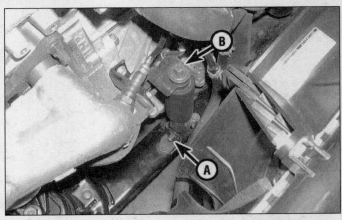

16.3 To replace the V6 mount shock absorber, remove the through-bolt (A) and the nut and bushing at the top (B)

14 Flywheel/driveplate - removal and installation

Refer to Chapter 2, Part A for this procedure, but be sure to use the torque specifications in this Part of Chapter 2 for the V6 engine.

15 Rear main oil seal - replacement

Refer to illustration 15.1

Refer to Chapter 2, Part A for this procedure, but note that the V6 engine doesn't have a gasket between the seal retainer and the engine block. Instead, apply a 2 to 3 mm wide bead of anaerobic sealant to the retainer flange **(see illustration)** before attaching the retainer to the block. Also, be sure to use the torque specifications in this Part of Chapter 2 for the V6 engine.

16 Engine mounts - check and replacement

Refer to illustrations 16.3 and 16.4

1 Refer to Chapter 2, Part A, but note that the V6 engine mounts are slightly different in ways that don't significantly affect the check and replacement procedures.
2 The basic arrangement and construction of the three engine mounts is very similar to the four-cylinder model, except that the engine movement control rod assembly has two stays, one mounted on either side of the bracket on the engine **(see illustration 8.9)**.
3 In addition to the front engine mount, V6 models have a shock-absorbing mount, mounted between the engine and chassis, to the left of the conventional front engine mount **(see illustration)**.
4 The rear mount on V6 models differs in that the engine mounting bracket is a casting that cradles the insulator, which is retained by a through-bolt to the chassis bracket **(see illustration)**.

16.4 Remove the through-bolt (not seen here) at the rear mount, raise the engine, remove the nuts (arrows) from below and remove the insulator

Chapter 2 Part C
General engine overhaul procedures

Contents

	Section
Balancer assembly (four-cylinder engine) - backlash check	21
Balancer assembly (four-cylinder engine) - installation	27
Balancer assembly (four-cylinder engine) - removal	13
CHECK ENGINE light	See Chapter 6
Crankshaft - inspection	19
Crankshaft - installation and main bearing oil clearance check	24
Crankshaft - removal	15
Cylinder head - cleaning and inspection	10
Cylinder head - disassembly	9
Cylinder head - reassembly	12
Cylinder compression check	3
Cylinder honing	17
Engine block - cleaning and inspection	16
Engine overhaul - disassembly sequence	8
Engine overhaul - reassembly sequence	22
Engine rebuilding alternatives	7
Engine - removal and installation	6
Engine removal - methods and precautions	5
General information - engine overhaul	1
Initial start-up and break-in after overhaul	28
Oil pressure check	2
Main and connecting rod bearings - inspection and selection	20
Pistons/connecting rods - inspection	18
Pistons/connecting rods - installation and rod bearing oil clearance check	26
Pistons/connecting rods - removal	14
Piston rings - installation	23
Rear main oil seal installation	25
Vacuum gauge diagnostic checks	4
Valves - servicing	11

Specifications

Four-cylinder engine

General

Engine designation	5S-FE
Displacement	134 cubic inches (2.2 liters)
Cylinder compression pressure	
Standard	178 psi
Minimum	142 psi
Oil pressure (engine warm)	
At idle	4.3 psi minimum
At 3000 rpm	36 to 71 psi

Cylinder head

Warpage limits	
Block surface	0.0020 inch
Manifold surfaces	0.0031 inch

Valves and related components

Valve margin width	
Standard	1/32 to 3/64 inch
Minimum	0.020 inch
Valve stem diameter	
Intake	0.2350 to 0.2356 inch
Exhaust	0.2348 to 0.2354 inch
Valve stem-to-guide clearance	
Intake	
Standard	0.0010 to 0.0024 inch
Service limit	0.0031 inch
Exhaust	
Standard	0.0012 to 0.0026 inch
Service limit	0.0039 inch

Four-cylinder engine (continued)

Valves and related components (continued)

Valve spring
- Out-of-square limit .... 0.079 inch
- Free length .... 1.6122 to 1.6850 inches
- Installed height .... 1.366 inches

Valve lifter
- Diameter .... 1.2191 to 1.2195 inches
- Lifter bore diameter .... 1.2205 to 1.2212 inch
- Lifter-to-bore clearance
  - Standard .... 0.0009 to 0.0020 inch
  - Service limit .... 0.0028 inch

Crankshaft and connecting rods

Connecting rod journal
- Diameter .... 2.0466 to 2.0472 inches
- Taper and out-of-round limits .... 0.0008 inch
- Bearing oil clearance
  - Standard .... 0.0009 to 0.0022 inch
  - Service limit .... 0.0031 inch

Connecting rod side clearance (endplay)
- Standard .... 0.0063 to 0.0123 inch
- Service limit .... 0.0138 inch

Main bearing journal
- Diameter .... 2.1653 to 2.1655 inches
- Taper and out-of-round limits .... 0.0008 inch
- Runout limit .... 0.0024 inch
- Bearing oil clearance (standard)
  - No. 3 (center) main .... 0.0010 to 0.0017 inch
  - All others .... 0.0006 to 0.0013 inch
  - Service limit .... 0.0031 inch

Crankshaft endplay
- Standard .... 0.0008 to 0.0087 inch
- Service limit .... 0.0118 inch
- Thrust washer thickness .... 0.0961 to 0.0980 inch

Balancer assembly

Balance shaft no.1-to-balance shaft no. 2 backlash (off-engine) .... 0.0002 to 0.0024 inch
Balance shaft no. 1-to-crankshaft backlash (on-engine) .... 0.0010 to 0.0035 inch
Balance shaft endplay (thrust clearance)
- Standard .... 0.0024 to 0.0043 inch
- Service Limit .... 0.0043 inch

Engine block

Deck warpage limit .... 0.0020 inch
Cylinder bore diameter
- Standard
  - Mark 1 .... 3.4252 to 3.4256 inches
  - Mark 2 .... 3.4256 to 3.4260 inches
  - Mark 3 .... 3.4260 to 3.4264 inches
- Service limit .... 3.4342 inches
Taper and out-of-round limits .... 0.0008 inch

Pistons and rings

Piston diameter (measured 0.8 inch from top of piston)
- Mark 1 .... 3.4193 to 3.4197 inches
- Mark 2 .... 3.4197 to 3.4201 inches
- Mark 3 .... 3.4201 to 3.4205 inches

Piston-to-bore clearance
- Standard .... 0.0055 to 0.0063 inch
- Service limit .... 0.0071 inch

Piston ring end gap
- No. 1 (top) compression ring
  - Standard .... 0.0106 to 0.0197 inch
  - Service limit .... 0.0433 inch
- No. 2 (middle) compression ring
  - Standard .... 0.0138 to 0.0236 inch
  - Service limit .... 0.0472 inch

Chapter 2 Part C General engine overhaul procedures

Piston ring end gap (continued)
  Oil ring
    Standard .... 0.0079 to 0.0217 inch
    Service limit .... 0.0453 inch
Piston ring groove clearance
  No. 1 (top) compression ring .... 0.0016 to 0.0031 inch
  No. 2 (middle) compression ring .... 0.0012 to 0.0028 inch

Torque specifications*

Ft-lbs (unless otherwise indicated)

Main bearing cap bolts 43
Connecting rod cap nuts
  Step 1 18
  Step 2 .... Tighten an additional 90-degrees (1/4-turn)
Engine balancer assembly to block 36

Note: Refer to Chapter 2 Part A for additional torque specifications

V6 engine

General

Engine designation 1MZ-FE
Displacement .... 183 cubic inches (3.0 liters)
Cylinder compression pressure
  Standard .... 218 psi
  Minimum .... 145 psi
Oil pressure (engine hot)
  At 3000 rpm .... 43 to 78 psi
  At idle .... 4.3 psi minimum

Cylinder head

Warpage limits
  Block surface .... 0.0039 inch
  Manifold surfaces .... 0.0031 inch

Valves and related components

Valve margin width
  Standard .... 0.039 inch
  Minimum .... 0.020 inch
Valve stem diameter
  Intake .... 0.2154 to 0.2159 inch
  Exhaust .... 0.2152 to 0.2157 inch
Valve stem-to-guide clearance
  Intake
    Standard .... 0.001 to 0.0024 inch
    Service limit .... 0.0031 inch
  Exhaust
    Standard .... 0.0012 to 0.0026 inch
    Service limit .... 0.0039 inch
Valve spring
  Out-of-square limit .... 0.079 inch
  Free length .... 1.7913 inches
  Installed height .... 1.331 inches
Valve lifter
  Diameter .... 1.2191 to 1.2195 inches
  Lifter bore diameter .... 1.2205 to 1.2211 inches
  Lifter-to-bore clearance
    Standard .... 0.0009 to 0.0020 inch
    Service limit .... 0.0028 inch

Crankshaft and connecting rods

Connecting rod journal
  Diameter .... 2.0863 to 2.0866 inches
  Taper and out-of-round limits .... 0.0008 inch
  Bearing oil clearance
    Standard .... 0.0015 to 0.0025 inch
    Service limit .... 0.0031 inch
Connecting rod side clearance (endplay)
  Standard .... 0.0059 to 0.0118 inch
  Service limit .... 0.0138 inch

V6 engine (continued)

Crankshaft and connecting rods (continued)

Main bearing journal
  Diameter .... 2.4011 to 2.4016 inches
  Taper and out-of-round limits .... 0.0008 inch
  Bearing oil clearance
    Standard
      Numbers 1 and 4 .... 0.0006 to 0.0014 inch
      Numbers 2 and 3 .... 0.0010 to 0.0019 inch
    Service limit
      Numbers 1 and 4 .... 0.0020 inch
      Numbers 2 and 3 .... 0.0024 inch
Crankshaft endplay
  Standard .... 0.0016 to 0.0095 inch
  Service limit .... 0.0118 inch
  Thrust washer thickness .... 0.0760 to 0.0780 inch

Engine block

Deck warpage limit .... 0.0028 inch
Cylinder bore diameter
  Standard .... 3.4449 to 3.4453 inches
  Service limit .... 3.4457 inches

Pistons and rings

Piston diameter (standard)
  Aisin-made (measured 0.9 inch from top of piston) .... 3.4412 to 3.4416 inches
  Mahle-made (measured 1.6 inch from top of piston) .... 3.4430 to 3.4436 inches
Piston-to-bore clearance
  Standard
    Aisin-made .... 0.0033 to 0.0042 inches
    Mahle-made .... 0.0013 to 0.0023 inches
  Service limit
    Aisin-made .... 0.0051 inch
    Mahle-made .... 0.0031 inch
Piston ring end gap
  No. 1 (top) compression ring
    Standard .... 0.0098 to 0.0138 inch
    Service limit .... 0.0374 inch
  No. 2 (middle) compression ring
    Standard .... 0.0138 to 0.0177 inch
    Service limit .... 0.0413 inch
  Oil ring
    Standard .... 0.0059 to 0.0157 inch
    Service limit .... 0.0394 inch
Piston ring groove clearance
  No. 1 (top) compression ring .... 0.0008 to 0.0028 inch
  No. 2 (middle) compression ring .... 0.0008 to 0.0024 inch

Torque specifications*   Ft-lbs (unless otherwise indicated)

Main bearing cap assembly bolts
  12-point bolts
    Step 1 16
    Step 2 .... Tighten an additional 90-degrees (1/4-turn)
  6-point bolts 20
Connecting rod cap nuts
  Step 1 18
  Step 2 .... Tighten an additional 90-degrees (1/4-turn)
Water seal plate nuts .... 120 in-lbs

*Note: *Refer to Chapter 2 Part B for additional torque specifications*

1 General information - engine overhaul

Included in this portion of Chapter 2 are the general overhaul procedures for the cylinder head and internal engine components.

The information ranges from advice concerning preparation for an overhaul and the purchase of replacement parts to detailed, step-by-step procedures covering Removal and installation of internal engine components and the inspection of parts.

The following Sections have been written based on the assumption that the engine has been removed from the vehicle. For information concerning in-vehicle engine repair, as well as removal and installation of the external components necessary for the overhaul, see Chapter 2A (four-cylinder models) or 2B (V6 models).

The Specifications included in this Part are only those necessary for the inspection and overhaul procedures which follow. Refer

Chapter 2 Part C General engine overhaul procedures

to Chapter 2, Part A or Part B for additional Specifications.

It's not always easy to determine when, or if, an engine should be completely overhauled, as a number of factors must be considered.

High mileage is not necessarily an indication that an overhaul is needed, while low mileage doesn't preclude the need for an overhaul. Frequency of servicing is probably the most important consideration. An engine that's had regular and frequent oil and filter changes, as well as other required maintenance, will most likely give many thousands of miles of reliable service. Conversely, a neglected engine may require an overhaul very early in its life.

Excessive oil consumption is an indication that piston rings, valve seals and/or valve guides are in need of attention. Make sure that oil leaks aren't responsible before deciding that the rings and/or guides are bad. Perform an oil pressure check and a cylinder compression check to determine the extent of the work required (see Sections 2 and 3). Also check the vacuum readings under various conditions (see Section 4).

Loss of power, rough running, knocking or metallic engine noises, excessive valve train noise and high fuel consumption rates may also point to the need for an overhaul, especially if they're all present at the same time. If a complete tune-up doesn't remedy the situation, major mechanical work is the only solution.

An engine overhaul involves restoring the internal parts to the specifications of a new engine. During an overhaul, the piston rings are replaced and the cylinder walls are reconditioned (re-bored and/or honed). If a re-bore is done by an automotive machine shop, new oversize pistons will also be installed. The main bearings, connecting rod bearings and camshaft bearings are generally replaced with new ones and, if necessary, the crankshaft may be reground to restore the journals. Generally, the valves are serviced as well, since they're usually in less-than-perfect condition at this point. While the engine is being overhauled, other components, such as the starter and alternator, can be rebuilt as well. The end result should be a like new engine that will give many trouble free miles. **Note:** *Critical cooling system components such as the hoses, drivebelts, thermostat and water pump should be replaced with new parts when an engine is overhauled. The radiator should be checked carefully to ensure that it isn't clogged or leaking (see Chapter 3). If you purchase a rebuilt engine or short block, some rebuilders will not warranty their engines unless the radiator has been professionally flushed. Also, be sure to check the oil pump carefully as described in Chapter 2A or 2B.*

Before beginning the engine overhaul, read through the entire procedure to familiarize yourself with the scope and requirements of the job. Overhauling an engine isn't difficult, but it is time-consuming. Plan on the vehicle being tied up for a minimum of two weeks, especially if parts must be taken to an automotive machine shop for repair or reconditioning. Check on availability of parts and make sure that any necessary special tools and equipment are obtained in advance. Most work can be done with typical hand tools, although a number of precision measuring tools are required for inspecting parts to determine if they must be replaced. Often an automotive machine shop will handle the inspection of parts and offer advice concerning reconditioning and replacement. **Note:** *Always wait until the engine has been completely disassembled and all components, especially the engine block, have been inspected before deciding what service and repair operations must be performed by an automotive machine shop. Since the block's condition will be the major factor to consider when determining whether to overhaul the original engine or buy a rebuilt one, never purchase parts or have machine work done on other components until the block has been thoroughly inspected. As a general rule, time is the primary cost of an overhaul, so it doesn't pay to install worn or substandard parts.*

As a final note, to ensure maximum life and minimum trouble from a rebuilt engine, everything must be assembled with care in a spotlessly-clean environment.

2 Oil pressure check

Refer to illustrations 2.2a and 2.2b

1 Low engine oil pressure can be a sign of an engine in need of rebuilding. A "low oil pressure" indicator (often called an "idiot light") is not a test of the oiling system. Such indicators only come on when the oil pressure is dangerously low. Even a factory oil pressure gauge in the instrument panel is only a relative indication, although much better for driver information than a warning light. A better test is with a mechanical (not electrical) oil pressure gauge.

2 Find the oil pressure indicator sending unit **(see illustrations)**.

3 Remove the oil pressure sending unit and install a fitting which will allow you to directly connect your hand-held, mechanical oil pressure gauge. Use Teflon tape or sealant on the threads of the adapter and the fitting on the end of your gauge's hose.

4 Connect an accurate tachometer to the engine, according to the tachometer manufacturer's instructions.

5 Check the oil pressure with the engine running (normal operating temperature) at the specified engine speed, and compare it to this Chapter's Specifications. If it's extremely low, the bearings and/or oil pump are probably worn out.

3 Cylinder compression check

Refer to illustration 3.6

1 A compression check will tell you what mechanical condition the upper end of your engine (pistons, rings, valves, head gaskets) is in. Specifically, it can tell you if the compression is down due to leakage caused by worn piston rings, defective valves and seats or a blown head gasket. **Note:** *The engine must be at normal operating temperature and the battery must be fully charged for this check.*

2 Begin by cleaning the area around the spark plugs before you remove them (compressed air should be used, if available). The idea is to prevent dirt from getting into the cylinders as the compression check is being done.

3 Remove all of the spark plugs from the engine (see Chapter 1).

4 Block the throttle wide open.

5 Disable the ignition system by disconnecting the primary (low voltage) electrical connectors from the coil packs (see Chapter 5). The fuel pump circuit should also be disabled (see Chapter 4).

2.2a The oil pressure can be checked by removing the sending unit and installing a pressure gauge in its place (four-cylinder model shown)

2.2b On V6 models, the oil pressure sending unit (arrow) is located at the front of the engine, above the crankshaft position sensor

3.6 A compression gauge with a threaded fitting for the spark plug hole is preferred over the type that requires hand pressure to maintain the seal - be sure to open the throttle valve as far as possible during the compression check! - arrow indicates number 1 plug hole

4.4 A simple vacuum gauge can be very handy in diagnosing engine condition and performance - arrow indicates source of manifold vacuum

6 Install the compression gauge in the spark plug hole **(see illustration)**.

7 Crank the engine over at least seven compression strokes and watch the gauge. The compression should build up quickly in a healthy engine. Low compression on the first stroke, followed by gradually increasing pressure on successive strokes, indicates worn piston rings. A low compression reading on the first stroke, which doesn't build up during successive strokes, indicates leaking valves or a blown head gasket (a cracked head could also be the cause). Deposits on the undersides of the valve heads can also cause low compression. Record the highest gauge reading obtained.

8 Repeat the procedure for the remaining cylinders and compare the results to this Chapter's Specifications.

9 Add some engine oil (about three squirts from a plunger-type oil can) to each cylinder, through the spark plug hole, and repeat the test.

10 If the compression increases after the oil is added, the piston rings are definitely worn. If the compression doesn't increase significantly, the leakage is occurring at the valves or head gasket. Leakage past the valves may be caused by burned valve seats and/or faces or warped, cracked or bent valves.

11 If two adjacent cylinders have equally low compression, there's a strong possibility that the head gasket between them is blown. The appearance of coolant in the combustion chambers or the crankcase would verify this condition.

12 If one cylinder is slightly lower than the others, and the engine has a slightly rough idle, a worn lobe on the camshaft could be the cause.

13 If the compression is unusually high, the combustion chambers are probably coated with carbon deposits. If that's the case, the cylinder head(s) should be removed and decarbonized.

14 If compression is way down or varies greatly between cylinders, it would be a good idea to have a leak-down test performed by an automotive repair shop. This test will pinpoint exactly where the leakage is occurring and how severe it is.

4 Vacuum gauge diagnostic checks

Refer to illustrations 4.4 and 4.6

A vacuum gauge provides valuable information about what is going on in the engine at a low-cost. You can check for worn rings or cylinder walls, leaking head or intake manifold gaskets, incorrect carburetor adjustments, restricted exhaust, stuck or burned valves, weak valve springs, improper ignition or valve timing and ignition problems.

Unfortunately, vacuum gauge readings are easy to misinterpret, so they should be used in conjunction with other tests to confirm the diagnosis.

Both the absolute readings and the rate of needle movement are important for accurate interpretation. Most gauges measure vacuum in inches of mercury (in-Hg). The following references to vacuum assume the diagnosis is being performed at sea level. As elevation increases (or atmospheric pressure decreases), the reading will decrease. For every 1,000 foot increase in elevation above approximately 2000 feet, the gauge readings will decrease about one inch of mercury.

Connect the vacuum gauge directly to intake manifold vacuum, not to ported (throttle body) vacuum **(see illustration)**. Be sure no hoses are left disconnected during the test or false readings will result.

Before you begin the test, allow the engine to warm up completely. Block the wheels and set the parking brake. With the transmission in Park, start the engine and allow it to run at normal idle speed. **Warning:** *Keep your hands and the vacuum gauge clear of the fans.*

Read the vacuum gauge; an average, healthy engine should normally produce about 17 to 22 in-Hg with a fairly steady needle **(see illustration)**. Refer to the following vacuum gauge readings and what they indicate about the engine's condition:

1 A low steady reading usually indicates a leaking gasket between the intake manifold and cylinder head(s) or throttle body, a leaky vacuum hose, late ignition timing or incorrect camshaft timing. Check ignition timing with a timing light and eliminate all other possible causes, utilizing the tests provided in this Chapter before you remove the timing chain cover to check the timing marks.

2 If the reading is three to eight inches below normal and it fluctuates at that low reading, suspect an intake manifold gasket leak at an intake port or a faulty fuel injector.

3 If the needle has regular drops of about two-to-four inches at a steady rate, the valves are probably leaking. Perform a compression check or leak-down test to confirm this.

4 An irregular drop or down-flick of the needle can be caused by a sticking valve or an ignition misfire. Perform a compression check or leak-down test and read the spark plugs.

5 A rapid vibration of about four in-Hg vibration at idle combined with exhaust smoke indicates worn valve guides. Perform a leak-down test to confirm this. If the rapid vibration occurs with an increase in engine speed, check for a leaking intake manifold gasket or head gasket, weak valve springs, burned valves or ignition misfire.

6 A slight fluctuation, say one inch up and down, may mean ignition problems. Check all the usual tune-up items and, if necessary, run the engine on an ignition analyzer.

7 If there is a large fluctuation, perform a compression or leak-down test to look for a weak or dead cylinder or a blown head gasket.

8 If the needle moves slowly through a

Chapter 2 Part C General engine overhaul procedures

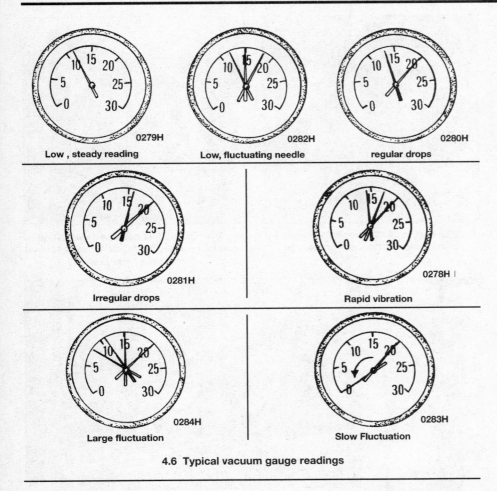

4.6 Typical vacuum gauge readings

wide range, check for a clogged PCV system, incorrect idle fuel mixture, carburetor/throttle body or intake manifold gasket leaks.

9 Check for a slow return after revving the engine by quickly snapping the throttle open until the engine reaches about 2,500 rpm and let it shut. Normally the reading should drop to near zero, rise above normal idle reading (about 5 in-Hg over) and then return to the previous idle reading. If the vacuum returns slowly and doesn't peak when the throttle is snapped shut, the rings may be worn. If there is a long delay, look for a restricted exhaust system (often the muffler or catalytic converter). An easy way to check this is to temporarily disconnect the exhaust ahead of the suspected part and redo the test.

5 Engine removal - methods and precautions

If you've decided that an engine must be removed for overhaul or major repair work, several preliminary steps should be taken.

Locating a suitable place to work is extremely important. Adequate work space, along with storage space for the vehicle, will be needed. If a shop or garage isn't available, at the very least a flat, level, clean work surface made of concrete or asphalt is required.

Cleaning the engine compartment and engine before beginning the removal procedure will help keep tools clean and organized.

An engine hoist or A-frame will also be necessary. Make sure the equipment is rated in excess of the combined weight of the engine and transaxle. Safety is of primary importance, considering the potential hazards involved in lifting the engine out of the vehicle.

If the engine is being removed by a novice, a helper should be available. Advice and aid from someone more experienced would also be helpful. There are many instances when one person cannot simultaneously perform all of the operations required when lifting the engine out of the vehicle.

Plan the operation ahead of time. Arrange for or obtain all of the tools and equipment you'll need prior to beginning the job. Some of the equipment necessary to perform engine removal and installation safely and with relative ease are (in addition to an engine hoist) a heavy duty floor jack, complete sets of wrenches and sockets as described in the front of this manual, wooden blocks and plenty of rags and cleaning solvent for mopping up spilled oil, coolant and gasoline. If the hoist must be rented, make sure that you arrange for it in advance and perform all of the operations possible without it beforehand. This will save you money and time.

Plan for the vehicle to be out of use for quite a while. A machine shop will be required to perform some of the work which the do-it-yourselfer can't accomplish without special equipment. These shops often have a busy schedule, so it would be a good idea to consult them before removing the engine in order to accurately estimate the amount of time required to rebuild or repair components that may need work.

Always be extremely careful when removing and installing the engine. Serious injury can result from careless actions. Plan ahead, take your time and a job of this nature, although major, can be accomplished successfully.

6 Engine - removal and installation

Refer to illustrations 6.7, 6.16, 6.18, 6.20, 6.21 and 6.26
Note: *Read through the entire Section before beginning this procedure. The manufacturer recommends removing the engine and transaxle from the top as a unit, then separating the engine from the transaxle on the shop floor. If the transaxle is not being serviced, it is possible to leave the transaxle in the vehicle and remove the engine from the top by itself, by removing the crankshaft pulley and tilting up the timing belt end of the engine for clearance. This second method may not be possible on V6 models, which have limited clearance between the timing belt end of the engine and the right side of the body.*

Removal

Warning 1: *The models covered by this manual are equipped with Supplemental Restraint systems (SRS), more commonly known as airbags. Always disable the airbag system before working in the vicinity of the impact sensors, steering column or instrument panel to avoid the possibility of accidental deployment of the airbag, which could cause personal injury (see Chapter 12).*
Warning 2: *Gasoline is extremely flammable, so take extra precautions when you work on any part of the fuel system. Don't smoke or allow open flames or bare light bulbs near the work area, and don't work in a garage where a natural gas-type appliance (such as a water heater or a clothes dryer) with a pilot light is present. Since gasoline is carcinogenic, wear latex gloves when there's a possibility of being exposed to fuel, and, if you spill any fuel on your skin, rinse it off immediately with soap and water. Mop up any spills immediately and do not store fuel-soaked rags where they could ignite. The fuel system is under constant pressure, so, if any fuel lines are to be disconnected, the fuel pressure in the system must be relieved first (see Chapter 4 for more information). When you perform any*

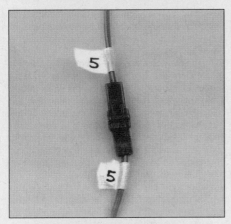
6.7 Label both ends of each wire and hose before disconnecting it

6.16 Unbolt the A/C compressor and use wire or rope to tie it out of the way

6.18 On models equipped with an oil cooler, remove the oil filter, cooler (arrow) and the cooler hoses

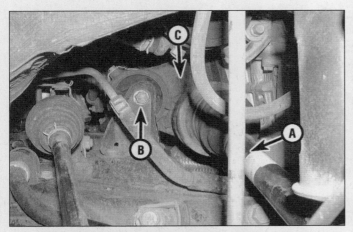
6.20 Remove the passenger-side driveaxle (A), the engine mount through-bolt (B), and the firewall-side engine mount bracket (C)

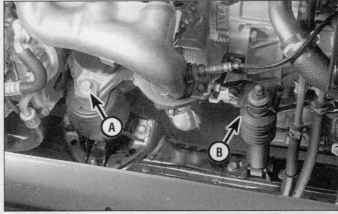
6.21 Remove the radiator-side engine mount bolt (A) - If the transaxle is coming out with the engine, unbolt the engine shock absorber (B)

kind of work on the fuel system, wear safety glasses and have a Class B type fire extinguisher on hand.

1 Relieve the fuel system pressure (see Chapter 4).
2 Disconnect the negative cable from the battery. **Caution:** *If the stereo in your vehicle is equipped with an anti-theft system, make sure you have the correct activation code before disconnecting the battery.*
3 Remove the battery and battery tray.
4 Place protective covers on the fenders and cowl and remove the hood (see Chapter 11).
5 Remove the air cleaner assembly (see Chapter 4). Remove the inner fender splash shields.
6 Raise the vehicle and support it securely on jackstands. Drain the cooling system and engine oil and remove the drivebelts (see Chapter 1).
7 Clearly label, then disconnect all vacuum lines, coolant and emissions hoses, wiring harness connectors, ground straps and fuel lines. Masking tape and/or a touch up paint applicator work well for marking items **(see illustration)**. Take instant photos or sketch the locations of components and brackets.
8 Remove the windshield washer tank and coolant reservoir tank.
9 Remove the cooling fan(s), shroud(s) and radiator (see Chapter 3).
10 Disconnect the heater hoses.
11 Release the residual fuel pressure in the tank by removing the gas cap, then undo the fuel lines connecting the engine to the chassis (see Chapter 4). Plug or cap all open fittings.
12 Disconnect the throttle linkage, transmission Throttle Valve (TV) linkage and speed control cable, if equipped, from the engine (see Chapter 4).
13 Disconnect the engine harness plugs and clips from the front and left side of the engine.
14 Disconnect the wire harness clips at the transaxle end of the engine and pull back the wiring on all sides to clear the engine.
15 On power steering equipped vehicles, unbolt the power steering pump. If clearance allows, tie the pump aside without disconnecting the hoses. If necessary, remove the pump (see Chapter 10).

16 On air-conditioned models, unbolt the compressor and set it aside **(see illustration)**. Do not disconnect the refrigerant hoses.
17 Detach the exhaust pipe(s) from the manifold(s) (see Chapter 4).
18 On four-cylinder models equipped with a oil cooler (small canister mounted between the oil filter and the block), remove the oil filter, hoses connected to the cooler, and the oil cooler **(see illustration)**. On Solara models, remove the front suspension upper brace between the tops of the front shock towers.
19 Attach a lifting sling to the engine. Position a hoist and connect the sling to it. Take up the slack until there is slight tension on the hoist.
20 Remove the passenger-side driveaxle (see Chapter 8), and remove the firewall-side engine mount bracket from the block and the through-bolt from the engine mount **(see illustration)**.
21 Remove the bolt securing the radiator-side engine mount to the chassis **(see illustration)**.
22 On automatic transaxle equipped models, detach the torque converter dust shield

Chapter 2 Part C General engine overhaul procedures

6.26 Lift the engine/transaxle high enough to clear the vehicle, then move it away and lower the hoist - this four-cylinder is being removed with the transaxle still attached

from the lower bellhousing. Remove the torque converter-to-driveplate fasteners (see Chapter 7B) and push the converter back slightly into the bellhousing.

23 If only the engine is being removed, remove the engine-to-transaxle bolts and separate the engine from the transaxle. The torque converter should remain in the transaxle. **Note:** *If the transaxle is to be removed at the same time, the driver's side engine mount should be disconnected, along with any wires, cables or hoses connected to the transaxle. The engine-to-transaxle bolts should remain in place at this time.*

24 Recheck to be sure nothing except the mounts are still connecting the engine to the vehicle or to the transaxle. Disconnect and label anything still remaining.

25 Support the transaxle with a floor jack. Place a block of wood on the jack head to prevent damage to the transaxle. Remove the bolts from the engine mounts, leaving those attached to the transaxle in place. **Warning:** *Do not place any part of your body under the engine/transaxle when it's supported only by a hoist or other lifting device.*

26 Slowly lift the engine (or engine/transaxle) out of the vehicle **(see illustration)**. It may be necessary to pry the mounts away from the frame brackets.

27 Move the engine away from the vehicle and carefully lower the hoist until the engine can be set on the floor; or remove the flywheel/driveplate and mount the engine on an engine stand. **Note:** *On automatic transaxle-equipped models, mark the front and rear spacer plates and keep them with the driveplate.*

Installation

28 Check the engine/transaxle mounts. If they're worn or damaged, replace them.
29 On manual transaxle equipped models, inspect the clutch components (see Chapter 8) and on automatic models inspect the converter seal and bushing.
30 On manual transaxle equipped vehicles, apply a dab of high temperature grease to the pilot bearing. On automatic transaxle equipped models, apply a dab of grease to the nose of the converter.
31 Carefully guide the transaxle into place, following the procedure outlined in Chapter 7B. **Caution:** *Do not use the bolts to force the engine and transaxle into alignment. It may crack or damage major components.*
32 Install the engine-to-transaxle bolts and tighten them securely.
33 Attach the hoist to the engine and carefully lower the engine/transaxle assembly into the engine compartment. **Note:** *If the engine was removed with the transaxle remaining in the car, lower the engine into the car until an assistant can help you line up the dowels pins on the block with the transaxle. Some twisting and angling of the engine and/or the transaxle will be necessary to secure proper alignment of the two.*
34 Install the mount bolts and tighten them securely.
35 Reinstall the remaining components and fasteners in the reverse order of removal.
36 Add coolant, oil, power steering and transmission fluids as needed (see Chapter 1).
37 Run the engine and check for proper operation and leaks. Shut off the engine and recheck the fluid levels.

7 Engine rebuilding alternatives

The do-it-yourselfer is faced with a number of options when performing an engine overhaul. The decision to replace the engine block, piston/connecting rod assemblies and crankshaft depends on a number of factors, with the number one consideration being the condition of the block. Other considerations are cost, access to machine shop facilities, parts availability, time required to complete the project and the extent of prior mechanical experience on the part of the do-it-yourselfer.

Some of the rebuilding alternatives include:

Individual parts - If the inspection procedures reveal that the engine block and most engine components are in reusable condition, purchasing individual parts may be the most economical alternative. The block, crankshaft and piston/connecting rod assemblies should all be inspected carefully. Even if the block shows little wear, the cylinder bores should be surface honed.

Short block - A short block consists of an engine block with a crankshaft and piston/connecting rod assemblies already installed. All new bearings are incorporated and all clearances will be correct. The existing camshaft, valve train components, cylinder head(s) and external parts can be bolted to the short block with little or no machine shop work necessary.

Long block - A long block consists of a short block plus an oil pump, oil pan, cylinder head(s), valve cover(s), camshaft and valve train components, timing sprockets and chain or gears and timing cover. All components are installed with new bearings, seals and gaskets incorporated throughout. The installation of manifolds and external parts is all that's necessary.

Used engine assembly - While overhaul provides the best assurance of a like-new engine, used engines available from wrecking yards and importers are often a very simple and economical solution. Many used engines come with warranties, but always give any engine a thorough diagnostic check-out before purchase. If possible, have the seller run the engine, ether in the vehicle or on a test stand so you can be sure it runs smoothly with no knocking or other noises. Check compression, vacuum and also for signs of oil leakage.

Give careful thought to which alternative is best for you and discuss the situation with local automotive machine shops, auto parts dealers and experienced rebuilders before ordering or purchasing replacement parts.

8 Engine overhaul - disassembly sequence

Refer to illustrations 8.5a and 8.5b

1 It's much easier to disassemble and work on the engine if it's mounted on a portable engine stand. A stand can often be rented quite cheaply from an equipment rental yard. Before the engine is mounted on a stand, the flywheel/driveplate and rear oil seal retainer should be removed from the engine.
2 If a stand isn't available, it's possible to disassemble the engine with it blocked up on the floor. Be extra careful not to tip or drop the engine when working without a stand.
3 If you're going to obtain a rebuilt engine, all external components must come off first, to be transferred to the replacement engine, just as they will if you're doing a complete engine overhaul yourself. These include:

Alternator and brackets
Emissions control components
Coil packs, spark plug wires and spark plugs
Thermostat and housing cover
Water pump
EFI components

2C-10 Chapter 2 Part C General engine overhaul procedures

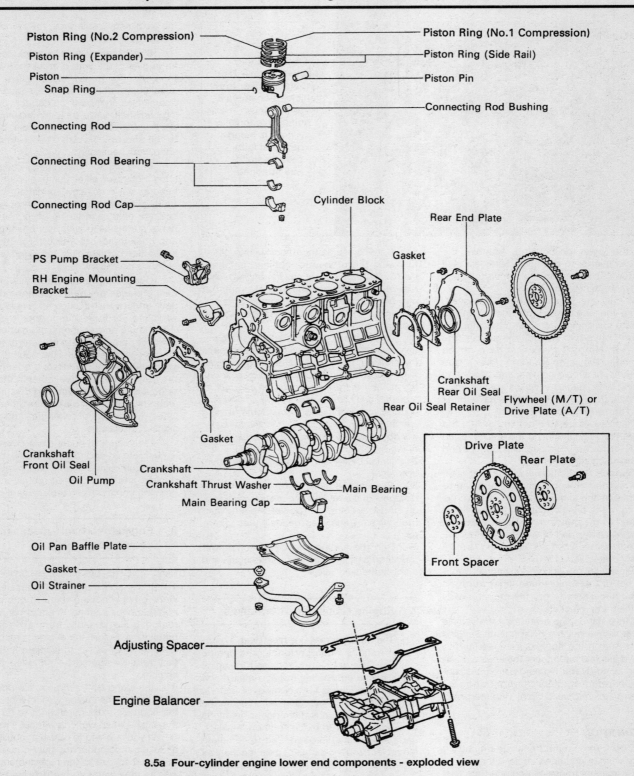

8.5a Four-cylinder engine lower end components - exploded view

Intake/exhaust manifolds
Oil filter
Engine mounts
Clutch and flywheel/driveplate
Engine rear plate

Note: When removing the external components from the engine, pay close attention to details that may be helpful or important during installation. Note the installed position of gaskets, seals, spacers, pins, brackets, washers, bolts and other small items.

4 If you're obtaining a short block, which consists of the engine block, crankshaft, pistons and connecting rods all assembled, then the cylinder head(s), oil pan and oil pump will have to be removed as well. See *Engine rebuilding alternatives* for additional information regarding the different possibilities to be considered.

5 If you're planning a complete overhaul, the engine must be disassembled and the

Chapter 2 Part C General engine overhaul procedures

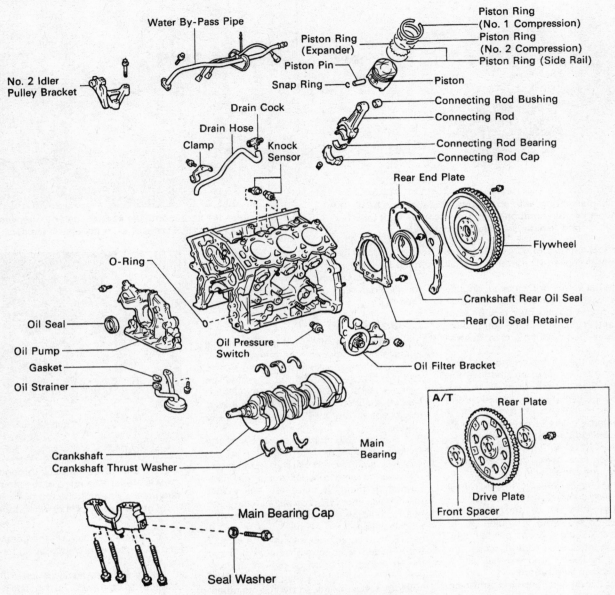

8.5b V6 engine lower end components - exploded view

internal components removed in the following order **(see illustrations)**.

Intake and exhaust manifolds
Valve cover(s)
Timing belt covers
Timing belt and sprockets
Cylinder head(s)
Oil pan
Oil pump
Engine balancer assembly (four-cylinder engines)
Piston/connecting rod assemblies
Crankshaft rear oil seal retainer
Crankshaft and main bearings

6 Before beginning the disassembly and overhaul procedures, make sure the following items are available. Also, refer to Section 22 for a list of tools and materials needed for engine reassembly.

Common hand tools
Small cardboard boxes or plastic bags for storing parts
Gasket scraper
Ridge reamer
Vibration damper puller
Micrometers
Telescoping gauges
Dial indicator set
Valve spring compressor
Cylinder surfacing hone
Piston ring groove cleaning tool
Electric drill motor
Tap and die set
Wire brushes
Oil gallery brushes
Cleaning solvent

9 Cylinder head - disassembly

Refer to illustrations 9.2, 9.3 and 9.4

Note: *New and rebuilt cylinder heads are commonly available for most engines at dealerships and auto parts stores. Due to the fact that some specialized tools are necessary for the disassembly and inspection procedures, and replacement parts may not be readily available, it may be more practical and economical for the home mechanic to purchase replacement head(s) rather than taking the time to disassemble, inspect and recondition the original(s).*

1 Cylinder head disassembly involves removal of the intake and exhaust valves and related components. It's assumed that the

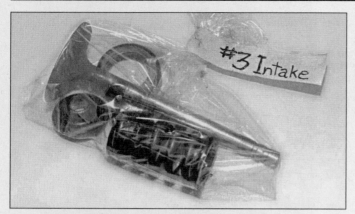

9.2 A small plastic bag, with an appropriate label, can be used to store the valve train components so they can be kept together and reinstalled in the correct guide

9.3 Compress the spring until the keepers can be removed with a small magnetic screwdriver or needle-nose pliers

lifters or rocker arms and camshaft(s) have already been removed (see Part A or B as needed).

2 Before the valves are removed, arrange to label and store them, along with their related components, so they can be kept separate and reinstalled in the same valve guides they are removed from **(see illustration)**.

3 Compress the springs on the first valve with a spring compressor and remove the keepers **(see illustration)**. Carefully release the valve spring compressor and remove the retainer, the spring and the spring seat (if used). **Caution:** *Be very careful not to nick or otherwise damage the lifter bores when compressing the valve springs.* **Note:** *If your spring compressor does not have an end such as the one shown with cutouts on the side, an adapter is available to use with a standard spring compressor.*

4 Pull the valve out of the head, then remove the oil seal from the guide. If the valve binds in the guide (won't pull through), push it back into the head and deburr the area around the keeper groove with a fine file or whetstone **(see illustration)**.

5 Repeat the procedure for the remaining valves. Remember to keep all the parts for each valve together so they can be reinstalled in the same locations.

6 Once the valves and related components have been removed and stored in an organized manner, the head should be thoroughly cleaned and inspected. If a complete engine overhaul is being done, finish the engine disassembly procedures before beginning the cylinder head cleaning and inspection process.

10 Cylinder head - cleaning and inspection

Refer to illustrations 10.12, 10.14, 10.15, 10.16, 10.17 and 10.18

1 Thorough cleaning of the cylinder head(s) and related valve train components, followed by a detailed inspection, will enable you to decide how much valve service work must be done during the engine overhaul. **Note:** *If the engine was severely overheated, the cylinder head is probably warped (see Step 12).*

Cleaning

2 Scrape all traces of old gasket material and sealing compound off the head gasket, intake manifold and exhaust manifold sealing surfaces. Be very careful not to gouge the cylinder head. Special gasket removal solvents that soften gaskets and make removal much easier are available at auto parts stores.

3 Remove all built up scale from the coolant passages.

4 Run a stiff wire brush through the various holes to remove deposits that may have formed in them. If there are heavy rust deposits in the water passages, the bare head should be professionally cleaned at a machine shop.

5 Run an appropriate size tap into each of the threaded holes to remove corrosion and thread sealant that may be present. If compressed air is available, use it to clear the holes of debris produced by this operation. **Warning:** *Wear eye protection when using compressed air!*

6 Clean the exhaust and intake manifold stud threads with a wire brush.

7 Clean the cylinder head with solvent and dry it thoroughly. Compressed air will speed the drying process and ensure that all holes and recessed areas are clean. **Note:** *Decarbonizing chemicals are available and may prove very useful when cleaning cylinder heads and valve train components. They are very caustic and should be used with caution. Be sure to follow the instructions on the container.*

8 Clean the lifters with solvent and dry them thoroughly (don't mix them up during the cleaning process). Compressed air will speed the drying process and can be used to clean out the oil passages.

9 Clean all the valve springs, spring seats, keepers and retainers with solvent and dry

9.4 If the valve won't pull through the guide, deburr the edge of the stem end and the area around the top of the keeper groove with a file or whetstone

them thoroughly. Work on the components from one valve at a time to avoid mixing up the parts.

10 Scrape off any heavy deposits that may have formed on the valves, then use a motorized wire brush to remove deposits from the valve heads and stems. Again, make sure the valves don't get mixed up.

Inspection

Note: *Be sure to perform all of the following inspection procedures before concluding that machine shop work is required. Make a list of the items that need attention. The inspection procedures for the lifters and the camshafts, can be found in Part A or B.*

Cylinder head

11 Inspect the head very carefully for cracks, evidence of coolant leakage and other damage. If cracks are found, check with an automotive machine shop concerning repair. If repair isn't possible, a new cylinder head should be obtained.

12 Using a straightedge and feeler gauge, check the head gasket mating surface for warpage **(see illustration)**. If the warpage

Chapter 2 Part C General engine overhaul procedures

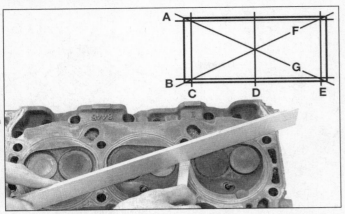

10.12 Check the cylinder head gasket surfaces for warpage by trying to slip a feeler gauge under the precision straightedge (see the Specifications for the maximum warpage allowed and use a feeler gauge of that thickness)

10.14 A dial indicator can be used to measure valve stem-to-guide clearance - move the valve stem back and forth as shown

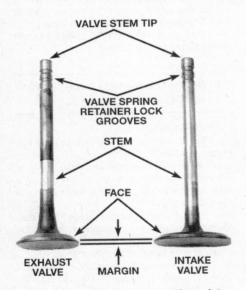

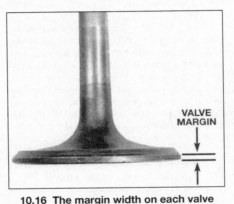

10.16 The margin width on each valve must be as specified (if no margin exists, the valve cannot be re-used)

10.17 Measure the free length of each valve spring with a dial or vernier caliper

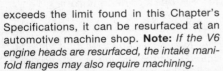

10.15 Check for valve wear at the points shown here

exceeds the limit found in this Chapter's Specifications, it can be resurfaced at an automotive machine shop. **Note:** *If the V6 engine heads are resurfaced, the intake manifold flanges may also require machining.*

13 Examine the valve seats in each of the combustion chambers. If they're pitted, cracked or burned, the head will require valve service that's beyond the scope of the home mechanic.

14 Check the valve stem-to-guide clearance with a dial indicator **(see illustration)**. Also, check the valve stem deflection with a dial indicator attached securely to the head. The valve must be in the guide and approximately 1/16-inch off the seat. The total valve stem movement indicated by the gauge needle must be noted, then divided by two to obtain the actual clearance value. If it exceeds the stem-to-guide clearance limit found in this Chapter's Specifications, the valve guides should be replaced. After this is done, if there's still some doubt regarding the condition of the valve guides they should be checked by an automotive machine shop (the cost should be minimal).

Valves

15 Carefully inspect each valve face for uneven wear, deformation, cracks, pits and burned areas **(see illustration)**. Check the valve stem for scuffing and galling and the neck for cracks. Rotate the valve and check for any obvious indication that it's bent. Look for pits and excessive wear on the end of the stem. The presence of any of these conditions indicates the need for valve service by an automotive machine shop.

16 Measure the margin width on each valve **(see illustration)**. Any valve with a margin narrower than that listed in this Chapter's Specifications will have to be replaced with a new one.

Valve components

17 Check each valve spring for wear (on the ends) and pits. Measure the free length and compare it to this Chapter's Specifications **(see illustration)**. Any springs that are shorter than specified have sagged and should not be re-used. The tension of all springs should be pressure checked with a special fixture before deciding that they're suitable for use in a rebuilt engine (take the springs to an automotive machine shop for this check).

18 Stand each spring on a flat surface and check it for squareness **(see illustration)**. If any of the springs are distorted or sagged, replace all of them with new parts.

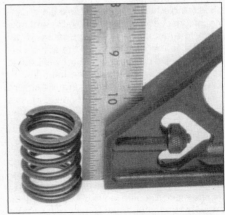

10.18 Check each valve spring for squareness

12.3 Gently tap the valve seals (arrow) into place with a deep socket and hammer - on four-cylinder engines, the intake seals are gray and exhaust seals are black, while on V6 engines the intake seals are light brown and the exhaust seals are gray

12.6 The small valve stem keepers are easier to position when coated with grease

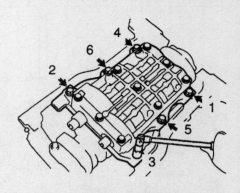

13.2 Balancer assembly bolt LOOSENING sequence

19 Check the spring retainers and keepers for obvious wear and cracks. Any questionable parts should be replaced with new ones, as extensive damage will occur if they fail during engine operation.
20 Any damaged or excessively worn parts must be replaced with new ones.
21 If the inspection process indicates that the valve components are in generally poor condition and worn beyond the limits specified, which is usually the case in an engine that's being overhauled, reassemble the valves in the cylinder head and refer to Section 12 for valve servicing recommendations.

11 Valves - servicing

1 Because of the complex nature of the job and the special tools and equipment needed, servicing of the valves, the valve seats and the valve guides, commonly known as a valve job, should be done by a professional.
2 The home mechanic can remove and disassemble the head(s), do the initial cleaning and inspection, then reassemble and take them to a dealer service department or an automotive machine shop for the actual service work. Doing the inspection will enable you to see what condition the head(s) and valvetrain components are in and will ensure that you know what work and new parts are required when dealing with an automotive machine shop.
3 The dealer service department, or automotive machine shop, will remove the valves and springs, recondition or replace the valves and valve seats, recondition the valve guides, check and replace the valve springs, spring retainers and keepers (as necessary), replace the valve seals with new ones, reassemble the valve components and make sure the installed spring height is correct. The cylinder head gasket surface will also be resurfaced if it's warped.
4 After the valve job has been performed by a professional, the head(s) will be in like new condition. When the heads are returned, be sure to clean them again before installation on the engine to remove any metal particles and abrasive grit that may still be present from the valve service or head resurfacing operations. Use compressed air, if available, to blow out all the oil holes and passages.

12 Cylinder head - reassembly

Refer to illustrations 12.3 and 12.6

1 Regardless of whether or not the head was sent to an automotive machine shop for valve servicing, make sure it's clean before beginning reassembly.
2 If the head was sent out for valve servicing, the valves and related components will already be in place. Begin the reassembly procedure with Step 8.
3 Install new seals on each of the valve guides. **Note:** *Intake and exhaust valves require different seals - DO NOT mix them up!* Gently tap each intake valve seal into place until it's seated on the guide **(see illustration)**. **Caution:** *Don't hammer on the valve seals once they're seated or you may damage them. Don't twist or cock the seals during installation or they won't seat properly on the valve stems.*
4 Beginning at one end of the head, lubricate and install the first valve. Apply moly-base grease or clean engine oil to the valve stem.
5 Drop the spring seat or shim(s) over the valve guide and set the valve spring and retainer in place.
6 Compress the springs with a valve spring compressor and carefully install the keepers in the upper groove, then slowly release the compressor and make sure the keepers seat properly. Apply a small dab of grease to each keeper to hold it in place if necessary **(see illustration)**.
7 Repeat the procedure for the remaining valves. Be sure to return the components to their original locations - don't mix them up!
8 Check the valve spring installed height with a dial or vernier caliper.

13 Balancer assembly (four-cylinder engine) - removal

Refer to illustration 13.2

1 Remove the oil pan, pick-up tube and oil pump (see Chapter 2A).
2 Loosen the balancer mounting bolts in the recommended sequence, 1/4 turn at a time, until you can unscrew them by hand **(see illustration)**.
3 Remove the balancer assembly. Keep the spacers with the balancer assembly by securing them with wire or twist-ties. **Caution:** *Do not disassemble the balancer.*
4 Be sure to check the balancer for excessive backlash as described in Section 21 before installing it.

14 Pistons/connecting rods - removal

Refer to illustrations 14.2, 14.4, 14.5 and 14.7
Note: *Prior to removing the piston/connecting rod assemblies, remove the cylinder head(s), the oil pan and the oil pump pick-up tube by referring to the appropriate Sections in Chapter 2A or 2B.*

1 On four-cylinder engines only, remove the engine balancer assembly if not already done (see Section 13).
2 Use your fingernail to feel if a ridge has formed at the upper limit of ring travel (about 1/4-inch down from the top of each cylinder). If carbon deposits or cylinder wear have produced ridges, they must be completely removed with a special tool **(see illustration)**. Follow the manufacturer's instructions provided with the tool. Failure to remove the ridges before attempting to remove the piston/connecting rod assemblies may result in

Chapter 2 Part C General engine overhaul procedures   2C-15

14.2 A ridge reamer is required to remove the ridge from the top of each cylinder - do this before removing the pistons!

14.4 Check the connecting rod side clearance (endplay) with a feeler gauge as shown here

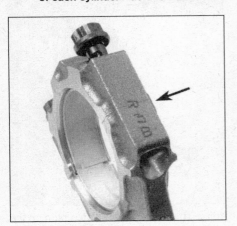

14.5 The connecting rods and caps should be marked to indicate which cylinder they're installed in - if they aren't, mark them with a center punch to avoid confusion during reassembly - do not confuse the markings shown here as rod numbers; these are bearing size identifications

14.7 To prevent damage to the crankshaft journals and cylinder walls, slip sections of hose over the bolts (four-cylinder models)

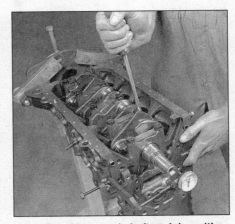

15.1 Checking crankshaft endplay with a dial indicator

piston damage.

3 After the cylinder ridges have been removed, turn the engine upside-down so the crankshaft is facing up.

4 Before the connecting rods are removed, check the endplay with feeler gauges. Slide them between the first connecting rod and the crankshaft throw until the play is removed (see illustration). The endplay is equal to the thickness of the feeler gauge(s). If the endplay exceeds the specified service limit, new connecting rods will be required. If new rods (or a new crankshaft) are installed, the endplay may fall under the service limit (if it does, the rods will have to be machined to restore it - consult an automotive machine shop for advice if necessary). Repeat the procedure for the remaining connecting rods.

5 Check the connecting rods and caps for identification marks. If they aren't plainly marked, use a small center punch to make the appropriate number of indentations on each rod and cap (1, 2, 3, etc., depending on the engine type and cylinder they're associated with) (see illustration).

6 Loosen each of the connecting rod cap nuts 1/2-turn at a time until they can be removed by hand. Remove the number one connecting rod cap and bearing insert. Don't drop the bearing insert out of the cap.

7 To protect the crankshaft journals on four-cylinder engines, slip short pieces of rubber hose over the rod bolts before removing the pistons (see illustration). Note: *The rods on V6 engines use rod bolts instead of nuts on bolts mounted in the rods. When removing or installing the piston/connecting rod assemblies on a V6, be careful not to scratch the crankshaft journals or cylinder walls.*

8 Push the connecting rod/piston assembly out through the top of the engine. Use a wooden hammer handle to push on the upper bearing surface in the connecting rod. If resistance is felt, double-check to make sure that all of the ridge was removed from the cylinder.

9 Repeat the procedure for the remaining cylinders. Note: *Turn the crankshaft as needed to put the rod to be removed close to parallel with the cylinder bore, i.e. don't try to drive it out while at a large angle to the bore.*

10 After removal, reassemble the connecting rod caps and bearing inserts in their respective connecting rods and install the cap nuts finger tight. Leaving the old bearing inserts in place until reassembly will help prevent the connecting rod bearing surfaces from being accidentally nicked or gouged.

11 Don't separate the pistons from the connecting rods (see Section 18 for additional information).

15 Crankshaft - removal

Refer to illustrations 15.1, 15.3 and 15.5

Note: *The crankshaft can be removed only after the engine has been removed from the vehicle. It's assumed that the flywheel or driveplate, vibration damper, timing belt, oil pan, oil pick-up tube, oil pump, balancer (four-cylinder models) and piston/connecting rod assemblies have already been removed. The rear main oil seal and retainer must be removed from the block before proceeding with crankshaft removal.*

1 Before the crankshaft is removed, check the endplay. Mount a dial indicator with the

Chapter 2 Part C General engine overhaul procedures

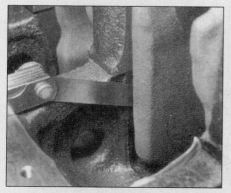

15.3 Checking crankshaft endplay with a feeler gauge

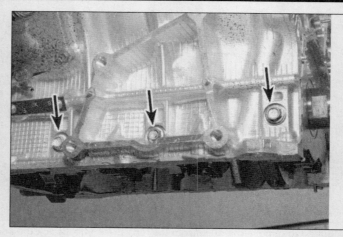

15.5 The V6 engine has side bolts (arrows) into the main caps (rear bolt on this side not seen here) - the engine mount must be removed to access the bolt in the center of this photo

16.1a A hammer and a large punch can be used to knock the core plugs sideways in their bores

16.1b Pull the core plugs from the block with pliers

16.1c The equivalent of a core plug on the V6 engine is this bolted-on plate, sealed with sealant

stem in line with the crankshaft and touching the nose of the crank or one of the crank throws **(see illustration)**.

2 Push the crankshaft all the way to the rear and zero the dial indicator. Next, pry the crankshaft to the front as far as possible and check the reading on the dial indicator. The distance that it moves is the endplay. If it's greater than specified, check the crankshaft thrust surfaces for wear. If no wear is evident, new thrust washers should correct the endplay.

3 If a dial indicator isn't available, feeler gauges can be used. Gently pry or push the crankshaft all the way to the front of the engine. Slip feeler gauges between the crankshaft and the front face of the thrust main bearing to determine the clearance **(see illustration)**. The thrust bearing on four-cylinder engines is the number three (center) bearing, while on the V6 engine it's the number two journal.

4 On four-cylinder engines, check the main bearing caps to see if they're marked to indicate their locations. They should be numbered consecutively from the front of the engine to the rear. If they aren't, mark them with number stamping dies or a center punch. Main bearing caps generally have a cast-in arrow, which points to the front of the engine. Loosen the main bearing cap bolts 1/4-turn at a time each, in the reverse order of the recommended tightening sequence **(see illustration 24.12a)**, until they can be removed by hand.

5 The V6 engine has main cap bolts along the sides of the block **(see illustration)**. Check the main bearing caps to see if they're marked to indicate their locations. If they aren't, mark them with number stamping dies or a center punch. Remove those first (with their sealing washers) in the reverse order of the tightening sequence **(see illustration 24.12d)**. Then remove the rest of the main cap bolts in the reverse order of the tightening sequence **(see illustration 24.12b)**. Note if any stud bolts are used and make sure they're returned to their original locations when the crankshaft is reinstalled.

6 Gently tap the caps with a soft-face hammer, then separate them from the engine block. If necessary, use the bolts as levers to remove the caps. Try not to drop the bearing inserts if they come out with the caps.

7 Carefully lift the crankshaft out of the engine. It may be a good idea to have an assistant available, since the crankshaft is quite heavy. With the bearing inserts in place in the engine block and main bearing caps or cap assembly, return the caps to their respective locations on the engine block and tighten the bolts finger tight.

16 Engine block - cleaning and inspection

Cleaning

Refer to illustrations 16.1a, 16.1b, 16.1c, 16.8 and 16.10

Caution: *The core plugs (also known as freeze or soft plugs) may be difficult or impossible to retrieve if they're driven completely into the block coolant passages.*

1 Using the blunt end of a punch, tap in on the outer edge of the core plug to turn the plug sideways in the bore. Then using pliers, pull the core plug from the engine block **(see illustrations)**. **Note:** *The V6 engine does not have conventional core plugs in its aluminum block, but rather bolted-on plates with RTV sealant* **(see illustration)**.

2 Using a gasket scraper, remove all traces of gasket material from the engine block. Be very careful not to nick or gouge the gasket sealing surfaces.

Chapter 2 Part C General engine overhaul procedures   2C-17

16.8 All bolt holes in the block - particularly the main bearing cap and head bolt holes - should be cleaned and restored with a tap (be sure to remove debris from the holes after this is done)

16.10 A large socket on an extension can be used to drive the new core plugs into the bores

3   Remove the main bearing caps or cap assembly and separate the bearing inserts from the caps and the engine block. Tag the bearings, indicating which cylinder they were removed from and whether they were in the cap or the block, then set them aside.

4   Remove all of the threaded oil gallery plugs from the block. The plugs are usually very tight - they may have to be drilled out and the holes retapped. Use new plugs when the engine is reassembled.

5   If the engine is extremely dirty, it should be taken to an automotive machine shop for cleaning.

6   After the block is returned, clean all oil holes and oil galleries one more time. Brushes specifically designed for this purpose are available at most auto parts stores. Flush the passages with warm water until the water runs clear, dry the block thoroughly and wipe all machined surfaces with a light, rust preventive oil. If you have access to compressed air, use it to speed the drying process and to blow out all the oil holes and galleries. **Warning:** *Wear eye protection when using compressed air!*

7   If the block isn't extremely dirty or sludged up, you can do an adequate cleaning job with hot soapy water and a stiff brush. Take plenty of time and do a thorough job. Regardless of the cleaning method used, be sure to clean all oil holes and galleries very thoroughly, dry the block completely and coat all machined surfaces with light oil.

8   The threaded holes in the block must be clean to ensure accurate torque readings during reassembly. Run the proper size tap into each of the holes to remove rust, corrosion, thread sealant or sludge and restore damaged threads **(see illustration)**. If possible, use compressed air to clear the holes of debris produced by this operation. Now is a good time to clean the threads on the head bolts and the main bearing cap bolts as well.

9   Reinstall the main bearing caps and tighten the bolts finger tight.

10 If you're working on a four-cylinder engine, coat the sealing surfaces of the new

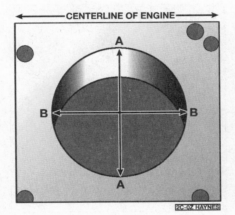

16.16a Measure the diameter of each cylinder at a right angle to the engine centerline (A), and parallel to engine centerline (B) - out-of-round is the difference between A and B, taper is the difference between A and B at the top of the cylinder and A and B at the bottom of the cylinder

core plugs with Permatex no. 1 sealant, then install them in the engine block **(see illustration)**. Make sure they're driven in straight and seated properly or leakage could result. Special tools are available for this purpose, but a large socket, with an outside diameter that will just slip into the core plug, a 1/2-inch drive extension and a hammer will work just as well. If you're working on a V6 engine, apply a 1/8 to 3/16-inch bead of RTV sealant to the water seal plate, then install the plate and tighten the nuts to the torque listed in this Chapter's Specifications.

11 Apply non-hardening sealant (such as Permatex no. 2 or Teflon pipe sealant) to the new oil gallery plugs and thread them into the holes in the block. Make sure they're tightened securely.

12 If the engine isn't going to be reassembled right away, cover it with a large plastic trash bag to keep it clean.

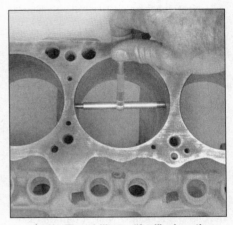

16.16b The ability to "feel" when the telescoping gauge is at the correct point will be developed over time, so work slowly and repeat the check until you're satisfied that the bore measurement is accurate

Inspection

Refer to illustrations 16.16a, 16.16b, 16.16c and 16.20

13 Before the block is inspected, it should be cleaned as described earlier in this Section.

14 Visually check the block for cracks, rust and corrosion. Look for stripped threads in the threaded holes. It's also a good idea to have the block checked for hidden cracks by an automotive machine shop that has the special equipment to do this type of work, especially if the vehicle had a history of overheating or using coolant. If defects are found, have the block repaired, if possible, or replaced.

15 Check the cylinder bores for scuffing and scoring.

16 Measure the diameter of each cylinder at the top (just under the ridge area), center and bottom of the cylinder bore, parallel to the crankshaft axis **(see illustrations)**.

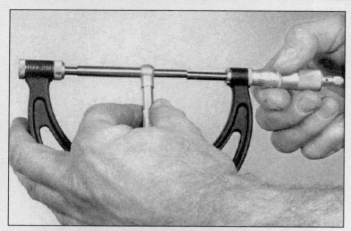

16.16c The gauge is then measured with a micrometer to determine the bore size

16.20 Check the block deck (both banks on a V6 engine) for distortion with a precision straightedge and feeler gauges - check straight along and diagonally

17   Next, measure each cylinder's diameter at the same three locations across the crankshaft axis. Compare the results to this Chapter's Specifications.

18   Repeat the procedure for the remaining pistons and cylinders.

19   If the cylinder walls are badly scuffed or scored, or if they're out-of-round or tapered beyond the limits given in this Chapter's Specifications, have the engine block rebored and honed at an automotive machine shop. If a rebore is done, oversize pistons and rings will be required.

20   Using a precision straightedge and feeler gauge, check the block deck (the surface that mates with the cylinder head[s]) for distortion **(see illustration)**. If it's distorted beyond the specified limit, it can be resurfaced by an automotive machine shop.

21   If the cylinders are in reasonably good condition and not worn to the outside of the limits, and if the piston-to-cylinder clearances can be maintained properly, then they don't have to be rebored. Honing is all that's necessary (refer to Section 17).

17   Cylinder honing

Refer to illustrations 17.3a and 17.3b

1   Prior to engine reassembly, the cylinder bores must be honed so the new piston rings will seat correctly and provide the best possible combustion chamber seal. **Note:** *If you don't have the tools or don't want to tackle the honing operation, most automotive machine shops will do it for a reasonable fee.*

2   Before honing the cylinders, install the main bearing caps or cap assembly (without bearing inserts) and tighten the bolts to the specified torque.

3   Two types of cylinder hones are commonly available - the flex hone or "bottle brush" type and the more traditional surfacing hone with spring-loaded stones. Both will do the job, but for the less experienced mechanic the "bottle brush" hone will probably be easier to use. You'll also need some

17.3a A "bottle brush" hone is the easiest type of hone to use

kerosene or honing oil, rags and an electric drill motor. The drill motor should be operated at a steady, slow speed. Use a large 1/2-inch drill or a 3/8-inch variable-speed drill. Proceed as follows:

a) Mount the hone in the drill motor, compress the stones and slip it into the first cylinder **(see illustration)**. *Warning: Be sure to wear safety goggles or a face shield!*

b) Lubricate the cylinder with plenty of honing oil, turn on the drill and move the hone up-and-down in the cylinder at a pace that will produce a fine crosshatch pattern on the cylinder walls. Ideally, the crosshatch lines should intersect at approximately a 60-degree angle **(see illustration)**. *Be sure to use plenty of lubricant and don't take off any more material than is absolutely necessary to produce the desired finish.* **Note:** *Piston ring manufacturers may specify a smaller crosshatch angle than the traditional 60-degrees - read and follow any instructions included with the new rings.*

c) Don't withdraw the hone from the cylin-

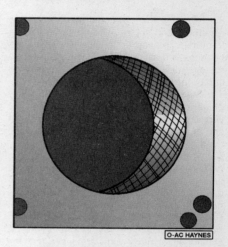

17.3b The cylinder hone should leave a smooth, crosshatch pattern with the lines intersecting at approximately a 60-degree angle

der while it's running. Instead, shut off the drill and continue moving the hone up-and-down in the cylinder until it comes to a complete stop, then compress the stones and withdraw the hone. If you're using a "bottle brush" type hone, stop the drill motor, then turn the chuck in the normal direction of rotation while withdrawing the hone from the cylinder.

d) *Wipe the oil out of the cylinder and repeat the procedure for the remaining cylinders.*

4   After the honing job is complete, chamfer the top edges of the cylinder bores with a small file so the rings won't catch when the pistons are installed. Be very careful not to nick the cylinder walls with the end of the file.

5   The entire engine block must be washed again very thoroughly with warm, soapy water to remove all traces of the abrasive grit produced during the honing operation. **Note:** *The bores can be considered clean when a lint-free white cloth - dampened with clean engine oil - used to wipe them out doesn't*

Chapter 2 Part C General engine overhaul procedures

18.4a The piston ring grooves can be cleaned with a special tool, as shown here . . .

18.4b . . . or a section of a broken ring

pick up any more honing residue, which will show up as gray areas on the cloth. Be sure to run a brush through all oil holes and galleries and flush them with running water.

6 After rinsing, dry the block and apply a coat of light rust preventive oil to all machined surfaces. Wrap the block in a plastic trash bag to keep it clean and set it aside until reassembly.

18 Pistons/connecting rods - inspection

Refer to illustrations 18.4a, 18.4b, 18.10 and 18.11

1 Before the inspection process can be carried out, the piston/connecting rod assemblies must be cleaned and the original piston rings removed from the pistons. **Note:** *Always use new piston rings when the engine is reassembled.*

2 Using a piston ring installation tool, carefully remove the rings from the pistons. Be careful not to nick or gouge the pistons in the process.

3 Scrape all traces of carbon from the top of the piston. A hand-held wire brush or a piece of fine emery cloth can be used once the majority of the deposits have been scraped away. Do not, under any circumstances, use a wire brush mounted in a drill motor to remove deposits from the pistons. The piston material is soft and may be eroded away by the wire brush.

4 Use a piston ring groove-cleaning tool to remove carbon deposits from the ring grooves. If a tool isn't available, a piece broken off the old ring will do the job. Be very careful to remove only the carbon deposits - don't remove any metal and do not nick or scratch the sides of the ring grooves **(see illustrations)**.

5 Once the deposits have been removed, clean the piston/rod assemblies with solvent and dry them with compressed air (if available). Make sure the oil return holes in the back sides of the ring grooves and the oil hole in the lower end of each rod are clear.

6 If the pistons and cylinder walls aren't damaged or worn excessively, and if the engine block is not rebored, new pistons won't be necessary. Normal piston wear appears as even vertical wear on the piston thrust surfaces and slight looseness of the top ring in its groove. New piston rings, however, should always be used when an engine is rebuilt.

7 Carefully inspect each piston for cracks around the skirt, at the pin bosses and at the ring lands.

8 Look for scoring and scuffing on the thrust faces of the skirt, holes in the piston crown and burned areas at the edge of the crown. If the skirt is scored or scuffed, the engine may have been suffering from overheating and/or abnormal combustion, which caused excessively high operating temperatures. The cooling and lubrication systems should be checked thoroughly. A hole in the piston crown is an indication that abnormal combustion (preignition) was occurring. Burned areas at the edge of the piston crown are usually evidence of spark knock (detonation). If any of the above problems exist, the causes must be corrected or the damage will occur again. The causes may include intake air leaks, incorrect air/fuel mixture, incorrect ignition timing and EGR system malfunctions.

9 Corrosion of the piston, in the form of small pits, indicates that coolant is leaking into the combustion chamber and/or the crankcase. Again, the cause must be corrected or the problem may persist in the rebuilt engine.

10 Measure the piston ring groove clearance by laying a new piston ring in each ring groove and slipping a feeler gauge in beside it **(see illustration)**. Check the clearance at three or four locations around each groove. Be sure to use the correct ring for each groove - they are different. If the clearance is greater than that listed in this Chapter's Specifications, new pistons will have to be used.

11 Check the piston-to-bore clearance by measuring the bore (see Section 16) and the piston diameter. Make sure the pistons and bores are correctly matched. Measure the piston across the skirt, at a 90-degree angle to the piston pin at the specified distance from the top of the piston **(see illustration)**. Subtract the piston diameter from the bore diameter to obtain the clearance. If it's greater than specified, the block will have to be rebored and new pistons and rings installed.

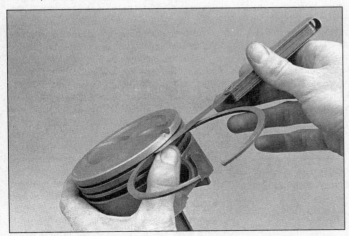

18.10 Check the ring groove clearance with a feeler gauge at several points around the groove

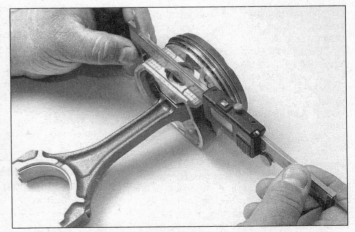

18.11 Measure the piston diameter at a 90-degree angle to the piston pin, at the specified distance from the top of the piston - a precision caliper may be used if a micrometer isn't available

19.5 Measure the diameter of each crankshaft journal at several points to detect taper and out-of-round conditions

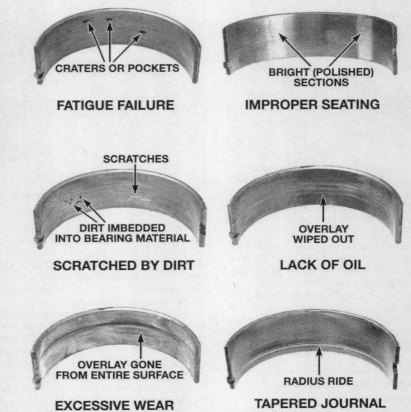

20.1 When inspecting the main and connecting rod bearings, look for these problems

12 Check the piston-to-rod clearance by twisting the piston and rod in opposite directions. Any noticeable play indicates excessive wear, which must be corrected.

13 If the pistons must be removed from the connecting rods for any reason, the rods should be taken to an automotive machine shop, to be checked for bend and twist, since automotive machine shops have special equipment for this purpose.

14 Check the connecting rods for cracks and other damage. Temporarily remove the rod caps, lift out the old bearing inserts, wipe the rod and cap bearing surfaces clean and inspect them for nicks, gouges and scratches. After checking the rods, replace the old bearings, slip the caps into place and tighten the nuts finger tight. **Note:** *If the engine is being rebuilt because of a connecting rod knock, be sure to install new rods.*

19 Crankshaft - inspection

Refer to illustration 19.5

1 Clean the crankshaft with solvent and dry it with compressed air (if available). Be sure to clean the oil holes with a stiff brush and flush them with solvent.

2 Check the main and connecting rod bearing journals for uneven wear, scoring, pits and cracks.

3 Remove all burrs from the crankshaft oil holes with a stone, file or scraper.

4 Check the rest of the crankshaft for cracks and other damage. It should be magnafluxed to reveal hidden cracks - an automotive machine shop will handle the procedure.

5 Using a micrometer, measure the diameter of the main and connecting rod journals and compare the results to this Chapter's Specifications **(see illustration)**. By measuring the diameter at a number of points around each journal's circumference, you'll be able to determine whether or not the journal is out-of-round. Take the measurement at each end of the journal, near the crank throws, to determine if the journal is tapered. Crankshaft runout should be checked also, but large V-blocks and a dial indicator are needed to do it correctly. If you don't have the equipment, have a machine shop check the runout.

6 If the crankshaft journals are damaged, tapered, out-of-round or worn beyond the limits given in the Specifications, have the crankshaft reground by an automotive machine shop. Be sure to use the correct size bearing inserts if the crankshaft is reconditioned.

7 Check the oil seal journals at each end of the crankshaft for wear and damage. If the seal has worn a groove in the journal, or if it's nicked or scratched, the new seal may leak when the engine is reassembled. In some cases, an automotive machine shop may be able to repair the journal by pressing on a thin sleeve. If repair isn't feasible, a new or different crankshaft should be installed.

8 Refer to Section 20 and examine the main and rod bearing inserts.

20 Main and connecting rod bearings - inspection and selection

Inspection

Refer to illustration 20.1

1 Even though the main and connecting rod bearings should be replaced with new ones during the engine overhaul, the old bearings should be retained for close examination, as they may reveal valuable information about the condition of the engine **(see illustration)**.

2 Bearing failure occurs because of lack of lubrication, the presence of dirt or other foreign particles, overloading the engine and corrosion. Regardless of the cause of bearing failure, it must be corrected before the engine is reassembled to prevent it from happening again.

3 When examining the bearings, remove them from the engine block, the main bearing caps, the connecting rods and the rod caps and lay them out on a clean surface in the same general position as their location in the engine. This will enable you to match any bearing problems with the corresponding crankshaft journal.

4 Dirt and other foreign particles get into the engine in a variety of ways. It may be left in the engine during assembly, or it may pass through filters or the PCV system. It may get into the oil, and from there into the bearings. Metal chips from machining operations and normal engine wear are often present. Abrasives are sometimes left in engine components after reconditioning, especially when parts are not thoroughly cleaned using the proper cleaning methods. Whatever the

Chapter 2 Part C General engine overhaul procedures

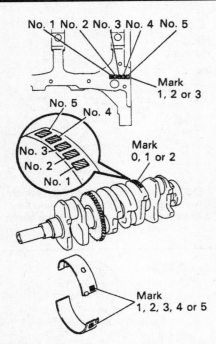

20.10a If the number on the original main bearing is not clear, install a new bearing with a number that matches the number stamped into the block - different journals may require different size bearings (four-cylinder grade number locations shown)

20.10b Grade numbers on the V6 engine are located on the front of the block above the crankshaft (A); and crankshaft journal grades are on the side of the first counterweight (B)

source, these foreign objects often end up embedded in the soft bearing material and are easily recognized. Large particles will not embed in the bearing and will score or gouge the bearing and journal. The best prevention for this cause of bearing failure is to clean all parts thoroughly and keep everything spotlessly clean during engine assembly. Frequent and regular engine oil and filter changes are also recommended.

5 Lack of lubrication (or lubrication breakdown) has a number of interrelated causes. Excessive heat (which thins the oil), overloading (which squeezes the oil from the bearing face) and oil leakage or throw off (from excessive bearing clearances, worn oil pump or high engine speeds) all contribute to lubrication breakdown. Blocked oil passages, which usually are the result of misaligned oil holes in a bearing shell, will also oil starve a bearing and destroy it. When lack of lubrication is the cause of bearing failure, the bearing material is wiped or extruded from the steel backing of the bearing. Temperatures may increase to the point where the steel backing turns blue from overheating.

6 Driving habits can have a definite effect on bearing life. Driving habits can have a definite effect on bearing life. Low speed operation in too high a gear (lugging the engine) puts very high loads on bearings, which tends to squeeze out the oil film. These loads cause the bearings to flex, which produces fine cracks in the bearing face (fatigue failure). Eventually the bearing material will loosen in pieces and tear away from the steel backing. Short trip driving leads to corrosion of bearings because insufficient engine heat is produced to drive off the condensed water and corrosive gases. These products collect in the engine oil, forming acid and sludge. As the oil is carried to the engine bearings, the acid attacks and corrodes the bearing material.

7 Incorrect bearing installation during engine assembly will lead to bearing failure as well. Tight-fitting bearings leave insufficient bearing oil clearance and will result in oil starvation. Dirt or foreign particles trapped behind a bearing insert result in high spots on the bearing which lead to failure.

Selection

Refer to illustrations 20.10a, 20.10b and 20.11

8 If the original bearings are worn or damaged, or if the oil clearances are incorrect (see Sections 24 or 26), the following procedures should be used to select the correct new bearings for engine reassembly. However, if the crankshaft has been reground, new undersize bearings must be installed - the following procedure should not be used if undersize bearings are required! The automotive machine shop that reconditions the crankshaft will provide or help you select the correct size bearings. Regardless of how the bearing sizes are determined, use the oil clearance, measured with Plastigage, as a guide to ensure the bearings are the right size.

Main bearings

9 If you need to use a STANDARD size main bearing, install one that has the same number as the original bearing (see illustrations 20.10a and 20.10b for the bearing number locations). There are five sizes of main bearings.

10 If the number on the original main bearing has been obscured, locate the main journal grade numbers stamped into the oil pan mating surface on the engine block and the crankshaft counterweights (see illustrations). On the V6 engine, the block numbers are on the front of the block, just above the first main bearing, and the crankshaft numbers are on the first counterweight (see illustration).

11 Adding the block number to the crank number for a particular journal will give the recommended bearing size for the four-cylinder engines. Use the accompanying chart to determine the correct bearings for each journal on the V6 engine (see illustration).

Number 1 and 4 Bearings

Cylinder block (A) + Crankshaft (B)	Total number				" ": Number mark
	0 – 5	6 – 11	12 – 17	18 – 23	24 – 28
Use bearing	"3"	"4"	"5"	"6"	"7"

EXAMPLE: Cylinder block "06" (A)
+ Crankshaft "08" (B)
= Total number 14 (Use bearing "5")

Number 2 and 3 Bearings

Cylinder block (A) + Crankshaft (B)	Total number				" ": Number mark
	0 – 5	6 – 11	12 – 17	18 – 23	24 – 28
Use bearing	"1"	"2"	"3"	"4"	"5"

EXAMPLE: Cylinder block "06" (A)
+ Crankshaft "08" (B)
= Total number 14 (Use bearing "3")

20.11 Select the proper bearings for the V6 engine based on this chart - add the crank and block numbers and select the bearing size listed underneath

Chapter 2 Part C General engine overhaul procedures

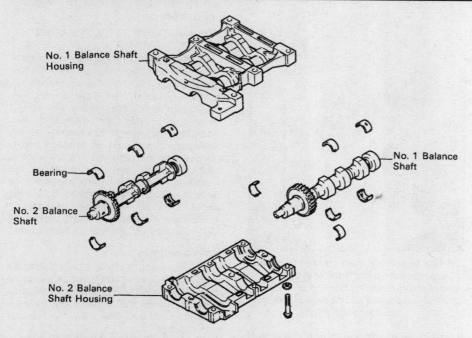

21.1 The balancer assembly is geared to the crankshaft - it is shown here in exploded view for information only (do NOT disassemble the unit)

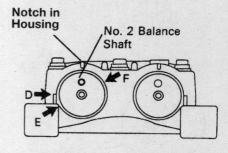

21.2a The balancer shafts have punch marks and the number 2 housing has alignment notches - start with the shafts aligned as shown . . .

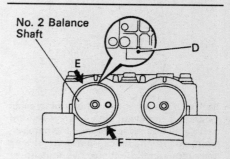

21.2b . . . then turn until mark D lines up with the notch in the housing

Connecting rod bearings

12 If you need to use a STANDARD size rod bearing, install one that has the same number as the number stamped into the connecting rod cap **(see illustration 13.5)**.

All bearings

13 Remember, the oil clearance is the final judge when selecting new bearing sizes. If you have any questions or are unsure which bearings to use, get help from an auto parts store or a dealer parts or service department.

21 Balancer assembly - backlash check (four-cylinder engine)

Refer to illustrations 21.1, 21.2a, 21.2b, 21.3a and 21.3b
Note: *This procedure checks the internal condition of the balancer assembly for worn gears, bearings and/or shafts. For balancer assembly installation and crankshaft-to-balancer backlash checks, see Section 27.*

1 The engine balancer assembly consists of two shafts contained in an upper and lower aluminum housing assembly. The shafts are geared together, and one of the shafts is geared to the crankshaft **(see illustration)**. The following procedure applies only with the balancer assembly removed from the engine. Do NOT disassemble the balancer assembly!
2 Start with the punch marks on the number 2 shaft lined up as indicated **(see illustration)**, then align the number 2 shaft so that mark D is aligned with the housing **(see illustration)**.
3 Using locking pliers to hold the number 2 shaft from turning, measure the backlash on the number 1 gear with a dial indicator **(see illustrations)**, while keeping forward pressure on the rear of both shafts to eliminate thrust play. Compare the backlash to this Chapter's Specifications.
4 Remove the dial indicator, and turn the number 2 shaft so that punch mark E now aligns with the case. Measure backlash again as in Step 3 and compare the results with this Chapter's Specifications.
5 Remove the dial indicator again, and turn the number 2 shaft so that punch mark F now aligns with the case. Measure backlash again as in Step 3 and compare the results with this Chapter's Specifications. **Note:** *If any of the three measurements exceeds the Specifications, it is suggested that the engine balance assembly be taken to a Toyota dealer or other qualified service facility for further disassembly and repair.*

22 Engine overhaul - reassembly sequence

1 Before beginning engine reassembly, make sure you have all the necessary new parts, gaskets and seals as well as the following items on hand:
    Common hand tools
    A 1/2-inch drive torque wrench
    Piston ring installation tool
    Piston ring compressor
    Short lengths of rubber or plastic hose
      to fit over connecting rod bolts

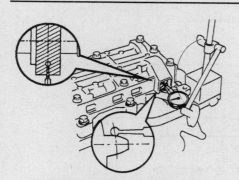

21.3a Set up the dial indicator as shown . . .

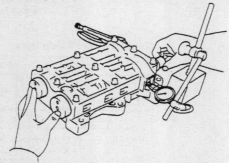

21.3b . . . and measure the backlash while applying mild hand pressure on the rear of both shafts

Chapter 2 Part C General engine overhaul procedures

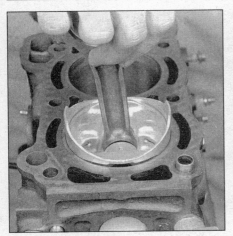

23.3 When checking piston ring end gap, the ring must be square in the cylinder bore (this is done by pushing the ring down with the top of a piston as shown)

Plastigage
Feeler gauges
A fine-tooth file
New engine oil
Engine assembly lube or moly-base grease
Gasket sealer
Thread locking compound

2   In order to save time and avoid problems, engine reassembly must be done in the following general order:

Four-cylinder engines

Piston rings (Part C)
Crankshaft and main bearings (Part C)
Piston/connecting rod assemblies (Part C)
Rear main (crankshaft) oil seal (Part C)
Engine balancer assembly (Part C)
Cylinder head and lifters (Part A)
Camshafts (Part A)
Timing belt and sprockets (Part A)
Timing belt covers (Part A)
Oil pump (Part A)
Oil pick-up (Part A)
Oil pan (Part A)
Intake and exhaust manifolds (Part A)
Valve cover (Part A)
Flywheel/driveplate (Part A)

V6 engines

Piston rings (Part C)
Crankshaft and main bearings (Part C)
Piston/connecting rod assemblies (Part C)
Rear main oil seal/retainer (Part C)
Oil pump (Part B)
Oil pan (Part B)
Cylinder heads (Part B)
Camshafts and lifters (Part B)
Timing belt and sprockets (Part B)
Timing belt covers (Part B)
Valve covers (Part B)
Intake and exhaust manifolds (Part B)
Flywheel/driveplate (Part B)

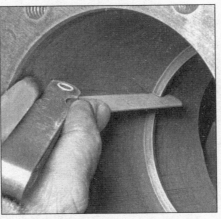

23.4 With the ring square in the cylinder, measure the end gap with a feeler gauge

23 Piston rings - installation

Refer to illustrations 23.3, 23.4, 23.9a, 23.9b and 23.12

1   Before installing the new piston rings, the ring end gaps must be checked. It's assumed that the piston ring groove clearance has been checked and verified correct (see Section 18).
2   Lay out the piston/connecting rod assemblies and the new ring sets so the ring sets will be matched with the same piston and cylinder during the end gap measurement and engine assembly.
3   Insert the top (number one) ring into the first cylinder and square it up with the cylinder walls by pushing it in with the top of the piston **(see illustration)**. The ring should be near the bottom of the cylinder, at the lower limit of ring travel.
4   To measure the end gap, slip feeler gauges between the ends of the ring until a gauge equal to the gap width is found **(see illustration)**. The feeler gauge should slide between the ring ends with a slight amount of drag. Compare the measurement to that found in this Chapter's Specifications. If the gap is larger or smaller than specified, double-check to make sure you have the correct rings before proceeding.
5   If the gap is too small, replace the rings - DO NOT file the ends to increase the clearance.
6   Excess end gap isn't critical unless it's greater than 0.040-inch. Again, double-check to make sure you have the correct rings for your engine.
7   Repeat the procedure for each ring that will be installed in the first cylinder and for each ring in the remaining cylinders. Remember to keep rings, pistons and cylinders matched up.
8   Once the ring end gaps have been checked/corrected, the rings can be installed on the pistons.
9   The oil control ring (lowest one on the piston) is usually installed first. It's composed of three separate components. Slip the

23.9a Install the spacer/expander in the oil control ring groove

23.9b DO NOT use a piston ring installation tool when installing the oil ring side rails

spacer/expander into the groove **(see illustration)**. If an anti-rotation tang is used, make sure it's inserted into the drilled hole in the ring groove. Refer to illustrations **26.5a and 26.5b** for the orientation of the expander gap to the piston. Next, install the lower side rail. Don't use a piston ring installation tool on the oil ring side rails, as they may be damaged. Instead, place one end of the side rail into the groove between the spacer/expander and the ring land, hold it firmly in place and slide a finger around the piston while pushing the rail into the groove **(see illustration)**. Next, install the upper side rail in the same manner.
10   After the three oil ring components have been installed, check to make sure that both the upper and lower side rails can be turned smoothly in the ring groove.
11   The number two (middle) ring is installed next. It's usually stamped with a mark which must face up, toward the top of the piston. **Note:** *Always follow the instructions printed on the ring package or box - different manufacturers may require different approaches. Do not mix up the top and middle rings, as they have different cross-sections.*
12   Use a piston ring installation tool and make sure the ring's identification mark is facing the top of the piston, then slip the ring

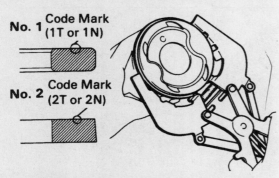

23.12 Install the compression rings with a ring expander - the mark must face up (four-cylinder engine shown - on V6 engines, top rings are marked 1R, G1 or T, second rings are marked 2R, G2 or T2)

24.10 Lay the Plastigage strips (arrow) on the main bearing journals, parallel to the crankshaft centerline

24.12a Main bearing cap bolt tightening sequence - four-cylinder engine

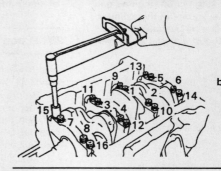

24.12b Main bearing cap bolt tightening sequence - V6 engine

into the middle groove on the piston **(see illustration)**. Don't expand the ring any more than necessary to slide it over the piston.

13 Install the number one (top) ring in the same manner. Make sure the mark is facing up. Be careful not to confuse the number one and number two rings.

14 Repeat the procedure for the remaining pistons and rings.

24 Crankshaft - installation and main bearing oil clearance check

Refer to illustrations 24.10, 24.12a, 24.12b, 24.12c, 24.12d, 24.14, 24.19 and 24.20

1 Crankshaft installation is the first major step in engine reassembly. It's assumed at this point that the engine block and crankshaft have been cleaned, inspected and repaired or reconditioned.

2 Position the engine with the bottom facing up.

3 Remove the main bearing cap bolts and lift out the caps or cap assembly. Lay the caps out in the proper order.

4 If they're still in place, remove the old bearing inserts from the block and the main bearing caps. Wipe the main bearing surfaces of the block and caps with a clean, lint free cloth. They must be kept spotlessly clean!

Main bearing oil clearance check

5 Clean the back sides of the new main bearing inserts and lay the bearing half with the oil groove in each main bearing saddle in the block. Lay the other bearing half from each bearing set in the corresponding main bearing cap. Make sure the tab on each bearing insert fits into the recess in the block or cap. Also, the oil holes in the block must line up with the oil holes in the bearing insert. **Caution:** *Do not hammer the bearings into place and don't nick or gouge the bearing faces. No lubrication should be used at this time.*

6 If you're working on a V6 engine, the thrust bearings (washers) must be installed in the number two cap and saddle. On four-cylinder engines, the thrust bearings (washers) must be installed in the number three (center) cap. Install the thrust bearings with the oil grooves facing out.

7 Clean the faces of the bearings in the block and the crankshaft main bearing journals with a clean, lint free cloth. Check or clean the oil holes in the crankshaft, as any dirt here can go only one way - straight through the new bearings.

8 Once you're certain the crankshaft is clean, carefully lay it in position in the main bearings.

9 Before the crankshaft can be permanently installed, the main bearing oil clearance must be checked.

10 Trim several pieces of the appropriate size Plastigage (they must be slightly shorter than the width of the main bearings) and place one piece on each crankshaft main bearing journal, parallel with the journal axis **(see illustration)**.

11 Clean the faces of the bearings in the caps and install the caps in their respective positions (don't mix them up) with the arrows pointing toward the front of the engine. Don't disturb the Plastigage. Apply a light coat of oil to the bolt threads and the under-sides of the bolt heads, then install them.

12 Following the recommended sequence **(see illustrations)**, tighten the main bearing cap bolts, in three steps, to the torque listed in this Chapter's Specifications. Don't rotate the crankshaft at any time during this operation! **Note:** *On V6 engines only, after reaching the specified torque on all of the bolts, tighten*

24.12c On V6 engines only, after reaching the specified torque, mark the front side of each bolt with paint as shown here, plus a paint mark on the socket, then turn the bolts an additional 1/4-turn - the paint marks should now be 90-degrees from their original position

Chapter 2 Part C General engine overhaul procedures

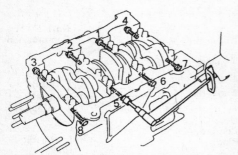

24.12d Main bearing cap side bolt tightening sequence - V6 engine (tighten side bolts after tightening cap bolts)

24.14 Compare the width of the crushed Plastigage to the scale on the envelope to determine the main bearing oil clearance (always take the measurement at the widest point of the Plastigage) - be sure to use the correct scale; standard and metric scales are included

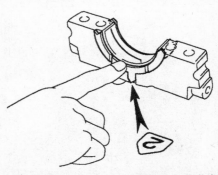

24.19 On four-cylinder engines, install the thrust washer in the number three cap and on V6 engines (shown), the tangs should fit into the slots on the number 2 cap - the grooves must face OUT

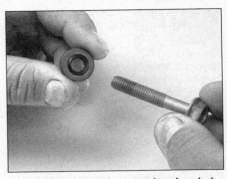

24.20 The V6 engine uses hex-headed bolts that secure the sides of the main caps - when final-assembling this engine use new sealing washers on these bolts

each bolt in sequence an additional 90-degrees (except the six-point bolts on the V6 engine) **(see illustration)**. The side bolts on V6 engines are torqued **after** the main cap bolts **(see illustration)**.

13 Remove the bolts and carefully lift off the main bearing caps or cap assembly. Keep them in order. Don't disturb the Plastigage or rotate the crankshaft. If any of the main bearing caps are difficult to remove, tap them gently from side-to-side with a soft-face hammer to loosen them.

14 Compare the width of the crushed Plastigage on each journal to the scale printed on the Plastigage envelope to obtain the main bearing oil clearance **(see illustration)**. Check the Specifications to make sure it's correct.

15 If the clearance is not as specified, the bearing inserts may be the wrong size (which means different ones will be required - see Section 20). Before deciding that different inserts are needed, make sure that no dirt or oil was between the bearing inserts and the caps or block when the clearance was measured. If the Plastigage is noticeably wider at one end than the other, the journal may be tapered (see Section 19).

16 Carefully scrape all traces of the Plastigage material off the main bearing journals and/or the bearing faces. Don't nick or scratch the bearing faces.

Final crankshaft installation

17 Carefully lift the crankshaft out of the engine. Clean the bearing faces in the block, then apply a thin, uniform layer of clean moly-base grease or engine assembly lube to each of the bearing surfaces. Coat the thrust washers as well.

18 Lubricate the crankshaft surfaces that contact the oil seals with multi-purpose grease or clean engine oil.

19 Make sure the crankshaft journals are clean, then lay the crankshaft back in place in the block. Clean the faces of the bearings in the caps or cap assembly, then apply lubricant to them. Install the caps in their respective positions with the arrows pointing toward the front of the engine. **Note:** *Be sure to install the thrust washers. On four-cylinder engines, the thrust washers go with the number 3 main journal, and the number 2 main journal on V6 engines. The upper (block side) thrust washers* can be rotated into position around the crank with the crank in the block, with the thrust washer grooves facing OUT. The tanged lower thrust washers should be placed on the caps with their grooves OUT and the tangs fitting into the cap slots **(see illustration)**.

20 Apply a light coat of oil to the bolt threads and the under sides of the bolt heads, then install them. Tighten all main bearing cap bolts to the torque listed in this Chapter's Specifications, following the recommended sequence. **Note:** *On the V6 engine, install the inner (closest to the crankshaft) 12-point bolts first, tap the caps down lightly with a plastic-faced hammer, then install the outer 12-point bolts and begin the torque sequence. The six-point (side) bolts are not installed and tightened until all of the 12-point bolts have been tightened to the proper torque* **(see illustrations 24.12b, 24.12c and 24.12d)**. *Be sure to use new sealing washers on the hex-head (side) bolts* **(see illustration)**.

21 On manual transmission equipped models, install a new pilot bearing in the end of the crankshaft.

22 Rotate the crankshaft a number of times by hand to check for any obvious binding.

23 Check the crankshaft endplay with a feeler gauge or a dial indicator as described in Section 15. The endplay should be correct if the crankshaft thrust faces aren't worn or damaged and new thrust washers have been installed.

24 Install a new rear main oil seal, then bolt the retainer to the block (see Section 25).

25 Rear main oil seal installation

Refer to illustrations 25.3 and 25.5

1 The crankshaft must be installed first and the main bearing caps or cap assembly bolted in place, then the new seal should be installed in the retainer and the retainer bolted to the block.

2 Check the seal contact surface on the crankshaft very carefully for scratches and nicks that could damage the new seal lip and cause oil leaks. If the crankshaft is damaged, the only alternative is a new or different crankshaft.

3 The old seal can be removed from the retainer by driving it out from the back side with a hammer and punch **(see illustration)**.

25.3 After removing the retainer from the block, support it on a couple of wood blocks and drive out the old seal with a punch and hammer

25.5 Drive the new seal into the retainer with a wood block or a section of pipe, if you have one large enough - make sure you don't cock the seal in the retainer bore

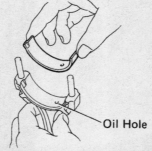

26.3 Align the oil hole (if equipped) in the bearing with the oil hole in the rod - on all rods, align the bearing tang with the notch in the rod

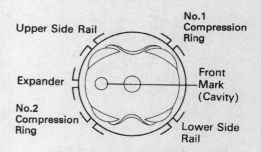

26.5a Stagger the ring end gaps around the piston as shown, before installing the pistons - four-cylinder engine

26.5b Stagger the ring end gaps around the piston as shown, before installing the pistons - V6 engine

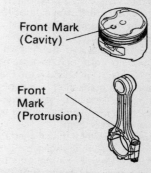

26.9 Check to be sure both the mark on the piston and the mark on the connecting rod are aligned as shown

Be sure to note how far it's recessed into the bore before removing it; the new seal will have to be recessed an equal amount. Be very careful not to scratch or otherwise damage the bore in the retainer or oil leaks could develop.

4 Make sure the retainer is clean, then apply a thin coat of engine oil to the outer edge of the new seal. The seal must be pressed squarely into the bore, so hammering it into place isn't recommended. If you don't have access to a press, sandwich the housing and seal between two smooth pieces of wood and press the seal into place with the jaws of a large vise. The pieces of wood must be thick enough to distribute the force evenly around the entire circumference of the seal. Work slowly and make sure the seal enters the bore squarely.

5 As a last resort, the seal can be tapped into the retainer with a hammer. Use a block of wood to distribute the force evenly and make sure the seal is driven in squarely **(see illustration)**.

6 The seal lips must be lubricated with clean engine oil or multi-purpose grease before the seal/retainer is slipped over the crankshaft and bolted to the block. On four-cylinder engines, use a new gasket - and sealant - and make sure the dowel pins are in place before installing the retainer. On V6 engines, no gasket is required. Instead, apply a 3/32-inch wide bead of anaerobic sealant to the retainer-to-block surface.

7 Tighten the bolts a little at a time to the torque listed in Chapter 2A or 2B Specifications.

26 Pistons/connecting rods - installation and rod bearing oil clearance check

Refer to illustrations 26.3, 26.5a, 26.5b, 26.9, 26.11, 26.13, 26.15 and 26.17

1 Before installing the piston/connecting rod assemblies, the cylinder walls must be perfectly clean, the top edge of each cylinder must be chamfered, and the crankshaft must be in place.

2 Remove the cap from the end of the number one connecting rod (refer to the marks made during removal). Remove the original bearing inserts and wipe the bearing surfaces of the connecting rod and cap with a clean, lint-free cloth. They must be kept spotlessly clean.

Connecting rod bearing oil clearance check

3 Clean the back side of the new upper bearing insert, then lay it in place in the connecting rod. Make sure the tab on the bearing fits into the recess in the rod so the oil holes line up **(see illustration)**. Don't hammer the bearing insert into place and be very careful not to nick or gouge the bearing face. Don't lubricate the bearing at this time.

4 Clean the back side of the other bearing insert and install it in the rod cap. Again, make sure the tab on the bearing fits into the recess in the cap, and don't apply any lubricant. It's critically important that the mating surfaces of the bearing and connecting rod are perfectly clean and oil free when they're assembled.

5 Position the piston ring gaps at staggered intervals around the piston **(see illustrations)**.

6 Slip a section of plastic or rubber hose over each connecting rod cap bolt (four-cylinder engines only).

7 Lubricate the piston and rings with clean engine oil and attach a piston ring compressor to the piston. Leave the skirt protruding about 1/4-inch to guide the piston into the cylinder. The rings must be compressed until they're flush with the piston.

8 Rotate the crankshaft until the number one connecting rod journal is at BDC (bottom dead center) and apply a coat of engine oil to the cylinder wall.

9 With the dimple on top of the piston **(see illustration)** facing the front of the engine, gently insert the piston/connecting rod assembly into the number one cylinder bore and rest the bottom edge of the ring compressor on the engine block.

10 Tap the top edge of the ring compressor to make sure it's contacting the block around its entire circumference.

11 Gently tap on the top of the piston with

Chapter 2 Part C General engine overhaul procedures

26.11 The piston can be driven (gently) into the cylinder bore with the end of a wooden or plastic hammer handle

26.13 Lay the Plastigage strips on each rod bearing journal, parallel to the crankshaft centerline

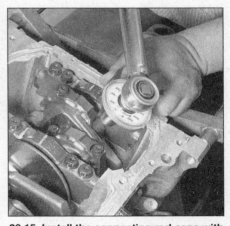

26.15 Install the connecting rod caps with the front mark facing the timing belt end of the engine, and torque to Specifications - for the 90-degree second step, use an angle gauge as shown, or use paint reference marks on the rod nuts/bolts and the socket

the end of a wooden hammer handle **(see illustration)** while guiding the end of the connecting rod into place on the crankshaft journal. The piston rings may try to pop out of the ring compressor just before entering the cylinder bore, so keep some downward pressure on the ring compressor. Work slowly, and if any resistance is felt as the piston enters the cylinder, stop immediately. Find out what's hanging up and fix it before proceeding. **Caution:** *Do not, for any reason, force the piston into the cylinder - you might break a ring and/or the piston.*

12 Once the piston/connecting rod assembly is installed, the connecting rod bearing oil clearance must be checked before the rod cap is permanently bolted in place.

13 Cut a piece of the appropriate size Plastigage slightly shorter than the width of the connecting rod bearing and lay it in place on the number one connecting rod journal, parallel with the journal axis **(see illustration)**.

14 Clean the connecting rod cap bearing face, remove the protective hoses from the connecting rod bolts and install the rod cap. Make sure the mating mark on the cap is on the same side as the mark on the connecting rod. Check the cap to make sure the front mark is facing the timing belt end of the engine.

15 Apply a light coat of oil to the under sides of the nuts, then install and tighten them to the torque listed in this Chapter's Specifications, working up to it in three steps. Use a thin-wall socket to avoid erroneous torque readings that can result if the socket is wedged between the rod cap and nut. If the socket tends to wedge itself between the nut and the cap, lift up on it slightly until it no longer contacts the cap. Do not rotate the crankshaft at any time during this operation. **Note:** *After reaching the specified torque, tighten each nut an additional 90-degrees (1/4-turn)* **(see illustration)**.

16 Remove the nuts and detach the rod cap, being very careful not to disturb the Plastigage.

17 Compare the width of the crushed Plastigage to the scale printed on the Plastigage

envelope to obtain the oil clearance **(see illustration)**. Compare it to this Chapter's Specifications to make sure the clearance is correct.

18 If the clearance is not as specified, the bearing inserts may be the wrong size (which means different ones will be required). Before deciding that different inserts are needed, make sure that no dirt or oil was between the bearing inserts and the connecting rod or cap when the clearance was measured. Also, recheck the journal diameter. If the Plastigage was wider at one end than the other, the journal may be tapered (refer to Section 19).

Final connecting rod installation

19 Carefully scrape all traces of the Plastigage material off the rod journal and/or bearing face. Be very careful not to scratch the bearing, use your fingernail or the edge of a credit card to remove the Plastigage.

20 Make sure the bearing faces are perfectly clean, then apply a uniform layer of clean moly-base grease or engine assembly lube to both of them. You'll have to push the piston higher into the cylinder to expose the face of the bearing insert in the connecting rod, be sure to slip the protective hoses over the rod bolts first.

21 Slide the connecting rod back into place on the journal, remove the protective hoses from the rod cap bolts, install the rod cap and tighten the nuts to the torque listed in this Chapter's Specifications. Again, work up to the torque in three steps.

22 Repeat the entire procedure for the remaining pistons/connecting rods.

23 The important points to remember are:

a) *Keep the back sides of the bearing inserts and the insides of the connecting rods and caps perfectly clean when assembling them.*
b) *Make sure you have the correct piston/rod assembly for each cylinder.*
c) *The dimple on the piston must face the front of the engine.*

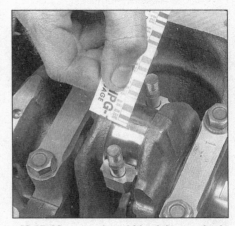

26.17 Measure the width of the crushed Plastigage to determine the rod bearing oil clearance (be sure to use the correct scale - standard and metric scales are included)

d) *Lubricate the cylinder walls with clean oil.*
e) *Lubricate the bearing faces when installing the rod caps after the oil clearance has been checked.*

24 After all the piston/connecting rod assemblies have been properly installed, rotate the crankshaft a number of times by hand to check for any obvious binding.

25 As a final step, the connecting rod endplay must be checked. Refer to Section 14 for this procedure.

26 Compare the measured endplay to this Chapter's Specifications to make sure it's correct. If it was correct before disassembly and the original crankshaft and rods were reinstalled, it should still be right. If new rods or a new crankshaft were installed, the endplay may be inadequate. If so, the rods will have to be removed and taken to an automotive machine shop for resizing.

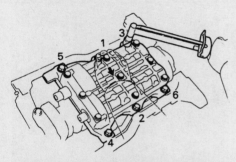

27.4 With the balance shafts aligned with the case, apply hand pressure as shown while tightening the bolts in sequence

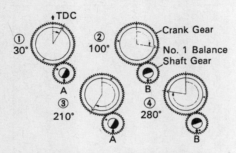

27.6 Check the balancer shaft gear-to-crankshaft gear backlash at the four locations shown

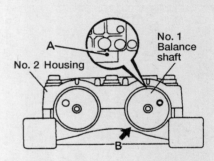

27.8 Check backlash with the balance shaft number 1 aligned as shown - punchmark A aligns with the notch in the housing

27 Balancer assembly installation (four-cylinder engines)

Refer to illustrations 27.4, 27.6, 27.8 and 27.9

1. After the crankshaft and pistons/rods have all been installed, torqued and checked, clean the surface of the block and the balancer assembly spacers, then place the spacers on the block.
2. Turn the crankshaft until number 1 piston is at TDC (see Chapter 2A or 2B).
3. Align the punch marks on the balance shafts with their corresponding marks on the housing (see illustration 21.2a). Place the balancer assembly on the block and install the bolts.
4. Apply hand pressure on the center of the assembly in the direction indicated and tighten the bolts to the torque listed in this Chapter's Specifications, in several steps, in the sequence shown (see illustration).
5. Position a dial indicator against the end of the number 1 balance shaft and check the shaft endplay while pushing the shaft back and forth by hand. Repeat the check for the number 2 balance shaft and compare your readings to this Chapter's Specifications. If the endplay is excessive, remove the balancer assembly and take it to a dealer service department or other qualified service facility for repair.
6. Because the backlash between the crankshaft gear and the number 1 balance shaft varies with rotation, backlash must be checked at the four locations indicated (see illustration).
7. Rotate the crankshaft several revolutions and return to TDC for number 1 piston. Check that the punch marks are still aligned (see illustration 21.2a).
8. Rotate the engine clockwise until punch mark A lines up with the notch in the balancer housing (see illustration). Position a lever-type dial indicator on the side of a tooth of the number 1 shaft gear (see illustration 21.3a). Lightly twist the number 1 shaft back and forth a few times by hand and, while pressing on the rear of the shaft to take up the endplay, lightly twist the shaft while observing the dial indicator. Compare the results to on-engine balancer backlash in this Chapter's Specifications.
9. Remove the dial indicator and rotate the engine clockwise until punch mark B on balance shaft 1 lines up with the notch in the housing (see illustration). Measure backlash at this position as in Step 8 and compare the results to on-engine balancer backlash in this Chapter's Specifications.
10. Repeat Steps 7 and 8 (punch mark A) and compare the results to on-engine balancer backlash in this Chapter's Specifications.
11. Repeat Step 9 (punch mark B) and compare the results to on-engine balancer backlash in this Chapter's Specifications.
12. If any of the readings exceed Specifications, change the spacer between the block and the balance assembly to obtain proper backlash. To increase backlash, select a thicker spacer; select a thinner spacer to reduce backlash. Changing the spacer by 0.0008-inch changes the backlash by 0.0006-inch. **Note:** *Various size spacers are available at your Toyota Dealership.*

28 Initial start-up and break-in after overhaul

Warning: *Have a fire extinguisher handy when starting the engine for the first time.*

1. Once the engine has been installed in the vehicle, double-check the engine oil and coolant levels.
2. With the spark plugs out of the engine and the ignition system and fuel pump disabled (see Section 3), crank the engine until oil pressure registers on the gauge or the light goes out.
3. Install the spark plugs, hook up the plug wires and restore the ignition system and fuel pump functions (see Section 3).
4. Start the engine. It may take a few moments for the fuel system to build up pressure, but the engine should start without a great deal of effort.
5. After the engine starts, it should be allowed to warm up to normal operating temperature. While the engine is warming up, make a thorough check for fuel, oil and coolant leaks.
6. Shut the engine off and recheck the engine oil and coolant levels.
7. Drive the vehicle to an area with minimum traffic, accelerate from 30 to 50 mph, then allow the vehicle to slow to 30 mph with the throttle closed. Repeat the procedure 10 or 12 times. This will load the piston rings and cause them to seat properly against the cylinder walls. Check again for oil and coolant leaks.
8. Drive the vehicle gently for the first 500 miles (no sustained high speeds) and keep a constant check on the oil level. It is not unusual for an engine to use oil during the break-in period.
9. At approximately 500 to 600 miles, change the oil and filter.
10. For the next few hundred miles, drive the vehicle normally. Do not pamper it or abuse it.
11. After 2000 miles, change the oil and filter again and consider the engine broken in.

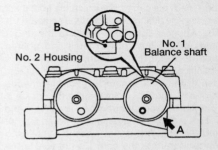

27.9 The second and fourth positions for checking balancer backlash are with balance shaft number 1 aligned as shown - punchmark B now aligns with the housing notch

Chapter 3
Cooling, heating and air conditioning systems

Contents

	Section
Air conditioning and heating system - check and maintenance	13
Air-conditioning compressor clutch circuit - check	14
Air conditioning compressor - removal and installation	16
Air conditioning condenser - removal and installation	17
Air-conditioning evaporator and expansion valve - removal and installation	18
Air conditioning receiver/drier - removal and installation	15
Antifreeze - general information	2
Blower motor circuit - check	9
Blower motor - removal and installation	10
CHECK ENGINE light	See Chapter 6
Coolant level check	See Chapter 1
Coolant temperature sending unit - check and replacement	8

	Section
Cooling system check	See Chapter 1
Cooling system servicing (draining, flushing and refilling)	See Chapter 1
Drivebelt check, adjustment and replacement	See Chapter 1
Engine cooling fans and circuit - check and replacement	4
General information	1
Heater and air conditioning control assembly - removal and installation, cable adjustment and electrical checks	11
Heater core - removal and installation	12
Radiator and coolant reservoir - removal and installation	5
Thermostat - check and replacement	3
Underhood hose check and replacement	See Chapter 1
Water pump - check	6
Water pump - removal and installation	7

Specifications

General

Radiator cap pressure rating
- Four-cylinder engine ..... 10.7 to 14.9 psi
- V6 engine ..... 12.1 to 16.4 psi

Thermostat rating
- Opens ..... 176 to 183-degrees F
- Fully open ..... 203-degrees F

Torque specifications
Ft-lbs (unless otherwise indicated)

Thermostat housing bolts
- Four-cylinder engine ..... 78 in-lbs
- V6 engine ..... 69 in-lbs

Water pump-to-block bolts
- Four-cylinder engine ..... 78 in-lbs
- V6 engine ..... 53 in-lbs

Chapter 3 Cooling, heating and air conditioning systems

1.1 Typical cooling system component locations (four-cylinder shown, V6 similar)

1. Radiator
2. Condenser
3. Cooling fans
4. Water pump
5. Thermostat
6. Coolant recovery tank

1 General information

Engine cooling system

Refer to illustrations 1.1 and 1.2

All vehicles covered by this manual employ a pressurized engine cooling system with thermostatically controlled coolant circulation **(see illustration)**. An impeller type water pump mounted on the front of the block pumps coolant through the engine. The coolant flows around each cylinder and toward the rear of the engine. Cast-in coolant passages direct coolant around the intake and exhaust ports, near the spark plug areas and in proximity to the exhaust valve guides.

A wax pellet-type thermostat is located in the thermostat housing near the front of the engine **(see illustration)**. During warm up, the closed thermostat prevents coolant from circulating through the radiator. When the engine reaches normal operating temperature, the thermostat opens and allows hot coolant to travel through the radiator, where it is cooled before returning to the engine.

The cooling system is sealed by a pressure-type radiator cap. This raises the boiling point of the coolant, and the higher boiling point of the coolant increases the cooling efficiency of the radiator. If the system pressure exceeds the cap pressure relief value, the excess pressure in the system forces the spring-loaded valve inside the cap off its seat and allows the coolant to escape through the overflow tube into a coolant reservoir. When the system cools, the excess coolant is automatically drawn from the reservoir back into the radiator.

The coolant reservoir serves as both the point at which fresh coolant is added to the cooling system to maintain the proper fluid level and as a holding tank for overheated coolant.

This type of cooling system is known as a closed design because coolant that escapes past the pressure cap is saved and reused.

Heating system

The heating system consists of a blower fan and heater core located within the heater box, the inlet and outlet hoses connecting the heater core to the engine cooling system and the heater/air conditioning control head on the dashboard. Hot engine coolant is circulated through the heater core. When the heater mode is activated, a flap door opens to expose the heater box to the passenger compartment. A fan switch on the control head activates the blower motor, which forces air through the core, heating the air.

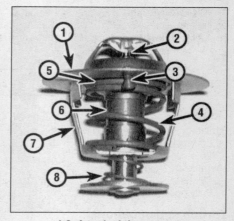

1.2 A typical thermostat

1. Flange
2. Piston
3. Jiggle valve
4. Main coil spring
5. Valve seat
6. Valve
7. Frame
8. Secondary coil spring

Chapter 3 Cooling, heating and air conditioning systems

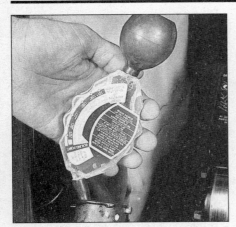

2.4 An inexpensive hydrometer can be used to test the condition of your coolant

Air conditioning system

The air conditioning system consists of a condenser mounted in front of the radiator, an evaporator mounted adjacent to the heater core under the dashboard, a compressor mounted on the engine, a filter-drier (accumulator) which contains a high pressure relief valve and the plumbing connecting all of the above.

A blower fan forces the warmer air of the passenger compartment through the evaporator core (sort of a radiator-in-reverse), transferring the heat from the air to the refrigerant. The liquid refrigerant boils off into low pressure vapor, taking the heat with it when it leaves the evaporator. The compressor keeps refrigerant circulating through the system, pumping the warmed coolant through the condenser where it is cooled and then circulated back to the evaporator.

Some models are equipped with an optional Automatic Air-Conditioning system. With this system, you select the desired interior temperature with a knob on the controls, similar to setting the temperature on a home heating/cooling thermostat, and the system automatically adds the right blend of cool or warm air to maintain this temperature. The system has sensors that detect both the interior and outside temperature.

2 Antifreeze - general information

Refer to illustration 2.4

Warning: *Do not allow antifreeze to come in contact with your skin or painted surfaces of the vehicle. Rinse off spills immediately with plenty of water. Antifreeze is highly toxic if ingested. Never leave antifreeze lying around in an open container or in puddles on the floor; children and pets are attracted by it's sweet smell and may drink it. Check with local authorities about disposing of used antifreeze. Many communities have collection centers which will see that antifreeze is disposed of safely. Never dump used antifreeze on the ground or into drains.*

Note: *Non-Toxic coolant is available at local auto parts stores. Although the coolant is non-toxic when fresh, proper disposal of used coolant is still required.*

The cooling system should be filled with a water/ethylene-glycol based antifreeze solution, which will prevent freezing down to at least -20 degrees F, or lower if local climate requires it. It also provides protection against corrosion and increases the coolant boiling point.

The cooling system should be drained, flushed and refilled at least every other year (see Chapter 1). The use of antifreeze solutions for periods of longer than two years is likely to cause damage and encourage the formation of rust and scale in the system. If your tap water is "hard", i.e. contains a lot of dissolved minerals, use distilled water with the antifreeze.

Before adding antifreeze to the system, check all hose connections, because antifreeze tends to leak through very minute openings. Engines do not normally consume coolant. Therefore, if the level goes down find the cause and correct it.

The exact mixture of antifreeze-to-water which you should use depends on the relative weather conditions. The mixture should contain at least 50 percent antifreeze, but should never contain more than 70 percent antifreeze. Consult the mixture ratio chart on the antifreeze container before adding coolant. Hydrometers are available at most auto parts stores to test the ratio of antifreeze to water **(see illustration)**. Use antifreeze which meets the vehicle manufacturer's specifications.

3 Thermostat - check and replacement

Warning: *Do not attempt to remove the radiator cap, coolant or thermostat until the engine has cooled completely.*

General check

1   Before assuming the thermostat is responsible for a cooling system problem, check the coolant level (Chapter 1), drivebelt tension (Chapter 1) and temperature gauge (or light) operation.
2   If the engine takes a long time to warm up (as indicated by the temperature gauge or heater operation), the thermostat is probably stuck open. Replace the thermostat with a new one.
3   If the engine runs hot, use your hand to check the temperature of the lower radiator hose. If the hose is not hot, but the engine is, the thermostat is probably stuck in the closed position, preventing the coolant inside the engine from escaping to the radiator. Replace the thermostat. **Caution:** *Do not drive the vehicle without a thermostat. The computer may stay in open loop and emissions and fuel economy will suffer.*
4   If the lower radiator hose is hot, it means that the coolant is flowing and the thermostat is open. Consult the Troubleshooting Section at the front of this manual for further diagnosis.

Thermostat test

Refer to illustration 3.7

5   A more thorough test of the thermostat can only be made when it is removed from the vehicle (see below). If the thermostat remains in the open position at room temperature, it is faulty and must be replaced.
6   To test it fully, suspend the (closed) thermostat on a length of string or wire in a container of cold water, with a thermometer (cooking type that reads beyond 212 degrees F). A clear Pyrex cooking container is easiest to use.
7   Heat the water on a stove while observing the temperature and the thermostat. Neither should contact the sides of the container **(see illustration)**.
8   Note the temperature when the thermostat begins to open and when it is fully open. Compare the temperatures to the Specifications in this Chapter. The number stamped into the thermostat is generally the fully-open temperature. Some manufacturers provide Specifications for the beginning-to-open temperature, the fully-open temperature, and sometimes the amount the valve should open.
9   If the thermostat doesn't open and close as specified, or sticks in any position, replace it.

Replacement

Refer to illustrations 3.12, 3.13a, 3.13b, 3.14 and 3.16

10   Disconnect the negative cable from the battery. **Caution:** *If the stereo in your vehicle is equipped with an anti-theft system, make sure you have the correct activation code before disconnecting the battery.*

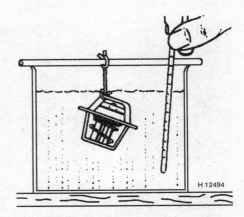

3.7 A thermostat can be accurately checked by heating it in a container of water with a thermostat and observing the opening and fully-open temperature

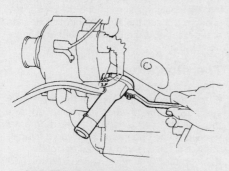

3.12 The thermostat cover on four-cylinder models is located behind the water pump - access to the nuts is easier with the oil filter removed

3.13a With air cleaner cover removed, remove the nuts (arrows indicate two of three) on the back of the V6 engine and move the harness for access to the thermostat cover bolts

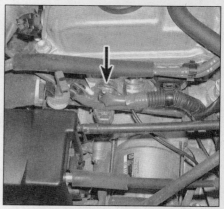

3.13b On V6 models, unbolt the water inlet pipe (arrow) from the cylinder head and remove it from the thermostat housing

3.14 The V6 thermostat housing (arrow) is at the back of the intake manifold, above the transaxle - remove the hoses, wiring plugs and the three nuts

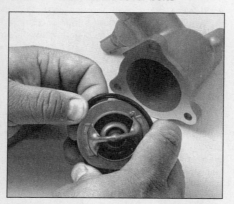

3.16 The thermostat gasket fits around the edge of the thermostat like a grooved sealing ring

11 Drain the coolant from the radiator (see Chapter 1). Disconnect the radiator hose from the thermostat housing or water inlet pipe.
12 On four-cylinder models remove the water inlet pipe from the thermostat housing (located on the back of the water pump housing) **(see illustration)**. **Note:** *For easier access to the lower bolt, remove the oil filter.*
13 On V6 models, unbolt the water inlet pipe and remove the pipe and O-ring **(see illustrations)**.

14 Remove the fasteners and electrical connectors from the thermostat housing and detach the housing from the engine **(see illustration)**. Be prepared for some coolant to spill as the gasket seal is broken. **Note:** *Access on V6 engines is easier with the air cleaner cover and hose removed.*
15 Remove the thermostat, noting the direction in which it was installed in the housing, and thoroughly clean the sealing surfaces.
16 Fit a new gasket onto the thermostat **(see illustration)**. Make sure it is evenly fitted all the way around.
17 Install the thermostat and housing, positioning the jiggle pin at the highest point.
18 Tighten the housing fasteners to the torque listed in this Chapter's Specifications and reinstall the remaining components in the reverse order of removal. On V6 models, use a new O-ring on the water inlet pipe and lubricate the O-ring with soapy water.
19 Refill the cooling system and on V6 models, bleed the air from the system (see Chapter 1). Run the engine and check for leaks and proper operation.

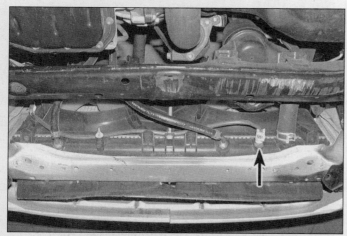

4.1a On all models, the number 1 coolant temperature switch (arrow) is located in the bottom of the radiator

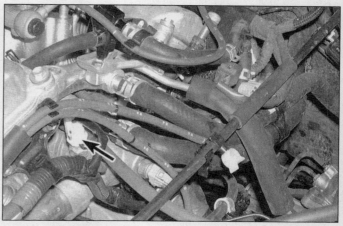

4.1b On V6 models, the number 2 coolant temperature switch (arrow) is on the thermostat housing at the driver's side of the intake manifold

Chapter 3 Cooling, heating and air conditioning systems

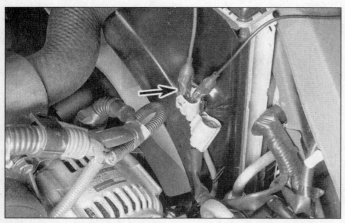

4.4 Disconnect the fan wiring connector and connect jumper wires directly to the positive and negative terminals of the battery

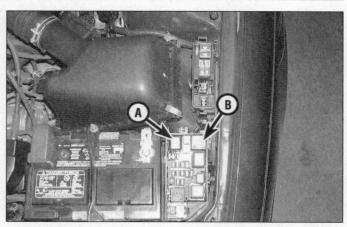

4.6a Main engine relay (A) and number 1 (B) cooling fan relay locations - Avalon model shown

4 Engine cooling fans and circuit - check and replacement

Warning: *To avoid possible injury, keep clear of the fan blades, as they may start turning at any time!*

Check

Refer to illustrations 4.1a, 4.1b and 4.4

1 If the radiator fan won't shut off when the engine is cool, disconnect the wiring connector from the coolant temperature switch(es) **(see illustrations)** and bridge the connector. If this shuts off the fan, replace the switch. **Note:** *Four-cylinder models have a number 1 coolant switch in the bottom of the radiator, while Camry/Avalon V6 models have two switches, each located on the thermostat housing. Solara V6 models have one switch in the radiator and one on the thermostat housing.*

2 Test the number 1 ECT (Engine Coolant Temperature switch) for continuity with an ohmmeter. When the engine is cold (below 190-degrees F) there should be continuity. When the engine is warm (above 208-degrees F), there should be NO continuity.

3 Check the number 2 ECT with an ohmmeter also. It should exhibit NO continuity when below 181-degrees F, and there should be continuity when above 190-degrees F. If either ECT fails these test replace the defective switch.

4 To test an inoperative fan motor (one that doesn't come on when the engine gets hot or when the air conditioner is on), first check the fuses and/or fusible links (see Chapter 12). Then disconnect the electrical connector at the motor and use fused jumper wires to connect the fan directly to the battery **(see illustration)**. If the fan still does not work, replace the fan motor. **Warning:** *Do not allow the test clips to contact each other or any metallic part of the vehicle.*

5 If the motor tested OK in the previous test but is still inoperative, then the fault lies in the relays, wiring or the cooling fan ECU. The fan relay and main engine relay can be tested as described below.

Relay check

Refer to illustrations 4.6a, 4.6b, 4.7 and 4.8

6 Locate the main relay box (located in the engine compartment, on the driver's side) **(see illustrations)**. Only air-conditioning-equipped models will have a cooling fan relay number 2, and not all will have a number 3

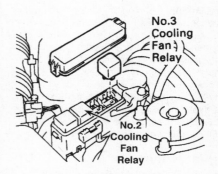

4.6b Cooling fan relay box - Avalon models

relay. If your vehicle has them, test them.

7 Remove the cooling fan no.1 relay, and, using an ohmmeter, test for continuity as shown **(see illustration)**. There should be continuity between terminals 1 and 2 (numbered 85 and 86 on Bosch relays), and between 3 and 4 (numbered 30 and 87a on Bosch relays). When battery voltage is applied across terminals 1 and 2, there should be NO continuity between terminals 3 and 4. On some Avalon models, Bosch relays may be found instead of Denso relays. These have different numbers on the terminals but use the same terminal layout and are tested the same.

8 If the fan relays check OK, remove the main engine relay and test it as shown **(see**

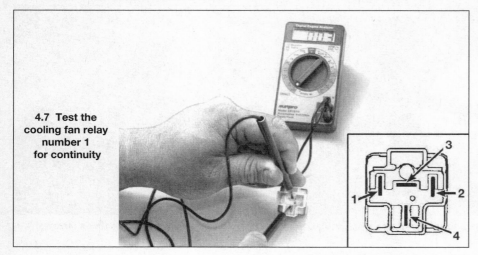

4.7 Test the cooling fan relay number 1 for continuity

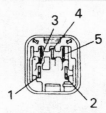

4.8 Test the main engine relay for continuity

3-6 Chapter 3 Cooling, heating and air conditioning systems

4.12 Unbolt the fan and shroud and remove the assembly from the radiator - arrows indicate the upper bolts at each fan; there are more at the bottom of the shroud

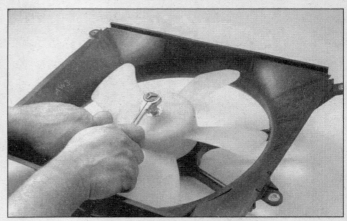

4.13 Remove the fan from the motor

illustration). There should be continuity between terminals 3 and 5, and between 2 and 4. There should be NO continuity between terminals 1 and 2. With battery voltage applied across terminals 3 and 5, there should be continuity between 1 and 2, and NO continuity between 2 and 4. If all relays test OK, we recommend you take the vehicle to a dealer or other qualified repair facility for further diagnosis, due to the complexity and variety of the circuits involved.

Replacement

Refer to illustrations 4.12, 4.13 and 4.14

9 Disconnect the negative battery cable.
Caution: *If the stereo in your vehicle is equipped with an anti-theft system, make sure you have the correct activation code before disconnecting the battery.*
10 Disconnect the wiring connector at the fan motor. **Note**: *All models have two fans. Both fans should operate all the time at low speed. When the air conditioning is On, both fans should operate at a higher speed.*
11 On models with cruise control, disconnect the cruise control actuator wire from the fan shroud for easier removal of fan number 1.
12 Unbolt the fan/shroud from the radiator

and lift it from the vehicle **(see illustration)**. On models with air-conditioning, the number 2 fan may have to be removed first to allow room for removal of the number 1 fan.
13 Hold the fan blades and remove the fan retaining clip or nut **(see illustration)**. **Note**: *The number 1 fan blade is attached with a nut, the number 2 with a clip.*
14 Unbolt the fan motor from the shroud **(see illustration)**.
15 Installation is the reverse of removal. When installing the blade to fan motor number 2, use a new clip, and orient the clip's open side to clear the projecting bump on the blade. If a new fan switch is installed, lubricate the O-ring with soapy water before installation.

---

5 Radiator and coolant reservoir - removal and installation

Refer to illustrations 5.6 and 5.7
Warning: *Do not start this procedure until the engine is completely cool. Do not allow antifreeze to come in contact with your skin or painted surfaces of the vehicle. Rinse off spills immediately with plenty of water.*

Antifreeze is highly toxic if ingested. Never leave antifreeze lying around in an open container or in puddles on the floor; children and pets are attracted by it's sweet smell and may drink it. Check with local authorities about disposing of used antifreeze. Many communities have collection centers which will see that antifreeze is disposed of safely. Never dump used antifreeze on the ground or into drains.
Note: *Non-Toxic coolant is available at local auto parts stores. Although the coolant is non-toxic when fresh, proper disposal of used coolant is still required.*

Removal

1 Disconnect the negative battery cable.
Caution: *If the stereo in your vehicle is equipped with an anti-theft system, make sure you have the correct activation code before disconnecting the battery.*
2 Drain the coolant into a container (see Chapter 1).
3 Remove both the upper and lower radiator hoses.
4 Disconnect the reservoir hose from the radiator filler neck.
5 Remove the cooling fan (see Section 4).

4.14 Remove the screws (arrows) and separate the motor from the shroud

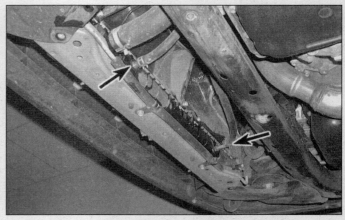

5.6 Remove the automatic transaxle cooler lines (arrows)

Chapter 3 Cooling, heating and air conditioning systems

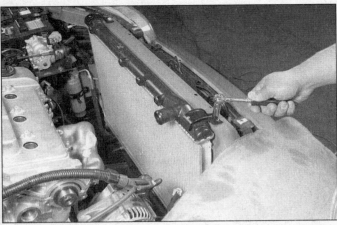

5.7 Remove the hold-down clamps from each end of the radiator

5.15 Typical coolant reservoir removal - remove the upper bolt (arrow)

6 If equipped with an automatic transaxle, disconnect the cooler lines from the radiator **(see illustration)**. Place a drip pan to catch the fluid and cap the fittings.
7 Remove the bolts that attach the radiator to its support **(see illustration)**.
8 Lift out the radiator. Be aware of dripping fluids and the sharp fins.
9 With the radiator removed, it can be inspected for leaks, damage and internal blockage. If in need of repairs, have a professional radiator shop or dealer service department perform the work, as special techniques are required.
10 Bugs and dirt can be cleaned from the radiator with compressed air and a soft brush. Don't bend the cooling fins as this is done. **Warning:** *Wear eye protection.*

Installation

11 Installation is the reverse of the removal procedure. Be sure the rubber mounts are in place.
12 After installation, fill the cooling system with the proper mixture of antifreeze and water. Refer to Chapter 1 if necessary.
13 Start the engine and check for leaks. Allow the engine to reach normal operating temperature, indicated by the upper radiator hose becoming hot. Recheck the coolant level and add more if required.
14 On automatic transmission equipped models, check and add fluid as needed.

Coolant reservoir, removal and installation

Refer to illustration 5.15

15 On most models, remove one bolt and the coolant reservoir simply pulls up and out of the bracket on the fenderwell **(see illustration)**.
16 Pour the coolant into a container. Wash out and inspect the reservoir for cracks and chafing. Replace it if damaged.
17 Installation is the reverse of removal.

6 Water pump - check

1 A failure in the water pump can cause serious engine damage due to overheating.
2 With the engine running and warmed to normal operating temperature, squeeze the upper radiator hose. If the water pump is working properly, a pressure surge should be felt as the hose is released. **Warning:** *Keep hands away from fan blades!*
3 Water pumps are equipped with weep or vent holes. If a failure occurs in the pump seal, coolant will leak from this hole. In most cases it will be necessary to use a flashlight to find the hole on the water pump by looking through the space behind the pulley just below the water pump shaft. A slight gray discoloration around the weep hole is normal, while dark brown stains indicate a problem.

4 If the water pump shaft bearings fail there may be a howling sound at the front of the engine while it is running. Bearing wear can be felt if the water pump pulley is rocked up and down. Do not mistake drivebelt slippage, which causes a squealing sound, for water pump failure. Spray automotive drivebelt dressing on the belts to eliminate the belt as a possible cause of the noise.

7 Water pump - removal and installation

Warning: *Do not start this procedure until the engine is completely cool. Do not allow antifreeze to come in contact with your skin or painted surfaces of the vehicle. Rinse off spills immediately with plenty of water. Antifreeze is highly toxic if ingested. Never leave antifreeze lying around in an open container or in puddles on the floor; children and pets are attracted by it's sweet smell and may drink it. Check with local authorities about disposing of used antifreeze. Many communities have collection centers which will see that antifreeze is disposed of safely. Never dump used antifreeze on the ground or into drains.*
Note: *Non-Toxic coolant is available at local auto parts stores. Although the coolant is non-toxic when fresh, proper disposal of used coolant is still required.*

Four-cylinder engine

Refer to illustrations 7.3, 7.4, 7.6, 7.9a and 7.9b

1 Disconnect the negative battery cable and drain the cooling system. **Caution:** *If the stereo in your vehicle is equipped with an anti-theft system, make sure you have the correct activation code before disconnecting the battery.*
2 Remove the timing belt, number 1 and number 2 idler pulleys (see Chapter 2A).
3 Remove the alternator adjusting bar **(see illustration)**. On models with an engine oil cooler, unbolt the air-conditioning compressor and set it aside without disconnect-

7.3 Remove the alternator adjusting bar (arrow)

Chapter 3 Cooling, heating and air conditioning systems

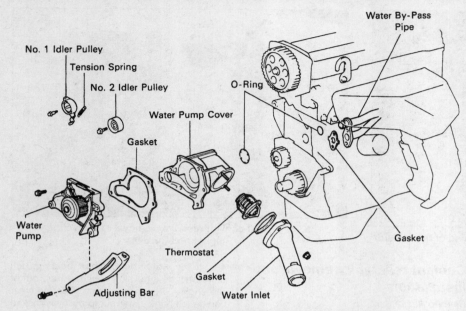

7.4 Typical four-cylinder engine water pump components

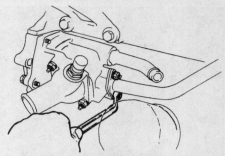

7.6 Disconnect the coolant bypass hose from the water neck - the heater pipe is attached with two nuts

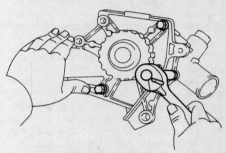

7.9a Install these bolts first . . .

ing the refrigerant lines (see Section 16). Disconnect the lower radiator hose.

4 Remove the bolts from the water pump **(see illustration)**, noting the locations of the different length bolts and the sequence of removal. Remove the pump and gasket. If necessary, tap the pump loose with a soft-face hammer.

5 It isn't necessary to remove to remove the pump cover (housing), but a thorough job would include removing it to replace the gaskets and O-rings.

6 Disconnect the coolant bypass hose from the water neck, then remove the two nuts and heater pipe **(see illustration)** and lift out the pump cover (housing).

7 Thoroughly clean all sealing surfaces, removing all traces of old gaskets, sealer and O-rings.

8 Be sure to use new O-rings between the pump cover and engine block and also a new gasket between the heater pipe and cover.

9 Using a new gasket, install the pump and bolts **(see illustrations)** and tighten them to the torque listed in this Chapter's Specifications.

10 Install the remaining parts in the reverse order of removal.

11 Refill the cooling system (see Chapter 1), run the engine and check for leaks and proper operation.

V6 engine

Refer to illustrations 7.14, 7.15 and 7.17

12 Disconnect the negative battery cable and drain the cooling system. **Caution:** *If the stereo in your vehicle is equipped with an anti-theft system, make sure you have the correct activation code before disconnecting the battery.*

13 Remove the timing belt and camshaft sprockets (see Chapter 2B).

14 Remove the bolt retaining the water inlet pipe to the alternator adjuster and remove the water inlet pipe and thermostat (see Section 3). Remove the number 3 timing belt cover **(see illustration)**.

15 Remove the water pump retaining bolts/nuts and separate the pump from the engine. **Note:** *On V6 models, there are two long studs (some engines may have bolts) that hold the engine mounting bracket and the water pump. The studs have a small hex-head that is used to remove the stud* **(see illustration)**. *If it is stuck, remove the stud with locking pliers and replace the stud with a new bolt of the same length. At least one of these long studs must be removed to allow the water pump to be removed.*

16 Thoroughly clean all sealing surfaces, removing all traces of old gaskets, sealer and

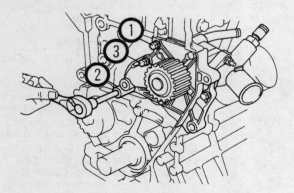

7.9b . . . then tighten these bolts in the order shown

7.14 Release the three wiring harness clips (arrows), pull the harness back and remove the number three belt cover

Chapter 3 Cooling, heating and air conditioning systems

7.15 V6 water pumps have two long studs (arrow) - at least one must be removed with a small wrench to allow water pump removal

7.17 On V6 models, a new metal/rubber gasket (arrow) is used when installing the water pump

8.3a Coolant temperature sending unit location - four-cylinder engine

8.3b The coolant temperature sending unit (arrow) on V6 engines is located in the water outlet at the passenger's side of the engine - use a deep socket to remove it

O-rings. Remove any traces of oil with acetone or lacquer thinner and a clean rag.
17 These models use a metal and rubber gasket **(see illustration)**.
18 Reinstall the pump and the remaining parts in the reverse order of removal. Tighten the water pump bolts in several steps to the torque listed in this Chapter's Specifications.
19 Refill the cooling system (see Chapter 1) and run the engine, checking for leaks and proper operation.

8 Coolant temperature sending unit - check and replacement

Warning: *Do not start this procedure until the engine is completely cool. Do not allow antifreeze to come in contact with your skin or painted surfaces of the vehicle. Rinse off spills immediately with plenty of water. Antifreeze is highly toxic if ingested. Never leave antifreeze lying around in an open container or in puddles on the floor; children and pets are attracted by it's sweet smell and may drink it.*

Check with local authorities about disposing of used antifreeze. Many communities have collection centers which will see that antifreeze is disposed of safely. Never dump used antifreeze on the ground or into drains.
Note: *Non-Toxic coolant is available at local auto parts stores. Although the coolant is non-toxic when fresh, proper disposal of used coolant is still required.*

Check

Refer to illustrations 8.3a and 8.3b
1 If the coolant temperature gauge is inoperative, check the fuses first (see Chapter 12).
2 If the temperature gauge indicates excessive temperature after running awhile, see the Troubleshooting Section in the front of the manual.
3 If the temperature gauge indicates Hot as soon as the engine is started cold, disconnect the wire at the coolant temperature sender **(see illustrations)**. If the gauge reading drops, replace the sending unit. If the reading remains high, the wire to the gauge may be shorted to ground or the gauge is faulty.

4 If the coolant temperature gauge fails to show any indication after the engine has been warmed up, (approx. 10 minutes) and the fuses checked out OK, shut off the engine. Disconnect the wire at the sending unit and, using a jumper wire, connect the wire to a clean ground on the engine. Briefly turn on the ignition without starting the engine. If the gauge now indicates Hot, replace the sending unit.
5 If the gauge fails to respond, the circuit may be open or the gauge may be faulty - see Chapter 12 for additional information.

Replacement

6 Drain the coolant (see Chapter 1).
7 Disconnect the wiring connector from the sending unit.
8 Using a deep socket or a wrench, remove the sending unit.
9 Install the new unit and tighten it securely. Do not use thread sealer as it may electrically insulate the sending unit.
10 Reconnect the wiring connector, refill the cooling system and check for coolant leakage and proper gauge function.

9 Blower motor circuit - check

Refer to illustrations 9.4, 9.8, 9.9 and 9.10

Note: *This procedure does not apply to models with Automatic Air Conditioning. For models with Automatic Air Conditioning, refer to Section 13, Steps 29 to 33.*

1 If the blower motor doesn't operate, first check all connections in the circuit for looseness and corrosion. Make sure the battery is fully charged. Check fuse 2 in the fuse panel below the driver's side of the dashboard. Also check the blower relay in the underhood fuse/relay box. Remove the relay and test it according to the procedure in Chapter 12.
2 Locate the blower motor relay in the underhood fuse/relay box. Remove the relay and test it for proper operation (see Chapter 12). If the relay fails the tests, replace it.
3 With the transmission in Park and the parking brake securely set, turn the ignition switch to the ON position. It isn't necessary to start the vehicle.
4 The blower motor is located under the glove compartment area of the dash, near the firewall **(see illustration)**.
5 Switch the heater controls to FLOOR and the blower speed to 4. Listen at the ducts to hear if the blower is operating. If it is, then switch the blower speed to 1 and listen again. Try all the speeds.
6 If the blower didn't operate, and you have checked the fuses and relay, test the blower. Locate the electrical connector at the blower motor **(see illustration 9.4)**. Backprobe the Blue/red wire terminal; there should be at least 10 volts with the mode switch in any position other than Off and the ignition switch On. If not, there is a problem in the circuit from the fuse panel to the heater/air conditioning control panel, or from the control panel to the blower.
7 If there is voltage at the feed wire, but the blower does not operate, backprobe the Black/white wire and connect it to a known good chassis ground with a jumper wire. If it still doesn't operate, replace the blower motor. If the blower now operates there is a problem in the circuit from the blower to the blower speed switch to the resistor to ground. The most likely problem is a bad ground connection. Check the connection between the blower resistor and ground and the blower switch and ground (White/Black wires)
8 If the blower operates, but not at all speeds, check the blower resistor. It's located on the blower housing under the right side of the dash **(see illustration)**. There are resistor elements mounted on the resistor board to provide blower speeds Lo, 2 and 3 (HI or 4 bypasses the resistor).
9 With the resistor removed from the vehicle, visually check the resistor for damage, indicated by the material melting out between the contacts. Check the resistor block with an ohmmeter between the terminals, as shown **(see illustration)**. If the resistance values are not as specified, replace the resistor.
10 If the blower operates, but not at all speeds and you have already checked the

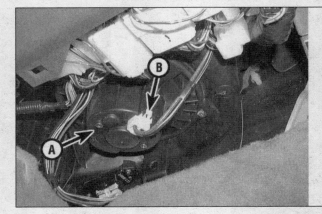

9.4 Blower motor location
- A Blower motor
- B Blower motor connector

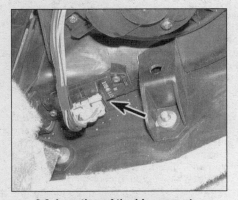

9.8 Location of the blower motor resistor (arrow)

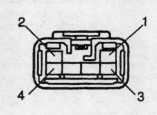

Test between terminals	Resistance (approximate)
2 and 4	0.38 ohms
2 and 3	1.12 ohms
1 and 3	1.74 ohms

9.9 Continuity diagram for the blower motor resistor

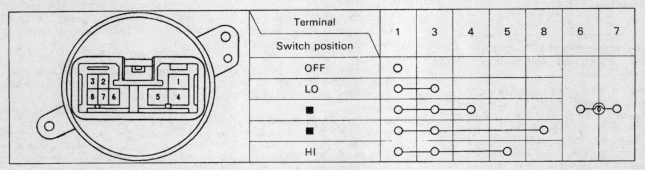

9.10 Continuity test for the blower speed switch

Chapter 3 Cooling, heating and air conditioning systems 3-11

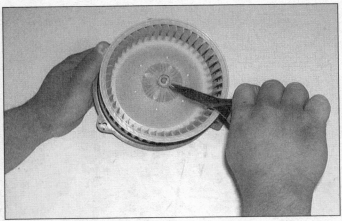

10.5 Use pliers to release and remove the clip, then lift the blower fan off the motor shaft

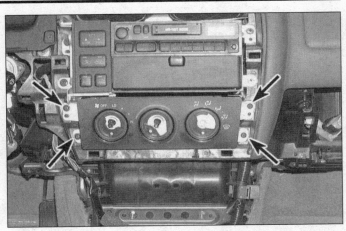

11.4 Remove these screws (arrows) to release the control assembly

blower resistor, refer to Section 12 and remove the heater/air conditioning control panel. Disconnect the electrical connector from the back of the blower speed switch and test the terminals for continuity **(see illustration)**. If the continuity is not as described, replace the blower speed switch.

10 Blower motor - removal and installation

Refer to illustration 10.5
Warning: *The models covered by this manual are equipped with Supplemental Restraint systems (SRS), more commonly known as airbags. Always disconnect the negative battery cable, then the positive battery cable and wait two minutes before working in the vicinity of the impact sensors, steering column or instrument panel to avoid the possibility of accidental deployment of the airbag, which could cause personal injury (see Chapter 12). The yellow wiring harnesses and connectors routed through the console and instrument panel are for this system. Do not use electrical test equipment on any of the airbag system wiring or tamper with them in any way.*
1   Disconnect the negative cable from the battery. **Caution:** *If the stereo in your vehicle is equipped with an anti-theft system, make sure you have the correct activation code before disconnecting the battery.*
2   The blower unit is located in the passenger compartment above the right front footwell.
3   Remove the glove compartment liner, the glove compartment door and the right lower dash panel (see Chapter 11).
4   To remove the blower, remove the blower unit retaining screws and lower the unit from the housing **(see illustration 9.4)**.
5   If the motor is being replaced, transfer the fan to the new motor prior to installation **(see illustration)**.
6   Installation is the reverse of removal. Check for proper operation.

11 Heater and air conditioning control assembly - removal and installation, cable adjustment and electrical checks

Refer to illustrations 11.4, 11.5a, 11.5b, 11.9, 11.10, 11.11 and 11.13
Warning: *The models covered by this manual are equipped with Supplemental Restraint systems (SRS), more commonly known as airbags. Always disconnect the negative battery cable, then the positive battery cable and wait two minutes before working in the vicinity of the impact sensors, steering column or instrument panel to avoid the possibility of accidental deployment of the airbag, which could cause personal injury (see Chapter 12). The yellow wiring harnesses and connectors routed through the console and instrument panel are for this system. Do not use electrical test equipment on any of the airbag system wiring or tamper with them in any way.*

Removal and installation

1   Disconnect the negative cable from the battery. **Caution:** *If the stereo in your vehicle is equipped with an anti-theft system, make sure you have the correct activation code before disconnecting the battery.*
2   Remove the center instrument trim panel (see Chapter 11). Remove the radio (see Chapter 12).
3   Pull off the control knobs (if equipped).
4   Remove the mounting screws located on the front of the control assembly **(see illustration)**.
5   Pull the control out slightly. Push the temperature control shaft (Camry models) through and leave it in position as the control

11.5a Disconnect the electrical connectors (arrows) from the control assembly (Avalon model shown)

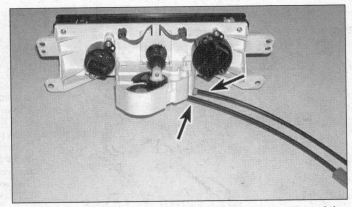

11.5b Disconnect the two cables (arrows) from the bottom of the control assembly (Avalon model shown)

3-12   Chapter 3   Cooling, heating and air conditioning systems

11.9 The air mix control cable (arrow) should be attached when the control is in the COOL position

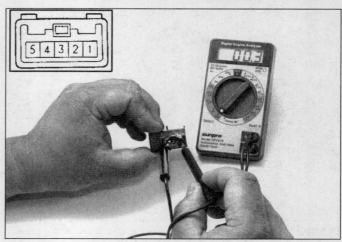

11.10 With the switch pushed IN, there should be continuity between terminals 2 and 5 and between 3 and 4 on the air-conditioning switch

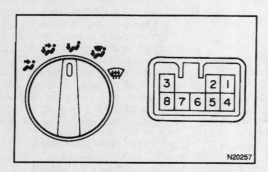

INSPECT MODE SWITCH CONTINUITY

Switch position	Tester connection	Specified condition
FACE	1 – 6	Continuity
B/L	1 – 5	Continuity
FOOT	1 – 4	Continuity
FOOT/DEF.	1 – 3	Continuity
DEF.	1 – 2 7 – 8	Continuity

11.11 Test the mode switch for continuity

panel is pulled forward **(see illustrations)**.
6 Installation is the reverse of the removal procedure.
7 Run the engine and check for proper functioning of the heater (and air conditioning, if equipped).

Cable adjustment (except Automatic Air Conditioning)

8 To adjust the air inlet damper control cable set the control lever to FRESH, install

the cable and clamp it in place.
9 To adjust the air mix control cable, set the air mix to COOL, install the cable and lock the clamp while applying slight pressure (away from the firewall) on the outer cable **(see illustration)**.

Electrical checks (except Automatic Air Conditioning)

10 Check the air-conditioning switch continuity between terminals 2 and 5. There

should be continuity when the switch is pushed in (ON) **(see illustration)**.
11 Using a self-powered continuity tester, test the terminals on the back of the control unit, starting with the mode control switch **(see illustration)**.
12 Test the blower speed control switch for continuity (see Section 9). Any tests which do not meet specifications indicate the control unit should be replaced.
13 To replace any control panel components, remove the control knobs, then the four screws on the face plate **(see illustration)**.

12 Heater core - removal and installation

Refer to illustrations 12.3, 12.5, 12.6 and 12.7
Warning 1: *The models covered by this manual are equipped with Supplemental Restraint systems (SRS), more commonly known as airbags. Always disconnect the negative battery cable, then the positive battery cable and wait two minutes before working in the vicinity of the impact sensors, steering column or instrument panel to avoid the possibility of accidental deployment of the airbag, which*

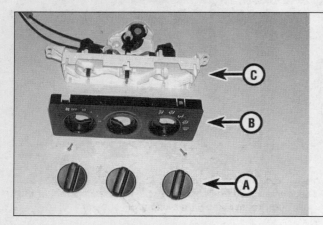

11.13 To replace any of the air-conditioning control panel switches, remove the knobs (A), then the screws and the face plate (B) - at each switch remove the mounting screws from the front of the base (C) and remove the switch from the back

Chapter 3 Cooling, heating and air conditioning systems   3-13

12.3 Squeeze the clamps and disconnect the heater hoses (arrows) at the firewall

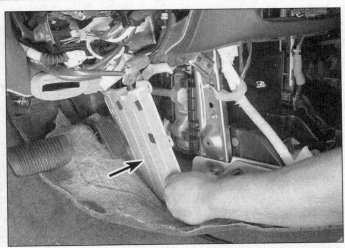

12.5 Pull back the carpeting on the driver's side, and remove the heater protector (arrow)

12.6 Disconnect the heater pipes at the junctions (arrows)

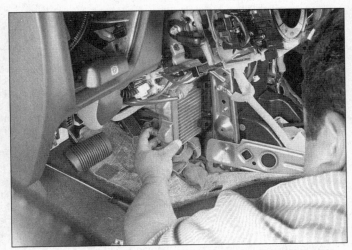

12.7 Slide the heater core out of heater/air conditioning unit

could cause personal injury (see Chapter 12). The yellow wiring harnesses and connectors routed through the console and instrument panel are for this system. Do not use electrical test equipment on any of the airbag system wiring or tamper with them in any way.
Warning 2: *Do not allow antifreeze to come in contact with your skin or painted surfaces of the vehicle. Rinse off spills immediately with plenty of water. Antifreeze is highly toxic if ingested. Never leave antifreeze lying around in an open container or in puddles on the floor; children and pets are attracted by it's sweet smell and may drink it. Check with local authorities about disposing of used antifreeze. Many communities have collection center s which will see that antifreeze is disposed of safely. Never dump used antifreeze on the ground or into drains.*
Note: *Non-Toxic coolant is available at local auto parts stores. Although the coolant is non-toxic when fresh, proper disposal of used coolant is still required.*

1 Disconnect the negative cable from the battery. **Caution:** *If the stereo in your vehicle is equipped with an anti-theft system, make sure you have the correct activation code before disconnecting the battery.*
2 Drain the cooling system (see Chapter 1).
3 Working in the engine compartment, disconnect the heater hoses where they enter the firewall, below the heater valve **(see illustration)**.
4 Pull back the carpeting from the driver's side of the console area.
5 Remove the two clips and the heater protector plate **(see illustration)**.
6 Remove the three screws and clamps and disconnect the heater pipes **(see illustration)**.
7 Pull the heater core out of the heater/air conditioning unit **(see illustration)**. **Note:** *Keep plenty of towels or rags on the carpeting to catch any coolant that may drip.*
8 Installation is the reverse order of removal. Use new O-rings to seal the heater core pipes.

9 Refill the cooling system, reconnect the battery and run the engine. Check for leaks and proper system operation.

13 Air conditioning and heating system - check and maintenance

Air conditioning system

Refer to illustration 13.1
Warning: *The air conditioning system is under high pressure. Do not loosen any hose fittings or remove any components until the system has been discharged. Air conditioning refrigerant should be properly discharged into an EPA-approved recovery/recycling unit by a dealer service department or an automotive air conditioning repair facility. Always wear eye protection when disconnecting air conditioning system fittings.*

1 The following maintenance checks should be performed on a regular basis to ensure that the air conditioner continues to

operate at peak efficiency **(see illustration)**.
  a) Inspect the condition of the compressor drivebelt. If it is worn or deteriorated, replace it (see Chapter 1).
  b) Check the drivebelt tension and, if necessary, adjust it (see Chapter 1).
  c) Inspect the system hoses. Look for cracks, bubbles, hardening and deterioration. Inspect the hoses and all fittings for oil bubbles or seepage. If there is any evidence of wear, damage or leakage, replace the hose(s).
  d) Inspect the condenser fins for leaves, bugs and any other foreign material that may have embedded itself in the fins. Use a "fin comb" or compressed air to remove debris from the condenser.
  e) Make sure the system has the correct refrigerant charge.

2 It's a good idea to operate the system for about ten minutes at least once a month. This is particularly important during the winter months because long term non-use can cause hardening, and subsequent failure, of the seals.

3 Because of the complexity of the air conditioning system and the special equipment necessary to service it, in-depth troubleshooting and repairs are beyond the scope of this manual. However, simple component replacement procedures are provided in this Chapter.

4 The most common cause of poor cooling is simply a low system refrigerant charge. If a noticeable drop in system cooling ability occurs, one on the following quick checks will help you determine whether the refrigerant level is low.

Check

Refer to illustration 13.8

5 Warm the engine up to normal operating temperature.

6 Place the air conditioning temperature selector at the coldest setting and put the blower at the highest setting. Open the doors (to make sure the air conditioning system doesn't cycle off as soon as it cools the passenger compartment).

7 After the system reaches operating temperature, feel the two pipes connected to the evaporator at the firewall.

8 The pipe (thinner tubing) leading from the condenser outlet to the evaporator should be cold, and the evaporator outlet line (the thicker tubing that leads back to the compressor) should be slightly colder (3 to 10 degrees F). If the evaporator outlet is considerably warmer than the inlet, the system needs a charge. Insert a thermometer in the center air distribution duct **(see illustration)** while operating the air conditioning system - the temperature of the output air should be 35 to 40 degrees F below the ambient air temperature (down to approximately 40 degrees F). If the ambient (outside) air temperature is very high, say 110-degrees F, the duct air temperature may be as high as 60 degrees F, but generally the air conditioning is 35 to 40 degrees F cooler than the ambient air. If the air isn't as cold as it used to be, the system probably needs a charge. Further inspection or testing of the system is beyond the scope of the home mechanic and should be left to a professional.

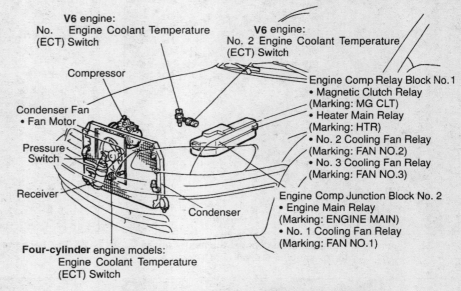

13.1 Basic components of the air conditioning system

Adding refrigerant

Refer to illustrations 13.10, 13.13 and 13.14

9 **Note:** *All models covered by this manual use refrigerant R-134a. When recharging or replacing air conditioning components, use only refrigerant, refrigerant oil and seals compatible with this system. The seals and compressor oil used with older, conventional R-12 refrigerant are not compatible with the components in this system.*

10 Buy an automotive charging kit at an auto parts store. A charging kit includes a 12-ounce can of R-134a refrigerant, a tap valve and a short section of hose that can be attached between the tap valve and the system low side service valve **(see illustration)**. Because one can of refrigerant may not be sufficient to bring the system charge up to the proper level, it's a good idea to buy a couple of additional cans. Try to find at least one can that contains red refrigerant dye. If the system is leaking, the red dye will leak out with the refrigerant and help you pinpoint the location of the leak.

11 Connect the charging kit by following the manufacturer's instructions.

12 Back off the valve handle on the charging kit and screw the kit onto the refrigerant can, making sure first that the O-ring or rubber seal inside the threaded portion of the kit is in place. **Warning:** *Wear protective eye wear when dealing with pressurized refrigerant cans.*

13.8 Place an accurate thermometer in the center dash vent, turn the air conditioning on and check the output temperature

13.10 A basic charging kit for R-134a systems is available at most auto parts stores - it must say R-134a (not R-12) and so must the can of refrigerant

Chapter 3 Cooling, heating and air conditioning systems

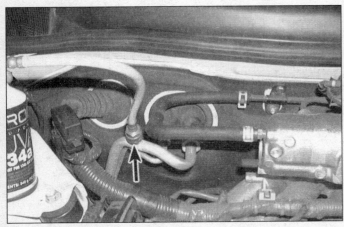

13.13 Add R-134a only to the low side port (arrow) - the procedure is easier if you wrap the can with a warm, wet towel to prevent icing

13.14 Pull the pressure switch connector off and temporarily bypass it with a jumper wire or paper clip (arrow) between terminals 1 and 4 - put the connector back on as soon as you are through charging the system

13   Remove the dust cap from the low-side charging port and attach the quick-connect fitting on the kit hose **(see illustration)**. **Warning:** *DO NOT hook the charging kit hose to the system high side! The fittings on the charging kit are designed to fit* **only** *on the low side of the system.*

14   Warm the engine to normal operating temperature and turn on the air conditioning. Keep the charging kit hose away from the fan and other moving parts. In some cases, if the refrigerant charge is low enough, the air-conditioning system pressure switch may prevent the compressor from operating. For purposes of charging the system only, pull the pressure switch connector off and connect both pins on the connector with a paper clip or jumper wire **(see illustration)**.

15   Turn the valve handle on the kit until the stem pierces the can, then back the handle out to release the refrigerant. You should be able to hear the rush of gas. Add refrigerant to the low side of the system until both the outlet and the evaporator inlet pipe feel about the same temperature. Allow stabilization time between each addition. **Warning:** *Never add more than two cans of refrigerant to the system.* The can may tend to frost up, slowing the procedure. Wet a shop towel with hot water and wrap it around the bottom of the can to keep it from frosting.

16   Put your thermometer back in the center register and check that the output air is getting colder.

17   When the can is empty, turn the valve handle to the closed position and release the connection from the low-side port. Replace the dust cap.

18   Remove the charging kit from the can and store the kit for future use with the piercing valve in the UP position, to prevent inadvertently piercing the can on the next use.

Heating systems

Refer to illustration 13.23

19   If the air coming out of the heater vents isn't hot, the problem could stem from any of the following causes:

a) *The thermostat is stuck open, preventing the engine coolant from warming up enough to carry heat to the heater core. Replace the thermostat (see Section 3).*
b) *A heater hose is blocked, preventing the flow of coolant through the heater core. Feel both heater hoses at the firewall. They should be hot. If one of them is cold, there is an obstruction in one of the hoses or in the heater core, or the heater control valve is shut. Detach the hoses and back flush the heater core with a water hose. If the heater core is clear but circulation is impeded, remove the two hoses and flush them out with a garden hose.*
c) *If flushing fails to remove the blockage from the heater core, the core must be replaced. (see Section 12).*

20   If the blower motor speed does not correspond to the setting selected on the blower switch, the problem could be a bad fuse, circuit, blower relay, speed switch or blower resistor (see Section 9).

21   If there isn't any air coming out of the vents:

a) *Turn the ignition ON and activate the fan control. Place your ear at the heating/air conditioning register (vent) and listen. Most motors are audible. Can you hear the motor running?*
b) *If you can't (and have already verified that the blower switch and the blower motor resistor are good), the blower motor itself is probably bad (see Section 10).*

22   If the carpet under the heater core is damp, or if antifreeze vapor or steam is coming through the vents, the heater core is leaking. Remove it (see Section 12) and install a new unit (most radiator shops will not repair a leaking heater core).

23   Inspect the drain hose from the heater/evaporator assembly at the right-cen-

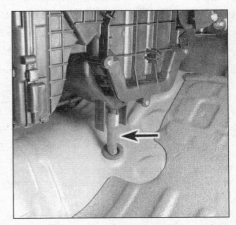

13.23 This drain hose from the heater/air conditioning unit should be kept clear to allow drainage of condensation

ter of the firewall, make sure it is not clogged **(see illustration)**. If there is a humid mist coming from the system ducts, this hose may be plugged with leaves or road debris.

Eliminating air-conditioning odors

Refer to illustration 13.27

24   Unpleasant odors that often develop in air-conditioning systems are caused by the growth of a fungus, usually on the surface of the evaporator core. The warm, humid environment there is a perfect breeding ground for mildew to develop.

25   The evaporator core on most vehicles is difficult to access, and factory dealerships have a lengthy, expensive process for eliminating the fungus by opening up the evaporator case and using a powerful disinfectant and rinse on the core until the fungus is gone. You can service your own system at home, but it takes something much stronger than basic household germ-killers or deodorizers.

26   Aerosol disinfectants for automotive air-

13.27 Remove the blower motor resistor and angle the spray nozzle of the disinfectant can towards the evaporator core to destroy mildew that causes air conditioning odors

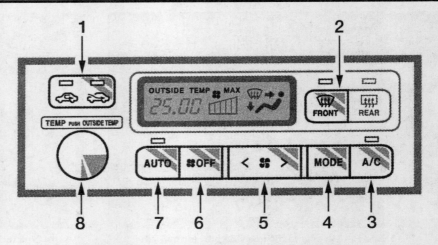

13.29 Controls for Automatic Air Conditioning system:

1. Recirculate/Fresh selector
2. Windshield airflow selector
3. Air conditioning button
4. Air flow mode selector
5. Blower fan speed selector
6. Off button
7. Automatic operation button
8. Temperature adjustment knob

conditioning systems are available in most auto parts stores, but remember when shopping for them that the most effective treatments are also the most expensive. The basic procedure for using these sprays is to start by running the system in the RECIRC mode for ten minutes with the blower on its highest speed. Use the highest heat mode to dry out the system and keep the compressor from engaging by disconnecting the wiring connector at the compressor (see Section 16).

27 The disinfectant can usually comes with a long spray hose. Remove the blower motor resistor (see Section 9), point the nozzle inside the hole and to the left towards the evaporator core, and spray according to the manufacturer's recommendations (see illustration). Try to cover the whole surface of the evaporator core, by aiming the spray up, down and sideways. Follow the manufacturer's recommendations for the length of spray and waiting time between applications.

28 Once the evaporator has been cleaned, the best way to prevent the mildew from coming back again is to make sure your evaporator housing drain tube is clear (see illustration 13.23) and to run the defrost cycle briefly to dry the evaporator out after a long drive with the air conditioning on.

Automatic Air Conditioning

Refer to illustrations 13.29 and 13.30

29 Solara and Avalon models may have an optional Automatic Air Conditioning system to control the temperature inside the vehicle. You set the desired temperature on the control panel and the climate control module (computer) blends the right amount of cooled or heated air to maintain this cabin temperature (see illustration). The speed of the blower motor in this system is electronically controlled by the microprocessor through a linear controller that replaces the conventional blower motor resistor.

30 Most repairs or diagnostics of the system are beyond the scope of the home mechanic, but there is an on-board diagnostics function in the computer that will display trouble codes relating to the climate-control system. The codes (see illustration) can indicate what area, if any, is malfunctioning.

31 To begin the self-test function (vehicle interior at normal temperature), push the RECIRC/FR and AUTO buttons at the same time, while turning the ignition key to On. If all is normal with the system, the four indicator lights will come on for one second then go off in sequence.

32 After the indicators display, the computer performs a search for Diagnostic Trouble Codes (DTC's). Any codes will appear on the control head where the temperature setting is usually displayed.

33 After the trouble code check, the system will operate all of the system's actuators at one-second intervals. The test mode can be exited any time by pressing the OFF button, but the computer will keep the codes in memory. If you want to clear the climate control DTC's, pull the fuse marked "ECU-B" from the underhood fuse/relay box and keep it disconnected for at least 10 seconds.

14 Air conditioning compressor clutch circuit - check

1 Proper operation of the compressor clutch is essential to the function of the air conditioning system. If your system doesn't seem to get cold, first check the clutch operation.

2 With the engine warmed up, set the air conditioning temperature selector on the coldest setting and the fan on high. Open the doors (to make sure the air conditioning system doesn't cycle off as soon as it cools the passenger compartment down).

3 Have an assistant push the AC button while you observe the front of the compressor. The clutch will make an audible click and the center of the clutch should rotate. If it doesn't, shut the engine off and disconnect the air conditioning system pressure switch. Insert a paper clip as a jumper and try the air conditioning again as in Steps 1 and 2 (see illustration 13.14). If it works now, the system pressure is too high or too low. Have your system tested by a dealer service department or air conditioning shop.

Trouble Code	Circuit
11	Room temperature sensor
12	Ambient temperature sensor
13	Evaporator temperature sensor
14	Engine coolant temperature sensor
21	Solar sensor
22	Compressor lock sensor
23	Pressure sensor
31	Air mix damper position sensor
32	Air intake damper position sensor
41	Air mix servomotor
42	Air intake servomotor

13.30 Trouble codes for Automatic Air Conditioning system

Chapter 3 Cooling, heating and air conditioning systems

4   If the clutch still didn't operate, check the appropriate fuses. Inspect the fuse marked "heater" in the interior fuse panel under the driver's side of the dash.
5   While at the underhood relay/fuse box, pull the "MG CLT" (compressor clutch) relay and test it (see Chapter 12). With the relay out and the ignition Off, check for battery power at the number 1 socket for the compressor clutch relay, which has the same terminals as the fan relays (see illustration 4.7). There should be battery power. Also check the large heater relay (see Section 9).
6   With the engine running and the air conditioning on, connect sockets 3 and 5 for the compressor clutch relay with a jumper wire and listen for the clutch to click as you make the connection.
7   If the clutch still doesn't operate, turn off the engine, disconnect the clutch connector at the compressor and attach a jumper wire long enough to allow use of an ohmmeter to check for continuity between the number 3 terminal of the clutch relay socket and the socket in the compressor clutch connector. If there is no continuity, check for a open in the circuit between the relay box and the compressor clutch connector. If it has continuity, check for an open in the circuit from the PCM to the relay box.

15 Air conditioning receiver/drier - removal and installation

Refer to illustration 15.3
Warning: *The air conditioning system is under high pressure. Do not loosen any hose fittings or remove any components until the system has been discharged. Air conditioning refrigerant should be properly discharged into an EPA-approved recovery/recycling unit by a dealer service department or an automotive air conditioning repair facility. Always wear eye protection when disconnecting air conditioning system fittings.*

1   Have the refrigerant discharged by an air conditioning technician.
2   Disconnect the negative battery cable. **Caution:** *If the stereo in your vehicle is equipped with an anti-theft system, make sure you have the correct activation code before disconnecting the battery.*
3   Disconnect the refrigerant lines (see illustration) from the receiver/drier and cap the open fittings to prevent entry of moisture.
4   Loosen the clamp bolt and slip the receiver/drier out of the bracket.
5   Installation is the reverse of removal.
6   Have the system evacuated, charged and leak tested by the shop that discharged it. If the receiver was replaced, have them add about 20 cc (0.71 oz.) refrigeration oil to the compressor. Use only compressor oil compatible with R-134a refrigerant.

16 Air conditioning compressor - removal and installation

Refer to illustrations 16.4a and 16.4b
Warning: *The air conditioning system is under high pressure. Do not loosen any hose fittings or remove any components until the system has been discharged. Air conditioning refrigerant should be properly discharged into an EPA-approved recovery/recycling unit by a dealer service department or an automotive air conditioning repair facility. Always wear eye protection when disconnecting air conditioning system fittings.*

1   Have the refrigerant discharged by an automotive air conditioning technician.
2   Disconnect the negative cable from the battery. **Caution:** *If the stereo in your vehicle is equipped with an anti-theft system, make sure you have the correct activation code before disconnecting the battery.*
3   Remove the drivebelt from the compressor (see Chapter 1). On V6 models, remove the alternator (see Chapter 5).
4   Detach the wiring connector and the

15.3 After the system has been discharged, unbolt the two refrigerant lines (arrows) from the receiver/drier and cap them

refrigerant lines (see illustrations).
5   Unbolt the compressor and lift it from the vehicle.
6   If a new or rebuilt compressor is being installed, follow the directions which come with it regarding the proper level of oil prior to installation.
7   Installation is the reverse of removal. Replace any O-rings with new ones specifically made for the purpose and lubricate them with refrigerant oil.
8   Have the system evacuated, recharged and leak tested by the shop that discharged it.

17 Air conditioning condenser - removal and installation

Refer to illustrations 17.4a, 17.4b and 17.5
Warning: *The air conditioning system is under high pressure. Do not loosen any hose fittings or remove any components until the*

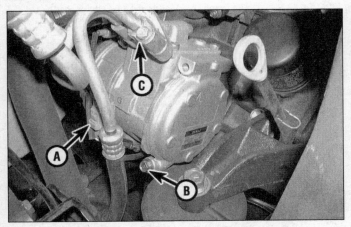

16.4a Compressor details - four-cylinder models
   A   Electrical connector
   B   Mounting bolts (one shown here)
   C   Refrigerant lines

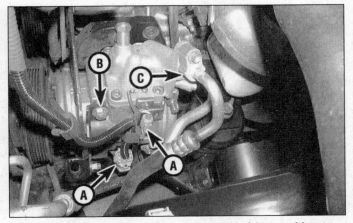

16.4b Compressor details - V6 models (shown with alternator removed)
   A   Electrical connectors
   B   Mounting bolts (one shown here)
   C   Refrigerant lines

system has been discharged. Air conditioning refrigerant should be properly discharged into an EPA-approved recovery/recycling unit by a dealer service department or an automotive air conditioning repair facility. Always wear eye protection when disconnecting air conditioning system fittings.

1 Have the refrigerant discharged by an air conditioning technician.
2 Remove the radiator upper mounts as described in Section 5. On Solara models, unbolt and set aside the right-hand underhood electrical relay box for clearance.
3 Remove the receiver/drier (see Section 15).
4 Disconnect the inlet and outlet fittings **(see illustrations)**. Cap the open fittings immediately to keep moisture and dirt out of the system.
5 Remove the mounting nuts **(see illustration)**. Push the radiator back toward the engine, then push the condenser rearward until it's free of the radiator support and can be pulled up and out if the vehicle.
6 Install the condenser, brackets and bolts, making sure the rubber cushions fit on the mounting points properly.
7 Reconnect the refrigerant lines, using new O-rings where needed.
8 Reinstall the remaining parts in the reverse order of removal.
Have the system evacuated, charged and leak tested by the shop that discharged it.

18 Air conditioning evaporator and expansion valve - removal and installation

Refer to illustrations 18.3a, 18.3b, 18.6, 18.7, 18.8 and 18.9

Warning 1: *The models covered by this manual are equipped with Supplemental Restraint systems (SRS), more commonly known as airbags. Always disconnect the negative battery cable, then the positive battery cable and wait two minutes before working in the vicin-*

17.4a Disconnect the liquid line (arrow) by reaching through the grille opening (left side) with a socket and long extension

ity of the impact sensors, steering column or instrument panel to avoid the possibility of accidental deployment of the airbag, which could cause personal injury (see Chapter 12). The yellow wiring harnesses and connectors routed through the console and instrument panel are for this system. Do not use electrical test equipment on any of the airbag system wiring or tamper with them in any way.
Warning 2: *The air conditioning system is under high pressure. Do not loosen any hose fittings or remove any components until after the system has been discharged. Air conditioning refrigerant should be properly discharged into an EPA-approved recovery/recycling unit at a dealer service department or an automotive air conditioning repair facility. Always wear eye protection when disconnecting air conditioning system fittings.*
Caution: *When replacing entire components, additional refrigerant oil should be added equal to the amount that is removed with the component being replaced. Be sure to read the can before adding any oil to the system, to make sure it is compatible with the R-134a system.*
Note: *This is a difficult procedure for the*

17.4b Remove the bolt (arrow) holding the refrigerant line to the right side of the condenser - on some models, the line is clamped to the side of the condenser as well

home mechanic, and involves removal of the dashboard. There are a number of hard-to-reach fasteners and many electrical connectors.

Removal

1 Have the air conditioning system discharged and recovered by an air conditioning technician.
2 Disconnect the cable from the negative battery cable. **Caution:** *The radio in your vehicle is equipped with an anti-theft system. Make sure you have the correct activation code before disconnecting the battery.*
3 Disconnect the receiver line and suction line from the evaporator **(see illustrations)**. Plug both lines to prevent contaminants and moisture from entering the air conditioning system.
4 Drain the coolant (see Chapter 1) and disconnect the heater hoses at the firewall **(see illustration 12.3)**.
5 Remove the dashboard trim panels, then the dashboard (see Chapter 11).

17.5 Remove the mounting nuts (arrows) on each side, near the top of the condenser

18.3a Using backup wrenches, disconnect the two refrigerant lines (arrows) at the firewall

Chapter 3 Cooling, heating and air conditioning systems   3-19

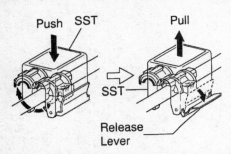

18.3b On 1999 models, a special clamp-disconnect tool is used on refrigerant line connections

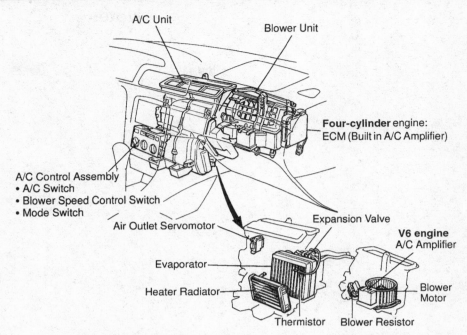

18.6 The AC unit (evaporator housing) is under the center area of the dashboard, with the blower unit to the right

6   Disconnect the electrical connectors from the blower housing and evaporator housing **(see illustration)**.

7   The blower housing must be removed before removing the evaporator housing. Remove the connecting duct between the two units, remove the mounting bolts and screws, and remove the blower unit **(see illustration)**.

8   Remove the bolts holding the expansion valve block to the evaporator core **(see illustration)**.

9   Remove the two rear heating ducts, then remove the fasteners retaining the evaporator housing to the firewall **(see illustration)**.

10   Disconnect the drain hose and remove the evaporator unit from the vehicle.

11   Remove the thermostat sensor from the evaporator.

12   Remove the screws, separate the housing and remove the evaporator.

Installation

13   Installation is the reverse of removal. Install new O-rings and coat them with R-134a refrigerant oil.

14   Have the system evacuated, charged and leak tested by an air conditioning technician. If a new evaporator was installed, add 1.4 ounces of new refrigerant oil.

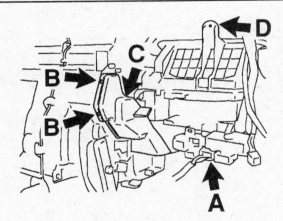

18.7 Remove the electrical connector-pack (A) from the bottom of the blower unit, exposing two screws to remove, remove the clips (B) and the duct (C), then remove the blower unit mounting bolts at D and those exposed when the duct is removed

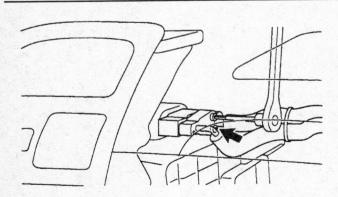

18.8 Remove the two socket-head bolts retaining the expansion valve (arrow) to the evaporator core

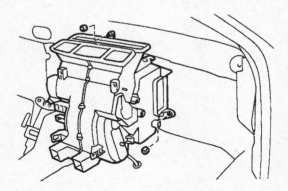

18.9 Remove the mounting nuts and take out the evaporator housing

Notes

Chapter 4
Fuel and exhaust systems

Contents

	Section		Section
Accelerator cable - removal, installation and adjustment	11	Fuel pressure relief	2
Air cleaner assembly - removal and installation	10	Fuel pump/fuel pressure - check	3
Air filter replacement	See Chapter 1	Fuel pump - removal and installation	5
CHECK ENGINE light	See Chapter 6	Fuel rail and injectors - removal and installation	15
Electronic fuel injection system - check	13	Fuel system check	See Chapter 1
Electronic fuel injection system - general information	12	Fuel tank cleaning and repair - general information	9
Exhaust system check	See Chapter 1	Fuel tank - removal and installation	8
Exhaust system servicing - general information	16	General information	1
Fuel level sending unit - check and replacement	7	Throttle body - check, removal and installation	14
Fuel lines and fittings - general information	4	Underhood hose check and replacement	See Chapter 1
Fuel pressure regulator - removal and installation	6		

Specifications

Fuel system
Fuel system pressure	44 to 50 psi
Fuel system hold pressure (after five minutes)	21 psi minimum
Fuel injector resistance (approximate)	13.4 to 14.2 ohms

Torque specifications
Ft-lbs (unless otherwise indicated)

Throttle body mounting nuts	
1997 and later Camry, 1997 through 1999 Avalon and 2000 and later Lexus ES 300	168 in-lbs
2000 and later Avalon and 1999 and later Lexus ES 300	
No. 1 intake air control valve-to-intake plenum nuts	22
Throttle body-to-No. 1 intake air control valve screws	61 in-lbs
Fuel rail mounting bolts	
Four-cylinder models	108 in-lbs
V6 models	84 in-lbs
Fuel pressure regulator mounting screw	17 in-lbs
In-tank fuel filter mounting screw	17 in-lbs

1 General information

Refer to illustrations 1.1a and 1.1b

The fuel system consists of a fuel tank, an electric fuel pump (located in the fuel tank), a fuel pressure regulator located next to the fuel pump, an EFI main relay, a fuel pump relay (circuit opening relay), the fuel rail and fuel injectors, an air cleaner assembly and a throttle body unit. All models are equipped with a Sequential Electronic Fuel Injection system **(see illustrations)**.

Sequential Electronic Fuel Injection system

Sequential Electronic Fuel Injection uses timed impulses to inject the fuel directly into the intake port of each cylinder according to its firing order. The injectors are controlled by the Powertrain Control Module (PCM). The PCM monitors various engine parameters and delivers the exact amount of fuel required into the intake ports. The throttle body serves only to control the amount of air passing into the system. Because each cylinder is equipped with its own injector, much better control of the fuel/air mixture ratio is possible.

Fuel pump and lines

Fuel is circulated from the fuel tank to the fuel injection system through a metal line running along the underside of the vehicle. An electric fuel pump and fuel level sending unit is located inside the fuel tank along with a fuel pressure regulator. A vapor return system routes all vapors back to the fuel tank through a separate return line.

The fuel pump relay is equipped with a primary and secondary voltage circuit. The primary circuit is controlled by the PCM and the secondary circuit is linked directly to the EFI main relay from the ignition switch. With the ignition switch ON (engine not running), the PCM will ground the relay for two seconds. During cranking, the PCM grounds the fuel pump relay as long as the camshaft position sensor sends its position signal (see Chapter 6). If there are no reference pulses, the fuel pump will shut off after two seconds.

Exhaust system

The exhaust system consists of exhaust manifolds (one on four-cylinder models, two on V6 models), exhaust pipes, catalytic converters, a muffler and a tail pipe. On some California models, two catalytic converters are used, in which case the front catalytic converter is an integral component of the exhaust manifold.

The catalytic converters are an emission control device added to the exhaust system to reduce pollutants. Refer to Chapter 6 for more information regarding the catalytic converters.

1.1a Fuel injection components - four-cylinder engine

1. Data Link Connector (DLC)
2. Fuel rail
3. Intake manifold
4. Throttle body
5. Idle Air Control (IAC) valve

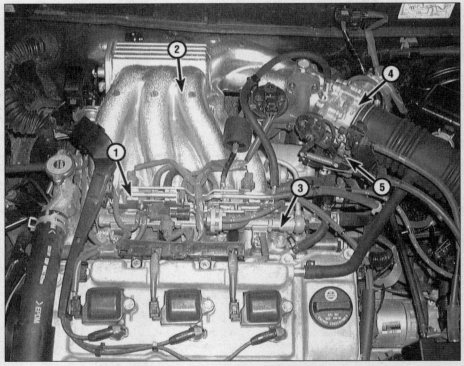

1.1b Fuel injection components - V6 engine

1. Fuel injector
2. Air intake plenum
3. Fuel rail
4. Throttle body
5. Idle Air Control (IAC) valve (below throttle body)

Chapter 4 Fuel and exhaust systems

2 Fuel pressure relief

Refer to illustration 2.3

Warning: *Gasoline is extremely flammable, so take extra precautions when you work on any part of the fuel system. Don't smoke or allow open flames or bare light bulbs near the work area, and don't work in a garage where a gas-type appliance (such as a water heater or a clothes dryer) is present. Since gasoline is carcinogenic, wear latex gloves when there's a possibility of being exposed to fuel, and, if you spill any fuel on your skin, rinse it off immediately with soap and water. Mop up any spills immediately and do not store fuel-soaked rags where they could ignite. The fuel system is under constant pressure, so, if any fuel lines are to be disconnected, the fuel pressure in the system must be relieved first. When you perform any kind of work on the fuel system, wear safety glasses and have a Class B type fire extinguisher on hand.*

1   Remove the rear seat and liner to access the fuel pump/sending unit access cover (see Section 5). Disconnect the harness connector.
2   Start the engine and allow it to run until it stops. Disconnect the cable from the negative terminal of the battery before working on the fuel system. **Caution:** *If the stereo in your vehicle is equipped with an anti-theft system, make sure you have the correct activation code before disconnecting the battery.*
3   Using an open-end wrench, loosen the fuel line fitting at the bottom of the fuel filter to relieve any extra fuel that is built up in the fuel lines **(see illustration)**. Be sure to place a shop rag around the bottom of the filter to catch the residual fuel as it bleeds off. Dispose of the fuel soaked rag in an approved safety container.
4   The fuel system pressure is now relieved. When you're finished working on the fuel system, tighten the fuel filter, reconnect the fuel pump/sending unit harness connector and connect the negative cable to the battery.

3 Fuel pump/fuel pressure - check

Warning: *Gasoline is extremely flammable, so take extra precautions when you work on any part of the fuel system. See the Warning in Section 2.*

General checks

1   Check that there is adequate fuel in the fuel tank.
2   Verify the fuel pump actually runs. Have an assistant turn the ignition switch to ON - you should hear a brief whirring noise (for approximately two seconds) as the pump comes on and pressurizes the system. **Note:** *The fuel pump is easily heard through the gas tank filler neck.* If there is no response from the fuel pump (makes no sound) proceed to Step 9 and check the fuel pump electrical circuit. If the fuel pump runs, but a fuel system problem is suspected, continue with the fuel pump pressure check.

Fuel pump pressure check

Refer to illustrations 3.3a and 3.3b

Note: *In order to perform the fuel pressure test, you will need to obtain a fuel pressure gauge capable of measuring high fuel pressure and the proper adapter set for the specific fuel injection system.*

3   Relieve the fuel system pressure (see Section 2). Disconnect the fuel filter outlet line and connect an adapter and fuel pressure gauge to the fuel filter and fuel line **(see illustrations)**. **Note:** *Early models are equipped with the bolt-and-washer fuel line connect, while later models are equipped with the quick-connect fuel lines. Each type will require a different adapter for the fuel pressure gauge assembly.*
4   Turn all the accessories Off and switch the ignition key On. The fuel pump should run for about two seconds; note the reading on the gauge. After the pump stops running, the pressure should hold steady. After five minutes it should not drop below the minimum listed in this Chapter's Specifications.
5   Start the engine and let it idle at normal operating temperature. The pressure should remain the same. If all the pressure readings are within the limits listed in this Chapter's Specifications, the system is operating properly.
6   If the fuel pressure is not within specifications, check the following:

a) If the pressure is higher than specified, replace the fuel pressure regulator (see Section 6).
b) If the pressure is lower than specified, check the fuel filter and fuel lines from the fuel rail to the fuel tank for restrictions or damage. Check the fuel injectors for leaks. If the pressure is still low, most likely the fuel pressure regulator and/or the fuel pump is defective. In this situation, it is recommended that both the fuel pressure regulator and fuel pump are replaced to prevent any future fuel pressure problems.

7   After the testing is done, relieve the fuel pressure (see Section 2) and remove the fuel pressure gauge.

2.3 After the fuel pump is disabled and the fuel pressure relieved, relieve the residual pressure by loosening the fuel line fitting at the bottom of the fuel filter (arrow)

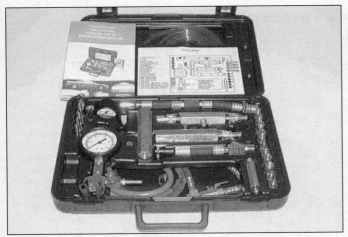

3.3a This aftermarket fuel pressure testing kit contains all the necessary fittings and adapters, along with the fuel pressure gauge, to test most automotive fuel systems

3.3b Attach the fuel pressure gauge to the fuel filter and fuel line; turn the ignition key ON and check the fuel pressure

Chapter 4 Fuel and exhaust systems

Fuel pump electrical circuit check

Refer to illustrations 3.8a, 3.8b, 3.9a and 3.9b

8 If the pump does not turn on (makes no sound) with the ignition switch in the ON position, check the IGN fuse and the EFI fuse located in the engine compartment fuse center. If either fuse is blown, replace the fuse and see if the pump works **(see illustration)**. If the pump now works, check for a short in the circuit between the fuel pump relay (circuit opening relay) and the fuel pump.

9 If the pump still does not work, remove the EFI main relay and check for battery voltage at the relay connector. One of the terminals should have battery voltage available at all times and another terminal should have battery voltage available with the ignition key On. If battery voltage does not exist, check the circuit from the fuse/relay box to the battery or ignition switch. If battery voltage exists at the relay socket, check the relays. **Note:** *These models are equipped with an EFI main relay and a fuel pump relay (circuit opening relay). On Camry and Solara models, the EFI main relay and the fuel pump relay are located in the engine compartment fuse/relay box while on Avalon models, the EFI main relay is located in the engine compartment fuse/relay box and the fuel pump relay (circuit opening relay) is located in the passenger side interior fuse/relay box* **(see illustrations)**.

10 If the relays are good and the fuel pump does not operate, check for power to the fuel pump at the fuel pump electrical connector. Also check for continuity to chassis ground at the connector. If voltage is present at the fuel pump connector, the ground circuit is good and the fuel pump does not operate when connected, replace the fuel pump.

3.8a The EFI fuse is located in the engine compartment fuse/relay box

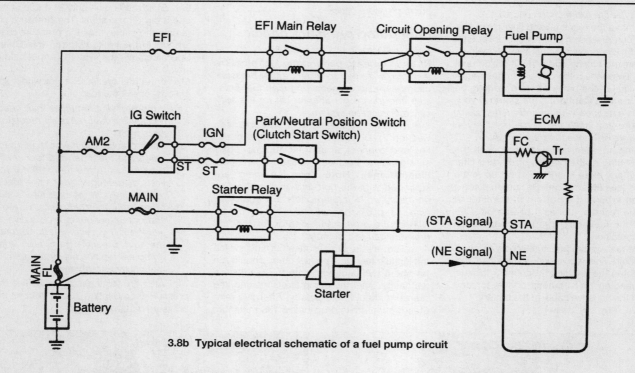

3.8b Typical electrical schematic of a fuel pump circuit

3.9a The EFI Main relay is located in the engine compartment fuse/relay box

3.9b The fuel pump relay (circuit opening relay) is located in the passenger side interior fuse/relay box on Avalon models

Chapter 4 Fuel and exhaust systems

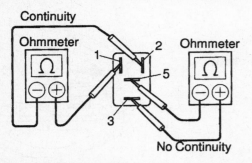

3.12a EFI main relay continuity check - Toyota type relay

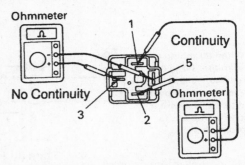

3.12b EFI main relay continuity check - Nippondenso type relay

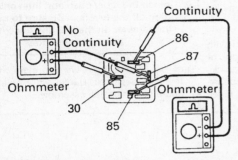

3.12c EFI main relay continuity check - Bosch type relay

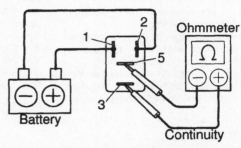

3.13a Apply battery voltage to terminals 1 and 2 and check for continuity between terminals 3 and 5 - Toyota type relay

Relay check

Refer to illustrations 3.12a, 3.12b, 3.12c, 3.13a, 3.13b and 3.13c

Note: *These models are equipped with three different type EFI main relays and fuel pump relays (circuit opening relays). Camry and Solara models use the Toyota type while the Avalon models use either the Nippondenso or the Bosch type. The following checks apply to both the EFI main relay and fuel pump relay (circuit opening relay).*

11 Remove the relay from the engine compartment fuse/relay box or interior fuse/relay box.

12 Using an ohmmeter, check for continuity between terminals 1 and 2 (Toyota or Nippondenso type) or 85 and 86 (Bosch type) **(see illustrations)**. If there's no continuity, replace the relay. Check that there is NO continuity between terminals 3 and 5 (Toyota or Nippondenso type) or 30 and 87 (Bosch type).

13 Connect battery voltage to terminals 1 and 2 (Toyota or Nippondenso type) or 85 and 86 (Bosch type) and verify there's continuity between terminals 3 and 5 (Toyota or Nippondenso type) or 30 and 87 (Bosch type) **(see illustrations)**. If there isn't, replace the relay.

4 Fuel lines and fittings - general information

Warning: *Gasoline is extremely flammable, so take extra precautions when you work on any part of the fuel system. See the* **Warning** *in Section 2.*

1 Always relieve the fuel pressure before servicing fuel lines or fittings (see Section 2).

2 The fuel line extends from the fuel tank to the engine compartment. The line is secured to the underbody with clip and screw assemblies. This line must be occasionally inspected for leaks, kinks and dents.

3 If evidence of dirt is found in the system or fuel filter during disassembly, the line should be disconnected and blown out. Check the fuel strainer on the fuel gauge sending unit (see Section 5) for damage and deterioration.

Steel tubing

4 If replacement of a fuel line or emission line is called for, use welded steel tubing meeting the manufacturer's specifications or its equivalent.

5 Don't use copper or aluminum tubing to replace steel tubing. These materials cannot withstand normal vehicle vibration.

6 Because fuel lines used on fuel-injected vehicles are under high pressure, they require special consideration.

7 Some fuel lines have threaded fittings

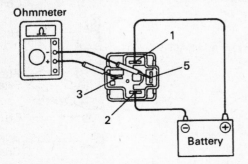

3.13b Apply battery voltage to terminals 1 and 2 and check for continuity between terminals 3 and 5 - Nippondenso type relay

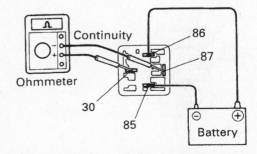

3.13c Apply battery voltage to terminals 85 and 86 and check for continuity between terminals 87 and 30 - Bosch type relay

Chapter 4 Fuel and exhaust systems

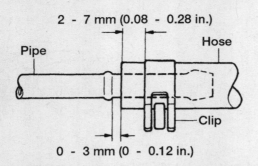

4.12a When attaching a section of rubber hose to a metal fuel line, be sure to overlap the hose as shown, secure it to the line with a new hose clamp of the proper type

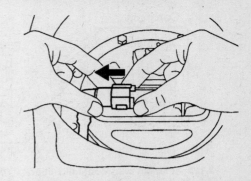

4.12b On quick-connect fuel line fittings, pinch the tabs, push and then pull the fitting off the fuel line. Be sure to replace the retainer on the fuel line once it has been disconnected

with O-rings. Any time the fittings are loosened to service or replace components:

a) Use a backup wrench while loosening and tightening the fittings.
b) Check all O-rings for cuts, cracks and deterioration. Replace any that appear hardened, worn or damaged.
c) If the lines are replaced, always use original equipment parts, or parts that meet the original equipment standards specified in this Section.

Flexible hose

Warning: *Use only original equipment replacement hoses or their equivalent. Others may fail from the high pressures of this system.*

8 Don't route fuel hose within four inches of any part of the exhaust system or within ten inches of the catalytic converter. Metal lines and rubber hoses must never be allowed to chafe against the frame. A minimum of 1/4-inch clearance must be maintained around a line or hose to prevent contact with the frame.

9 Some models may be equipped with nylon fuel line and quick-connect fittings at the fuel filter and/or fuel pump. The quick-connect fittings cannot be serviced separately. Do not attempt to service these types of fuel lines in the event the retainer tabs or the line becomes damaged. Replace the entire fuel line as an assembly.

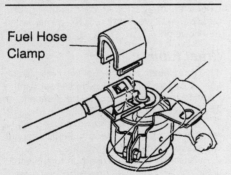

4.12c On late models with quick-connect fittings on the fuel filter, remove the fuel hose clamp (cover) . . .

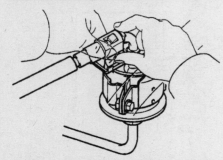

4.12d . . . then pull the tab down to unlock the fuel line fitting

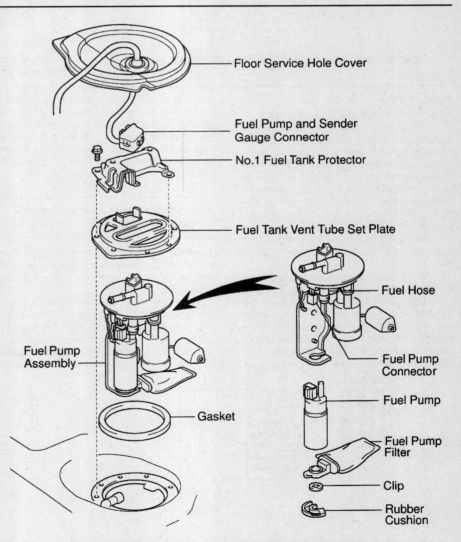

5.4a Exploded view of the Type A fuel pump assembly

Chapter 4 Fuel and exhaust systems

Replacement

Refer to illustrations 4.12a, 4.12b, 4.12c and 4.12d

10 In the event of any fuel line damage (metal or flexible lines) it is necessary to replace the damaged lines with factory replacement parts. Others may fail from the high pressures of this system.
11 Relieve the fuel pressure.
12 Remove all fasteners attaching the lines to the vehicle body. On fuel lines so equipped, detach the clamp(s) that attach the fuel hoses to the metal lines, then pull the hose off the fitting **(see illustration)**. Twisting the hoses back and forth will allow them to separate more easily. If equipped with quick-connect fittings, hold the connector with one hand and depress the retaining tabs with the other hand, then separate the connector from the pipe **(see illustrations)**.
13 Installation is the reverse of removal. Be sure to use new O-rings at the threaded fittings (if equipped). On quick-connect fittings, align the retainer locking pawls with the connector grooves. Push the connector onto the pipe until both retaining pawls lock with a clicking sound.

5 Fuel pump - removal and installation

Warning: *Gasoline is extremely flammable, so take extra precautions when you work on any part of the fuel system. See the* **Warning** *in Section 2.*

Removal

Refer to illustrations 5.4a, 5.4b, 5.5a, 5.5b, 5.8, 5.9a, 5.9b, 5.9c, 5.9d, 5.9e and 5.10

1 Relieve the fuel system pressure (see Section 2) and remove the fuel tank cap.
2 Disconnect the cable from the negative terminal of the battery. **Caution:** *If the stereo in your vehicle is equipped with an anti-theft system, make sure you have the correct activation code before disconnecting the battery.*
3 Remove the rear seat from inside the passenger compartment (see Chapter 11).
4 Remove the fuel pump/sending unit floor service hole cover **(see illustrations)**. **Note:** *There are two different types of fuel pumps in these models. Type A uses a bracket to mount the fuel pump onto the fuel pump assembly while Type B uses a housing*

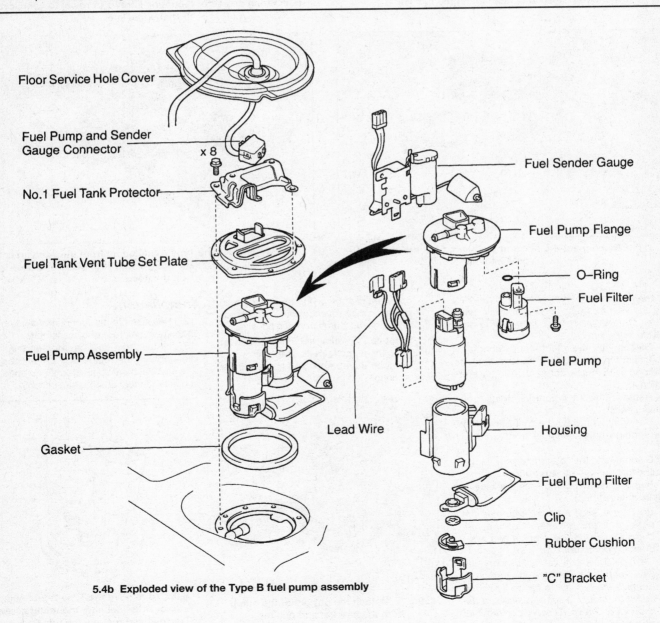

5.4b Exploded view of the Type B fuel pump assembly

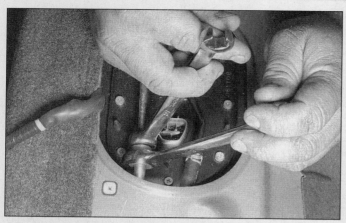

5.5a If equipped with a flared fitting, use a flare nut wrench and a back-up wrench to disconnect the fuel line

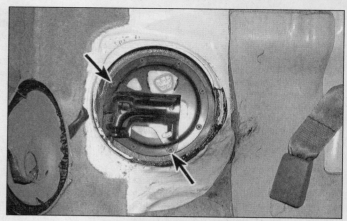

5.5b If equipped with a quick-connect fitting, remove the fuel tank fitting protector mounting screws (arrows) to gain access to the fuel line

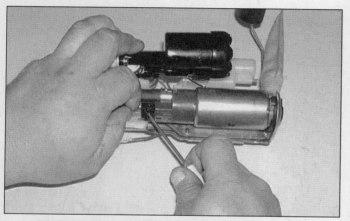

5.8 Disconnect the fuel pump connector at the pump (Type A shown)

5.9a To remove the Type A fuel pump from the sending unit assembly, squeeze the hose clamp with pliers and slide it up the hose . . .

to contain the fuel pump. Follow the procedure and refer to the particular exploded view that illustrates the correct type fuel pump installed in the vehicle.

5   Disconnect the electrical connector. Disconnect the fuel line **(see illustrations)**. If equipped with quick-connect fittings, see Section 4.
6   Remove the fuel pump/sending unit retaining bolts.
7   Carefully withdraw the fuel pump/fuel level sending unit assembly from the fuel tank.
8   Disconnect the electrical connector from the fuel pump **(see illustration)**.
9   Remove the fuel pump from the assembly **(see illustrations)**.

a) On Type A, slide the hose clamp up the hose, pull the lower end of the fuel pump loose from the bracket and withdraw the pump from the hose.
b) On Type B, remove the fuel sending unit (see Section 7). Remove the fuel filter. Remove the fuel pump flange by prying up on the housing snap-retainers in the sequence shown. Disconnect the fuel pump electrical connector. Remove the upper bracket by prying on the snap-retainers in the sequence shown and separating the fuel pump from the housing.

10   Remove the clip securing the inlet strainer to the pump **(see illustration)**.
11   Remove the strainer and inspect it for contamination. If it is dirty, replace it.

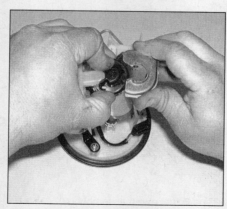

5.9b . . . detach the bottom of the pump from the bracket and pull the pump off the hose

Installation

12   Reassemble the fuel pump/sending unit in the reverse order of disassembly.
13   Install the fuel pump/sending unit assembly in the fuel tank. Connect the fuel line and electrical connector. If equipped with a quick-connect fitting, see Section 4.

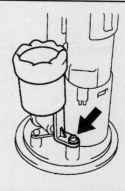

5.9c To remove the Type B fuel pump, remove the fuel filter mounting screw (arrow) and remove the fuel filter . . .

Chapter 4 Fuel and exhaust systems

4-9

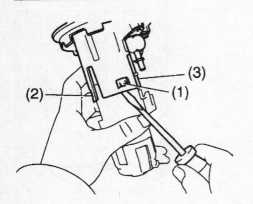

5.9d ... carefully pry the snap-retainers on the fuel pump flange in sequence ...

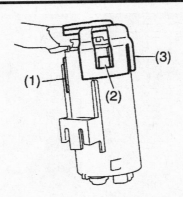

5.9e ... pry the snap-retainers on the bracket in sequence to release the housing from the upper section of the fuel pump assembly

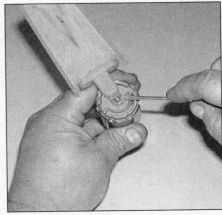

5.10 Remove the clip and the strainer from the fuel pump

14 The remainder of installation is the reverse of removal.

6 Fuel pressure regulator - removal and installation

Refer to illustrations 6.2, 6.4a and 6.4b
Warning: *Gasoline is extremely flammable, so take extra precautions when you work on any part of the fuel system. See the* **Warning** *in Section 2.*
1 Relieve the fuel system pressure (see Section 2). Disconnect the negative battery cable. **Caution:** *If the stereo in your vehicle is equipped with an anti-theft system, make sure you have the correct activation code before disconnecting the battery.*
2 Remove the fuel pump/fuel level sending unit from the fuel tank (see Section 5). Remove the fuel filter retaining screw **(see illustration)**.
3 Remove the fuel filter from the fuel pump/sending unit assembly.
4 Remove the fuel pressure regulator mounting screw and separate the regulator from the assembly **(see illustrations)**.
5 Installation is the reverse of removal. Be sure to install new O-rings on the fuel pressure regulator and fuel filter.

7 Fuel level sending unit - check and replacement

Warning: *Gasoline is extremely flammable, so take extra precautions when you work on any part of the fuel system. See the* **Warning** *in Section 2.*

Check

Refer to illustration 7.3
1 Relieve the fuel system pressure (see Section 2). Disconnect the negative battery cable. **Caution:** *If the stereo in your vehicle is equipped with an anti-theft system, make sure you have the correct activation code before disconnecting the battery.*
2 Remove the fuel pump/fuel level sending unit from the fuel tank (see Section 5).
3 Disconnect the sending unit connector and connect the probes of an ohmmeter to

6.2 Remove the fuel filter mounting screw (arrow) and remove the filter from the fuel pump/sending unit assembly

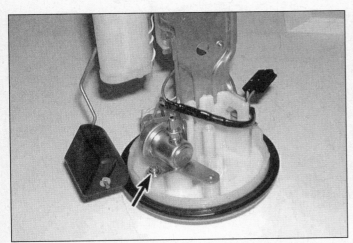

6.4a Remove the fuel pressure regulator mounting screw (arrow) and remove the pressure regulator from the fuel pump/sending unit assembly

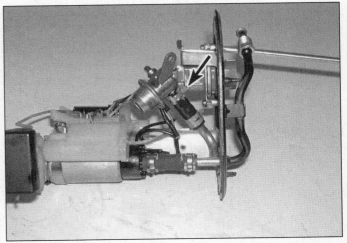

6.4b On some models, squeeze the hose clamp (arrow) with pliers and pull the fuel pressure regulator off the hose

4-10 Chapter 4 Fuel and exhaust systems

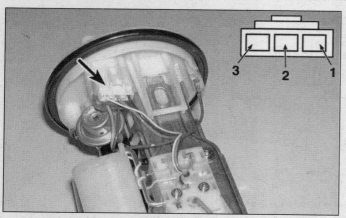

7.3 An accurate check of the sending unit can be made by removing the fuel pump/fuel level sending unit assembly from the fuel tank and checking the sending unit resistance

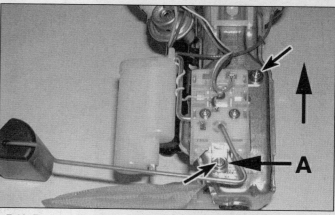

7.10 Remove the mounting screws (arrows) and push down on Tab A while pushing UP to separate the sending unit from the bracket

terminals 2 and 3 **(see illustration)**. Check the resistance of the sending unit as you move the float arm.

4   With the float arm down (tank empty), the resistance of the sending unit should be approximately 110 to 115 ohms.

5   With the float arm up (tank full), the resistance should be approximately 2 to 3 ohms.

6   The resistance should change smoothly from empty to full. If the readings are incorrect, replace the sending unit.

Replacement

Refer to illustration 7.10

7   Remove the fuel pump/fuel level sending unit assembly from the fuel tank (see Section 5).

8   Carefully angle the sending unit out of the opening without damaging the fuel level float located at the bottom of the assembly.

9   Disconnect the electrical connectors from the sending unit.

10   Remove the mounting screws **(see illustration)** and separate the sending unit

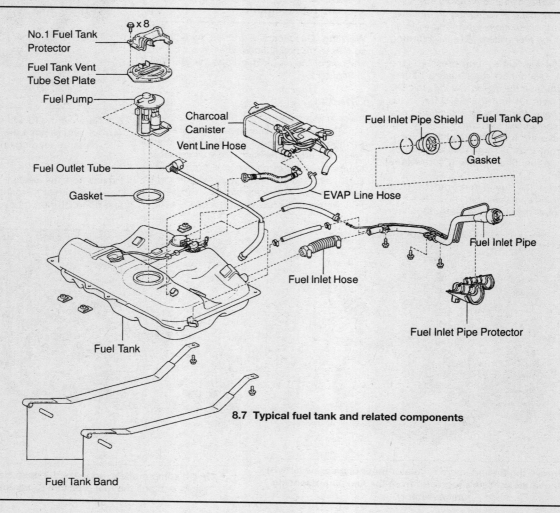

8.7 Typical fuel tank and related components

Chapter 4 Fuel and exhaust systems

4-11

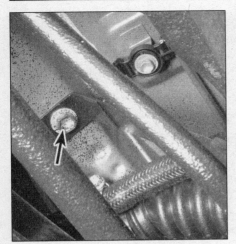

8.9 Remove the tank strap bolt (arrow) from the body

10.2a To remove the air intake duct on 1997 and later Camry, 1997 through 1999 Avalon and 1997 and 1998 Lexus ES 300 models, loosen the air intake duct clamps (arrows)

from the assembly.
11 Installation is the reverse of removal.

8 Fuel tank - removal and installation

Refer to illustrations 8.7 and 8.9
Warning: *Gasoline is extremely flammable, so take extra precautions when you work on any part of the fuel system. See the Warning in Section 2.*
1 Relieve the fuel system pressure (see Section 2).
2 Remove the fuel filler cap to relieve fuel tank pressure.
3 Detach the cable from the negative terminal of the battery. **Caution:** *If the stereo in your vehicle is equipped with an anti-theft system, make sure you have the correct activation code before disconnecting the battery.*
4 If the tank is full or nearly full, siphon the fuel into an approved container using a siphoning kit (available at most auto parts stores). **Warning:** *Do not start the siphoning action by mouth!*
5 Raise the vehicle and place it securely on jackstands.
6 Remove the exhaust system from the center exhaust pipe completely to the rear of the vehicle.
7 Disconnect the fuel lines, the vapor return line and the fuel inlet pipe from the fuel tank fittings **(see illustration)**. **Note:** *Be sure to plug the hoses to prevent leakage and contamination of the fuel system.*
8 Support the fuel tank with a floor jack. Place a sturdy plank between the jack head and the fuel tank to protect the tank.
9 Remove the bolts from the fuel tank retaining straps **(see illustration)**.
10 Lower the tank enough to disconnect the electrical connector from the fuel pump/fuel gauge sending unit.
11 Remove the tank from the vehicle.
12 Installation is the reverse of removal.

9 Fuel tank cleaning and repair - general information

1 Any repairs to the fuel tank or filler neck should be carried out by a professional who has experience in this critical and potentially dangerous work. Even after cleaning and flushing of the fuel system, explosive fumes can remain and ignite during repair of the tank.
2 If the fuel tank is removed from the vehicle, it should not be placed in an area where sparks or open flames could ignite the fumes coming out of the tank. Be especially careful inside garages where a natural gas-type appliance is located, because the pilot light could cause an explosion.

10 Air cleaner assembly - removal and installation

Refer to illustrations 10.2a, 10.2b and 10.3
1 Detach the clips, lift up the air cleaner cover and remove the filter element (see Chapter 1). On V6 models, disconnect the MAF sensor electrical connector.
2 Disconnect the air intake hose from the cover **(see illustrations)**.
3 Remove the three bolts and remove the air cleaner assembly from the engine compartment **(see illustration)**.
4 Installation is the reverse of removal.

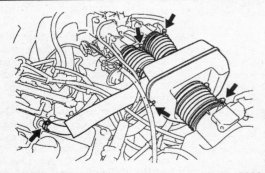

10.2b To remove the air intake duct and resonator on 2000 and later Avalon and 1999 and later Lexus ES 300 models, detach the PCV hose and the (active engine control mount) VSV vacuum hose from the resonator, and then loosen the three intake duct hose clamps (arrows)

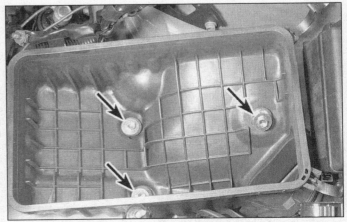

10.3 Remove the three bolts (arrows) from the air cleaner assembly and lift the assembly from the compartment

11.2 Loosen the locknuts on the accelerator cable using two wrenches

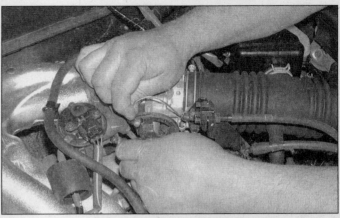

11.3 Rotate the throttle lever and remove the cable end from the slot

11 Accelerator cable - removal, installation and adjustment

Refer to illustrations 11.2, 11.3 and 11.4

Removal and installation

1 Detach the cable from the negative terminal of the battery. **Caution:** *If the stereo in your vehicle is equipped with an anti-theft system, make sure you have the correct activation code before disconnecting the battery.*
2 Loosen the locknut on the threaded portion of the accelerator cable at the throttle body **(see illustration)**.
3 Rotate the throttle lever, then slip the accelerator cable end out of the slot in the lever **(see illustration)**.
4 Detach the accelerator cable from the accelerator pedal **(see illustration)**.
5 Pull the cable and casing out of the firewall into the engine compartment.
6 Installation is the reverse of removal.

Adjustment

7 To adjust the cable:
 a) Lift up on the cable to remove any slack.
 b) Turn the adjusting nut until it is 1/8-inch (3 mm) away from the cable bracket.

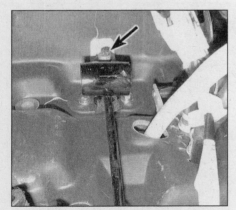

11.4 Detach the cable from the accelerator pedal arm and slide the cable end (arrow) out of the slot

 c) Tighten the locknut and check cable deflection between the throttle lever and the cable casing. Deflection should be 3/8 to 1/2-inch. If deflection is not within specifications, loosen the locknut and turn the adjusting nut until the deflection is as specified.
 d) After you have adjusted the accelerator cable, have an assistant help you verify that the throttle valve opens all the way when you depress the accelerator pedal to the floor and that it returns to the idle position when you release the accelerator. Verify the cable operates smoothly. It must not bind or stick.

12 Electronic fuel injection system - general information

The Electronic Fuel Injection (EFI) system consists of three sub-systems: air intake, electronic control and fuel delivery. The system uses a Powertrain Control Module (PCM) along with several sensors to determine the proper air/fuel ratio under all operating conditions. **Refer to illustrations 1.1a and 1.1b** for component locations.

The fuel injection system and the emissions control system are closely linked in function and design. For additional information, refer to Chapter 6.

Air intake system

The air intake system consists of the air cleaner, the air intake ducts, the throttle body, the idle control system and the intake manifold. V6 engines are equipped with an intake manifold (lower) and an intake air plenum (upper). Four-cylinder engines are equipped with a single intake manifold. Refer to Chapters 2A or 2B for the replacement procedures.

The throttle body is a single barrel, side-draft design. The lower portion of the throttle body is heated by engine coolant to prevent icing in cold weather. The idle adjusting screw is located on top of the throttle body. A throttle position sensor is attached to the throttle shaft to monitor changes in the throttle opening.

When the engine is idling, the air/fuel ratio is controlled by the idle air control system, which consists of the Powertrain Control Module (PCM), the Idle Air Control (IAC) valve and the various other sensors (ECT, IAT, TPS, MAP, MAF etc.) working in conjunction with the EFI system. The IAC valve is activated by the PCM depending upon the running conditions of the engine (air conditioning on, power steering demand, cold or warm temperature etc.). This valve regulates the amount of airflow bypassing the throttle plate and into the intake manifold. The PCM receives information from the sensors and adjusts the idle according to the demands of the engine and driver. Finally, to prevent rough running after the engine starts, the starting valve is opened during cranking and immediately after starting to provide additional air into the intake manifold.

Electronic control system

The electronic control system, Powertrain Control Module and sensors are described in detail in Chapter 6.

Fuel delivery system

The fuel delivery system consists of these components: The fuel pump, fuel pressure regulator, fuel filter, fuel lines, fuel rail and the fuel injectors.

The fuel pump is an electric in-line type. Fuel is drawn through an inlet strainer into the pump, flows through the fuel pressure regulator, passes through the fuel filter and is delivered to the injectors. The fuel pressure regulator maintains a constant fuel pressure to the injectors.

The injectors are solenoid-actuated, constant stroke, pintle types consisting of a solenoid, plunger, needle valve and housing. When current is applied to the solenoid coil, the needle valve raises and pressurized fuel fills the injector housing and squirts out the nozzle. The injection quantity is determined by the length of time the valve is open (the length of time during which current is supplied to the solenoid coils). Because it determines opening and closing intervals - which

Chapter 4 Fuel and exhaust systems   4-13

13.7 Use a stethoscope or a screwdriver to determine if the injectors are working properly - they should make a steady clicking sound that rises and falls with engine speed changes

13.8 Install the "noid" light into the fuel injector electrical connector and check to see that it blinks with the engine running

13.9 Using an ohmmeter, measure the resistance across both terminals of the injector

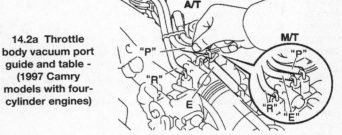

14.2a Throttle body vacuum port guide and table - (1997 Camry models with four-cylinder engines)

Port name	At idle	Other than idle
Mark "P" for M/T	No vacuum	Vacuum
E for A/T	Vacuum	Vacuum
Mark "E"	No vacuum	Vacuum
Mark "R"	No vacuum	No vacuum

in turn determines the air-fuel mixture ratio - injector timing must be quite accurate.

The EFI main relay, located in the engine compartment relay/fuse box, supplies power to the fuel pump relay (circuit opening relay) from the ignition key. The PCM controls the grounding signal to the fuel pump in response to the starting and camshaft position signals at start-up.

13 Electronic fuel injection system - check

Refer to illustrations 13.7, 13.8 and 13.9
Warning: *Gasoline is extremely flammable, so take extra precautions when you work on any part of the fuel system. See the Warning in Section 2.*

1   Check all electrical connectors, especially ground connections, for the system. Loose connectors and poor grounds can cause many engine control system problems.
2   Verify that the battery is fully charged because the Powertrain Control Module (PCM) and sensors cannot operate properly without adequate supply voltage.
3   Refer to Chapter 1 and check the air filter element. A dirty or partially blocked filter

will reduce performance and economy.
4   Check fuel pump operation (Section 3). If the fuel pump fuse is blown, replace it and see if it blows again. If it does, refer to Chapter 12 and the wiring diagrams and look for a short in the wiring harness to the fuel pump.
5   Inspect the vacuum hoses connected to the intake manifold for damage, deterioration and leakage.
6   Remove the air intake duct from the throttle body and check for dirt, carbon, varnish, or other residue in the throttle body, particularly around the throttle plate. If it's dirty, refer to Chapter 6 and troubleshoot the PCV and EGR systems for the cause of excessive varnish buildup.
7   With the engine running, place an automotive stethoscope against each injector, one at a time, and listen for a clicking sound that indicates operation **(see illustration)**. If you don't have a stethoscope, you can place the tip of a long screwdriver against the injector and listen through the handle.
8   If an injector does not seem to be operating electrically (not clicking), purchase a special injector test light (sometimes called a "noid" light) and install it into the injector wiring harness connector **(see illustration)**. Start the engine and see if the noid light

flashes. If it does, the injector is receiving proper voltage. If it doesn't flash, further diagnosis is necessary. You might want to have it checked by a dealership service department or other qualified repair shop.
9   With the engine off and the fuel injector electrical connectors disconnected, measure the resistance of each injector with an ohmmeter **(see illustration)**. Refer to the Specifications at the beginning of this chapter for the correct resistance.
10   Refer to Chapter 6 for other system checks.

14 Throttle body - check, removal and installation

Warning: *Gasoline is extremely flammable, so take extra precautions when you work on any part of the fuel system. See the Warning in Section 2.*

Check

Refer to illustrations 14.2a, 14.2b, 14.3a and 14.3b
1   Verify that the throttle linkage operates smoothly.
2   Start the engine, detach each vacuum

4-14 Chapter 4 Fuel and exhaust systems

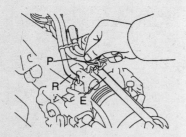

Port name	At idle	Other than idle
Mark E	Vacuum	Vacuum
Mark P	No vacuum	Vacuum
Mark R	No vacuum	No vacuum

14.2b Throttle body vacuum port guide and table (1998 and later Camry models with four-cylinder engines)

14.3a The area inside the throttle body near the throttle plate (arrow) suffers from sludge build-up because the PCV hose vents vapor from the crankcase here

hose and, using your finger, check the vacuum at each port on the throttle body with the engine at idle and above idle, then compare your observations with the vacuum table **(see illustrations)**. **Note:** *The throttle body ports on V6 models are plugged and not used in this application.*

3   Remove the air intake duct from the throttle body and check for carbon and residue build-up. If it is dirty, clean it with aerosol carburetor cleaner (make sure the can specifically states that it is safe with oxygen sensor systems and catalytic converters) and a tooth brush **(see illustrations)**. **Caution:** *Do not clean the throttle position sensor (TPS) or Idle Air Control (IAC) valve with the solvent.*

Removal and installation

1997 and 1998 Camry and Lexus ES 300 models, 1997 through 1999 Avalon

Refer to illustrations 14.8a, 14.8b and 14.10
Warning: *Wait until the engine is completely cool before beginning this procedure.*

14.3b With the engine off, use aerosol carburetor cleaner (make sure it is safe for use with catalytic converters and oxygen sensors), a toothbrush and a rag to clean the throttle body - open the throttle plate so you can clean behind it

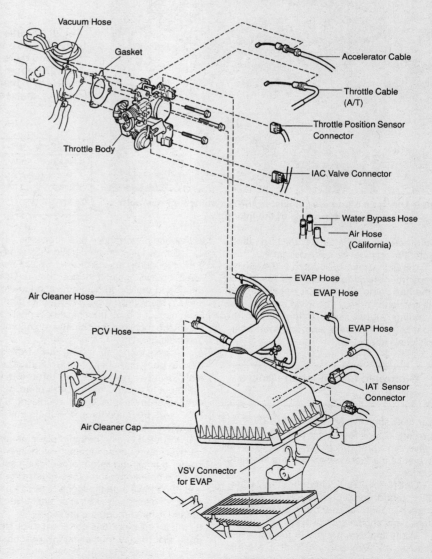

14.8a An exploded view of a typical throttle body assembly (1997 and later Camry four-cylinder models)

Chapter 4 Fuel and exhaust systems

4-15

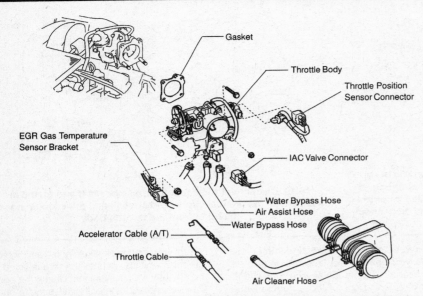

14.8b An exploded view of a typical throttle body assembly (1997 and later Camry V6 models, 1997 through 1999 Avalon models and 1997 and 1998 Lexus ES 300 models)

14.10 The throttle body on four-cylinder models is retained by three bolts (arrows)

4 Detach the cable from the negative terminal of the battery. **Caution:** *If the stereo in your vehicle is equipped with an anti-theft system, make sure you have the correct activation code before disconnecting the battery.*
5 Loosen the hose clamps and remove the air intake duct.
6 Detach the accelerator cable from the throttle lever (see Section 10).
7 If your vehicle is equipped with an automatic transmission, detach the throttle valve (TV) cable from the throttle lever (see Chapter 7B).
8 Clearly label, then detach, all vacuum and coolant hoses from the throttle body **(see illustrations)**. Plug the coolant hoses to prevent coolant leakage.
9 Disconnect the electrical connector from the Throttle Position Sensor (TPS).
10 Remove the throttle body mounting nuts/bolts **(see illustration)**.
11 Detach the throttle body and gasket from the intake manifold **(see illustration 14.8a or 14.8b)**.
12 Installation of the throttle body is the reverse of removal. Be sure to use a new gasket between the throttle body and the intake manifold.
13 Be sure to tighten the throttle body mounting bolts to the torque listed in this Chapter's Specifications.
14 Check the coolant level and add some, if necessary, to bring it to the appropriate level (see Chapter 1).

2000 and later Avalon and 1999 and later Lexus ES 300 models

Refer to illustrations 14.15, 14.18, 14.19 and 14.20

Warning: *Wait until the engine is completely cool before beginning this procedure.*

15 Remove the three V-bank cover cap nuts and remove the V-bank cover **(see illustration)**.
16 Remove the intake duct and resonator (see Section 10).
17 Disconnect the accelerator cable (see Section 11).

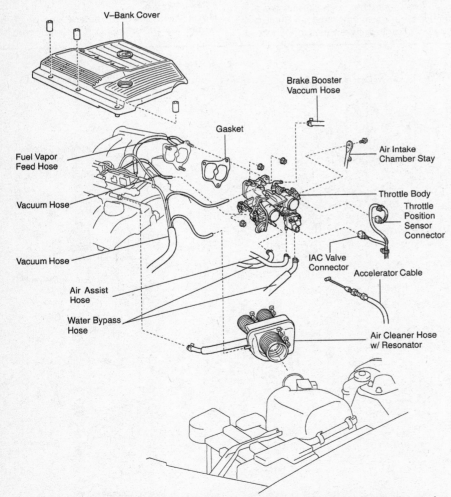

14.15 Throttle body removal and installation details (2000 and later Avalon and 1999 and later Lexus ES 300 models)

Chapter 4 Fuel and exhaust systems

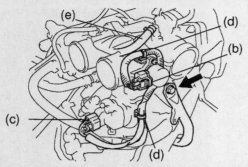

14.18 Remove the bolt (arrow) which secures the throttle body to the throttle body bracket, disconnect the electrical connectors from the throttle position sensor (b) and from the idle air control valve (c), detach the two wire harness clamps (d) and disconnect the brake booster vacuum hose (e)

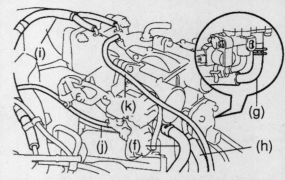

14.19 Disconnect the coolant bypass hoses (f and g), the air assist hose (h), the fuel vapor feed hose (i) and the vacuum hoses (j and k) from the throttle body

18 Remove the bolt which secures the throttle body to the throttle body bracket, disconnect the electrical connectors from the throttle position sensor and from the idle air control valve, detach the two wire harness clamps and disconnect the brake booster vacuum hose **(see illustration)**.

19 Disconnect the coolant bypass hoses, the air assist hose, the fuel vapor feed hose and the vacuum hoses from the throttle body **(see illustration)**. Plug the coolant hoses to prevent coolant leakage.

20 Remove the three nuts that secure the No. 1 intake air control valve to the air intake plenum **(see illustration)**.

21 It's not necessary to detach the throttle bodies from the No. 1 intake air control valve unless you need to replace either throttle body or the No. 1 intake air control valve. If you are replacing one of these components, refer to Steps 23 through 38).

22 Before installing the throttle body/No. 1 intake air control valve assembly, make sure that the two throttle bodies are correctly balanced (see Steps 31 through 38). Installation is otherwise the reverse of removal. Be sure to use new gaskets between the No 1 intake air control valve and the intake plenum and tighten the three retaining nuts to the torque listed in this Chapter's Specifications. Check the coolant level and add some, if necessary, to bring it to the appropriate level (see Chapter 1).

Replacement

Refer to illustrations 14.25a, 14.25b, 14.26, 14.27 and 14.28

Note: *If you need to replace one or both throttle bodies, or the No. 1 intake air control*

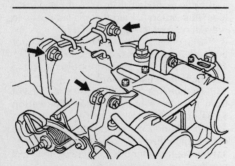

14.20 Remove the three nuts (arrows) that secure the No. 1 intake air control valve to the air intake plenum

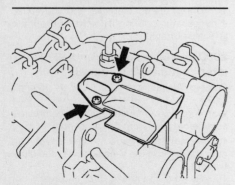

14.25a To detach the protector, remove these two screws (arrows)

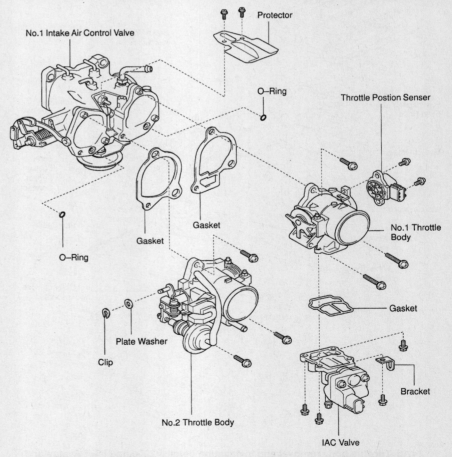

14.25b Throttle body and No. 1 intake air control valve assembly details

Chapter 4 Fuel and exhaust systems

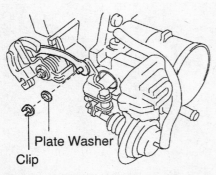

14.26 To disconnect the accelerator link, remove the clip and washer

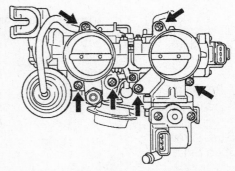

14.27 To detach the throttle bodies from the No. 1 intake air control valve, remove these six screws (arrows)

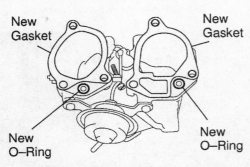

14.28 Be sure to use new O-rings and gaskets when reattaching the throttle bodies to the No. 1 intake air control valve and make sure that the O-rings are positioned correctly

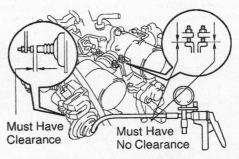

14.31 Using a hand-operated vacuum pump, apply 15.7 in-Hg vacuum to the throttle opener and verify that there's no clearance between each throttle stop screw and the throttle lever, but that there is clearance between each adjusting screw and the throttle lever

valve on these models, the two throttle bodies must be separated from the No. 1 intake air control valve.

23 Remove the throttle body/No. 1 intake air control valve assembly (see Steps 14 through 20).

24 Remove the throttle position sensor and the IAC valve (see Chapter 6).

25 Remove the two protector screws and remove the protector **(see illustrations)**.

26 Remove the clip and washer and disconnect the accelerator link **(see illustration)**.

27 Remove the six throttle body retaining screws (three in each throttle body) **(see illustration)** and then detach the throttle bodies from the No. 1 intake air control valve. Discard the old throttle body gaskets and O-rings.

28 Installation is the reverse of removal. Be sure to use new O-rings and gaskets **(see illustration)** and tighten the six throttle body retaining screws to the torque listed in this Chapter's Specifications.

29 Before installing the throttle body/No. 1 intake air control valve assembly on the air intake plenum, check and, if necessary, balance the throttle bodies as follows.

30 After the throttle bodies are balanced, install the throttle body/No. 1 intake air control valve assembly on the air intake plenum and tighten the retaining nuts to the torque listed in this Chapter's Specifications. Installation is otherwise the reverse of removal.

Adjustment

Refer to illustrations 14.31, 14.32, 14.33 and 14.35

Note: *Any time the throttle bodies are detached from the No. 1 intake air control valve, they must be balanced before being reinstalled.*

31 Using a hand-operated vacuum pump, apply 15.7 in-Hg vacuum to the throttle opener and verify that there's no clearance between each throttle stop screw and the throttle lever **(see illustration)**. If there's any clearance between either throttle stop screw and the

throttle lever, it will be removed in Step 35.

Caution: *The throttle stop screw is adjusted at the factory. Do NOT attempt to adjust it.*

32 On the No. 1 throttle body (the one to which the TPS and IAC valve are attached), measure the clearance between each adjusting screw and the throttle lever **(see illustration)**. On the close side, it should be 0.0051 inch; on the open side it should be 0.0087 inch. If the clearance is incorrect, it will be corrected in Step 35.

33 Fully open the throttle lever on the No. 2 throttle body by hand and measure the clear-

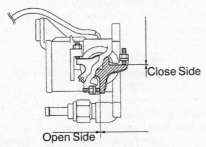

14.32 On the No. 1 throttle body (the one to which the TPS and IAC valve are attached), measure the clearance between each adjusting screw and the throttle lever and compare your measurement to the specified clearance

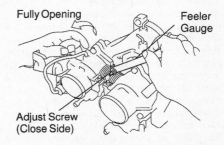

14.33 Fully open the throttle lever on the No. 2 throttle body and measure the clearance between the adjusting screw on the close side and the throttle lever on the No. 1 throttle body

Chapter 4 Fuel and exhaust systems

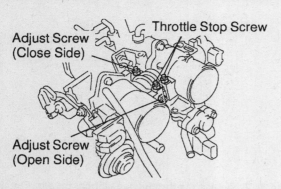

14.35 To adjust either adjusting screw (open or close side), loosen the locknut, back off the screw until there's some clearance between the screw tip and the throttle lever on the No. 1 throttle body, turn the screw back in until it's just touching the throttle lever, back off the screw another 1/4-turn and tighten the locknut

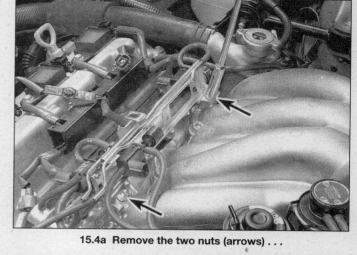

15.4a Remove the two nuts (arrows) . . .

15.4b . . . and position the VSV assembly off to the side

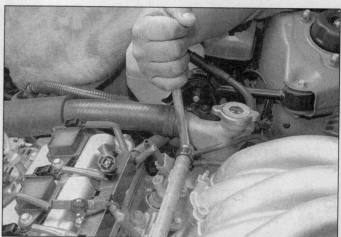

15.6 Disconnect the fuel lines from the fuel rail

ance between the adjusting screw on the close side and the throttle lever on the No. 1 throttle body **(see illustration)**. The clearance should be between 0.0098 and 0.0177 inch. If the clearance is out of range, go to the next step.

34 If any of the clearances in the previous three steps are incorrect, proceed as follows.

35 First, adjust the adjusting screw on the open side **(see illustration)**. Loosen the locknut and back off the screw until there's some clearance between the screw tip and the throttle lever on the No. 1 throttle body. Then, turn the screw back in until it's just touching the throttle lever. **Caution:** *Don't turn in the screw too much or you'll push the throttle lever off the throttle stop screw, which will make it impossible to complete the adjustment procedure.* Back off the screw another 1/4-turn and tighten the locknut.

36 Next, adjust the adjusting screw on the close side **(see illustration 14.35)**. Loosen the locknut and back off the screw until there's some clearance between the screw tip and the throttle lever on the No. 1 throttle body. Then, turn the screw back in until it's just touching the throttle lever. **Caution:**

Don't turn in the screw too much or you'll push the throttle lever off the throttle stop screw, which will make it impossible to complete the adjustment procedure. Back off the screw another 1/2 to 3/4-turn and tighten the locknut.

37 Recheck the clearance between the adjusting screw and the throttle lever on the No. 1 throttle body (see Step 32).

38 Install the throttle body/No. 1 intake air control valve assembly (see Step 30).

15 Fuel rail and injectors - removal and installation

Warning: *Gasoline is extremely flammable, so take extra precautions when you work on any part of the fuel system. See the* **Warning** *in Section 2.*

Removal

Refer to illustrations 15.4a, 15.4b, 15.6, 15.7a, 15.7b, 15.8, 15.9 and 15.10

1 Relieve the fuel pressure (see Section 2).

2 Detach the cable from the negative terminal of the battery. **Caution:** *If the stereo in your vehicle is equipped with an anti-theft system, make sure you have the correct activation code before disconnecting the battery.*

3 Remove the PCV hose.

4 On V6 models, remove the assembly that retains the Vacuum Switching Valve (VSV) control solenoids **(see illustrations)**. On V6 models, remove the air intake plenum to gain access to the rear (right bank) fuel rail (see Chapter 2B).

5 Disconnect the fuel injector electrical connectors and set the injector wire harness aside.

6 Disconnect the fuel lines from the fuel rail **(see illustration)**.

7 Remove the fuel rail mounting bolts **(see illustrations)**.

8 Remove the fuel rail with the fuel injectors attached **(see illustration)**.

9 Remove the fuel injector(s) from the fuel rail **(see illustration)** and set them aside in a clearly labeled storage container.

10 If you intend to re-use the same injectors, replace the grommets and O-rings **(see illustration)**.

Chapter 4 Fuel and exhaust systems

15.7a Remove the fuel rail mounting bolts (arrows) (four-cylinder model shown)

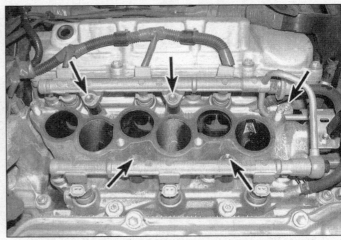

15.7b Location of the fuel rail mounting bolts (arrows) on V6 models

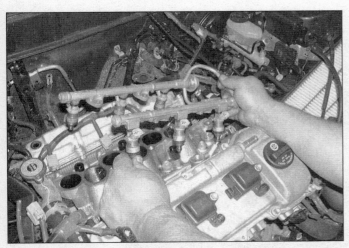

15.8 Carefully lift the fuel rail/injector assembly out of the engine compartment (V6 engine shown)

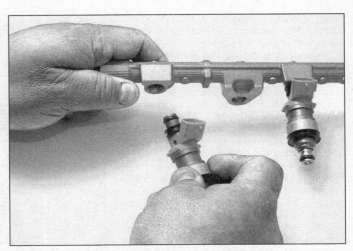

15.9 Simultaneously twist and pull the injector to remove it from the fuel rail

Installation

11 Installation of the fuel injectors is the reverse of removal.
12 Tighten the fuel rail mounting bolts to the torque listed in this Chapter's Specifications.

16 Exhaust system servicing - general information

Refer to illustrations 16.1a, 16.1b, 16.1c and 16.4

Warning: *Inspection and repair of exhaust system components should be done only after the system components have cooled completely.*

1 The exhaust system consists of the exhaust manifold, catalytic converter, the muffler, the tailpipe and all connecting pipes, brackets, hangers and clamps. The exhaust system is attached to the body with mounting brackets and rubber hangers **(see illustra-**

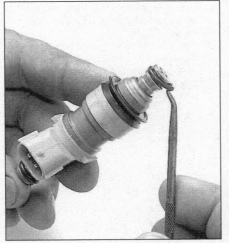

15.10 If you plan to reinstall the original injectors, remove and discard the O-rings and grommets and replace them with new ones

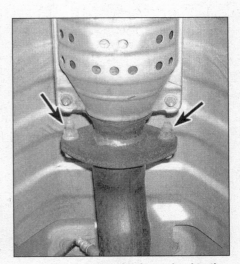

16.1a The exhaust pipe is retained to the catalytic converter with two flange bolts (arrows)

16.1b On V6 models, the front exhaust pipe is retained to the body with two bracket mounting bolts (arrows)

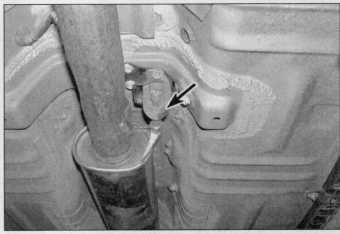

16.1c Inspect the muffler hanger (arrow) for cracks and deterioration of the rubber

tions). If any of these parts are damaged or deteriorated, excessive noise and vibration will be transmitted to the body.

2 Conducting regular inspections of the exhaust system will keep it safe and quiet. Look for any damaged or bent parts, open seams, holes, loose connections, excessive corrosion or other defects which could allow exhaust fumes to enter the vehicle. Deteriorated exhaust system components should not be repaired - they should be replaced with new parts.

3 If the exhaust system components are extremely corroded or rusted together, they will probably have to be cut from the exhaust system. The convenient way to accomplish this is to have a muffler repair shop remove the corroded sections with a cutting torch. If, however, you want to save money by doing it yourself and you don't have an oxy/acetylene welding outfit with a cutting torch, simply cut off the old components with a hacksaw. If you have compressed air, special pneumatic cutting chisels can also be used. If you do decide to tackle the job at home, be sure to wear eye protection to protect your eyes from metal chips and work gloves to protect your hands.

4 Here are some simple guidelines to apply when repairing the exhaust system:

a) Work from the back to the front when removing exhaust system components.

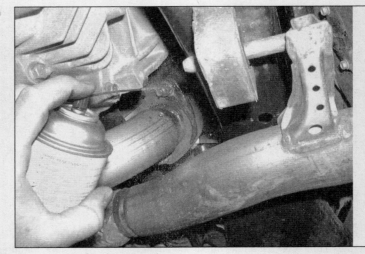

16.4 Lubricate the exhaust system fasteners with penetrating oil before attempting to loosen them

b) Apply penetrating oil to the exhaust system component fasteners to make them easier to remove (see illustration).

c) Use new gaskets, hangers and clamps when installing exhaust system components.

d) Apply anti-seize compound to the threads of all exhaust system fasteners during reassembly. Be sure to allow sufficient clearance between newly installed parts and all points on the underbody to avoid overheating the floor pan and possibly damaging the interior carpet and insulation. Pay particularly close attention to the catalytic converter and its heat shield. **Warning:** *The catalytic converter operates at very high temperatures and takes a long time to cool. Wait until it's completely cool before attempting to remove the converter. Failure to do so could result in serious burns.*

Chapter 5
Engine electrical systems

Contents

	Section		Section
Alternator - removal and installation	12	Ignition system - check	6
Battery cables - check and replacement	4	Ignition system - general information and precautions	5
Battery - check and replacement	3	Ignition timing - check	8
Battery check, maintenance and charging	See Chapter 1	Spark plug replacement	See Chapter 1
Battery - emergency jump starting	2	Spark plug wire check and replacement	See Chapter 1
Charging system - check	11	Starter motor and circuit - check	15
Charging system - general information and precautions	10	Starter motor - removal and installation	16
CHECK ENGINE light	See Chapter 6	Starter solenoid - removal and installation	17
Drivebelt check, adjustment and replacement	See Chapter 1	Starting system - general information and precautions	14
General information	1	Voltage regulator and alternator brushes - replacement	13
Igniter (V6 models) - check and replacement	9		
Ignition coil(s) - check and replacement	7		

Specifications

Idle speed (cooling fan and all accessories off)*

Camry
    1997 through 1999
        Automatic transaxle.. 750 rpm
        Manual transaxle.. 700 rpm
    2000 .... 650 to 750 rpm
Avalon and Lexus .... 650 to 750 rpm

If the idle speed specified on the VECI label is different, it supersedes the information provided here.

Ignition timing (transmission in Neutral)

Test terminals TE1 and E1 connected	8 to 12 degrees BTDC
Test terminals not connected	
Camry	0 to 10 degrees BTDC
Avalon	7 to 24 degrees BTDC
Lexus ES 300	10 to 25 degrees BTDC

Ignition coil

Four-cylinder models	
Primary resistance	Not available
Secondary resistance	
Cold	9.7 to 16.7 k-ohms
Hot	12.4 to 19.6 k-ohms
V6 models (except 1999 through 2001 Lexus ES 300 models)	
Primary resistance	
Cold	0.70 to 0.94 ohms
Hot	0.85 to 1.10 ohms
Secondary resistance	
Aisan coils	
Cold	10.8 to 14.9 k-ohms
Hot	13.1 to 17.5 k-ohms
Diamond coils	
Cold	6.8 to 11.7 k-ohms
Hot	8.6 to 13.7 k-ohms

Charging system

Charging voltage	13.5 to 15.0 volts
Standard amperage	
No load	10 amps or less
With load	30 amps or more
Alternator brush length	
Standard	0.413 inch
Minimum	0.059 inch

5-2 Chapter 5 Engine electrical systems

1.1a Charging and ignition system components - four-cylinder engine

1 Alternator
2 Spark plug wires
3 Ignition coil/igniter assembly
4 Starter
5 Battery

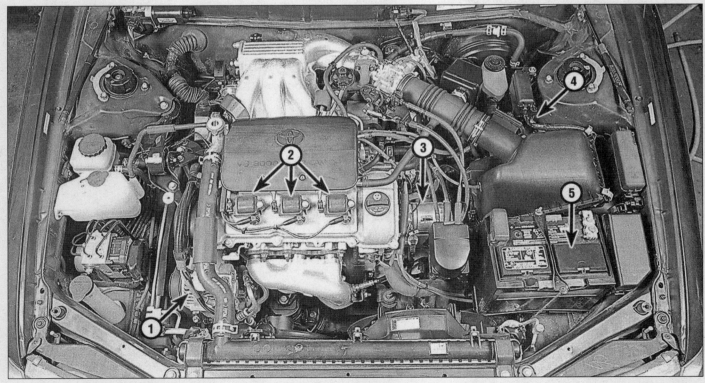

1.1b Charging and ignition system components - V6 engine

1 Alternator
2 Ignition coils
3 Starter
4 Igniter (under relay box)
5 Battery

Chapter 5 Engine electrical systems

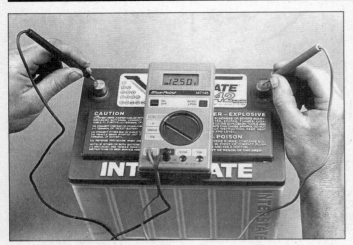

3.2 To test the open circuit voltage of the battery, simply touch the black probe of the voltmeter to the negative terminal and the red probe to the positive terminal of the battery - a fully charged battery should read between 11.5 to 12.5 volts depending on the outside air temperature

3.3 Some battery load testers are equipped with an ammeter which enables the battery load to be precisely dialed in, as shown - less expensive testers have a load switch and voltmeter only

1 General information

Refer to illustrations 1.1a and 1.1b

The engine electrical systems include all ignition, charging and starting components **(see illustrations)**. Because of their engine-related functions, these components are considered separately from chassis electrical devices like the lights, instruments, etc.

Be very careful when working on the engine electrical components. They are easily damaged if checked, connected or handled improperly. The alternator is driven by an engine drivebelt which could cause serious injury if your hands, hair or clothes become entangled in it with the engine running. Both the starter and alternator are connected directly to the battery and could arc or even cause a fire if mishandled, overloaded or shorted out.

Never leave the ignition switch on for long periods of time with the engine off. Don't disconnect the battery cables while the engine is running. Correct polarity must be maintained when connecting battery cables from another source, such as another vehicle, during jump starting. Always disconnect the negative cable first and hook it up last or the battery may be shorted by the tool being used to loosen the cable clamps.

Additional safety related information on the engine electrical systems can be found in *Safety first* near the front of this manual. It should be referred to before beginning any operation included in this Chapter.

2 Battery - emergency jump starting

Refer to the *Booster battery (jump) starting* procedure at the front of this manual.

3 Battery - check and replacement

Check

Refer to illustrations 3.2 and 3.3

1 Disconnect the negative battery cable, then the positive cable from the battery. **Caution:** *If the radio in your vehicle is equipped with an anti-theft system, make sure you have the correct activation code before disconnecting the battery.*
2 Check the battery state of charge. Visually inspect the indicator eye on the top of the battery; if the indicator eye is black in color charge the battery as described in Chapter 1. Next perform an open voltage circuit test using a digital voltmeter **(see illustration)**. **Note:** *The battery's surface charge must be removed before accurate voltage measurements can be made. Turn On the high beams for ten seconds, then turn them Off, let the vehicle stand for two minutes.* With the engine and all accessories Off, touch the negative probe of the voltmeter to the negative terminal of the battery and the positive probe to the positive terminal of the battery. The battery voltage should be 11.5 to 12.5 volts or slightly above. If the battery is less than the specified voltage, charge the battery before proceeding to the next test. Do not proceed with the battery load test unless the battery charge is correct.
3 Perform a battery load test. An accurate check of the battery condition can only be performed with a load tester (available at most auto parts stores). This test evaluates the ability of the battery to operate the starter and other accessories during periods of heavy amperage draw (load). Install a special battery load testing tool onto the terminals **(see illustration)**. Load test the battery according to the manufacturer's instructions for the particular tool. This tool utilizes a carbon pile to increase the load demand (amperage draw) on the battery. Maintain the load on the battery for 15 seconds or less and observe that the battery voltage does not drop below 9.6 volts. If the battery condition is weak or defective, the tool will indicate this condition immediately. **Note:** *Cold temperatures will cause the minimum voltage requirements to drop slightly. Follow the chart given in the manufacturer's instructions to compensate for cold climates. Minimum load voltage for freezing temperatures (32 degrees F) should be approximately 9.1 volts.*

3.5 Remove the two nuts (arrows) and detach the hold-down clamps

Replacement

Refer to illustration 3.5

4 Disconnect the negative battery cable, then the positive cable from the battery. **Caution:** *Always disconnect the negative cable first and hook it up last or the battery may be shorted by the tool being used to loosen the cable clamps.*
5 Remove the battery hold-down clamp **(see illustration)**.

5-4   Chapter 5   Engine electrical systems

4.4a The negative battery cable is attached to the chassis (arrow) as well as the engine to ensure a proper ground

4.4b Note the routing of the cable and replace it as originally installed

6   Lift out the battery. Be careful - it's heavy. **Note:** *Battery straps and handlers are available at most auto parts stores for a reasonable price. They make it easier to remove and carry the battery.*

7   While the battery is out, inspect the battery tray for corrosion.

8   If corrosion exists on the battery tray, detach the bolts and remove the tray from the engine compartment. Clean the deposits from the metal to prevent the battery tray from further corrosion.

9   If you are replacing the battery, make sure you get one that's identical, with the same dimensions, amperage rating, cold cranking rating, etc.

10 Installation is the reverse of removal.

4   Battery cables - check and replacement

Refer to illustrations 4.4a and 4.4b

1   Periodically inspect the entire length of each battery cable for damage, cracked or burned insulation and corrosion. Poor battery cable connections can cause starting problems and decreased engine performance.

2   Check the cable-to-terminal connections at the ends of the cables for cracks, loose wire strands and corrosion (see Chapter 1). The presence of white, fluffy deposits under the insulation at the cable terminal connection is a sign that the cable is corroded and should be replaced. Check the terminals for distortion, missing mounting bolts and corrosion.

3   When removing the cables, always disconnect the negative cable first and hook it up last or the battery may be shorted by the tool used to loosen the cable clamps. Even if only the positive cable is being replaced, be sure to disconnect the negative cable from the battery first. **Caution:** *If the radio in your vehicle is equipped with an anti-theft system, make sure you have the correct activation code before disconnecting the battery.*

4   Disconnect the old cables from the battery, then trace each of them to their opposite ends and detach them from the starter solenoid and ground terminals. Note the routing of each cable to ensure correct installation **(see illustrations)**.

5   If you are replacing either or both of the old cables, take them with you when buying new cables. It is vitally important that you replace the cables with identical parts. Cables have characteristics that make them easy to identify: positive cables are usually red and larger in cross-section; ground cables are usually black and smaller in cross section.

6   Clean the threads of the solenoid or ground connection with a wire brush to remove rust and corrosion. Apply a light coat of battery terminal corrosion inhibitor, or petroleum jelly, to the threads to prevent future corrosion.

7   Attach the cable to the solenoid or ground connection and tighten the mounting nut/bolt securely.

8   Before connecting a new cable to the battery, make sure that it reaches the battery post without having to be stretched.

9   Connect the positive cable first, followed by the negative cable.

5   Ignition system - general information and precautions

Refer to illustration 5.1

1   All models covered by this manual are equipped with a Distributorless Ignition Sys-

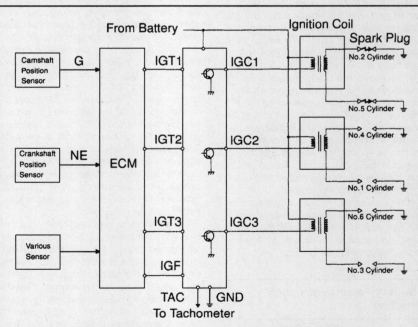

5.1 Typical ignition system electrical schematic (V6 models)

Chapter 5 Engine electrical systems

6.1 To use a calibrated ignition tester, simply disconnect a spark plug wire, connect it to the tester, clip the tester to a convenient ground and operate the starter with the ignition ON - if there is enough power to fire the plug, sparks will be visible between the electrode tip and the tester body

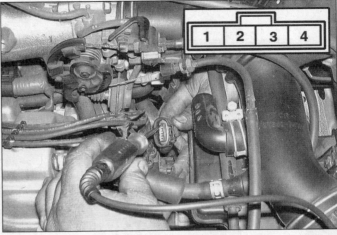

6.4a On four-cylinder models, disconnect the electrical connector at each ignition coil and check for battery voltage at terminal no. 1 with the ignition key turned to ON (engine not running)

tem (DIS) **(see illustration)**. The DIS system uses a "waste spark" method of spark distribution. Each cylinder is paired with its companion cylinder in the firing order (1-4, 2-5, 3-6 on V6 models or 1-4, 3-2 on four-cylinder models); the cylinder under compression fires simultaneously with its companion cylinder, which is on the exhaust stroke. Since the cylinder on the exhaust stroke requires very little of the available voltage to fire its spark plug, most of the voltage is used to fire the plug of the cylinder on the compression stroke.

2 The DIS system includes the camshaft position sensor, the crankshaft position sensor, coils (one coil for a pair of cylinders), igniter and the PCM (computer). The PCM generates cylinder identification signals which allow the igniter to trigger the correct coil. The igniter distributes the signal to the proper coil driver circuit and determines dwell period based on coil primary current flow. Each coil is fired independently and in the proper sequence.

3 On four-cylinder models, the coils are mounted adjacent to the cylinder head and spark plug wires run to each spark plug. On V6 models, the front (left bank) spark plug connectors are integrated into the ignition coils. Spark plug wires connect the rear (right bank) spark plugs to the coils. Both types incorporate an igniter (module) in the ignition system.

4 On four-cylinder models, the igniter is incorporated into the coil packs which are mounted on the engine. The ignition coil and igniter on four-cylinder models must be replaced as a single unit if found to be defective. On 1997 Camry V6 models, 1997 through 1999 Avalon models and 1997 through 1998 Lexus ES 300 models, the igniter is mounted on the fenderwell under the fan relay assembly near the master cylinder. On 2000 and later Avalon models and on 1999 and later Lexus ES 300 models, each

coil has its own integral igniter.

5 When working on the ignition system, take the following precautions:

   a) *Do not keep the ignition switch on for more than 10 seconds if the engine will not start.*
   b) *Always connect a tachometer in accordance with the manufacturer's instructions. Some tachometers may be incompatible with this ignition system. Consult an auto parts counterperson before buying a tachometer for use with this vehicle.*
   c) *Never allow the ignition coil terminals to touch ground. Grounding the coil could result in damage to the igniter and/or the ignition coil.*
   d) *Do not disconnect the battery when the engine is running.*
   e) *Make sure that the igniter is properly grounded.*

6 Ignition system - check

Refer to illustrations 6.1, 6.4a, 6.4b, 6.4c, 6.6a, 6.6b, 6.6c and 6.7

Warning: *Because of the high voltage generated by the ignition system, extreme care should be taken whenever an operation is performed involving ignition components. This not only includes the igniter, coil and spark plug wires, but related components such as plug connectors, tachometer and other test equipment as well.*

1 If the engine turns over but won't start, disconnect the spark plug wire from any spark plug and attach it to a calibrated ignition tester (available at most auto parts stores) **(see illustration)**. Make sure the tester is designed for a distributorless ignition system if a universal tester isn't available.

2 Connect the clip on the tester to a bolt or metal bracket on the engine, crank the

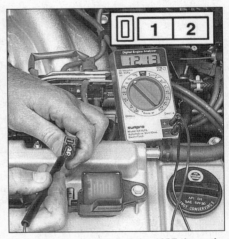

6.4b On Camry V6 models, 1997 through 1999 Avalon models and 1997 and 1998 Lexus ES 300 models, disconnect the electrical connector at each ignition coil and check for battery voltage at terminal no. 1 with the ignition key turned to ON (engine not running)

engine and watch the end of the tester to see if bright blue, well-defined sparks occur.

3 If sparks occur, sufficient voltage is reaching the spark plug to fire it (repeat the check at the remaining spark plug wires to verify that all the ignition coils are functioning). However, the plugs themselves may be fouled, so remove and check them as described in Chapter 1 or install new ones. **Note:** *On V6 models, use the spark plug wires leading to the rear bank spark plugs to check the coils. Since the waste spark system is used, it can be assumed the coils are firing the front bank spark plugs if the rear bank plug wires are firing.*

4 If no sparks or intermittent sparks occur, check for battery voltage to the ignition coils **(see illustrations)**. If battery voltage is pre-

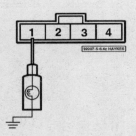

6.4c On 2000 and later Avalon models and 1999 and later Lexus ES 300 models, unplug the black four-pin electrical connector at each ignition coil/igniter and check for battery voltage at terminal no. 1 with the key turned to ON (engine not running)

6.6a On four-cylinder models, check for 5.0 volts from the PCM at terminal no. 3 with the ignition key turned to ON - check for a fluctuating voltage signal at terminal no. 2 as an assistant cranks the engine

6.6b On Camry V6 models, 1997 through 1999 Avalon models and 1997 and 1998 Lexus ES 300 models, use an LED test light to check for a trigger signal from the igniter on terminal no. 2 as an assistant cranks the engine

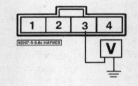

6.6c On 2000 and later Avalon models and 1999 and later Lexus ES 300 models, unplug the black four-pin electrical connector at each ignition coil/igniter and, using a digital voltmeter, measure the trigger signal from the PCM at terminal no. 3 while an assistant cranks the engine; it should be more than 0.1 volts and less than 4.5 volts

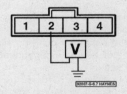

6.7 On 2000 and later Avalon models, and on 1999 and later Lexus ES 300 models, unplug the black four-pin electrical connector at each ignition coil/igniter unit and measure the ignition confirmation voltage between terminal no. 2 and ground; it should be between 4.5 and 5.5 volts

sent, check the coil resistance (see Section 7).

5 Check the camshaft and crankshaft position sensors (see Chapter 6).

6 Check for a coil trigger signal to the coil packs from the PCM (four-cylinder models) or igniter (V6 models) as follows:

a) On four-cylinder models, measure the voltage at terminal no. 3 **(see illustration)** with the ignition key turned to ON. It should be about 5.0 volts. Measure the voltage at terminal no. 2 as an assistant cranks the engine - it should fluctuate between 0.1 and 4.5 volts. Perform these tests on each coil pack connector. If the test results are incorrect, have the PCM checked by a dealership or other properly equipped repair facility. Also check for continuity to a good chassis ground on terminal no. 4. If the test results are correct, the coil/igniter unit is probably faulty.

b) On 1997 and later V6 models, 1997 through 1999 Avalon models and 1997 and 1998 Lexus ES 300 models, disconnect the electrical connector from each coil pack and attach a 12 volt test light to the battery positive terminal. Make sure it is a light emitting diode (LED) type test light. Touch the test light probe to terminal no. 2 **(see illustration)**. Have an assistant crank the engine over - the test light should blink on and off as the igniter completes the circuit to ground. Make this test for each coil pack. If the test results are incorrect, check the igniter (see Section 10).

c) On 2000 and later Avalon models, and on 1999 and later Lexus ES 300 models, disconnect the black four-terminal electrical connector from each of the six ignition coil/igniter units and measure the voltage between terminal no. 3 **(see illustration)** and ground. It should be more than 0.1 and less than 4.5 volts. If it's out of range, troubleshoot the wiring harness between the ECM and the ignition coil/igniter unit. If the wiring is okay, the ECM may be defective. Have the ECM checked by a Toyota or Lexus service department.

7 On 2000 and later Avalon models, and on 1999 and later Lexus ES 300 models, the igniter units also send an ignition confirmation signal (IGF) as a fail-safe measure to the ECM. To verify that this signal is present and correct, unplug the black four-pin electrical connector at each of the six ignition coil/igniter units and measure the voltage between terminal no. 2 **(see illustration)** and ground. It should be between 4.5 and 5.5 volts. If it isn't within this range, replace the ignition coil/igniter unit.

Chapter 5 Engine electrical systems

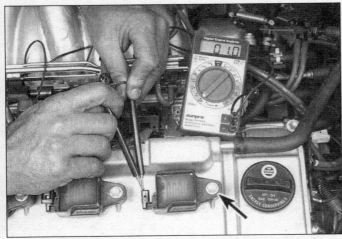

7.1 On V6 models, disconnect the electrical connectors and measure the coil primary resistance across the two terminals at each coil; to remove a coil, remove the retaining bolt (arrow)

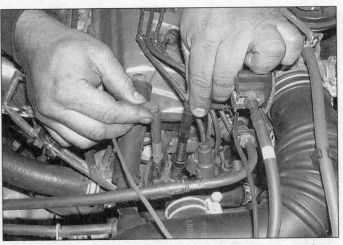

7.2a Remove the spark plug wires and measure the secondary resistance across the two spark plug wire terminals (four-cylinder models)

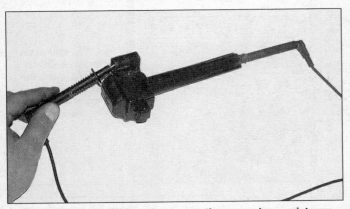

7.2b Remove the coils and measure the secondary resistance across the two spark plug wire terminals (Camry V6 models, 1997 through 1999 Avalon models and 1997 and 1998 Lexus ES 300 models)

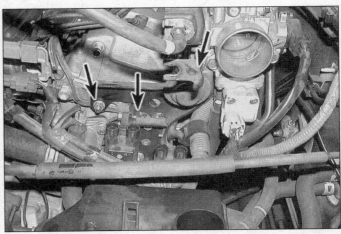

7.6 Ignition coil/igniter assembly mounting bolts (arrows) (four-cylinder models)

7 Ignition coil(s) - check and replacement

Check

Refer to illustrations 7.1, 7.2a and 7.2b

Note 1: *The following coil primary and secondary resistance checks do not apply to the ignition coil/igniter units used on 2000 and later Avalon models and 1999 and later Lexus ES 300 models. Toyota and Lexus do not publish primary or secondary resistance specifications for the ignition coil/igniter units used on these models. The only way to check them is to substitute a known good unit. But this option isn't feasible for a home mechanic because you can't return electrical components to a parts department once you have purchased them. If you have already eliminated all other possible causes of an ignition malfunction, a defective ignition coil/igniter unit is the likely cause of the problem, but the only way to verify this is to have a dealer substitute a known good unit.*

Note 2: *The following checks should be made with the engine cold. If the engine is hot the resistance will be greater.*

1 Check the primary resistance of *each* ignition coil on Camry V6 models, 1997 through 1999 Avalon models and 1997 and 1998 Lexus ES 300 models. With the ignition key OFF, disconnect the electrical harness connector(s) from each coil. Connect an ohmmeter across the coil primary terminals **(see illustration)**. The resistance should be as listed in this Chapter's Specifications. If not, replace the coil. **Note:** *Because of the internal circuitry of the ignition coil/igniter assembly, the primary resistance cannot be checked on four-cylinder models.*

2 Check the secondary resistance of each ignition coil. With the ignition key OFF, label and detach the spark plug wires from each coil. On Camry V6 models, 1997 through 1999 Avalon models and 1997 and 1998 Lexus ES 300 models, remove the coil for this test. Connect an ohmmeter across the two secondary terminals of each coil **(see illustrations)**. The resistance should be as listed in this Chapter's Specifications. If not, replace the coil.

Replacement

3 Disconnect the negative cable from the battery. **Caution:** *If the stereo in your vehicle is equipped with an anti-theft system, make sure you have the correct activation code before disconnecting the battery.*

Four-cylinder models

Refer to illustration 7.6

4 Remove the throttle body from the intake manifold (see Chapter 4).
5 Disconnect the ignition coil harness electrical connector(s) from each individual coil pack. Label each connector so they don't get mixed up. Label and detach the spark plug wires.
6 Remove the bolts from the mounting bracket and separate the bracket/coil pack assembly from the engine **(see illustration)**.
7 Remove the bolts securing the ignition coil to the mounting bracket and remove the individual coils.

8.2 Attach a jumper wire between terminals E1 and TE1 on the test connector

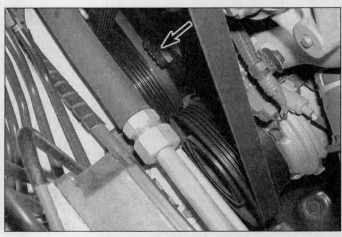

8.4 The timing marks are clearly molded into the timing bolt cover

8 Installation is the reverse of the removal procedure.

V6 models

9 On 2000 and later Avalon models, and on 1999 and later Lexus ES 300 models, remove the V-bank cover (see Section 30 in Chapter 1).

10 Disconnect the ignition coil harness electrical connector(s) from each individual coil pack. Label each connector so they don't get mixed up. Label and detach the spark plug wires for the rear cylinder bank spark plugs.

11 Remove the bolt securing the ignition coil to the engine **(see illustration 7.1)** and remove the coil.

12 Installation is the reverse of the removal procedure.

8 Ignition timing - check

Refer to illustrations 8.2 and 8.4

Note: *The following ignition timing procedure should apply to most models covered by this manual. However, if the procedure specified on the VECI label of your vehicle differs from this one, use the procedure found on the VECI label. The ignition timing cannot be adjusted on these models. The distributorless ignition system is controlled by the PCM (computer) during operation. If the ignition timing is incorrect, diagnose the ignition system components (see Section 6).*

1 On 2000 and later Avalon models and on 1999 and later Lexus ES 300 models, remove the V-bank cover (see Section 30 in Chapter 1).

2 Locate the diagnostic electrical connector and insert a jumper wire between terminals E1 and TE1 **(see illustration)**.

3 With the ignition switch off, connect a timing light and a tachometer according to the tool manufacturer's instructions.

4 Locate the timing marks on the timing belt cover and the crankshaft pulley **(see illustration)**.

5 Start the engine and allow it to warm up to normal operating temperature (upper radiator hose hot). Verify that the engine idle is correct (refer to the Specifications listed at the beginning of this Chapter).

6 Idle speed cannot be adjusted. If the idle speed is incorrect, inspect the intake system (see Chapter 4) for an air leak and check the operation of the idle air control valve (see Chapter 6). If the intake system and the idle air control valve are okay, the PCM might be defective. Have the vehicle diagnosed by a dealer service department or a qualified repair shop.

7 Aim the timing light at the timing scale on the front engine cover. The notch on the crankshaft pulley should line up with the pointer on the scale on the timing cover. Refer to the Specifications listed at the beginning of this Chapter.

8 Ignition timing cannot be adjusted. If the timing is incorrect, the PCM might be defective. Have the vehicle diagnosed by a dealer service department or other qualified repair shop.

9 Remove the jumper wire from the diagnostic connector and recheck the ignition timing. Refer to the Specifications listed at the beginning of this Chapter. Again, if the timing is incorrect, have the vehicle diagnosed by a dealer or shop.

10 Turn off the engine and remove the tachometer and the timing light.

9.1 The igniter on the V6 models is located under the fan relay housing near the master cylinder

9 Igniter (V6 models) - check and replacement

Refer to illustration 9.1

Note 1: *The following procedure does not apply to the integral ignition coil/igniter units used on 2000 and later Avalon models and on 1999 and later Lexus ES 300 models. Testing the igniter function of these units is beyond the scope of the home mechanic. You can check the ignition confirmation signal of each igniter (see Section 6, Step 7), but other than that, identifying a bad igniter is simply a process of elimination. If you suspect a bad igniter on one of these models, have the ignition coil/igniter unit tested by a dealer service department or a qualified repair shop.*

Note 2: *Refer to the wiring diagrams at the end of Chapter 12 when performing the following checks.*

1 Disconnect the electrical connector from the igniter and perform the following electrical checks at the igniter harness connector with the ignition switch On **(see illustration)**:

a) *Check for battery voltage on terminal no. 9. If battery voltage is not present check the circuit from the igniter to the battery.*

b) *Check for approximately 5.0 volts at terminal no. 4. If 5.0 volts are not present,*

Chapter 5 Engine electrical systems

11.3 Connect a voltmeter to the battery terminals and check the battery voltage with the engine Off and again with the engine running

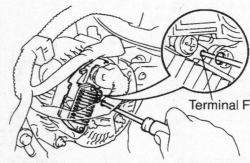

11.6 If the alternator is not charging, ground terminal F to the alternator body, start the engine and check the voltage at the battery – if the reading is now within the specified range, replace the regulator; if the reading is less than the minimum charging voltage, replace the alternator

have the PCM checked by a dealership service department or other properly equipped repair facility.
c) Check for a fluctuating 0.1 to 4.5 volt signal at terminal no. 5, 6 and 7 in turn, as an assistant cranks the engine. If a 0.1 to 4.5 volt signal at terminal no. 5, 6 and 7 is not present as an assistant cranks the engine, have the PCM checked by a dealership service department or other properly equipped repair facility.
d) Check for continuity to a good chassis ground at terminal no. 3.

2 If the above checks are good, turn the ignition switch Off and reconnect the electrical connector to the igniter.

3 Disconnect the electrical connectors at each ignition coil. Using an LED test light, check for an ignition coil trigger signal at each ignition coil connector as an assistant cranks the engine (see Section 6, Step 6). If no ignition coil trigger signal is present, check the circuits from the igniter to the ignition coils for continuity. If the circuits are good, replace the igniter.

10 Charging system - general information and precautions

The charging system includes the alternator, an internal voltage regulator, a charge indicator, the battery, a fusible link and the wiring between all the components. The charging system supplies electrical power for the ignition system, the lights, the radio, etc. The alternator is driven by a drivebelt at the front of the engine.

The purpose of the voltage regulator is to limit the alternator's voltage to a preset value. This prevents power surges, circuit overloads, etc., during peak voltage output.

The charging system doesn't ordinarily require periodic maintenance. However, the drivebelt, battery and wires and connections should be inspected at the intervals outlined in Chapter 1.

The dashboard warning light should come on when the ignition key is turned to Start, then should go off immediately. If it remains on, there is a malfunction in the charging system. Some vehicles are also equipped with a voltage gauge. If the voltage gauge indicates abnormally high or low voltage, check the charging system (see Section 11).

Be very careful when making electrical circuit connections to a vehicle equipped with an alternator and note the following:

a) When reconnecting wires to the alternator from the battery, be sure to note the polarity.
b) Before using arc welding equipment to repair any part of the vehicle, disconnect the wires from the alternator and the battery terminals.
c) Never start the engine with a battery charger connected.
d) Always disconnect both battery leads before using a battery charger.
e) The alternator is driven by an engine drivebelt which could cause serious injury if your hand, hair or clothes become entangled in it with the engine running.
f) Because the alternator is connected directly to the battery, it could arc or cause a fire if overloaded or shorted out.
g) Wrap a plastic bag over the alternator and secure it with rubber bands before steam cleaning the engine.

11 Charging system - check

Refer to illustrations 11.3 and 11.6

1 If a malfunction occurs in the charging circuit, do not immediately assume that the alternator is causing the problem. First, check the following items:

a) Make sure the battery cable clamps, where they connect to the battery, are clean and tight.
b) Test the condition of the battery (see Section 2). If it does not pass all the tests, replace it with a new battery.
c) Check the external alternator wiring and connections.
d) Check the drivebelt condition and tension (see Chapter 1).
e) Check the alternator mounting bolts for tightness.
f) Run the engine and check the alternator for abnormal noise.
g) Check the fusible links (if equipped) in the engine compartment fuse box (see Chapter 12). If they're burned, determine the cause and repair the circuit.
h) Check the charge light on the dash. It should illuminate when the ignition key is turned ON (engine not running). If it does not, check the circuit from the alternator to the charge light on the dash.
i) Check all the fuses that are in series with the charging system circuit. The location of these fuses and fusible links may vary from year and model but the designations are generally the same. Refer to the wiring schematics at the end of Chapter 12 for additional information.

2 With the ignition key off, check the battery voltage with no accessories operating. It should be approximately 12.5 volts. It may be slightly higher if the engine had been operating within the last hour.

3 Connect an ammeter to the charging system following the tool manufacturer's instructions. Start the engine, and check the battery voltage and amperage. It should now be approximately 13.2 to 15.0 volts at 10 amps or less **(see illustration)**.

4 Load the battery by turning on the high beam headlights and the air conditioning (if equipped and place the blower fan on HIGH. Raise the engine speed to 2,000 rpm and check the voltage and amperage. If the charging system is working properly the voltage should stay above 13.5 volts and the amperage should be 30 amps or more (depending on the condition of the battery - it could be less than 30 amps).

5 If the voltage rises above 15.0 volts in either test, the regulator is defective.

6 If the charging voltage is below the minimum charging voltage, turn the accessories and the engine off and use a small screwdriver to full field the alternator circuit **(see**

5-10 Chapter 5 Engine electrical systems

illustration). Start the engine and check the charging voltage. If the charging voltage is now within the specified range, the voltage regulator is defective. If the charging voltage is still low, the alternator is defective.

12 Alternator - removal and installation

Removal

Refer to illustrations 12.3, 12.4a and 12.4b

1 Detach the cable from the negative terminal of the battery. **Caution:** *If the stereo in your vehicle is equipped with an anti-theft system, make sure you have the correct activation code before disconnecting the battery.*
2 Detach the electrical connectors from the alternator.
3 Loosen the alternator adjustment and pivot bolts **(see illustration)** and detach the drivebelt.

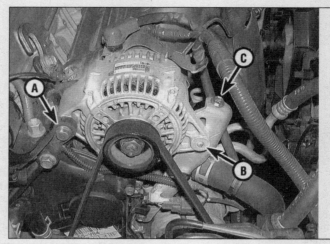

12.3 First loosen the pivot bolt (A), loosen the lock bolt (B) and then turn the adjustment bolt (C) counterclockwise to release the tension on the drivebelt

4 Remove the adjustment and pivot bolts and remove the alternator from the engine bracket and the adjustment bracket **(see illustrations)**.

Installation

5 If you are replacing the alternator, take the old alternator with you when purchasing a replacement unit. Make sure that the new/rebuilt unit is identical to the old alternator. Look at the terminals - they should be the same in number, size and locations as the terminals on the old alternator. Finally, look at the identification markings - they will be stamped in the housing or printed on a tag or plaque affixed to the housing. Make sure that these numbers are the same on both alternators.
6 Many new/rebuilt alternators do not have a pulley installed, so you may have to switch the pulley from the old unit to the new/rebuilt one. When buying an alternator, find out the shop's policy regarding installation of pulleys - some shops will perform this service free of charge.
7 Installation is the reverse of removal.
8 After the alternator is installed, adjust the drivebelt tension (see Chapter 1).
9 Check the charging voltage to verify proper operation of the alternator (see Section 11).

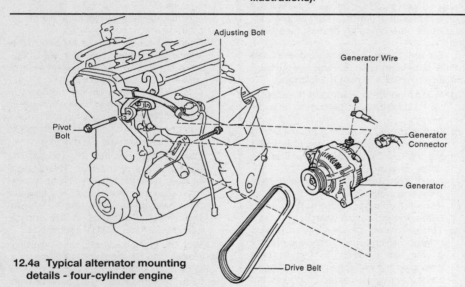

12.4a Typical alternator mounting details - four-cylinder engine

13 Voltage regulator and alternator brushes - replacement

Refer to illustrations 13.2a, 13.2b, 13.3, 13.5, 13.6 and 13.9

1 Remove the alternator (see Section 12) and place it on a clean workbench.
2 Remove the output terminal nut and insulator. Remove the rear cover nuts and remove the rear cover **(see illustrations)**.
3 Remove the two brush holder retaining screws **(see illustration)**.
4 Remove the brush holder.
5 Remove the three voltage regulator retaining screws and remove the voltage regulator **(see illustration)**.
6 Measure the exposed length of each brush **(see illustration)** and compare it to the specified minimum length. If the length of the brush is less than the minimum listed in this Chapter's Specifications, replace the brush

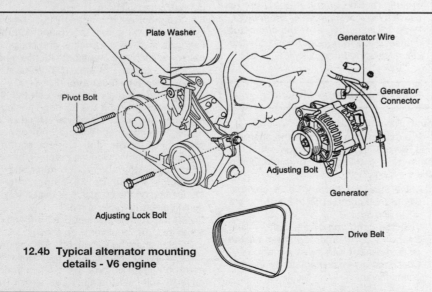

12.4b Typical alternator mounting details - V6 engine

Chapter 5 Engine electrical systems

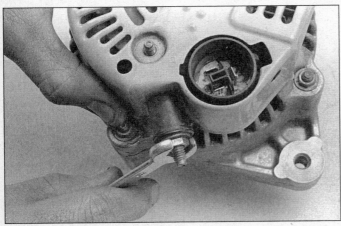

13.2a Remove the nut, washer and insulator from the output terminal

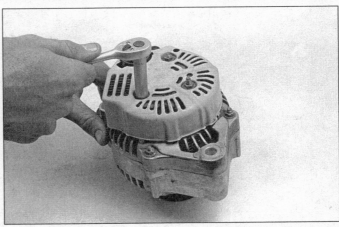

13.2b Remove the nuts and detach the rear cover from the alternator

13.3 Remove the two screws (arrows) that retain the brush holder

13.5 Remove the three screws (arrows) that retain the voltage regulator

assembly.
7 Make sure that each brush moves smoothly in the brush holder.
8 Install the voltage regulator.
9 Install the brush holder by depressing the brush with a small screwdriver to clear the shaft (see illustration).
10 Install the brush holder screws into the rear frame.
11 Install the rear cover and tighten the nuts securely.
12 Install the output terminal insulator and tighten it with the nut.
13 Install the alternator.

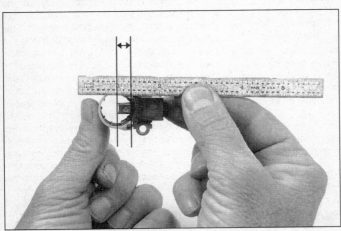

13.6 Measure the exposed length of the brushes and compare your measurements to the specified minimum length to determine if they should be replaced

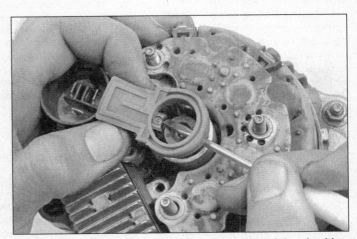

13.9 When installing the brush holder, depress each brush with a small screwdriver to clear the shaft

Chapter 5 Engine electrical systems

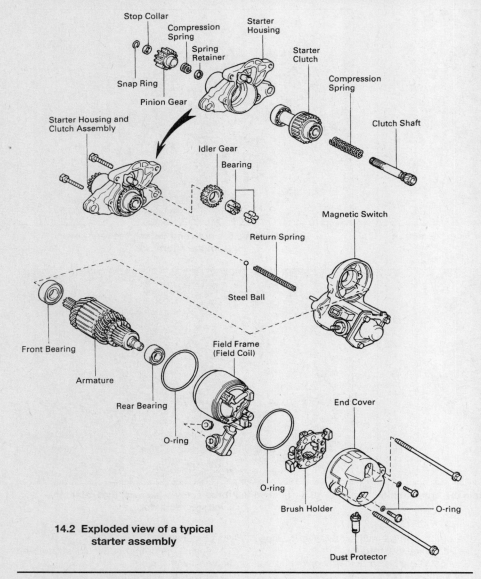

14.2 Exploded view of a typical starter assembly

14 Starting system - general information and precautions

Refer to illustration 14.2

The starting system consists of the battery, the starter motor, the starter solenoid and the electrical circuit connecting the components **(see illustration)**. The solenoid is mounted directly on the starter motor.

The solenoid/starter motor assembly is installed on the upper part of the engine, next to the transaxle bellhousing.

When the ignition key is turned to the START position, the starter solenoid is actuated through the starter control circuit. The starter solenoid then connects the battery to the starter. The battery supplies the electrical energy to the starter motor, which does the actual work of cranking the engine.

The starter motor on a vehicle equipped with a manual transaxle can be operated only when the clutch pedal is depressed; the starter on a vehicle equipped with an automatic transaxle can be operated only when the transaxle selector lever is in Park or Neutral.

Always observe the following precautions when working on the starting system:

a) *Excessive cranking of the starter motor can overheat it and cause serious damage. Never operate the starter motor for more than 15 seconds at a time without pausing to allow it to cool for at least two minutes.*
b) *The starter is connected directly to the battery and could arc or cause a fire if mishandled, overloaded or short circuited.*
c) *Always detach the cable from the negative terminal of the battery before working on the starting system.* **Caution:** *If the stereo in your vehicle is equipped with an anti-theft system, make sure you have the correct activation code before disconnecting the battery.*

15 Starter motor and circuit - check

Refer to illustration 15.4

1 If a malfunction occurs in the starting circuit, do not immediately assume that the starter is causing the problem. First, check the following items:

a) *Make sure the battery cable clamps, where they connect to the battery, are clean and tight.*
b) *Check the condition of the battery cables (see Section 4). Replace any defective battery cables with new parts.*
c) *Test the condition of the battery (see Section 3). If it does not pass all the tests, replace it with a new battery.*
d) *Check the starter solenoid wiring and connections. Refer to the wiring diagrams at the end of Chapter 12.*
e) *Check the starter mounting bolts for tightness.*
f) *Check the fusible links (if equipped) exiting the engine compartment fuse box (see Chapter 12). If they're burned, determine the cause and repair the circuit. Also, check the ignition switch circuit for correct operation (see Chapter 12).*
g) *Check the operation of the Park/Neutral switch (automatic transaxle) or clutch start switch (manual transaxle). Make sure the shift lever is in PARK or NEUTRAL. (automatic transaxle) or the clutch pedal is pressed (manual transaxle). Refer to Chapter 7 for the Park/Neutral switch check and adjustment procedure. Refer to Chapter 12 wiring diagrams, if necessary, when performing circuit checks. These systems must operate correctly to provide battery voltage to the ignition solenoid.*
h) *Check the operation of the starter relay. The starter relay is located in the fuse/relay box inside the engine compartment. Refer to Chapter 12 for the testing procedure.*

2 If the starter does not actuate when the ignition switch is turned to the start position, check for battery voltage to the solenoid. This will determine if the solenoid is receiving the correct voltage signal from the ignition switch. Connect a test light or voltmeter to the starter solenoid positive terminal and while an assistant turns the ignition switch to the start position. If voltage is not available, refer to the wiring diagrams in Chapter 12 and check all the fuses and relays in series with the starting system. If voltage is available but the starter motor does not operate, remove the starter from the engine compartment (see Section 16) and bench test the starter (see Step 4).

3 If the starter turns over slowly, check the starter cranking voltage and the current draw from the battery. This test must be performed with the starter assembly on the engine. Crank the engine over (for 10 seconds or less) and observe the battery voltage. It

Chapter 5 Engine electrical systems

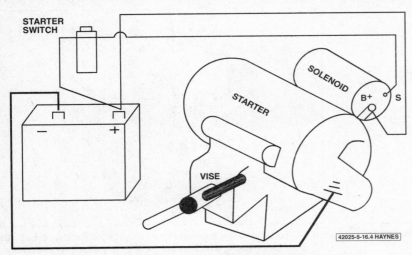

15.4 Starter motor bench testing details

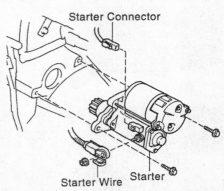

16.5a Starter motor installation details (four-cylinder engine)

should not drop below 8.0 volts on manual transaxle models or 8.5 volts on automatic transaxle models. Also, observe the current draw using an amp meter. It should not exceed 400 amps or drop below 250 amps. If the starter motor cranking amp values are not within the correct range, replace it with a new unit. There are several conditions that may affect the starter cranking potential. The battery must be in good condition and the battery cold-cranking rating must not be underrated for the particular application. Be sure to check the battery specifications carefully. The battery terminals and cables must be clean and not corroded. Also, in cases of extreme cold temperatures, make sure the battery and/or engine block is warmed before performing the tests.

4 If the starter is receiving voltage but does not activate, remove and check the starter/solenoid assembly on the bench. Most likely the solenoid is defective. In some rare cases, the engine may be seized so be sure to try and rotate the crankshaft pulley (see Chapter 2A or 2B) before proceeding. With the starter/solenoid assembly mounted in a vise on the bench, install one jumper cable from the negative battery terminal to the body of the starter **(see illustration)**. Install the other jumper cable from the positive battery terminal to the B+ terminal on the starter. Install a starter switch and apply battery voltage to the solenoid S terminal (for 10 seconds or less) and see if the solenoid plunger, shift lever and overrunning clutch extends and rotates the pinion drive. If the pinion drive extends but does not rotate, the solenoid is operating but the starter motor is defective. If there is no movement but the solenoid clicks, the solenoid and/or the starter motor is defective. If the solenoid plunger extends and rotates the pinion drive, the starter/solenoid assembly is working properly.

16 Starter motor - removal and installation

Refer to illustrations 16.5a and 16.5b

1 Detach the cable from the negative terminal of the battery. **Caution:** *If the stereo in your vehicle is equipped with an anti-theft system, make sure you have the correct activation code before disconnecting the battery.*
2 Remove the battery from the engine compartment.
3 Disconnect and remove the cruise control actuator from the engine compartment, if equipped.
4 Detach the electrical connectors from the starter/solenoid assembly.
5 Remove the starter motor mounting bolts **(see illustrations)**.
6 Remove the bracket from the upper section of the starter/solenoid assembly.
Note: *It is necessary to loosen one or two of the bracket bolts to allow the starter/solenoid assembly to partially drop down to gain access to the remaining bracket assembly bolts and hardware.*
7 Installation is the reverse of removal.

17 Starter solenoid - removal and installation

Refer to illustrations 17.2, 17.3, 17.4, 17.5, 17.6a and 17.6b

1 Remove the starter motor (see Section 16).
2 Scribe or paint a mark across the starter

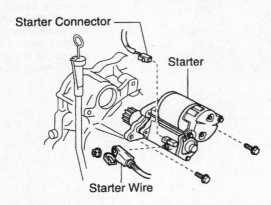

16.5b Starter motor installation details (V6 engine)

17.2 Before disassembling the starter motor, solenoid and gear reduction assembly, scribe or paint an alignment mark across the starter motor and the gear reduction assembly

Chapter 5 Engine electrical systems

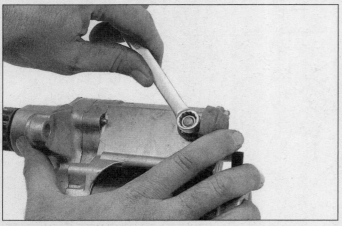

17.3 To disconnect the strap that connects the starter to the solenoid, remove this nut

17.4 To detach the solenoid from the starter motor, remove the screws (arrows) which secure the gear reduction assembly to the solenoid . . .

17.5 . . . and remove the through-bolts (arrows) which secure the starter motor to the gear reduction assembly

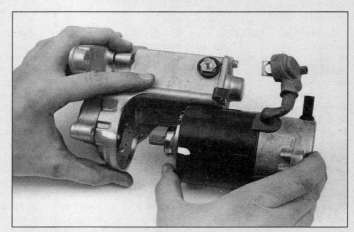

17.6a Separate the starter from the gear reduction assembly . . .

motor and gear reduction assembly **(see illustration)**.

3   Disconnect the strap from the solenoid to the starter motor terminal **(see illustration)**.

4   Remove the screws which secure the gear reduction assembly to the solenoid **(see illustration)**.

5   Remove the through-bolts which secure the starter motor to the gear reduction assembly **(see illustration)**.

6   Separate the motor from the gear reduction and solenoid assembly then remove the solenoid from the gear reduction assembly **(see illustrations)**.

7   Installation is the reverse of removal. Be sure to align the paint or scribe mark.

17.6b . . . and separate the solenoid from the gear reduction assembly (note the return spring protruding from the solenoid assembly - make sure that this spring is installed before reassembling the solenoid and the gear reduction assembly)

form
Chapter 6
Emissions and engine control systems

Contents

	Section		Section
Acoustic Control Induction System (ACIS) (V6 models only)	16	Knock sensor - check and replacement	13
Camshaft Position sensor - check and replacement	10	Manifold Absolute Pressure (MAP) sensor (four-cylinder models) - check and replacement	5
Catalytic converter	20		
Crankshaft Position sensor - check and replacement	9	Mass Airflow (MAF) sensor (V6 models) - check and replacement	6
Engine Coolant Temperature (ECT) sensor - check and replacement	8	On-Board Diagnostic (OBD) system and trouble codes	2
Exhaust Gas Recirculation (EGR) system - description, check and component replacement	18	Oxygen sensor - general information and precautions	11
		Oxygen sensor and air/fuel sensor - check and replacement	12
Evaporative emissions control (EVAP) system - description, check and component replacement	19	Positive Crankcase Ventilation (PCV) system	17
General information	1	Powertrain Control Module (PCM) - removal and installation	3
Idle Air Control (IAC) valve - check and replacement	15	Throttle Position Sensor (TPS) - check and replacement	4
Intake Air Temperature (IAT) sensor - check and replacement	7	Vehicle Speed Sensor (VSS) - check and replacement	14

Specifications

Air/fuel sensor heater resistance
  Four-cylinder engine (Camry, at 68 degrees F)
    1997 .... 11 to 16 ohms
    1998 and later .... 0.8 to 1.4 ohms
  V6 engine (1998 on)
    1998 (at 68 degrees F)
      Camry and Avalon .... 0.8 to 1.4 ohms
      Lexus ES 300 .... 11 to 16 ohms
    1999 and later
      Cold .... 0.8 to 1.4 ohms
      Hot .... 1.8 to 3.2 ohms
Oxygen sensor heater resistance
  1997 through 1999 (at 68 degrees F) .... 11 to 16 ohms
  2000 and later
    Cold .... 11 to 16 ohms
    Hot .... 23 to 32 ohms
Oxygen sensor voltage
  Open loop .... 0.1 to 0.2 volt
  Closed loop
    Camry, Avalon and 1997 Lexus ES 300 .... 0.1 to 0.9 volt
    1998 and later Lexus ES 300 .... 0.05 to 0.96 volt
Crankshaft position sensor resistance
  Four-cylinder models
    Cold .... 985 to 1,600 ohms
    Hot .... 1,265 to 1,890 ohms
  V6 engine
    Cold .... 1,630 to 2,740 ohms
    Hot .... 2,065 to 3,225 ohms

6-1

Camshaft position sensor resistance
    Four-cylinder models
        Cold .... 835 to 1,400 ohms
        Hot .... 1,060 to 1,645 ohms
    V6 models
        Nippondenso
            Cold .... 835 to 1,400 ohms
            Hot .... 1,060 to 1,645 ohms
        Wabash
            Cold .... 1,690 to 2,560 ohms
            Hot .... 2,145 to 3,010 ohms
Idle Air Control valve resistance
    Camry
        1997 and 1998
            Cold .... 17 to 24.5 ohms
            Hot .... 21.5 to 28.5 ohms
        1999
            Four-cylinder engine
                Cold .... 17 to 24.5 ohms
                Hot .... 21.5 to 28.5 ohms
            V6 engine
                Cold .... 17 to 25 ohms
                Hot .... 21.5 to 29.5 ohms
        2000 and later
            Four-cylinder engine.. N/A
            V6 engine
                Cold .... 17 to 25 ohms
                Hot .... 21.5 to 29.5 ohms
    Avalon
        1997 through 1999
            Cold .... 17 to 24.5 ohms
            Hot .... 21.5 to 28.5 ohms
        2000 and later... N/A
    Lexus ES 300
        1997 and 1998
            Cold .... 17 to 24.5 ohms
            Hot .... 21.5 to 28.5 ohms
        1999 and later... N/A
EGR system (Camry V6, 1997 through 1999 Avalon and 1997 and 1998 Lexus ES 300)
    EGR gas temperature sensor resistance
        122-degrees F.. 64 to 97 k-ohms
        212-degrees F.. 11 to 16 k-ohms
        302-degrees F.. 2 to 4 k-ohms
    EGR valve position sensor resistance.. 1.5 to 4.3 k-ohms

Torque specifications

Ft-lbs (unless otherwise noted)

Intake air control valve/No. 2 intake air control valve retaining nuts .... 120 in-lbs
No. 1 intake air control valve-to-intake plenum retaining nuts 22

1 General information

Refer to illustrations 1.1a, 1.1b, 1.6a and 1.6b

To prevent pollution of the atmosphere from incompletely burned and evaporating gases, and to maintain good driveability and fuel economy, a number of emission control systems are incorporated **(see illustrations)**. They include the:

    On-Board Diagnostic (OBD) system
    Electronic engine controls (EFI) system
    Exhaust Gas Recirculation (EGR) system
    Evaporative Emissions Control (EVAP) system
    Positive Crankcase Ventilation (PCV) system
    Acoustic Control Induction System (ACIS) (V6 models)
    Catalytic converter

The Sections in this Chapter include general descriptions, checking procedures within the scope of the home mechanic and component replacement procedures (when possible) for each of the systems listed above.

Before assuming that an emissions control system is malfunctioning, check the fuel and ignition systems carefully. The diagnosis of some emission control devices requires specialized tools, equipment and training. If checking and servicing become too difficult or if a procedure is beyond your ability, consult a dealer service department or other repair shop. Remember, the most frequent cause of emissions problems is simply a loose or broken wire or vacuum hose, so always check the hose and wiring connections first.

This doesn't mean, however, that emissions control systems are particularly difficult to maintain and repair. You can quickly and easily perform many checks and do most of the regular maintenance at home with common tune-up and hand tools. **Note:** *Because of a Federally mandated warranty which covers the emissions control system components, check with your dealer about warranty coverage before working on any emissions-related systems. Once the warranty has expired, you may wish to perform some of the component checks and/or replacement procedures in this Chapter to save money.*

Pay close attention to any special precautions outlined in this Chapter. It should be noted that the illustrations of the various systems may not exactly match the system installed on your vehicle because of changes made by the manufacturer during production or from year-to-year.

Chapter 6 Emissions and engine control systems

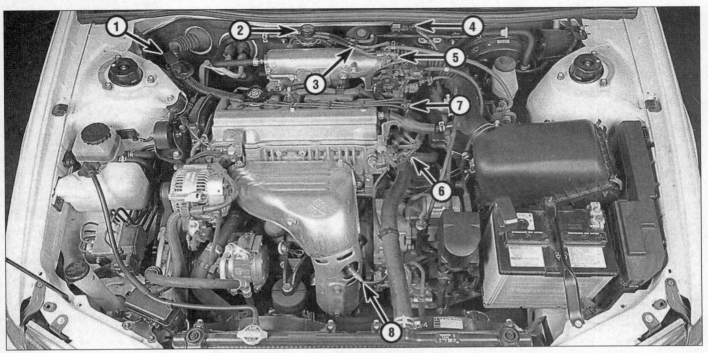

1.1a Typical emission and engine control system components - four-cylinder engine

1. Diagnostic test connector
2. Exhaust Gas Recirculation (EGR) vacuum modulator
3. Exhaust Gas Recirculation (EGR) valve
4. Manifold Absolute Pressure (MAP) sensor
5. Throttle Position Sensor (TPS) (behind throttle body)
6. Coolant temperature sensor
7. Coil packs
8. Oxygen sensor

1.1b Typical emission and engine control system components - V6 engine

1. Diagnostic test connector
2. Exhaust Gas Recirculation (EGR) valve
3. Idle Air Control (IAC) valve (under throttle body)
4. Throttle Position Sensor (TPS)
5. Mass Airflow (MAF) sensor
6. Igniter (under relay housing)
7. Oxygen sensor
8. Coil packs
9. Coolant temperature sensor

Chapter 6 Emissions and engine control systems

1.6a The Vehicle Emission Control Information (VECI) label contains such essential information as the types of emission control systems installed on the engine and the idle speed and ignition timing specifications

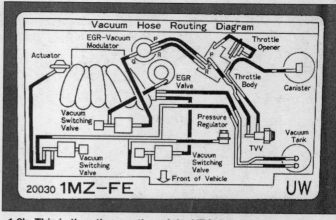

1.6b This is the other section of the VECI label that contains the vacuum hose routing

A Vehicle Emissions Control Information (VECI) label is attached to the underside of the hood **(see illustration)**. This label contains important emissions specifications and adjustment information. Part of this label, the Vacuum Hose Routing Diagram **(see illustration)**, provides a vacuum hose schematic with emissions components identified. When servicing the engine or emissions systems, the VECI label and the vacuum hose routing diagram in your particular vehicle should always be checked for up-to-date information.

2 On-Board Diagnostic (OBD) system and trouble codes

Diagnostic tool information

Refer to illustrations 2.1 and 2.2

1 A digital multimeter is necessary for checking fuel injection and emission related components **(see illustration)**. A digital volt-ohmmeter is preferred over the older style analog multimeter for several reasons. The analog multimeter cannot display the volts-ohms or amps measurement in hundredths and thousandths increments. When working with electronic circuits which are often very low voltage, this accurate reading is most important. Another good reason for the digital multimeter is the high impedance circuit. The digital multimeter is equipped with a high resistance internal circuitry (10 million ohms). Because a voltmeter is hooked up in parallel with the circuit when testing, it is vital that none of the voltage being measured should be allowed to travel the parallel path set up by the meter itself. This dilemma does not show itself when measuring larger amounts of voltage (9 to 12 volt circuits) but if you are measuring a low voltage circuit such as the oxygen sensor signal voltage, a fraction of a volt may be a significant amount when diagnosing a problem.

2 Hand-held scanners are the most powerful and versatile tools for analyzing engine management systems used on later model vehicles **(see illustration)**. Early model scanners handle codes and some diagnostics for many systems. Each brand scan tool must be examined carefully to match the year, make and model of the vehicle you are working on. Often, interchangeable cartridges are available to access the particular manufacturer (Ford, GM, Chrysler, Asian etc.). Some manufacturers will specify by continent (Asia, Europe, USA, etc.).

3 With the arrival of the Federally mandated emission control system (OBD-II), a specially designed scanner has been developed. Several tool manufacturers have released OBD-II scan tools for the home mechanic. Ask the parts salesman at a local auto parts store for additional information concerning dates and costs.

OBD system general description

4 As specified by CARB and EPA regulations, automobile manufacturers have designed a second generation self diagnosis system called On Board Diagnosis-II (OBD-II). This system incorporates a series of diagnostic monitors that detect and identify emissions systems faults and store the information in the computer memory. This updated system also tests sensors and output actuators, diagnoses drive cycles, freezes data and clears codes.

5 This powerful diagnostic computer must be accessed using the new OBD-II SCAN tool and 16 pin Data Link Connector (DLC) located under the driver's dash area. All engines and powertrain combinations described in this manual are equipped with the On Board Diagnosis II (OBD-II) system. This system consists of an onboard computer, known as the Powertrain Control Module (PCM), and information sensors, which monitor various functions of the engine and send data to the PCM. Based on the data and the information programmed into the computer's memory, the PCM generates output signals to control various engine functions via control relays, solenoids and other output actuators.

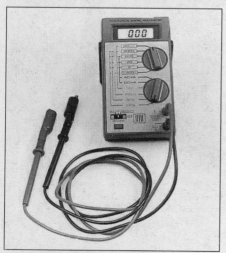

2.1 Digital multimeters can be used for testing all types of circuits; because of their high impedance, they are much more accurate than analog meters for measuring low-voltage computer circuits

2.2 Scanners like the Actron Scantool and the AutoXray XP240 are powerful diagnostic aids - programmed with comprehensive diagnostic information, they can tell you just about anything you want to know about your engine management system

Chapter 6 Emissions and engine control systems

6 The PCM is the "brain" of the electronically controlled fuel and emissions system. It receives data from a number of sensors and other electronic components (switches, relays, etc.). Based on the information it receives, the PCM generates output signals to control various relays, solenoids and other actuators. The PCM is specifically calibrated to optimize the emissions, fuel economy and driveability of the vehicle.

7 Because of a Federally mandated warranty which covers the emissions system components and because any owner-induced damage to the PCM, the sensors and/or the control devices may void the warranty, it isn't a good idea to attempt diagnosis or replacement of the PCM at home while the vehicle is under warranty. Take the vehicle to a dealer service department if the PCM or a system component malfunctions.

Information sensors

8 **Oxygen sensors (O2S)** - The O2S generates a voltage signal that varies with the difference between the oxygen content of the exhaust and the oxygen in the surrounding air.

9 **Crankshaft Position (CKP) sensor** - The CKP sensor provides information on crankshaft position and the engine speed signal to the PCM.

10 **Camshaft Position (CMP) sensor** - The CMP sensor produces a signal which the PCM uses to identify number 1 cylinder and to time the sequential fuel injection.

11 **Air/Fuel Sensor (California models)** - California models with an automatic transaxle are equipped with an air/fuel ratio sensor mounted upstream of the catalytic converter. These sensors work similar to the O2 sensors.

12 **Engine Coolant Temperature (ECT) sensor** - The ECT sensor monitors engine coolant temperature and sends the PCM a voltage signal that affects PCM control of the fuel mixture, ignition timing, and EGR operation.

13 **Intake Air Temperature (IAT) sensor** - The IAT provides the PCM with intake air temperature information. The PCM uses this information to control fuel flow, ignition timing, and EGR system operation.

14 **Throttle Position Sensor (TPS)** - The TPS senses throttle movement and position, then transmits a voltage signal to the PCM. This signal enables the PCM to determine when the throttle is closed, in a cruise position, or wide open.

15 **Manifold Absolute Pressure (MAP) sensor (four-cylinder models)** - The MAP sensor measures the amount (volume) of the intake airflow entering the engine. The MAP sensor, along with the IAT sensor, provide airflow volume and air temperature information for the most precise fuel metering.

16 **Mass Airflow Meter (MAF) (V6 models)** - The MAF sensor measures the mass of the intake air by detecting volume and weight of the air from samples passing over the hot wire element.

17 **Vehicle Speed Sensor (VSS)** - The vehicle speed sensor provides information to the PCM to indicate vehicle speed.

18 **EGR valve position sensor** - The EGR valve position sensor monitors the position of the EGR pintle in relation to the operating conditions of the EGR system.

19 **Vapor pressure sensor** - The fuel tank pressure sensor is part of the evaporative emission control system and is used to monitor vapor pressure in the fuel tank. The PCM uses this information to turn on and off the vacuum switching valves (VSV) of the evaporative emission system.

20 **Power Steering Pressure (PSP) switch** - The PSP sensor is used to increase engine idle speed during low-speed vehicle maneuvers.

21 **Transaxle sensors** - In addition to the vehicle speed sensor, the PCM receives input signals from the following sensors inside the transaxle or connected to it: (a) the direct clutch speed sensor (b) the vehicle speed sensor.

Output actuators

22 **EFI main relay** - The EFI main relay activates power to the fuel pump relay (circuit opening relay). It is activated by the ignition switch and supplies battery power to the PCM and the EFI system when the switch is in the Start or Run position. Refer to Chapter 4 or your owner's manual for more information on relay location.

23 **Fuel injectors** - The PCM opens the fuel injectors individually in firing order sequence. The PCM also controls the time the injector is open, called the "pulse width." The pulse width of the injector (measured in milliseconds) determines the amount of fuel delivered. For more information on the fuel delivery system and the fuel injectors, including injector replacement, refer to Chapter 4.

24 **Igniter** - The igniter triggers the ignition coil and determines proper spark advance based on inputs from the PCM. V6 models mount the igniter below the relay housing next to the brake master cylinder. Four-cylinder models incorporate the igniter into the coil assembly. Refer to Chapter 5 for more information on the igniter (V6 models) or the ignition coil/igniter assembly (four-cylinder models).

25 **Idle air control (IAC) valve** - The IAC valve controls the amount of air to bypass the throttle plate when the throttle valve is closed or at idle position. The IAC valve opening and the resulting airflow is controlled by the PCM. Refer to Chapter 4 for more information on the IAC valve.

26 **EVAP vacuum switching valve (VSV)** - The EVAP vacuum switching valve is a solenoid valve, operated by the PCM to purge the fuel vapor canister and route fuel vapor to the intake manifold for combustion.

27 **Vapor Pressure Sensor vacuum switching valve (VSV)** - The Vapor Pressure Sensor vacuum switching valve is operated by the PCM during the OBD-II evaporative emission monitor and during an emission test of the evaporative system.

Obtaining OBD-II system trouble codes

Refer to illustration 2.29

28 The PCM will illuminate the CHECK ENGINE light (also called the Malfunction Indicator Light) on the dash if it recognizes a component fault for two consecutive drive cycles. It will continue to set the light until the PCM does not detect any malfunction for three or more consecutive drive cycles.

29 The diagnostic codes for the OBD-II system can be extracted from the PCM by plugging a generic OBD-II scan tool **(see illustration 2.2)** into the PCM's data link connector **(see illustration)**, which is located under the left end of the dash. **Note:** *An aftermarket generic scanner should work with any model covered by this manual. However, some early OBD-II models, although technically classified as OBD-II compliant by the manufacturer and by the federal government, might not be fully compliant with all SAE standards for OBD-II. Some generic scanners are unable to extract all "P0" codes from these early OBD-II models. Before purchasing a generic scan tool, contact the manufacturer of the scanner you're planning to buy and verify that it will work properly with the OBD-II system you want to scan. If necessary, of course, you can always have the codes extracted by a dealer service department or an independent repair shop with a professional scan tool that's capable of extracting the codes on all Toyota and Lexus models.*

30 Plug the scan tool into the 16-pin data link connector (DLC), and then turn the ignition key ON (engine not running) and make sure the CHECK ENGINE light on the instru-

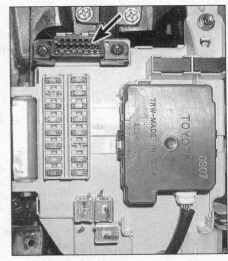

2.29 16-pin Data Link Connector (DLC) (arrow)

Chapter 6 Emissions and engine control systems

ment panel is on. If the light is off, either the bulb is burned out or there is a problem with the PCM or related circuit. Using the instructions included with the scan tool, extract any code(s) stored in the PCM.

Clearing the codes

31 After you have extracted the diagnostic trouble codes and made any necessary repairs, make sure that you "clear" (erase) them before operating the vehicle. Also, always clear any stored codes from the PCM before replacing any electronic emission control component (information sensor or output actuator) that is part of the engine management system, i.e. computer controlled.

Because the PCM stores trouble codes during sensor malfunctions, it will also store a code if a new sensor is installed before any codes regarding the old sensor have been erased. Clearing the old codes allows the computer to relearn the new operating parameters relayed by the new component. During the computer relearning process, the engine may experience a rough idle or slight driveability changes. This period of time, however, should last no longer than 15 to 20 minutes.

32 Plug the scan tool into the 16-pin DLC, scroll the menu for the "CLEARING CODES" mode and follow the manufacturer's instructions.

Diagnostic Trouble Codes

Note: *The following table lists all trouble codes used on all 1997 and later Camry, Avalon and Lexus ES 300 models. But not all codes apply to all models. Some codes apply only to four-cylinder models, others only to V6 models; some codes apply only to models with an automatic transaxle, others only to models with a manual transaxle; etc. Also, some of the following codes apply to all model years, while other apply only to some model years. But a generic scan tool can only extract those codes which have been stored in the PCM of the vehicle that it's scanning, so it's impossible to access a code that's doesn't apply to the vehicle you're diagnosing.*

Code	Identification	Possible problem area
P0100	Mass airflow (MAF) sensor sensor circuit malfunction	Open or short in MAF sensor circuit MAF sensor malfunction PCM malfunction
P0101	Mass airflow (MAF) circuit range/performance problem	MAF sensor malfunction
P0105	Manifold Absolute Pressure (MAP) sensor circuit malfunction	Open or short in MAP sensor circuit MAP sensor malfunction PCM malfunction
P0106	Manifold Absolute Pressure (MAP) sensor malfunction	Defective MAP sensor Loose, torn or missing vacuum line
P0110	Intake Air Temperature (IAT) sensor circuit malfunction	Open or short in IAT sensor circuit IAT sensor malfunction PCM malfunction
P0115	Engine coolant temperature (ECT) sensor circuit malfunction	Open or short in ECT sensor circuit ECT sensor malfunction PCM malfunction
P0116	Engine coolant temperature (ECT) circuit range/performance problem	ECT sensor malfunction Cooling system problem
P0120	Throttle/pedal position sensor (TPS) circuit malfunction	Open or short in TPS circuit TPS malfunction PCM malfunction
P0121	Throttle/pedal position sensor (TPS) range/performance problem	TPS malfunction
P0125	Insufficient coolant temperature for closed loop fuel control (non-California)	Open or short in heated oxygen sensor (bank 1, sensor 1) circuit Heated oxygen sensor (bank 1, sensor 1) malfunction PCM
P0125	Insufficient coolant temperature for closed loop fuel control (California)	Open or short in air/fuel (A/F) sensor circuit A/F sensor malfunction PCM malfunction
P0128	Thermostat malfunction	Thermostat malfunction
P0130	Heated oxygen sensor circuit malfunction (bank 1, sensor 1)	Heated oxygen sensor malfunction Fuel trim malfunction
P0133	Heated oxygen sensor slow	Heated oxygen sensor malfunction
P0135	Heated oxygen sensor heater circuit malfunction (bank 1, sensor 1)	Open or short in heated oxygen sensor heater circuit Heated oxygen sensor heater malfunction PCM malfunction

Chapter 6 Emissions and engine control systems

Code	Identification	Possible problem area
P0136	Heated oxygen sensor circuit malfunction (bank 1, sensor 2)	Heated oxygen sensor malfunction
P0141	Heated oxygen sensor heater circuit malfunction (bank 1, sensor 2)	Open or short in heated oxygen sensor heater circuit Heated oxygen sensor heater malfunction PCM malfunction
P0150	Heated oxygen sensor circuit malfunction (bank 2, sensor 1)	Heated oxygen sensor malfunction Fuel trim malfunction
P0153	Heated oxygen sensor circuit slow response (bank 2, sensor 1)	Heated oxygen sensor malfunction
P0155	Heated oxygen sensor heater circuit malfunction (bank 2, sensor 1)	Open or short in heated oxygen sensor heater circuit Heated oxygen sensor heater malfunction PCM malfunction
P0171	System too lean	Loose air intake hose or duct Incorrect fuel line pressure Obstruction in injector Heated oxygen sensor malfunction MAP or MAF sensor malfunction ECT sensor malfunction
P0172	System too rich	Incorrect fuel line pressure Leaking injector Heated oxygen sensor (bank 1, sensor 1) malfunction MAP or MAF sensor malfunction ECT sensor malfunction
P0174	System too lean (air/fuel lean malfunction, bank 2)	Same as P0171
P0175	System too lean (air/fuel rich malfunction, bank 2)	Same as P0172
P0300	Random or multiple cylinder misfire Detected	Ignition system malfunction Injector malfunction Incorrect fuel line pressure EGR malfunction Incorrect compression pressure Incorrect valve clearance Incorrect valve timing MAP or MAF sensor malfunction ECT sensor malfunction
P0301	Cylinder no. 1 misfire detected	Same as P0300
P0302	Cylinder no. 2 misfire detected	Same as P0300
P0303	Cylinder no. 3 misfire detected	Same as P0300
P0304	Cylinder no. 4 misfire detected	Same as P0300
P0305	Cylinder no. 5 misfire detected (V6 models)	Same as P0300
P0306	Cylinder no. 6 misfire detected (V6 models)	Same as P0300
P0325	Knock sensor 1 circuit malfunction (bank 1 on V6 models)	Open or short in knock sensor 1 circuit Loose knock sensor 1 PCM malfunction
P0330	Knock sensor 2 circuit malfunction (bank 2, V6 models only)	Open or short in knock sensor 2 circuit Loose knock sensor 2 PCM malfunction

Code	Identification	Possible problem area
P0335	Crankshaft position sensor A circuit Malfunction	Open or short in crankshaft position sensor circuit Crankshaft position sensor malfunction Starter malfunction PCM malfunction
P0340	Camshaft position sensor circuit Malfunction	Open or short in camshaft position sensor circuit Camshaft position sensor malfunction Starter malfunction PCM malfunction
P0401	EGR insufficient flow detected (four-cylinder engine)	EGR valve stuck closed Open or short in VSV circuit for EGR Vacuum or EGR hose disconnected, torn or missing MAP sensor malfunction EGR VSV stuck open or closed PCM malfunction
P0401	EGR insufficient flow detected (V6 engine)	EGR valve stuck closed Open or short in EGR gas temperature sensor circuit EGR gas temperature sensor malfunction Open in VSV circuit for EGR EGR VSV malfunction Vacuum control valve malfunction Vacuum hose disconnected or blocked PCM malfunction
P0402	EGR excessive flow detected (four-cylinder engine)	EGR valve stuck open Vacuum or EGR hose connected to wrong pipe MAP sensor malfunction PCM malfunction
P0402	EGR excessive flow detected (V6 engine)	EGR valve stuck open EGR VSV stuck open Short in EGR VSV circuit Open or short in EGR valve position sensor circuit EGR valve position sensor malfunction PCM malfunction
P0420	Catalyst system efficiency below threshold (four-cylinder engine)	Three-way catalytic converter malfunction Open or short in heated oxygen sensor (bank 1, sensor 2) circuit Heated oxygen sensor (bank 1, sensor 2) malfunction Open or short in heated oxygen sensor (bank 1, sensor 1) or in A/F sensor 2 circuit
P0420	Catalyst system efficiency below threshold (V6 engine)	Three-way catalytic converter malfunction Open or short in heated oxygen sensor circuit Heated oxygen sensor malfunction
P0440	EVAP system malfunction	Vapor pressure sensor malfunction Fuel tank cap cracked, damaged or incorrectly installed Vacuum hose blocked, cracked, torn or disconnected Hose or tube blocked, cracked, torn or disconnected Fuel tank damaged Charcoal canister damaged
P0441	EVAP control system - incorrect purge flow	Open or short in VSV circuit for EVAP VSV for EVAP Open or short in vapor pressure sensor circuit Vapor pressure sensor malfunction Open or short in VSV circuit for vapor pressure sensor Malfunction in VSV for vapor pressure circuit Vacuum hose blocked, damaged or disconnected Charcoal canister malfunction
P0442	EVAP system - small leak detected	Same as P0440
P0446	EVAP system - vent control malfunction	Same as P0441

Chapter 6 Emissions and engine control systems

Code	Identification	Possible problem area
P0450	EVAP system - pressure sensor Malfunction	Open or short in vapor pressure sensor circuit Vapor pressure sensor malfunction PCM malfunction
P0451	EVAP system - pressure sensor range/performance problem	Same as P0450
P0500	Vehicle speed sensor malfunction	Open or short in speed sensor circuit Vehicle speed sensor malfunction Instrument cluster malfunction PCM malfunction
P0505	Idle control system malfunction	Idle air control (IAC) valve stuck or closed Open or short in IAC valve circuit Open or short in A/C switch circuit Air intake duct or hose loose PCM malfunction

3 Powertrain Control Module (PCM) - removal and installation

Refer to illustrations 3.1a, 3.1b and 3.1c

Warning: *The models covered by this manual are equipped with Supplemental Restraint systems (SRS), more commonly known as airbags. Always disable the airbag system before working in the vicinity of any airbag system components to avoid the possibility of accidental deployment of the airbag, which could cause personal injury (see Chapter 12).*

Caution: *To avoid electrostatic discharge damage to the PCM, handle the PCM only by its case. Do not touch the electrical terminals during removal and installation. If available, ground yourself to the vehicle with a anti-static ground strap, available at computer supply stores.*

Note: *On earlier models, Toyota refers to the PCM as the Electronic Control Module (ECM).*

1 The Powertrain Control Module (PCM) is located inside the passenger compartment in the vicinity of the glovebox. On four-cylinder Camry models and on Avalon models, the PCM is mounted to the right of the glovebox, behind the cowl side trim **(see illustrations)**. On V6 Camry models and on Lexus ES 300 models, the PCM is mounted closer to the

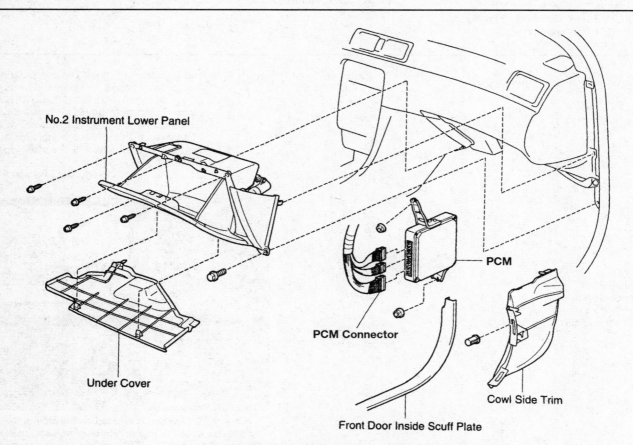

3.1a Typical PCM mounting details (four-cylinder Camry models)

6-10　Chapter 6　Emissions and engine control systems

center console **(see illustration)**.

2　Disconnect the cable from the negative battery terminal. **Caution:** *If the stereo in your vehicle is equipped with an anti-theft system, make sure you have the correct activation code before disconnecting the battery.*

3　Remove the glovebox and, if necessary, the right scuff plate and cowl side trim (see Section 23 in Chapter 12).

4　Unplug the electrical connectors from the PCM. **Caution:** *The ignition switch must be turned OFF when pulling out or plugging in the electrical connectors to prevent damage to the PCM.*

5　Remove the retaining bolts from the PCM bracket.

6　Carefully remove the PCM. **Note:** *Avoid any static electricity damage to the computer by grounding yourself to the body before touching the PCM and using a special anti-static pad to store the PCM on once it is removed.*

7　Installation is the reverse of removal.

4　Throttle Position Sensor (TPS) - check and replacement

Check

Refer to illustration 4.2

1　The Throttle Position Sensor (TPS) is located on the end of the throttle shaft on the throttle body. By monitoring the output voltage from the TPS, the PCM can determine fuel delivery based on throttle valve angle (driver demand). A broken or loose TPS can cause intermittent bursts of fuel from the injectors and an unstable idle because the PCM thinks the throttle is moving. A problem with the TPS circuits will set a diagnostic trouble code (see Section 2).

2　First, check the reference voltage from

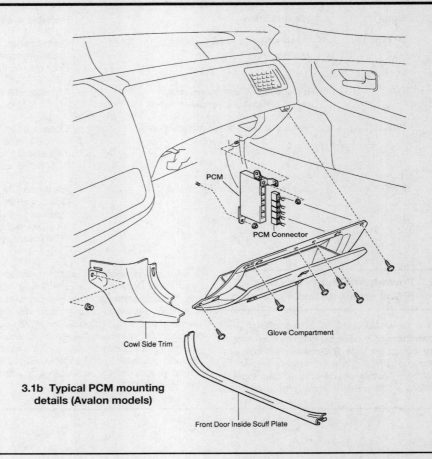

3.1b Typical PCM mounting details (Avalon models)

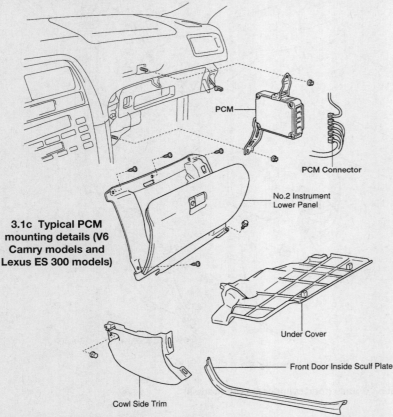

3.1c Typical PCM mounting details (V6 Camry models and Lexus ES 300 models)

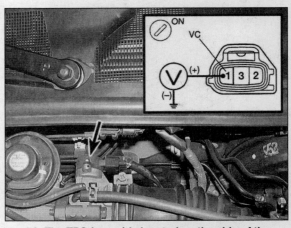

4.2 The TPS (arrow) is located on the side of the throttle body; to check the reference voltage, unplug the connector and touch the probes of a voltmeter to terminal 1 and to ground; to check the signal voltage, touch the voltmeter probes to terminals 2 and 3

Chapter 6 Emissions and engine control systems

4.6 Remove the screws and detach the TPS from the throttle body

the PCM to the TPS. Disconnect the TPS electrical connector and touch the probes of a voltmeter to terminal number 1 (on the *harness* side of the connector, not the TPS side) and ground **(see illustration)**. With the ignition key turned to ON (engine not running), it should read approximately 4.5 to 5.5 volts. If the reference voltage is too low or too high, or if there is no reference voltage, have the PCM checked at a dealership service department or at a properly equipped independent repair facility.

3   Next, check the TPS signal voltage. Reconnect the electrical connector to the TPS. Using suitable probes, backprobe terminals 2 and 3 of the TPS electrical connector. Be very careful not to damage the terminals. With the ignition key turned to ON (engine not running) and the throttle fully closed, the voltage should read approximately 0.3 to 1.0 volt. Then, slowly open the throttle. The voltage should increase to between 3.2 and 4.9 volts on a four-cylinder engine, or between 2.7 and 5.2 volts on a V6 engine. If the indicated readings are below or above the specified range, replace the TPS.

Replacement

Refer to illustration 4.6

4   Make sure the ignition key is in the OFF position.
5   Disconnect the electrical connector from the TPS.
6   Remove the screws that retain the TPS to the throttle body and remove the TPS **(see illustration)**.
7   Installation is the reverse of removal.

5   Manifold Absolute Pressure (MAP) sensor (four-cylinder models) - check and replacement

Check

Refer to illustration 5.2

1   The Manifold Absolute Pressure (MAP) sensor monitors the intake manifold pressure changes resulting from changes in engine load and speed and converts the information into a voltage output. The PCM uses the MAP sensor to control fuel delivery and ignition timing. The PCM will receive information as a voltage signal that will vary from 1.9 to 2.1 volts at closed throttle (high vacuum) and 0.3 to 0.5 volt at wide open throttle (low vacuum). The voltage range values will vary slightly according to changes in altitude. The MAP sensor is attached to bracket mounted on the engine compartment firewall. A problem in any of the MAP sensor circuits will set a diagnostic trouble code (see Section 2).

2   First, check the reference voltage from the PCM to the MAP sensor. Disconnect the MAP sensor harness connector and connect the positive probe of a voltmeter to terminal 3 and the negative probe to terminal 1 of the *harness* side of the connector, not the MAP sensor side **(see illustration)**. With the ignition key turned to ON (engine not running), it should read approximately 4.5 to 5.5 volts. If the reference voltage is too low or too high, or if there is no reference voltage, have the PCM checked at a dealership service department or a properly equipped independent repair facility.

3   Next, check the MAP signal voltage under simulated operating conditions. Reconnect the MAP sensor electrical connector and, using a voltmeter and suitable probes, backprobe connector terminal number 2 with the positive probe and terminal number 1 with the negative probe. Disconnect the vacuum hose at the sensor and connect a hand-held vacuum pump to the sensor. Turn the ignition key to ON (engine not running).

4   Apply vacuum to the sensor and verify that the voltage from the signal wire increases as the vacuum increases. The MAP sensor should exhibit the following voltage readings:

    0.3 to 0.5 volt at 3.9 in-Hg
    0.7 to 0.9 volt at 7.8 in-Hg
    1.1 to 1.3 volts at 11.8 in-Hg
    1.5 to 1.7 volts at 15.75 in-Hg
    1.9 to 2.1 volts at 19.7 in-Hg

If the readings are incorrect, replace the MAP sensor.

Replacement

Refer to illustration 5.7

5   Make sure the ignition key is in the OFF position.
6   Disconnect the electrical connector and vacuum hose from the MAP sensor.
7   Remove the bolt that retains the MAP sensor to the firewall and remove the MAP sensor **(see illustration)**.
8   Installation is the reverse of removal.

6   Mass Airflow (MAF) sensor (V6 models) - check and replacement

Check

Refer to illustration 6.2

1   The Mass Airflow (MAF) sensor is located on the air intake duct. The MAF sys-

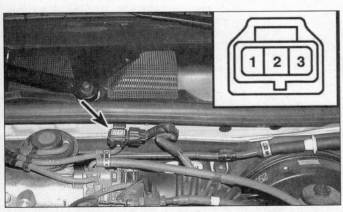

5.2 The MAP sensor (arrow) is located on the firewall at the rear of the engine compartment on four-cylinder models. Disconnect the electrical connector and using a voltmeter, touch the positive probe to terminal 3 and the negative probe to terminal 1, then check the reference voltage with the ignition key ON (engine not running). It should be 5.0 volts

5.7 To detach the MAP sensor from the firewall, remove the sensor mounting screw (arrow)

Chapter 6 Emissions and engine control systems

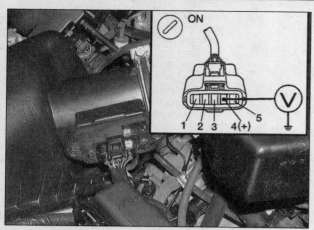

6.2 To check for battery power to the MAF sensor, hook up a test light between terminal number 4 (on the harness side of the connector) and ground, with the ignition key ON (engine not running)

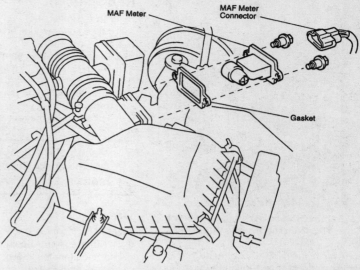

6.7a Typical MAF sensor mounting details (1997 and later Camry models, 1997 through 1999 Avalon models and 1997 and 1998 Lexus ES 300 models)

tem circuit consists of a platinum hot wire, a thermistor and a control circuit inside a plastic housing. The platinum hot wire and the thermistor detect even the slightest change in the temperature of the intake air passing over them. As the throttle opens, more air passes over the hot wire, which cools the wire. The MAF sensor circuit is designed to maintain the hot wire at a constant preset temperature by controlling the current flow through the hot wire. So, as the wire cools, the PCM increases the flow of current through the hot wire in order to maintain the wire at a constant temperature. The output voltage signal of the MAF sensor varies in accordance with this current flow and is measured by the PCM, which converts this signal into a digital wave form, compares it to the map, calculates the requisite fuel injector pulse width (duration) and turns the injectors on and off accordingly. A problem in the MAF sensor circuit will set a diagnostic trouble code (see Section 2).

2   First, check for power to the MAF sensor. Disconnect the MAF sensor electrical connector. Working on the harness side, with the ignition key turned to ON (engine not running), check for battery voltage (9 to 14 volts is acceptable) between terminal number 4 and ground **(see illustration)**. If there's no voltage, check the circuit between the EFI main relay and the MAF sensor.

3   Next, check the MAF sensor voltage signal to the PCM. Reconnect the electrical connector and backprobe the MAF sensor connector as follows: On Camry models, 1997 through 1999 Avalon models and 1997 and 1998 Lexus ES 300 models, insert the positive probe into the backside of terminal 5 and the negative probe into the backside of terminal 3 **(see illustration 6.2)**. Start the engine and allow it to idle, the voltage should be approximately 1.1 to 1.5 volts. On 1999 and later Lexus models and 2000 and later Avalon models, insert the positive probe into the backside of terminal 3 and the negative probe into the backside of terminal 2 **(see illustration 6.2)**. Watching the indicated signal voltage on the voltmeter, raise and lower the engine speed. The signal voltage from the MAF sensor should fluctuate between 1.1 and 1.5 volts in accordance with the changing engine speed.

4   If the indicated signal voltage is out of range, check the wiring harness for a bad connection, an open circuit or any other damage (see Chapter 12).

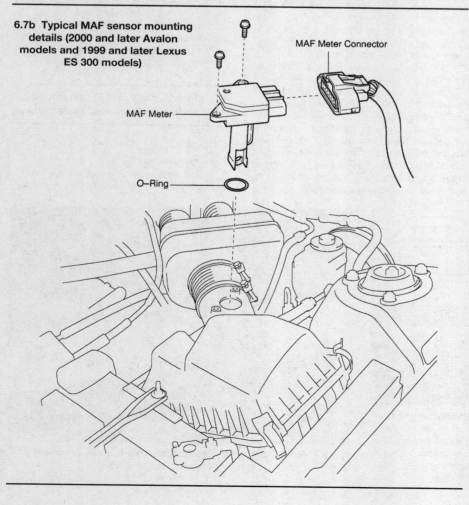

6.7b Typical MAF sensor mounting details (2000 and later Avalon models and 1999 and later Lexus ES 300 models)

Chapter 6 Emissions and engine control systems

7.2a Location of the Intake Air Temperature (IAT) sensor (four-cylinder model)

Temperature (degrees - F)	Resistance (ohms)
212	125
194	250
176	300
158	400
140	500
122	625
112	750
104	875
95	1000
86	1500
76	2000
68	2500
58	3000
50	3500
40	4000
32	4500

7.2b Coolant temperature and intake air temperature sensors approximate temperature vs. resistance values

5 If the wiring and connectors are in good shape, the MAF sensor is probably bad. However, this is a very expensive component, so it's a good idea, before replacing it, to have the MAF sensor circuit checked at a dealership service department or a properly equipped independent repair facility.

Replacement

Refer to illustrations 6.7a and 6.7b

6 Make sure the ignition key is in the OFF position.
7 Disconnect the electrical connector from the MAF sensor **(see illustrations)**.
8 On Camry models, 1997 through 1999 Avalon models and 2000 and later Lexus ES 300 models, remove the air cleaner assembly (see Chapter 4).
9 On Camry models, 1997 through 1999 Avalon models and 2000 and later Lexus ES 300 models, remove the two sensor retaining bolts and remove the MAF sensor and gasket **(see illustration 6.7a)**.
10 On 1999 and later Lexus ES 300 models and 2000 and later Avalon models, remove the two sensor retaining screws and remove the MAF sensor and O-ring **(see illustration 6.7b)**.
11 Installation is the reverse of removal. Be sure to install a new gasket or O-ring between the MAF sensor and the intake duct.

7 Intake Air Temperature (IAT) sensor - check and replacement

Check

Refer to illustrations 7.2a, 7.2b, 7.3a and 7.3b

1 The intake air temperature (IAT) sensor is a thermistor (a resistor which varies the value of its resistance in accordance with temperature changes). The change in the resistance values will directly affect the voltage signal from the sensor to the PCM. As the sensor temperature DECREASES, the resistance values will INCREASE. As the sensor temperature INCREASES, the resistance values will DECREASE. A problem in any of the IAT sensor circuits will set a diagnostic trouble code. Refer to Section 2 for the code accessing procedure.
2 On four-cylinder models, disconnect the electrical connector from the IAT sensor which is located in the air cleaner assembly **(see illustration)**. Using an ohmmeter, measure the resistance between the two terminals on the sensor. Compare your measurements with the resistance values listed in the accompanying chart **(see illustration)**. Warm the sensor with a hair dryer or heat gun while monitoring the resistance. If you don't have a heat gun or hair dryer, you can also heat up the sensor in a pan of water. If the resistance doesn't decrease proportionally to the temperature increase, replace the IAT sensor.
3 On V6 models, disconnect the electrical connector from the MAF sensor (see Section 6) and measure the resistance across the terminals as shown **(see illustrations)**. If the resistance values are not as specified, replace the MAF sensor (see Section 6).

Replacement (four-cylinder models)

Note: On V6 models, the IAT sensor is incorporated into the MAF sensor (see Section 6).

4 Make sure the ignition key is in the OFF position.

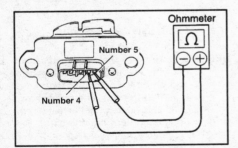

7.3a To check the IAT sensor on a V6 model, touch the negative probe of an ohmmeter to terminal 5 and the positive probe to terminal 4 of the MAF sensor connector and measure the resistance at the specified temperatures (1997 Camry V6 models, 2000 and later Avalon models and 1997 and 1999 and later Lexus ES 300 models)

Terminals	Resistance	Temperature
5 - 4	13.6 – 18.4 kΩ	–20°C (–4°F)
5 - 4	2.21 – 2.69 kΩ	20°C (68°F)
5 - 4	0.49 – 0.67 kΩ	60°C (140°F)

Terminals	Resistance	Temperature
5 - 4	14.6 – 17.8 kΩ	–20°C (–4°F)
5 - 4	2.21 – 2.69 kΩ	20°C (68°F)
5 - 4	0.29 – 0.35 kΩ	60°C (140°F)

7.3b The procedure for checking the IAT sensor on other V6 models is the same, but the resistance range at 140 degrees F. is different (1998 and later Camry V6 models, 1998 and 1999 Avalon models and 1998 Lexus ES 300 models)

6-14   Chapter 6 Emissions and engine control systems

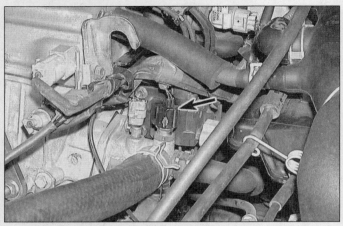

8.2a Location of the ECT sensor on a four-cylinder model

8.2b Location of the ECT sensor on a V6 model

5   Disconnect the electrical connector from the IAT sensor.
6   Remove the IAT sensor from the air cleaner.
7   Installation is the reverse of removal.

8   Engine Coolant Temperature (ECT) sensor - check and replacement

Check

Refer to illustrations 8.2a and 8.2b

1   The engine coolant temperature (ECT) sensor is a thermistor (a resistor which varies the value of its resistance in accordance with temperature changes). The change in the resistance values will directly affect the voltage signal from the sensor to the PCM. As the sensor temperature DECREASES, the resistance values will INCREASE. As the sensor temperature INCREASES, the resistance values will DECREASE. A problem in any of the ECT sensor circuits will set a diagnostic trouble code. Refer to Section 2 for the code accessing procedure.
2   Check the ECT sensor resistance. Disconnect the ECT electrical connector **(see illustrations)**. Using an ohmmeter, measure the resistance between the two terminals on the sensor. It should be between 2,500 and 3,000 ohms at room temperature (68-degrees F) **(see illustration 7.2b)**.
3   Start the engine and monitor the resistance, or remove the sensor and warm it in a pan of water. The resistance should decrease to approximately 250 to 300 ohms at 180-degrees F. If the test results are incorrect, replace the ECT sensor.

Replacement

Warning: *Wait until the engine has cooled completely before beginning this procedure.*

4   Make sure the ignition key is in the OFF position.
5   Drain approximately one gallon from the cooling system (see Chapter 1).

9.2 The crankshaft sensor harness can be disconnected near the alternator and checked using an ohmmeter (four-cylinder model)

6   Wrap the threads of the new sensor with Teflon sealing tape to prevent leakage and thread corrosion.
7   Disconnect the electrical connector and carefully unscrew the sensor.
8   Installation is the reverse of removal.
Caution: *Handle the coolant sensor with care. Damage to this sensor will affect the operation of the entire fuel injection system.*
9   Refill the cooling system.

9   Crankshaft Position sensor - check and replacement

Check

1   The crankshaft position sensor (CKP) determines the timing for the fuel injection and ignition on each cylinder. On four-cylinder models, the crankshaft sensor is mounted under the timing belt cover. On V6 models, the crankshaft sensor is mounted outside the timing belt cover, above the crankshaft pulley. A problem in the crankshaft sensor circuits will set a diagnostic trouble code (see Section 2).

9.4 Location of the crankshaft sensor on a V6 model

Four-cylinder models

Refer to illustration 9.2

2   To check the crankshaft sensor, disconnect the electrical connector at the sensor and probe the terminals with an ohmmeter **(see illustration)**. The crankshaft sensor is difficult to access but the harness can be disconnected near the alternator. Refer to the resistance values listed in this Chapter's Specifications.
3   This test checks only the crankshaft sensor. If the test results are correct, there may be a problem with the harness or the PCM. If necessary, have the PCM diagnosed by a dealer service department or other properly equipped repair facility.

V6 models

Refer to illustration 9.4

4   To check the crankshaft sensor, remove the crankshaft sensor (see Steps 6 through 9) and probe the terminals with an ohmmeter **(see illustration)**. Because the crankshaft sensor is difficult to access using ohmmeter probes, it is best to remove the sensor and check it on the bench. Refer to the resistance values listed in this Chapter's Specifications.
5   This test checks only the crankshaft

Chapter 6 Emissions and engine control systems   6-15

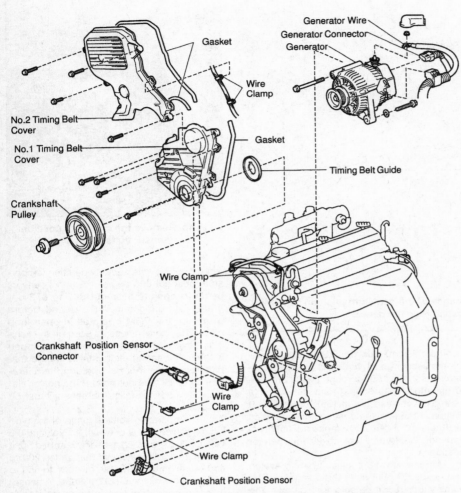

9.7 The crankshaft position sensor on four-cylinder models is located under the timing belt cover near the front pulley

9.8 Location of the crankshaft position sensor on V6 models

sensor. If the test results are correct, there may be a problem with the harness or the PCM. If necessary, have the PCM diagnosed by a dealer service department or other properly equipped repair facility.

Replacement

Refer to illustrations 9.7 and 9.8

6 Disconnect the cable from the negative battery terminal. **Caution:** *If the stereo in your vehicle is equipped with an anti-theft system, make sure you have the correct activation code before disconnecting the battery.*

7 The timing belt cover must be removed to access the crankshaft sensor on four-cylinder models **(see illustration)**. Refer to Chapter 2A for timing belt cover removal procedures.

8 The inner fender shield (see Chapter 11) must be removed to access the crankshaft sensor on the V6 engine **(see illustration)**.

9 Disconnect the crankshaft position sensor electrical connector.

10 Remove the bolt and detach the sensor.

11 Installation is the reverse of removal.

10 Camshaft Position sensor - check and replacement

Check

Refer to illustration 10.2

1 The camshaft position sensor determines the position of the cylinder for sequential fuel injection to each cylinder. The camshaft sensor is mounted on the backside of the cylinder head on four-cylinder engines while the V6 engines mount the sensor on the cylinder head toward the front. A problem in the camshaft sensor circuits will set a diagnostic trouble code. Refer to Section 2 for the code accessing procedure.

2 To check the camshaft sensor, disconnect the electrical connector at the sensor and probe the two terminals with an ohmmeter **(see illustration)**. Refer to the resistance values listed in this Chapter's Specifications.

10.2 Check the resistance of the camshaft position sensor on V6 models installed on the cylinder head

Chapter 6 Emissions and engine control systems

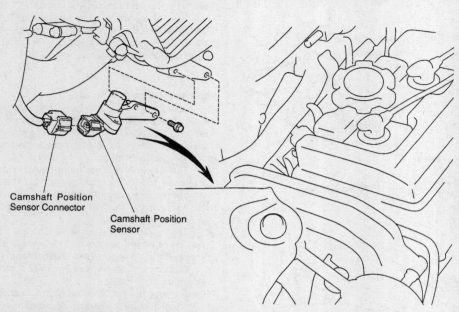

10.4a On four-cylinder models, check the resistance of the camshaft position sensor by removing the sensor and testing it on the bench

10.4b Remove the camshaft position sensor bolts (arrows)

The camshaft sensor on the four-cylinder engine is difficult to access, so remove the sensor and check it on the bench **(see illustration 10.4a)**.

Replacement

Refer to illustrations 10.4a and 10.4b

3 Make sure the ignition key is in the OFF position.

4 Disconnect the harness connector, remove the mounting screws and remove the camshaft sensor from the cylinder head **(see illustrations)**.

11 Oxygen sensor and air/fuel sensor - general information and precautions

General information

1 All vehicles covered by this manual have On-Board Diagnostics II (OBD-II) engine management systems, which means, among other things, that they have the ability to verify the accuracy of the basic feedback loop between the oxygen sensor and the PCM. They accomplish this by using an oxygen sensor or air/fuel sensor ahead of the catalytic converter and an oxygen sensor behind the catalyst. By sampling the exhaust gas before and after the catalytic converter, the PCM can determine the efficiency of the converter (and can even predict when it will fail).

2 Camry models with four-cylinder engines use two oxygen sensors or one oxygen sensor and one air/fuel sensor. The primary (upstream) oxygen sensor is located in the exhaust manifold and the secondary (downstream) oxygen sensor is located behind the catalytic converter The downstream sensor on all models is a heated oxygen sensor. On 1997 through 2000 non-California models, the upstream sensor is a heated oxygen sensor; on 1997 through 2000 California models, the upstream sensor is an air/fuel sensor. All 2001 models (California and non-California) use an air/fuel sensor upstream.

3 Models with V6 engines (Camry, Avalon and Lexus ES 300) have three sensors. The primary (upstream) sensors are located in the exhaust manifolds and the secondary (downstream) sensor is located after the catalytic converter. On 1997 models, all three sensors are heated oxygen sensors. On 1998 through 2000 California models with an automatic transaxle, the upstream sensors are air/fuel sensors; on 1998 and later California models with a manual transaxle and on 1998 through 2000 non-California models, the upstream sensors are heated oxygen sensors. On 2001 California and non-California models with an automatic transaxle, the upstream sensors are air/fuel sensors and the downstream sensor is a heated oxygen sensor.

4 Don't confuse oxygen sensors and air/fuel sensors. They're similar in appearance, but they operate differently and have different operating characteristics.

Oxygen sensor

5 An oxygen sensor is a galvanic battery that reacts to the oxygen content in the exhaust stream to produce a voltage output between 200 and 800 millivolts (mV) that the PCM monitors in order to determine the ratio of oxygen to fuel in the mixture and make the necessary correction.

6 The oxygen sensor produces no voltage when it is below its normal operating temperature of about 600-degrees F. (most sensors begin to operate at about 482 to 572 degrees F). During this initial period before warm-up, the PCM operates in open loop mode. The oxygen sensors used on the vehicles covered by this manual are equipped with small heating elements to ensure that they reach their normal operating temperature as quickly as possible. If the heater element in a sensor fails, the sensor must be replaced.

7 The operating voltage range of the typical oxygen sensor is between 0.1 volt and 0.9 volt. Once the oxygen sensor is warmed up, it produces a voltage signal that varies 200 to 450 mV (high oxygen, lean mixture) to 550 to 800 mV (low oxygen, rich mixture). A sensor in good condition responds so quickly to changes in the air/fuel ratio that it's constantly switching back and forth from lean to rich, rich to lean, etc. as the air-to-fuel mixture ratio varies from the ideal 14.7 to 1. Although this constantly switching voltage is, technically speaking, a variable output, it switches back and forth so quickly (many times a second) that it looks more like a digital signal, i.e. a switch turning on and off. The number of times that an oxygen sensor switches back and forth is known as its *cross-count*. The PCM responds to the cross-count by constantly altering the pulse width (open time) of the fuel injectors, which alters the air/fuel ratio. An air-to-fuel mixture ratio of 14.7 (air) to 1 (fuel) is the ideal mixture ratio for allowing the catalytic converter to operate at maximum efficiency, thus minimizing exhaust emissions The PCM and the oxygen sensor(s) attempt to maintain this ratio of 14.7 to 1 at all times.

8 The oxygen sensor will operate correctly only under the following four conditions:

a) *Electrical - The low voltage generated by the sensor depends upon good, clean connections which should be checked whenever a malfunction of the sensor is suspected or indicated.*

Chapter 6 Emissions and engine control systems

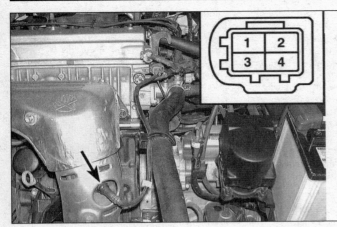

12.1 Location of the upstream heated oxygen sensor (non-California four-cylinder model shown); to check voltage, insert probes into the backside of the appropriate terminals (terminal guide shows the harness, i.e. PCM, side of unplugged sensor connector)

b) *Outside air supply* - The sensor is designed to allow air circulation to the internal portion of the sensor. Whenever the sensor is removed and installed or replaced, make sure the air passages are not restricted.

c) *Correct operating temperature* - The PCM will not react to the sensor signal until the sensor reaches approximately 600-degrees F. This factor must be taken into consideration when evaluating the performance of the sensor.

d) *Unleaded fuel* - The use of unleaded fuel is essential for proper operation of the sensor. Make sure the fuel you are using is of this type.

9 The PCM will set a diagnostic trouble code if the oxygen sensor continues to produce a steady signal voltage below 0.45-volts at 1,500 or more rpm after the engine reaches normal operating temperature and/or has been running for two or more minutes. The PCM will also set other codes if it detects any problem with the sensor's heater circuit (see Section 2).

10 When there is a problem with the oxygen sensor or its circuit, the PCM operates in the open loop mode - that is, it controls fuel delivery in accordance with a programmed default value instead of feedback information from the oxygen sensor. For help in diagnosing an oxygen sensor problem, refer to the next Section.

Air/fuel (A/F) sensor

11 The PCM uses the air/fuel (A/F) sensor as a feedback device to verify the accuracy of the alterations it makes to the injector pulse width in response to feedback from the oxygen sensor.

12 Like an oxygen sensor, the A/F sensor provides a variable voltage output to the PCM that's proportional to the air/fuel mixture ratio in the exhaust stream. But the A/F sensor doesn't "switch" back and forth like an oxygen sensor at the 14.7 to 1 stoichiometric threshold. Instead, it alters a PCM-controlled voltage between 3.3 volts (at the positive PCM terminal for the A/F sensor) and 3.0 volts (at the negative PCM terminal for the A/F sensor) in direct proportion to the amount of oxygen in the exhaust. In other words, as the air/fuel mixture in the exhaust becomes leaner, i.e. has a higher ratio of oxygen to fuel, the A/F sensor voltage increases (within its operating range of 3.0 to 3.3 volts).

13 Like an oxygen sensor, the A/F sensor doesn't operate correctly until it's warmed up. Also, like an oxygen sensor, the A/F sensor has a heating element which enables it to warm up quickly.

14 Because the operating voltage range of the A/F sensor is controlled by the PCM within such a narrow range (0.3 volt), it's impossible, according to Toyota and Lexus, to check A/F sensor output accurately by backprobing the PCM terminals of the A/F sensor electrical connector. This information can only be accessed by a special Toyota OBD-II scan tool or hand-held tester. After eliminating other possible problem areas that might cause the PCM to display diagnostic trouble code P0125, P0171 or P0172 (see Section 2), take the vehicle to a Toyota or Lexus dealer to have the A/F sensor diagnosed. However, you can check the resistance of the heater element in the A/F sensor and you can replace the A/F sensor (see Section 12).

Precautions

15 Special care must be taken whenever a sensor is serviced.

a) Oxygen sensors and air/fuel sensors have a permanently attached pigtail and electrical connector which should not be removed from the sensor. Damage or removal of the pigtail or electrical connector can adversely affect operation of the sensor.

b) Grease, dirt and other contaminants should be kept away from the electrical connector and the louvered end of the sensor.

c) Do not use cleaning solvents of any kind on an oxygen sensor or air/fuel ratio sensor.

d) Do not drop or roughly handle and oxygen sensor or air/fuel ratio sensor.

e) The silicone boot must be installed in the correct position to prevent the boot from being melted and to allow the sensor to operate properly.

12 Oxygen sensor and air/fuel sensor - check and replacement

Oxygen sensor

Check

Refer to illustrations 12.1 and 12.4

1 Locate the oxygen sensor electrical connector and without disconnecting it, insert a long pin into the oxygen sensor connector signal terminal number 3 (+) and another pin into ground terminal number 4 (-) **(see illustration)**. Install the positive and negative probes of a voltmeter onto the corresponding pins. Apply the parking brake, shift the transaxle into Park (automatic) or Neutral (manual), raise the front of the vehicle and place it securely on jackstands. Primary O2 sensors are fairly easy to access because they are mounted in the front exhaust manifold but the secondary O2 sensors are difficult to access because they are mounted under the vehicle behind the catalytic converter. Remove the front seat to access the harness connector.

2 Check the signal voltage from the oxygen sensor during operating conditions. Start the engine and monitor the voltage signal as the engine warms up. **Caution:** *Be extremely careful of hot exhaust components when performing this procedure.* **Note:** *The secondary oxygen sensor will produce much slower fluctuating voltage values to reflect the results of the catalyzed exhaust gas.*

3 The oxygen sensor will produce a steady voltage signal at first (open loop) of approximately 0.1 to 0.2 volts with the engine cold. After a period of approximately two minutes, the engine will reach operating temperature and the oxygen sensor should start to fluctuate between 0.1 to 0.9 volts (closed loop). If the oxygen sensor fails to reach the closed loop mode or there is a very long period of time until it does switch into closed loop mode, replace the oxygen sensor.

4 Also inspect the oxygen sensor heater. Disconnect the oxygen sensor electrical connector and connect an ohmmeter between the HT and B+ terminals **(see illustration)**. Compare your measurement to the sensor heater resistance listed in this Chapter's Specifications. If the indicated heater resis-

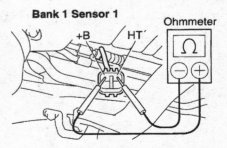

12.4 Ohmmeter hook-up for testing the oxygen sensor heater resistance

6-18 Chapter 6 Emissions and engine control systems

tance is incorrect, replace the sensor. Locations for the oxygen sensor connector are as follows:

a) **Four-cylinder models - upstream sensor:** In the engine compartment, attached to the left side of the cylinder head
b) **Four-cylinder models - downstream sensor:** Beneath the passenger's seat - seat must normally be removed for good access to the connector.
c) **V6 models - front manifold sensor:** Near the front of the engine compartment. Follow the wires from the sensor to the connector.
d) **V6 models - rear manifold sensor:** Beneath the rear of the engine compartment. Raise the vehicle, place it securely on jackstands, locate the sensor attached to the rear exhaust manifold and follow the wires to the connector.
e) **V6 models - downstream sensor:** On the driver's side floor of the passenger compartment. It is normally necessary to remove the driver's seat and carpet for access.

5 Verify that the heater is receiving battery voltage. Working on the PCM side of the connector, measure voltage between the B+ terminal and chassis ground. There should be battery voltage with the ignition key turned to ON (engine not running). If there is no voltage, check the circuit between the main relay, the PCM and the sensor. Refer to the wiring diagrams at the end of Chapter 12 for additional information.

6 If the oxygen sensor fails any of these tests, replace it with a new part.

Replacement

Refer to illustrations 12.11a and 12.11b

Note: *Because it is installed in the exhaust manifold or pipe, which contracts when cool, the oxygen sensor may be very difficult to loosen when the engine is cold. Rather than risk damage to the sensor (assuming you are planning to reuse it in another manifold or pipe), start and run the engine for a minute or two, then shut it off. Be careful not to burn yourself during the following procedure.*

7 Disconnect the cable from the negative terminal of the battery. **Caution:** *If the stereo in your vehicle is equipped with an anti-theft system, make sure you have the correct activation code before disconnecting the battery.*

8 If you're replacing the downstream sensor, remove the seat and, if necessary, the carpet, and then unplug the electrical connector.

9 If you're replacing the upstream sensor for the rear cylinder bank on a V6 model, or the downstream sensor on any model, raise the vehicle and place it securely on jackstands (the upstream sensor on four-cylinder models and the upstream sensor for the front cylinder bank on V6 models can be replaced without raising the vehicle).

10 If you're replacing an upstream sensor, unplug the sensor electrical connector. If

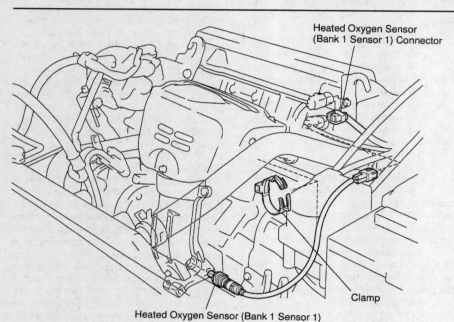

12.11a To replace the upstream heated oxygen sensor in a four-cylinder model (shown), or the upstream sensor for the front cylinder bank on a V6 model, simply unplug the connector and unscrew the sensor

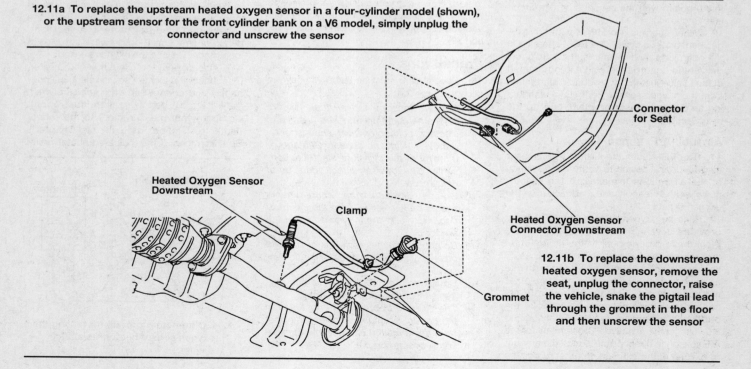

12.11b To replace the downstream heated oxygen sensor, remove the seat, unplug the connector, raise the vehicle, snake the pigtail lead through the grommet in the floor and then unscrew the sensor

Chapter 6 Emissions and engine control systems

you're replacing a downstream sensor, carefully snake the connector and pigtail lead through the grommet in the floor.
11 Unscrew the sensor from the exhaust manifold or exhaust pipe **(see illustrations)**.
Note: *The best tool for removing an oxygen sensor is a special slotted socket, especially if you're planning to reuse a sensor. If you don't have this tool, and you plan to reuse the sensor, be extremely careful when unscrewing the sensor.*
12 Apply anti-seize compound to the threads of the sensor to facilitate future removal. The threads of new sensors should already be coated with this compound, but if you're planning to reuse an old sensor, recoat the threads. Install the sensor and tighten it securely.
13 Reconnect the electrical connector of the pigtail lead to the main wiring harness.
14 Lower the vehicle (if it was raised), test drive the car and verify that no trouble codes have been set.

Air/fuel (A/F) sensor

Check

15 The air/fuel (A/F) sensor voltage cannot be checked without a special Toyota OBD-II scan tool or hand-held tester. However, the resistance of the heating element in the A/F sensor can be checked.
16 Unplug the electrical connector for the A/F sensor and, using an ohmmeter, measure the resistance between terminals B+ and HT **(see illustration 12.4)**. Compare your measurement to the resistance listed in this Chapter's Specifications. If the indicated resistance is too low or too high, replace the A/F sensor.

Replacement

Note 1: *Because it is installed in the exhaust manifold or pipe, which contracts when cool, the A/F sensor may be very difficult to loosen when the engine is cold. Rather than risk damage to the sensor (assuming you are planning to reuse it in another manifold or pipe), start and run the engine for a minute or two, then shut it off. Be careful not to burn yourself during the following procedure.*
Note 2: *Air/fuel sensors are installed in the same locations and look the same as oxygen sensors. For information on sensor location, refer to Steps 4 and 11.*
17 Disconnect the cable from the negative terminal of the battery. **Caution:** *If the stereo in your vehicle is equipped with an anti-theft system, make sure you have the correct activation code before disconnecting the battery.*
18 If you're replacing the A/F sensor for the rear cylinder bank on a V6 model, raise the vehicle and place it securely on jackstands (the A/F sensor on four-cylinder models and the A/F sensor for the front cylinder bank on V6 models can be replaced without raising the vehicle).
19 Unplug the sensor electrical connector from the sensor.
20 Carefully unscrew the sensor from the exhaust manifold.
21 Apply anti-seize compound to the threads of the sensor to facilitate future removal. The threads of new sensors should already be coated with this compound, but if you're planning to reuse an old sensor, recoat the threads.
22 Install the sensor and tighten it securely.
23 Reconnect the electrical connector to the main wiring harness.
24 Lower the vehicle (if it was raised), test drive the car and verify that no trouble codes have been set.

13 Knock sensor - check and replacement

Check

Refer to illustration 13.2

1 The knock control system is designed to reduce spark knock during periods of heavy detonation. This allows the engine to use optimal spark advance to improve driveability. The knock sensor detects abnormal vibration in the engine and produces a voltage output which increases with the severity of the knock. The voltage signal is monitored by the PCM, which retards ignition timing until the detonation ceases. The knock sensor on four-cylinder models is located on the backside of the engine block, directly below the cylinder head (facing toward the rear of the engine compartment). V6 models use two knock sensors located under the intake manifold.
2 To check a knock sensor, disconnect the electrical connector **(see illustration)** and use an ohmmeter to check continuity between the knock sensor terminal and the body of the sensor. Continuity should NOT exist. If continuity exists, replace it with a new sensor.

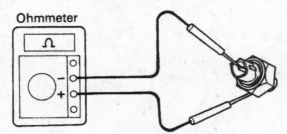

13.2 Verify that there is no continuity between the sensor terminal and the body of the sensor

Replacement

Refer to illustrations 13.6a and 13.6b
Warning: *Wait for the engine to cool completely before performing this procedure.*

3 Disconnect the cable from the negative terminal of the battery. **Caution:** *If the stereo in your vehicle is equipped with an anti-theft system, make sure you have the correct activation code before disconnecting the battery.*
4 Drain the cooling system (see Chapter 1).
5 On V6 models, remove the intake manifold (see Chapter 2B).
6 Disconnect the electrical connector and remove the knock sensor **(see illustrations)**.

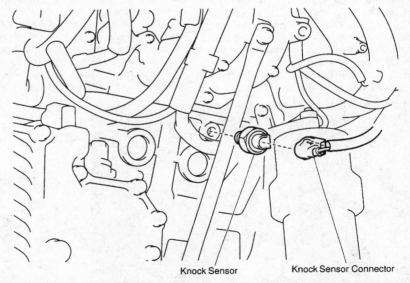

13.6a On four-cylinder engines, the knock sensor is located on the rear of the block (facing the firewall)

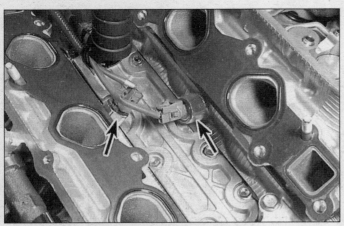

13.6b On V6 engines, the knock sensors (arrows) are located under the intake manifold

14.2a Location of the VSS (arrow) on four-cylinder models

7   If you're going to reuse the old sensor, coat the threads with thread sealant. New sensors are pre-coated with thread sealant, do not apply any additional sealant or the operation of the sensor may be affected.

8   Install the knock sensor and tighten it securely (approximately 30 ft-lbs). Don't overtighten the sensor or damage may occur. Plug in the electrical connector, refill the cooling system and check for leaks.

14 Vehicle Speed Sensor (VSS) - check and replacement

Check

Refer to illustrations 14.2a and 14.2b

Note 1: *The VSS is operating normally if the speedometer is working.*

Note 2: *The VSS is used on all manual and most automatic transaxles used with the vehicles covered by this manual The following procedure applies only to the conventional VSS used on 1997 and later Camry models, 1997 through 1999 Avalon models and 1997 through 2000 Lexus ES 300 models. The VSS to which this procedure applies has three wires (blue, pink and red/blue).*

Note 3: *Some automatic transaxles are also equipped with other speed sensor devices, in addition to or instead of, the VSS. This procedure does not apply to those devices.*

Note 4: *The following procedure does not apply to the overdrive (O/D) direct clutch speed sensor, which is used on 1997 and later Avalon models. The O/D direct clutch speed sensor has two wires: either a blue wire and a yellow wire (1997 through 1999 models) or a green wire and a red wire (2000 and later models). This sensor is controlled by the PCM, and it can only be checked by a dealer service department with a special Toyota OBD-II scan tool or hand-held tester.*

Note 5: *The following procedure does not apply to the other vehicle speed sensor (the one with a red wire and a green wire) used on 1997 through 2000 Lexus ES 300 models, nor does it apply to the counter gear speed sensor (red wire and green wire) and the input turbine speed sensor (blue wire and green wire) used on 2001 Lexus models. These sensors are controlled by the PCM, and they can only be checked by a dealer service department with a special Toyota OBD-II scan tool or hand-held tester.*

1   The Vehicle Speed Sensor (VSS) is located on top of the transaxle. This sensor is an electronic component that produces a pulsing voltage signal whenever the sensor shaft is rotated. These voltage pulses are monitored by the PCM, which uses this information to help control the fuel and ignition systems and transaxle shifting.

2   To check the vehicle speed sensor, disconnect the electrical connector from the sensor **(see illustration)**. Using a voltmeter, check for voltage on the harness side of the electrical connector by inserting the positive probe into the terminal for the red/blue wire and touching the negative probe to ground. There should be battery voltage. If there is no voltage, check for an open circuit between the VSS and the 10A GAUGE fuse.

3   Raise the front of the vehicle and place it securely on jackstands. Block the rear wheels and place the transaxle in Neutral. Reconnect the electrical connector to the VSS and back-probe the VSS connector blue wire and pink wire with suitable probes. Insert the positive probe into the terminal for the blue wire and the negative probe into the terminal for the pink wire. Turn the ignition to ON (engine not running). While holding one wheel steady, rotate the other wheel by hand. The voltmeter should pulse between zero and 11 volts as the sensor rotates. If it doesn't, replace the sensor.

Replacement

Refer to illustration 14.6

4   Disconnect the electrical connector from the VSS.

14.2b Location of the VSS (arrow) on V6 models

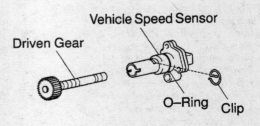

14.6 An exploded view of a typical VSS; be sure to replace the O-ring

Chapter 6 Emissions and engine control systems   6-21

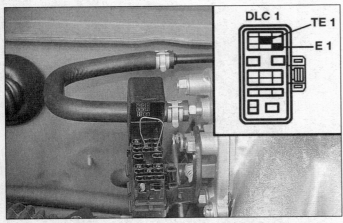

15.2a Test terminal guide (all except 1999 and later Camry four-cylinder models)

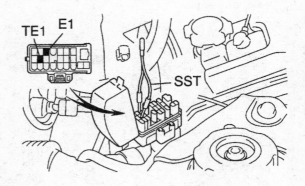

15.2b Test terminal guide (1999 and later Camry four-cylinder models)

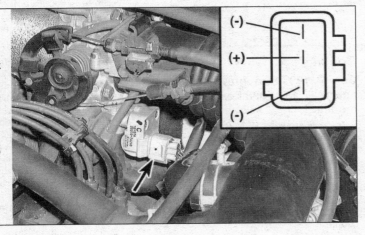

15.6 Using an ohmmeter, check the resistance between the middle terminal and each of the outer terminals (four-cylinder model shown; V6 models similar)

15.9 The IAC valve mounting screws (arrows) are located under the throttle body (V6 model shown, four-cylinder models similar)

5   Remove the VSS hold-down bolt and pull the VSS out of the transaxle.
6   Replace the O-ring **(see illustration)**.
7   Installation is the reverse of removal.

15 Idle Air Control (IAC) valve - check and replacement

Note: *The minimum idle speed is pre-set at the factory and should not require adjustment under normal operating conditions; however if the throttle body has been replaced or you suspect the minimum idle speed has been tampered with (for example, if the idle speed screw was removed from the throttle body) have the vehicle checked by a dealer service department or other qualified automotive repair shop.*

Check

Refer to illustrations 15.2a, 15.2b and 15.6

1   The engine idle speed is controlled by the Idle Air Control (IAC) valve. The IAC valve controls the amount of air that bypasses the throttle plate into the intake manifold. The IAC valve is controlled by the PCM in accordance with the demands on the engine (air conditioning, power steering) and the operating conditions (cold or warmed up).
2   Apply the parking brake, shift the transaxle to Neutral (manual) or Park (automatic) and block the drive wheels. Install a tachometer in accordance with the manufacturer's instructions (see Chapter 1). Start the engine and allow it to reach normal operating temperature. Check the idle speed. Using a jumper wire, bridge terminals TE1 and E1 of the diagnostic test connector **(see illustrations)**.
3   The engine speed should increase to approximately 900 to 1,300 rpm (four-cylinder models) or 1000 rpm (V6 models) for five seconds then return to normal idle speed.
   a) *If the engine speed changes as described, the IAC valve is okay.*
   b) *If the engine speed does not change as described, measure the IAC valve resistance.*
4   Remove the jumper wire.
5   Disconnect the IAC valve electrical connector.
6   Measure the resistance between the middle terminal and each of the two outer terminals **(see illustration)**. Compare your results to the IAC valve resistance in this Chapter's Specifications.
   a) *If the resistance is not as specified, replace the IAC valve.*
   b) *If the resistance is as specified, the IAC valve is okay, but there may be a problem with the wiring or the PCM.*
7   To determine whether there's a problem with the wiring or the PCM, remove the IAC valve (see below), and then apply battery voltage to the valve connector. Connect the positive battery terminal to the center terminal and ground each outer terminal, one at a time. The valve should open with one of the terminals grounded and close with the other terminal grounded.
   a) *If the IAC operates as described, check the wiring harness between the PCM and the IAC valve.*
   b) *If the wiring is okay, the PCM is probably defective. Have it checked by a dealer service department.*

Replacement

Refer to illustration 15.9

8   Remove the throttle body (see Chapter 4).
9   Remove the mounting screws and detach the IAC valve and gasket **(see illustration)**.

6-22   Chapter 6 Emissions and engine control systems

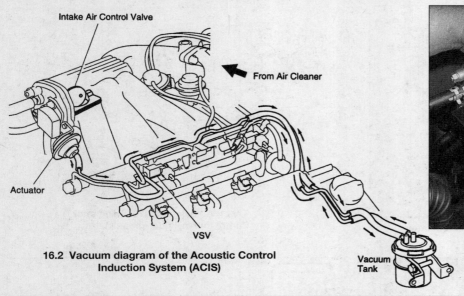

16.2 Vacuum diagram of the Acoustic Control Induction System (ACIS)

16.3 Using a T-fitting, install a vacuum gauge between the vacuum actuator and the vacuum source and see if vacuum increases slightly with an increase in the engine rpm

10 Installation is the reverse of removal. Be sure to use a new gasket when installing the IAC valve.

16 Acoustic Control Induction System (ACIS) (V6 models only)

General information

Refer to illustration 16.2

1 The Acoustic Control Induction System (ACIS) diverts the path of intake air to one of two paths through the intake manifold in order to match the intake volume to the engine speed. Engine performance is enhanced by switching the intake air control valve to the closed position for high torque at low rpm or to the open position for maximum horsepower at high rpm.

2 The ACIS actuator mechanism that opens and closes the butterfly valve inside the intake manifold is known as an "intake air control valve." On Camry models, 1997 through 1999 Avalon models and 1997 through 1998 Lexus ES 300 models, there is one intake air control valve, located on the opposite end of the intake plenum from the throttle body. On 2000 and later Avalon models, and on 1999 and later Lexus ES 300 models, there is an additional intake air control valve between the two throttle bodies and the other end of the air intake plenum. On these models, this other intake air control valve is referred to as the "No. 1 intake air control valve," because it's upstream in relation to the intake air control valve, which is referred to as the "No. 2 intake air control valve." The actuator mechanism(s) is/are controlled by vacuum from the intake manifold **(see illustration)**. A malfunction in the system can result in poor driveability. The following simple checks will help you diagnose any system defects. **Note:** *There are no diagnostic trouble codes for this system.*

Check

Camry, 1997 through 1999 Avalon and 1997 and 1998 Lexus ES 300 models

Refer to illustrations 16.3, 16.5, 16.7, 16.8, 16.9, 16.11 and 16.12

3 Install a T-fitting between the vacuum actuator and the vacuum hose and connect a vacuum gauge **(see illustration)** into the T-fitting.

4 Start the engine and allow it to idle. There should be no vacuum indicated by the gauge.

5 Quickly depress the accelerator pedal to

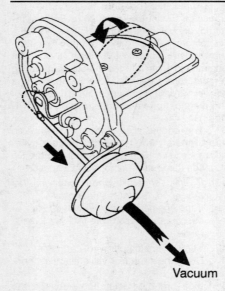

16.5 When you depress the accelerator pedal to its wide open position, the vacuum actuator should pull the rod out, which opens the butterfly valve inside the plenum (mechanism removed for clarity)

its wide open position and verify that the gauge indicates a vacuum of about 8 in-Hg. The actuator should pull the rod out (toward the actuator) **(see illustration)**.

6 If no vacuum was indicated by the gauge, check for a blocked vacuum port, a collapsed or broken vacuum hose to the control valve or a defective vacuum switching valve (VSV) (see Step 9).

7 Inspect the vacuum tank. Locate the vacuum tank under the battery tray and apply pressure to port B. Air should pass through port B and out port A. Apply air pressure to port A and observe that air does not flow out port B **(see illustration)**.

8 Plug port B with the tip of your finger and apply vacuum to port A and observe that vacuum does not leak down after one minute **(see illustration)**.

9 Check the VSV resistance. Touch the probes of an ohmmeter to the terminals and check the resistance **(see illustration)**. It should be between 33 and 39 ohms. If the resistance is incorrect, replace the VSV.

10 Check the VSV ground. Using the ohmmeter, verify that there's no continuity

16.7 When air pressure is applied to port B, air should flow out of port A; when pressure is applied to port A, air should NOT flow out port B

Chapter 6 Emissions and engine control systems

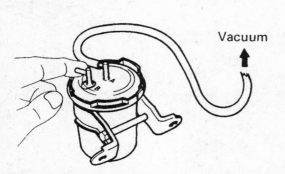

16.8 Check the vacuum tank for a tight seal. Apply vacuum to port A while closing off port B with the tip of your finger. The tank should hold vacuum for at least one minute

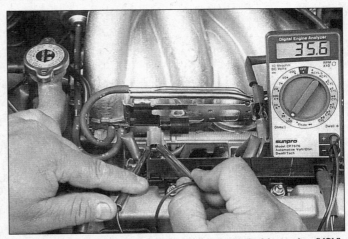

16.9 Measure the resistance of the vacuum switching valve (VSV). It should be between 33 and 39 ohms at 68-degrees F

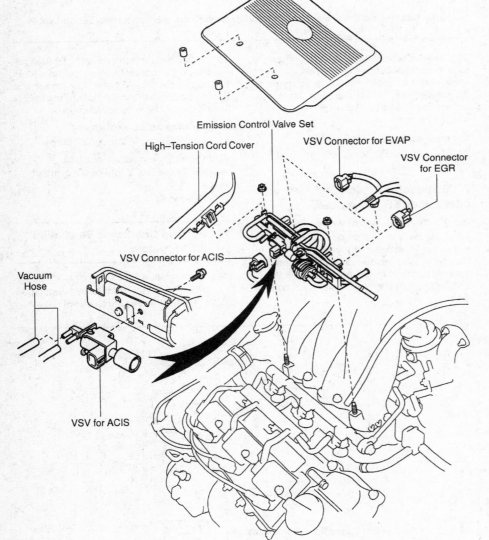

between each terminal and the body of the valve.

11 Remove the V-bank cover **(see illustration)**. Remove the emission control valve set retaining nuts and remove the emission control valve set. Remove the high-tension cord cover. Clearly mark the two vacuum hoses to the VSV and then detach them from the VSV, remove the VSV retaining screw and remove the VSV from the emission control valve set.

12 Blow air through port E of the VSV and verify that it comes out the filter **(see illustration)**.

13 Apply battery voltage across the electrical terminals of the VSV, blow through port E and verify that air comes out port F (the small port directly adjacent to Port E.

14 If the VSV fails any of the above tests, replace it.

15 Install the VSV on the emission control valve set and tighten the retaining screw securely. Reattach the vacuum hoses to the VSV (make sure you didn't accidentally switch them). Install the high-tension cord cover and install the emission control valve set. Tighten the emission control valve nuts securely. Install the V-bank cover.

16.11 Typical ACIS system vacuum and electrical connections on 1997 and later Camry, 1997 through 1999 Avalon and 1997 and 1998 Lexus ES 300 models. Other models are similar but use two VSV's

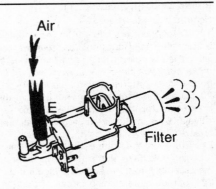

16.12 Blow air through port E of the VSV and verify that it comes out the filter

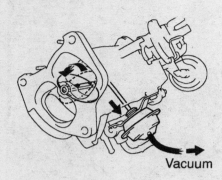

16.16 Quickly depress the accelerator pedal to its wide open position and verify that the gauge indicates a vacuum of about 8 in-Hg. The actuator should pull the rod out (toward the actuator) (No. 1 intake air control valve assembly removed for clarity)

2000 and later Avalon and 1999 and later Lexus ES 300 models

Refer to illustration 16.16

16 To check the No. 1 intake air control valve, follow Steps 3 through 6 above but refer to the accompanying illustration **(see illustration)**. To check the No. 2 intake air control valve, follow Steps 3 through 6 and refer to the illustrations accompanying those steps **(see illustrations 16.3 and 16.5)**.

17 To check the vacuum tank, follow Step 7 above.

18 There are two vacuum switching valves (VSV) on these models, one for the No. 1 intake air control valve, another for the No. 2 intake air control valve. Both VSVs are identical - and they're identical to the single VSV used on the above models. To check either one, follow Steps 8 through 15.

Replacement

Camry, 1997 through 1999 Avalon and 1997 and 1998 Lexus ES 300 models

Intake air control valve

Refer to illustration 16.21

19 Disconnect the cable from the negative battery terminal. **Caution:** *If the stereo in your vehicle is equipped with an anti-theft system, make sure you have the correct activation code before disconnecting the battery.*

20 Disconnect the vacuum hoses from the brake booster, the air conditioning idle-up valve and the intake air control valve vacuum hose.

21 Remove the intake air control valve retaining bolts **(see illustration)** and separate the assembly from the air intake plenum.

22 Scrape all remaining traces of the gasket material from the air intake plenum and valve body without damaging the aluminum material.

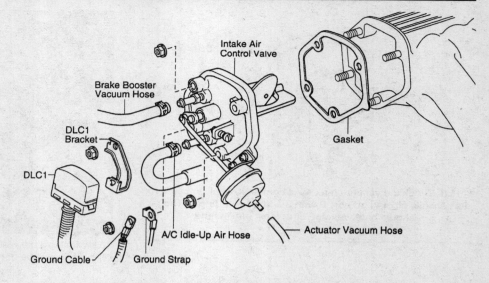

16.21 Intake air control valve installation details. Some models have a second valve

23 Installation is the reverse of removal. Be sure to coat the rubber O-ring seal with light engine oil and apply sealant to the ends of the rubber seal. Also, be sure to use a new gasket between the intake air control valve and the plenum.

Vacuum switching valve (VSV)

24 To replace the VSV, refer to Steps 11 and 15.

2000 and later Avalon and 1999 and later Lexus ES 300 models

No. 1 intake air control valve

25 Remove the throttle body/No. 1 intake air control valve assembly (see Section 14 in Chapter 4). This section also includes the disassembly procedure for separating the throttle bodies from the No. 1 intake air control valve.

26 Install the throttle body/No. 1 intake air control valve assembly (see Section 14 in Chapter 4). Be sure to use a new gasket and tighten the three mounting nuts to the torque listed in this Chapter's Specifications.

No. 2 intake air control valve

27 Follow Steps 19 through 23 above.

VSV for No. 1 intake air control valve/VSV for No. 2 intake air control valve

28 Refer to Step 24.

17 Positive Crankcase Ventilation (PCV) system

Refer to illustration 17.1

1 The Positive Crankcase Ventilation (PCV) system **(see illustration)** reduces

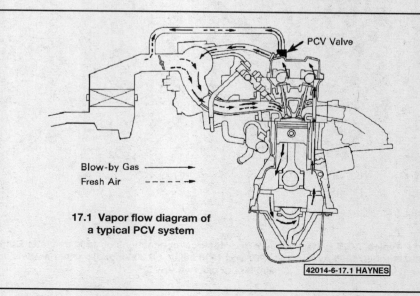

17.1 Vapor flow diagram of a typical PCV system

Chapter 6 Emissions and engine control systems

18.4 Apply vacuum to the EGR valve and see if the valve opens and allows exhaust gases to circulate. Once it is activated, the EGR valve should hold steady (no loss in vacuum) (four-cylinder model)

18.6 Clearly mark the vacuum hoses at ports P, Q and R, disconnect all three hoses, block ports P and R with your finger and then blow air into port Q and verify that air freely passes through the modulator to the air filter side

hydrocarbon emissions by scavenging crankcase vapors. It does this by circulating fresh air from the air cleaner through the crankcase, where it mixes with blow-by gases and is then rerouted through a PCV valve to the intake manifold.

2 The main components of the PCV system are the PCV valve, a blow-by filter and the vacuum hoses connecting these two components with the engine.

3 To maintain idle quality, the PCV valve restricts the flow when the intake manifold vacuum is high. If abnormal operating conditions (such as piston ring problems) arise, the system is designed to allow excessive amounts of blow-by gases to flow back through the crankcase vent tube into the air cleaner to be consumed by normal combustion.

4 Checking and replacement of the PCV valve is covered in Chapter 1.

18 Exhaust Gas Recirculation (EGR) system - description, check and component replacement

General description

1 To reduce oxides of nitrogen (NOx) emissions, some of the exhaust gases are recirculated through the EGR valve to the intake manifold to lower combustion temperatures.

2 The two EGR systems on these models vary slightly in design. The EGR system on four-cylinder models consists of an EGR valve, an EGR modulator, a vacuum switching valve (VSV) and the vacuum lines to the throttle body. The EGR system on V6 models consists of an EGR position sensor (EGR valve and sensor), a vacuum control valve (VCV), a vacuum switching valve (VSV) and the EGR gas temperature sensor. On both systems, the position of the EGR valve is controlled by vacuum which is controlled by the Powertrain Control Module (PCM). On V6 models, the position of the EGR valve is monitored by the EGR position sensor mounted on top of the EGR assembly.

Check

Four-cylinder models

EGR valve

Refer to illustration 18.4

3 Start the engine and allow it to idle.
4 Detach the vacuum hose from the EGR valve and attach a hand vacuum pump in its place **(see illustration)**.
5 Apply vacuum to the EGR valve. Vacuum should remain steady and the engine should run poorly.

   a) If the vacuum doesn't remain steady and the engine doesn't run poorly, replace the EGR valve and recheck it.
   b) If the vacuum remains steady but the engine doesn't run poorly, remove the EGR valve and check the valve and the intake manifold for blockage. Clean or replace parts as necessary and recheck.

EGR vacuum modulator

Refer to illustration 18.6

6 Clearly mark the vacuum hoses at ports P, Q and R, disconnect all three hoses, block ports P and R with your finger and then blow air into port Q and verify that air freely passes through the modulator to the air filter side **(see illustration)**.
7 Start the engine, allow it to warm up, bring the engine speed up to 2500 rpm, repeat the test in Step 6 and verify that there's a strong resistance to air flow.
8 If the modulator doesn't operate as described, replace it. If the modulator operates correctly, reconnect the vacuum hoses to their corresponding ports.

EGR system

Refer to illustration 18.9

9 Disconnect the hose from the EGR valve and install a T-fitting and vacuum gauge between the EGR valve and the hose from the Vacuum Switching Valve (VSV) **(see illustration)**. **Note:** *The VSV itself is located on the backside of the engine block, so if you have difficulty locating the correct hose, raise the front of the vehicle, place it securely on jackstands and trace the vacuum hose from the VSV to the EGR valve.*

10 Start the engine and connect terminals TE1 and E1 on the test terminal **(see illustration 15.2a or 15.2b)**.

   a) *With the coolant temperature below 131-degrees F (cold), verify that the vacuum gauge indicates zero (no vacuum) at 2,500 rpm.*
   b) *With the engine warm, verify that the vacuum gauge indicates low vacuum at 2,500 rpm.*
   c) *Disconnect the vacuum hose from the R port of the EGR vacuum modulator and connect the R port directly to the intake manifold using an extension (see illustration 18.6). Raise the rpm to 2,500 and verify the vacuum gauge indicates high vacuum.*

18.9 Install a vacuum gauge between the vacuum switching valve (VSV) and the EGR valve using a three way adapter

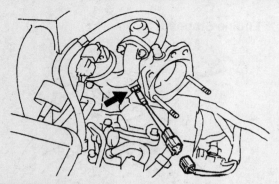

18.13 The EGR gas temperature sensor is located in the inlet pipe (arrow) directly ahead of the EGR valve

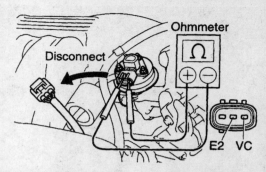

18.14 Disconnect the harness connector and check the resistance of the EGR valve position sensor on terminals E2 and VC

11 Check the operation of the VSV.
   a) The VSV on four-cylinder engines is located on the back of the engine under the intake manifold. This VSV is of a similar design to the VSV used on the ACIS system **(see illustration 16.12)**.
   b) Using an ohmmeter, measure the resistance of the VSV (between the two terminals of its electrical connector. It should be between 33 and 39 ohms at 68-degrees F.
   c) Verify that there is no continuity between either VSV terminal and ground.
   d) Blow air into the port nearest the filter (Port E) and verify that it flows out the port nearest the electrical connector (Port G).
   e) Apply battery voltage to the VSV terminals and verify that air flows from Port E through the filter.
   f) If the VSV fails any of the above tests, replace it.

V6 models

Note: The following tests apply to Camry V6, 1997 through 1999 Avalon and 1997 and 1998 Lexus ES 300 models. 2000 and later Avalon and 1999 and later Lexus ES 300 models do not have an EGR system.

EGR valve

12 Refer to Steps 3 through 5.

EGR gas temperature sensor

Refer to illustration 18.13

Note: You'll need a digital pyrometer to perform this test. If you don't have access to this tool, the only other way to check the EGR gas temperature sensor is to suspend it in a pot of oil (you can't use water because it vaporizes at the boiling point) and use a cooking thermometer to determine the temperature. However, because of the danger of being burned by hot oil, we don't recommend this method.

13 Disconnect the harness connector for the EGR gas temperature sensor **(see illustration)** and, using an ohmmeter, measure the resistance across the sensor terminals at the three temperatures listed in this Chapter's Specifications and compare your measurement with the resistance values listed in this Chapter's Specifications.

EGR valve position sensor

Refer to illustration 18.14

14 Disconnect the harness connector for the EGR valve position sensor and, using an ohmmeter, measure the resistance across the indicated sensor terminals **(see illustration)**. Compare your measurement to the correct range of resistance listed in this Chapter's Specifications. If the resistance is out of range, replace the EGR valve position sensor.

15 Reconnect the sensor connector and, using a voltmeter and a pair of backprobes, measure output voltage as follows. Disconnect the vacuum hose from the EGR valve, turn the ignition switch to ON and backprobe terminals E2 (brown wire) and VC (yellow wire on Camry and Lexus models, blue/red wire on Avalon models). (These are the same two terminals between which you measured the resistance in the last Step, but now you're backprobing the closed connector, so don't confuse the terminals.) The voltage should be between 4.5 and 5.5 volts. If the indicated voltage is out of range, replace the EGR valve position sensor.

16 Disconnect the vacuum hose from the EGR valve position sensor and hook up a hand-operated vacuum pump in its place. Backprobe terminals E2 and EGLS (white/green wire on Camry and Lexus models, red/white wire on Avalon models). (Terminal EGLS is the other terminal in the three-wire sensor connector.) Apply a 5.1 in-Hg vacuum to the sensor and measure the voltage. It should be between 3.2 and 5.1 volts. Release the vacuum from the EGR valve and read the voltage again. It should be between 0.4 and 1.6 volts. If the indicated voltage is incorrect with or without vacuum, replace the EGR valve position sensor.

Vacuum control valve (VCV)

Refer to illustration 18.18

17 Clearly mark the vacuum hoses, disconnect them from the vacuum control valve (VCV) and remove the VCV.
18 Hook up a hand-operated vacuum pump to port S on the VCV and plug up port Z with a finger **(see illustration)**.
19 Apply vacuum to the VCV by squeezing the hand pump three times (you must apply a

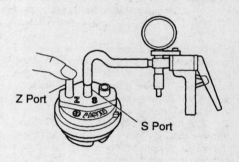

18.18 To test the VCV, hook up a hand-operated vacuum pump to port S on the VCV and plug up port Z with a finger, apply 11.88 in-Hg vacuum to the VCV, wait 10 seconds and note whether the VCV still holds a vacuum between 4.4 and 7.1 in-Hg

vacuum of about 10.88 in-Hg). Wait 10 seconds and then note the remaining vacuum. It should be between 4.4 and 7.1 in-Hg. If the indicated vacuum is lower than this, replace the VCV.
20 Install the VCV and reattach the vacuum hoses. Make sure that the hoses are reconnected to the correct ports.

Vacuum switching valve (VSV)

Refer to illustration 18.21 and 18.24

21 To get to the VSV assembly **(see illustration)**, remove the V-Bank cover. The VSV is located on the intake manifold side of the emission control valve set. The following tests can be performed with the VSV installed or removed. If you're going to remove the VSV, make sure you clearly mark the vacuum hoses before disconnecting them from the VSV.
22 Using an ohmmeter, measure the resistance across the VSV electrical connector terminals. It should be between 33 and 39 ohms at 68-degrees F. If the indicated resistance is out of range, replace the VSV.
23 Verify that there is no continuity between either of the VSV terminals and ground. If there is continuity between either

Chapter 6 Emissions and engine control systems 6-27

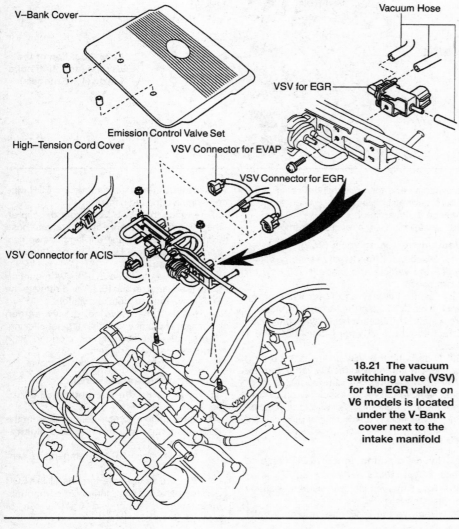

18.21 The vacuum switching valve (VSV) for the EGR valve on V6 models is located under the V-Bank cover next to the intake manifold

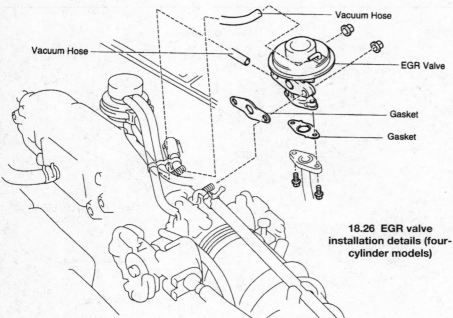

18.26 EGR valve installation details (four-cylinder models)

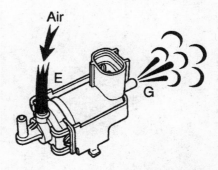

18.24 Blow air into port E and verify that it flows out port G

terminal and ground, replace the VSV.
24 Blow air into port E and verify that it flows out port G **(see illustration)**. If it doesn't, replace the VSV.
25 Apply battery voltage to the VSV terminals, blow through port E and verify it flows out port F. If it doesn't, replace the VSV.

Replacement

Four-cylinder models

EGR valve

Refer to illustrations 18.26

26 Detach the vacuum hose from the EGR valve. Remove the EGR valve mounting bolts and remove the EGR valve from the intake manifold **(see illustration)**. Check the valve for sticking and heavy carbon deposits. If the valve is sticking or clogged with deposits, clean or replace it.
27 Installation is the reverse of removal. Be sure to use new gaskets.

EGR vacuum modulator and modulator filter

Refer to illustrations 18.29a and 18.29b

28 Clearly label the vacuum hoses to the EGR vacuum modulator and then disconnect them. Remove the EGR vacuum modulator from its bracket.
29 If you're planning to reuse the old modulator, pull the cover off and check the filters **(see illustrations)**. Clean them with com-

18.29a To remove the EGR vacuum modulator filters for cleaning, remove the cap . . .

18.29b ... pull out the two filters and blow them out with compressed air; make sure the coarse side of the outer filter faces out when reinstalling the filters

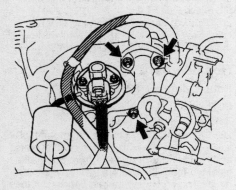

18.37 Detach the EVAP hose (1) from the clip on the EGR valve, disconnect the vacuum hose (2) from the EGR valve, unplug the electrical connector (3), then remove the EGR valve mounting bolts (arrows) (V6 models)

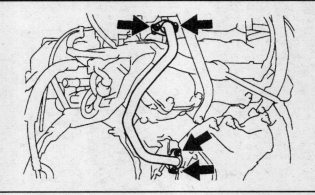

18.36 Location of the EGR pipe mounting bolts (arrows) (V6 models)

pressed air and then reinstall the cover. If the filters cannot be cleaned, replace them or replace the modulator.

30 Installation is the reverse of removal.

Vacuum switching valve (VSV)

31 Disconnect the negative battery cable. Raise the vehicle and place it securely on jackstands.
32 Locate the VSV on the backside of the engine block. Unplug the electrical connector from the VSV. Clearly label the vacuum hoses attached to the VSV and then disconnect them.
33 Separate the VSV from its mounting bracket. If you have difficulty removing the VSV from the bracket, remove the bracket bolt, remove the entire assembly and separate the VSV from the bracket off the vehicle.
34 Installation is the reverse of removal.

V6 models

EGR valve position sensor/EGR valve

Refer to illustrations 18.36 and 18.37

35 Raise the vehicle and place it securely on jackstands.
36 Remove the EGR pipe retaining bolts **(see illustration)**, then remove the EGR pipe and discard the old gaskets.
37 Detach the EVAP hose from the clip on the EGR valve, disconnect the vacuum hose from the EGR valve and unplug the electrical connector from the EGR valve position sensor **(see illustration)**.
38 Remove the three EGR valve mounting bolts **(see illustration 18.37)** and remove the EGR valve from the intake manifold.
39 Inspect the valve for heavy carbon deposits, and make sure that it's not sticking. If the valve is sticking or clogged with deposits, clean or replace it.
40 Installation is the reverse of removal.

EGR gas temperature sensor

41 Remove the throttle body (see Chapter 4).
42 Unplug the electrical connector from the EGR gas temperature sensor **(see illustration 18.13)**.
43 Unscrew the EGR gas temperature sensor.
44 Be sure to coat the threads of the EGR gas temperature with anti-seize compound. Installation is otherwise the reverse of removal.

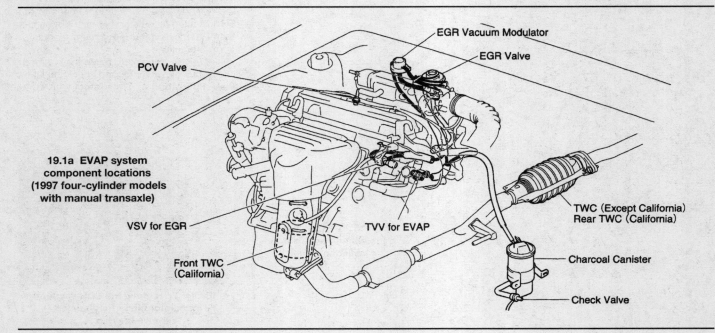

19.1a EVAP system component locations (1997 four-cylinder models with manual transaxle)

Chapter 6 Emissions and engine control systems

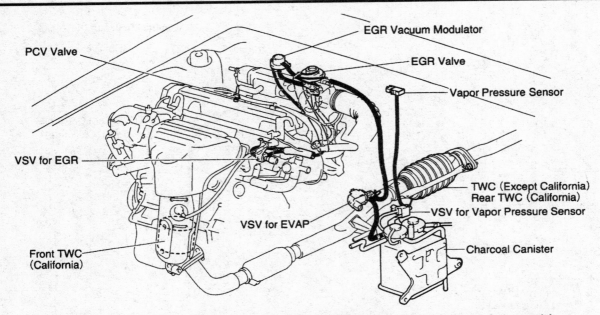

19.1b EVAP system component locations (1997 four-cylinder models with automatic transaxle)

Vacuum control valve (VCV)
45 Clearly label the vacuum hoses and then detach them from the VCV. Remove the VCV.
46 Installation is the reverse of removal. Make sure that the hoses are reconnected to the correct ports.

Vacuum switching valve (VSV)
47 To remove the VSV, see Step 21.
48 Installation is the reverse of removal.

19 Evaporative emissions control (EVAP) system - description, check and component replacement

General description
Refer to illustrations 19.1a, 19.1b, 19.1c, 19.1d and 19.1e
1 The fuel evaporative emissions control (EVAP) system absorbs fuel vapors and, during engine operation, releases them into the engine intake where they mix with the incoming air-fuel mixture. On 1997 four-cylinder models and on 1997 and 1998 V6 models, the charcoal canister is mounted in the engine compartment **(see illustrations)**. On 1998 and later four-cylinder models and on 1999 and later V6 models, the charcoal canister is mounted behind the fuel tank **(see illustrations)**.

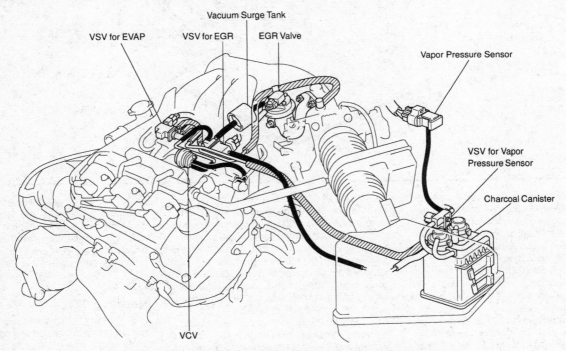

19.1c EVAP system component locations (1997 and 1998 V6 models)

6-30 Chapter 6 Emissions and engine control systems

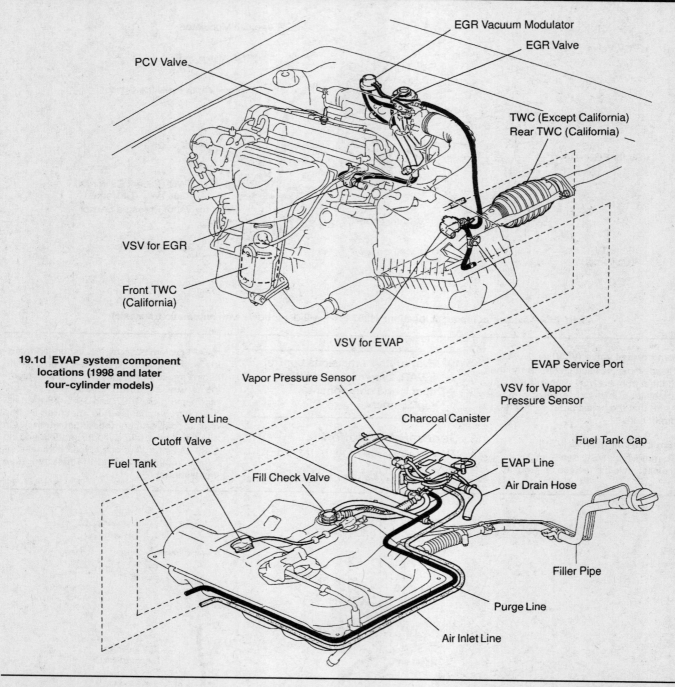

19.1d EVAP system component locations (1998 and later four-cylinder models)

2 The EVAP system employs a canister filled with activated charcoal to absorb fuel vapors. The following description of a typical system for the models covered by this manual should provide you enough information to understand the system on your vehicle.

3 The fuel filler cap is fitted with a two-way valve as a safety device. The valve vents fuel vapors to the atmosphere if the EVAP system fails.

4 When the engine is off, fuel vapors from the fuel tank and air cleaner housing are routed through hoses to the EVAP canister. The activated charcoal in the canister absorbs and stores these vapors.

5 When the engine is running and warmed up to a pre-set temperature, a thermal vacuum valve (TVV) (1997 four-cylinder models with a manual transaxle) or vacuum switching valve (VSV) (all other models) opens, allowing intake manifold vacuum to draw the fuel vapors from the canister to the intake manifold, where they're mixed with intake air before being burned with the air/fuel mixture inside the combustion chambers.

6 On 1997 four-cylinder models with a manual transaxle, a check valve allows vapors from the fuel tank to migrate to the canister, but prevent them from returning back to the tank. On all other models a fuel tank vapor pressure sensor monitors changes in pressure inside the tank and, when the pressure exceeds a preset threshold, opens a vacuum switching valve (VSV), which allows a purge port in the canister to admit fuel tank vapors into the canister.

Check

Hoses

7 Always inspect the hoses first. A disconnected, damaged or missing hose is the most likely cause of a malfunctioning EVAP system. Refer to the *Vacuum Hose Routing Diagram* (attached to the underside of the hood)

Chapter 6 Emissions and engine control systems

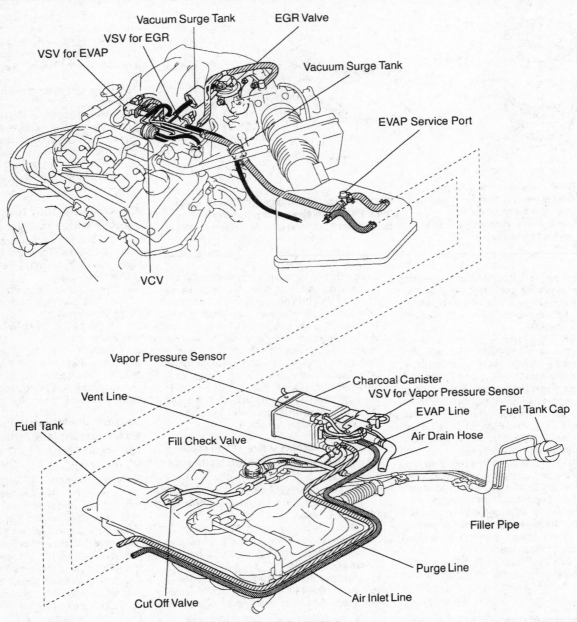

19.1e EVAP system component locations (1999 and later V6 models)

to determine whether the hoses are correctly routed and attached. Repair damaged hoses and replace missing hoses as necessary. If the diagram is missing, it can be ordered from a dealer service department.

Canister

Refer to illustrations 19.10a, 19.10b, 19.11, 19.12a and 19.12b

8   Remove the canister from the vehicle (see Steps 54 through 56) and then check it as follows.
9   When the EVAP system is functioning, some parts of it, including the canister, are slightly pressurized, so inspect the canister and all ports for leaks, damage or cracks that could allow vapors to leak out during EVAP system operation. If the canister is obviously damaged, replace it. If the canister appears to be in satisfactory condition, leak test it as follows (these tests will also pinpoint a clogged filter or a stuck check valve or diaphragm).
10   On 1997 four-cylinder models with a manual transaxle (cylinder-shaped canister), blow low-pressure air (pressure from your mouth should be enough) into Port A and verify that air comes out the other three ports **(see illustration)**. Blow low-pressure air into port B and verify that no air comes out the other three ports. If the canister doesn't operate as described, replace it. If the canister operates as described, clean the filter by blowing 43 psi of compressed air into port A

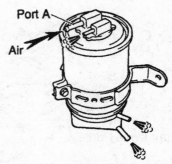

19.10a Apply low-pressure (0.68 psi) compressed air to the indicated port and verify that air comes out the other three ports (1997 four-cylinder models with a manual transaxle)

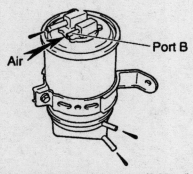

19.10b Blow low-pressure (0.68 psi) air into port B and verify that no air comes out the other three ports (1997 four-cylinder models with a manual transaxle)

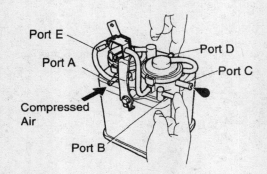

19.11 Port identification for the rectangular-type EVAP canister mounted in the engine compartment (second test described in text is being performed)

while plugging port B with your finger (do NOT try to wash the canister!). No activated charcoal should come out when you clean the filter with compressed air. If it does, replace the canister.

11 On 1997 four-cylinder models with an automatic transaxle and 1997 V6 models with an automatic transaxle (rectangular canister in engine compartment). Plug port E and port B with your fingers (if that's awkward, plug one of the ports with a rubber plug), blow gently into port A and verify that air flows out port D. Next, plug ports B and D, blow gently into port A and verify that no air flows out port C **(see illustration)**. Then, apply a 1.01 in-Hg vacuum to port B, and then verify that the vacuum does not decrease when port C is closed, and verify that vacuum decreases when port C is opened. Finally, plug port C, apply a vacuum of 0.36 in-Hg (four-cylinder models) or 1.01 in-Hg (V6 models) to port A and verify that air flows into port B. If the canister fails any of these four checks, replace it.

12 On 1998 and later four-cylinder and most V6 models, plug the vent port with a cap, plug the purge port with a finger, blow gently into the EVAP port and verify that air flows out the air drain port. Next, plug the purge port and the air drain port, blow gently into the EVAP port and verify that air does not flow out the air inlet port **(see illustration)**. Then, apply a vacuum of 1.01 in-Hg vacuum to the purge port and verify that the vacuum doesn't decrease when the air inlet port is closed, and that the vacuum does decrease when the air inlet port is opened. Finally, plug the air inlet port, apply 1.01 in-Hg vacuum to the EVAP port and verify that air flows into the purge port. If the canister fails one or more of these tests, replace it.

Thermal vacuum valve (TVV) (1997 four-cylinder models with a manual transaxle)

13 Drain the engine coolant (see Chapter 1). Locate the TVV. Refer to **illustration 19.1a**, if necessary.

14 Disconnect the hoses from the water outlet to which the TVV is attached. Remove the water outlet retaining nuts and remove the water outlet.

15 Remove the TVV from the water outlet.

16 Place the TVV in a container of water with a temperature of less than 95 degrees F, blow a little air into the lower port and verify that air does not flow out of the upper port.

17 Heat the container of water to at least 129 degrees F, blow some air into the upper port and verify that air flows out the lower port.

18 If the TVV doesn't operate as described, replace it.

19 Before installing the TVV, be sure to apply teflon tape or thread sealant to the threads of the TVV. Install the TVV so that the ports face up, at a 45-degree angle to the water outlet gasket surface.

20 Installation is otherwise the reverse of removal.

Check valve (1997 four-cylinder models with a manual transaxle)

21 Locate the check valve in the hose between the fuel tank and the canister and remove the check valve. Refer to illustration 19.1a, if necessary.

22 Verify that air flows from the yellow port to the black port but not from the black port to the yellow port.

23 If the check valve doesn't operate as described, replace it.

24 Installation is the reverse of removal. Make sure that you install the check valve so that it's oriented in the same direction as before (yellow port faces toward the fuel tank, black port faces toward the canister).

Vacuum switching valve (VSV) for EVAP system (1997 through 1999 four-cylinder and 1997 and later V6 models)

Refer to illustrations 19.25a and 19.25b

25 Locate the VSV. On 1997 through 1999 four-cylinder models, it's under the intake duct **(see illustration)**. On V6 models, it's under the V-bank cover, on the emission control valve set **(see illustration)**; on these models, remove the V-bank cover. On all models, unplug the electrical connector, clearly label and then detach the EVAP hoses, remove the retaining bolt and detach the VSV from its mounting bracket.

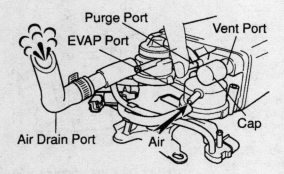

19.12a Plug the vent port with a cap, plug the purge port with a finger, blow air into the EVAP port and verify that air flows out the air drain port (1998 and later models)

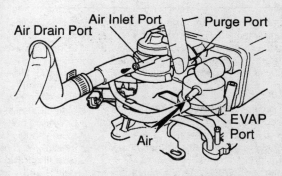

19.12b Plug the purge port and the air drain port, blow air into the EVAP port and verify that air does not flow out the air inlet port (1998 and later models)

Chapter 6 Emissions and engine control systems

26 Using an ohmmeter, measure the resistance of the VSV at the terminals. On 1997 through 1999 four-cylinder models, it should be between 30 and 34 ohms; on V6 models, it should be between 27 and 33 ohms. If the resistance is incorrect, replace the VSV.

27 Verify that there is no continuity between the VSV terminal and ground. If there is, replace the VSV.

28 Blow some low-pressure air (no more than 13.5 psi) into the port closest to the electrical connector and verify that it doesn't flow out of the other (outer) port. If it does, replace the VSV.

29 Apply battery voltage to the VSV terminals, blow some low-pressure air into the port closest to the electrical connector and verify that it does flow out of the other (outer) port.

30 Installation is the reverse of removal.

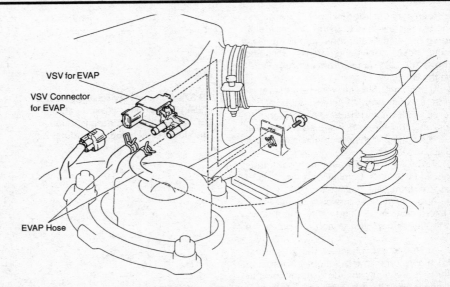

19.25a Vacuum switching valve (VSV) for EVAP system - installation details (1997 through 1999 four-cylinder models)

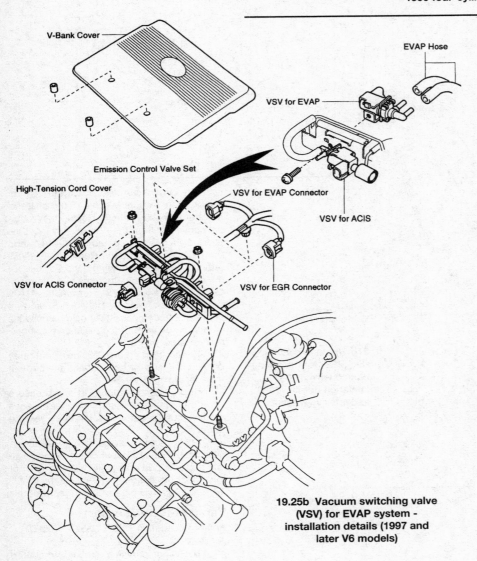

19.25b Vacuum switching valve (VSV) for EVAP system - installation details (1997 and later V6 models)

Vacuum switching valve (VSV) for vapor pressure sensor (1997 through 1999 models)

31 Locate the VSV for the vapor pressure sensor, referring to illustrations 19.1a through 19.1e. On 1997 four-cylinder and on 1997 and 1998 V6 models, it's in the engine compartment, near the canister. On 1998 and 1999 four-cylinder models, raise the rear of the vehicle and place it securely on jackstands, then locate the VSV at the charcoal canister, behind the fuel tank. On all models, unplug the electrical connector from the VSV, clearly label the vacuum hoses connected to the VSV and then disconnect them. Remove the VSV retaining bolt and remove the VSV.

32 Using an ohmmeter, measure the resistance between the VSV terminals. It should be between 33 and 39 ohms. If it isn't, replace the VSV.

33 Verify that there is no continuity between either VSV terminal and the body of the valve. If there is, replace the VSV.

34 There are three ports on the VSV: a closely grouped pair of ports and another port at the other end of the VSV, near the electrical connector. Apply a small amount of low-pressure air to the inner of the closely-grouped pair of ports and verify it flows out the port closest to the electrical connector. If it doesn't, replace the VSV.

35 Apply battery voltage to the VSV terminals, blow a small amount of low-pressure air into the inner of the closely-grouped pair of ports and verify that it comes out the other port of the pair. If it doesn't, replace the VSV.

36 Installation is the reverse of removal.

Vacuum switching valve (VSV) for canister closed valve (CCV) (2000 and later models)

Refer to illustrations 19.37a and 19.37b

37 Locate the VSV on the air cleaner hous-

ing (four-cylinder models) or on a bracket near the air cleaner housing (V6 models), unplug the electrical connector, detach the vacuum hose, remove the retaining bolts and remove the VSV **(see illustrations)**. Don't lose the O-ring.

38 Using an ohmmeter, measure the resistance across the VSV terminals. At 68 degrees F, it should be between 25 and 30 ohms; at 248 degrees F, it should be between 33 and 42 ohms. If the resistance is incorrect, replace the VSV.

39 Verify that there is no continuity between either VSV terminal and the body of the valve. If there is, replace the VSV.

40 Apply some low-pressure compressed air into the outer port and verify that it flows out the inner port. If it doesn't, replace the VSV.

41 Apply battery voltage to the VSV terminals, blow some low-pressure air into the outer port and verify that it does not flow out the inner port. If it does, replace the VSV.

42 Installation is the reverse of removal.

Vacuum switching valve (VSV) for pressure switching valve (2000 and later models)

Refer to illustration 19.43

43 Raise the rear of the vehicle and place it

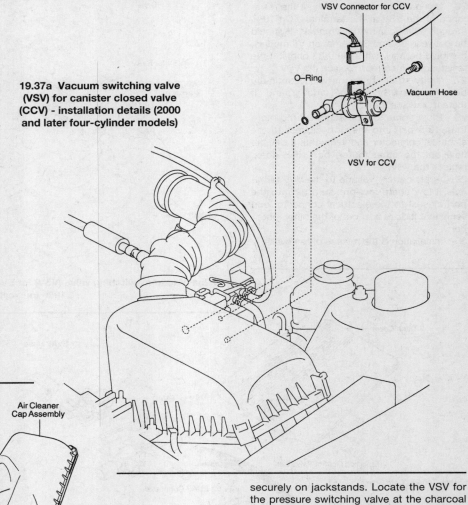

19.37a Vacuum switching valve (VSV) for canister closed valve (CCV) - installation details (2000 and later four-cylinder models)

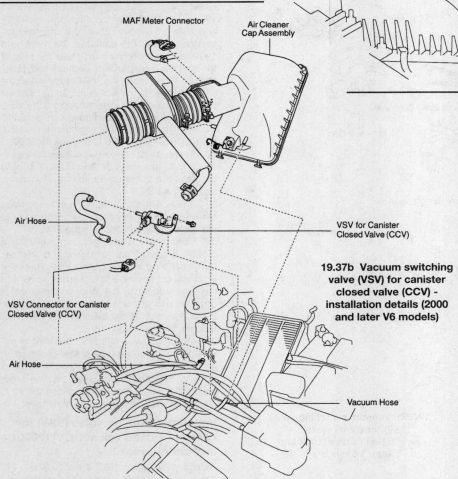

19.37b Vacuum switching valve (VSV) for canister closed valve (CCV) - installation details (2000 and later V6 models)

securely on jackstands. Locate the VSV for the pressure switching valve at the charcoal canister **(see illustration)**. Unplug the electrical connector from the VSV, clearly label the vacuum hoses connected to the VSV and then disconnect them. Remove the VSV retaining bolt and remove the VSV.

44 Using an ohmmeter, measure the resistance between the VSV terminals. At 68 degrees F, it should be between 37 and 44 ohms; at 248 degrees F, it should be between 51 and 62 degrees. If the resistance is incorrect, replace the VSV.

45 Verify that there is no continuity between either VSV terminal and the body. If there is, replace the VSV.

46 Apply a small amount of low-pressure air into the inner port and verify that it does not flow out the outer port. If it does, replace the VSV.

47 Apply battery voltage to the VSV terminals, blow a small amount of low-pressure air into the inner port and verify that it comes out the outer port. If it doesn't, replace the VSV.

48 Installation is the reverse of removal.

Vapor pressure sensor

Refer to illustration 19.51, 19.52a, 19.52b and 19.52c

49 The vapor pressure sensor is mounted either in the engine compartment, on the fire-

Chapter 6 Emissions and engine control systems

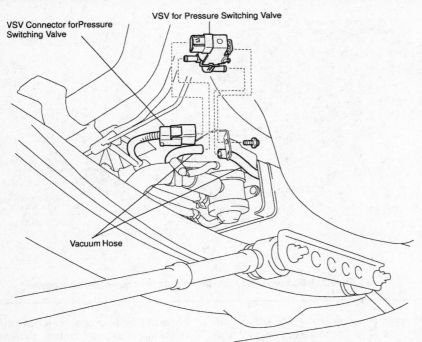

19.43 Vacuum switching valve (VSV) for the pressure switching valve - installation details (2000 and later four-cylinder models)

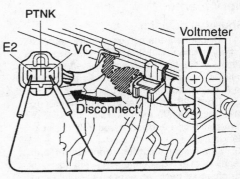

19.51 Vapor pressure sensor terminal guide

wall (on 1997 four-cylinder and 1997 and 1998 V6 models) or at the charcoal canister, behind the fuel tank (on 1998 and later four-cylinder and 1999 and later V6 models). Refer to illustrations 19.1a through 19.1e, if necessary.

50 If you're checking the vapor pressure sensor on a 1998 and later four-cylinder or a 1999 and later V6 model, raise the rear of the vehicle and place it securely on jackstands.

51 Unplug the electrical connector from the vapor pressure sensor, touch the probes of a voltmeter to the indicated terminals, turn the ignition key to ON (engine not running) and measure the voltage supply to the vapor pressure sensor **(see illustration)**. It should be between 4.5 and 5.5 volts. Reconnect the electrical connector. If the voltage supply is out of range, look for a short, open or bad connection in the wiring harness (see Wiring Diagrams at the end of Chapter 12).

52 Using a voltmeter and a pair of back-probes, measure the variation in voltage when vacuum is applied to the vapor pressure sensor **(see illustrations)**. Backprobe terminals PTNK and E2 **(see illustration 19.51)**; connect the positive probe to terminal PTNK and the negative probe to terminal E2. Turn the ignition key ON (engine not running).

  a) Using a hand-held vacuum pump, apply a vacuum of 0.59 in-Hg to the vapor pressure sensor; the voltage should be between 1.3 and 2.1 volts.
  b) Release the vacuum and verify that the voltage is between 3.0 and 3.6 volts.
  c) Apply a pressure of 0.22 psi to the vapor pressure sensor and verify that the voltage is between 4.2 and 4.8 volts.

Turn the ignition key to OFF and pull the probes out of the sensor connector.
53 If the vapor pressure sensor does not operate as described, replace it.

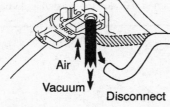

19.52b To test the vapor pressure sensor output voltage on 1998 and 1999 four-cylinder and 1999 V6 models, determine whether you have a Type A or Type B vapor pressure sensor, and then detach the indicated vacuum line and hook up a hand-operated vacuum pump in its place

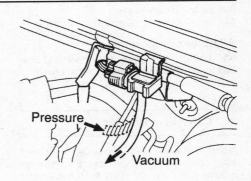

19.52a To test the vapor pressure sensor output voltage on 1997 four-cylinder and 1997 and 1998 V6 models, detach this vacuum line and hook up a hand-operated vacuum pump in its place

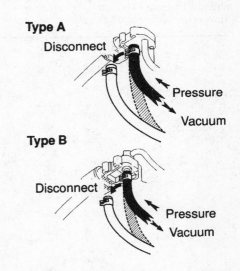

19.52c To test the vapor pressure sensor output voltage on 2000 and later models, determine whether you have a Type A or Type B vapor pressure sensor, and then detach the indicated vacuum line and hook up a hand-operated vacuum pump in its place

Replacement

Charcoal canister

54 Disconnect the cable from the negative battery terminal. **Caution:** *If the stereo in your vehicle is equipped with an anti-theft system, make sure you have the correct activation code before disconnecting the battery.*

55 On 1998 and later four-cylinder and 1999 and later V6 models, raise the rear of the vehicle and support it securely on jackstands.

56 Unplug all electrical connectors and clearly label and disconnect the vent hoses to the charcoal canister, remove the bolts and separate the canister from the engine compartment or the underside of the vehicle. Refer to the illustrations at the beginning of this section if necessary.

57 Installation is the reverse of removal.

All other components

58 Refer to the appropriate checking procedure and the illustrations that accompany that procedure.

20 Catalytic converter

Note: *Because of a Federally mandated extended warranty which covers emissions-related components such as the catalytic converter, check with a dealer service department before replacing the converter at your own expense.*

General description

1 The catalytic converter is an emission control device added to the exhaust system to reduce pollutants from the exhaust gas stream. There are two types of converters. The conventional oxidation catalyst reduces the levels of hydrocarbon (HC) and carbon monoxide (CO). The three-way catalyst lowers the levels of oxides of nitrogen (NOx) as well as hydrocarbons (HC) and carbon monoxide (CO).

Check

2 The test equipment for a catalytic converter is expensive and highly sophisticated. If you suspect that the converter on your vehicle is malfunctioning, take it to a dealer or authorized emissions inspection facility for diagnosis and repair.

3 Whenever the vehicle is raised for servicing of underbody components, check the converter for leaks, corrosion, dents and other damage. Check the welds/flange bolts that attach the front and rear ends of the converter to the exhaust system. If damage is discovered, the converter should be replaced.

4 Although catalytic converters don't break too often, they can become plugged. The easiest way to check for a restricted converter is to use a vacuum gauge to diagnose the effect of a blocked exhaust on intake vacuum.

  a) Connect a vacuum gauge to an intake manifold vacuum source (see Chapter 2C).
  b) Warm the engine to operating temperature, place the transaxle in Park (automatic) or Neutral (manual) and apply the parking brake.
  c) Note and record the vacuum reading at idle.
  d) Quickly open the throttle to near full throttle and release it shut. Note and record the vacuum reading.
  e) Perform the test three more times, recording the reading after each test.
  f) If the reading after the fourth test is more than one in-Hg lower than the reading recorded at idle, the exhaust system may be restricted (the catalytic converter could be plugged or an exhaust pipe or muffler could be restricted).

Replacement

Refer to illustrations 20.5a and 20.5b

Note: *On some California models, the front catalytic converter is incorporated into the exhaust manifold. Refer to Chapter 2A or Chapter 2B for the exhaust manifold replacement procedure.*

5 Be sure to spray the nuts on the exhaust flange studs before removing them from the catalytic converter **(see illustrations)**.

6 Remove the nuts and separate the catalytic converter from the exhaust system.

7 Installation is the reverse of removal.

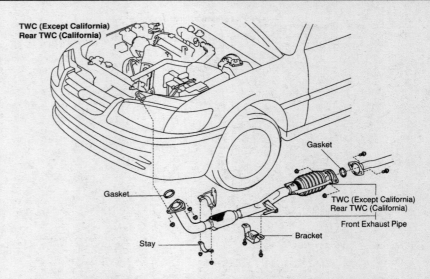

20.5a Catalytic converter installation details - four-cylinder models

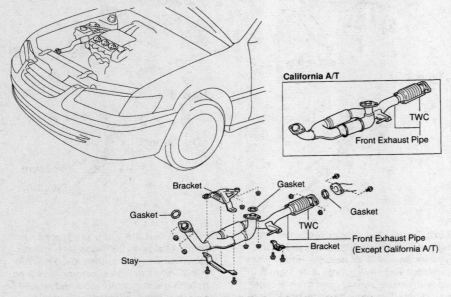

20.5b Catalytic converter installation details - V6 models

Chapter 7 Part A
Manual transaxle

Contents

	Section
Back-up light switch - check and replacement	4
Engine mounts - check and replacement	See Chapter 2A or 2B
General information	1
Manual transaxle lubricant change	See Chapter 1
Manual transaxle lubricant level check	See Chapter 1
Manual transaxle overhaul - general information	6
Manual transaxle - removal and installation	5
Shift and select cables - replacement	2
Shift lever - removal and installation	3

Specifications

Torque specifications

Ft-lbs (unless otherwise indicated)

Back-up light switch	33
Exhaust manifold brace	
Four cylinder transaxles	31
V6 transaxles	14
Power steering gear assembly	134
Subframe assembly **(see illustration 5.32)**	
Bolt A	134
Bolt B	24
Nut C	27
Stiffener plate	
Left side	27
Right side	29
Rear end plate	
Four cylinder models	82 in-lbs
V6 models **(see illustration 5.36b)**	
Bolt A	34
Bolt B	27
Transaxle-to-engine bolts	
14 mm bolts	34
17 mm bolts	47

1 General information

The vehicles covered by this manual are equipped with either a 5-speed manual or a 4-speed automatic transaxle. There are two different-type manual transaxles: one for the four cylinder engine (S51) and one for the V6 engine (E-153). The transaxles vary in oil capacity, bearing and gear size to compensate for the engine horsepower and torque ratings. Information on the manual transaxle is included in this Part of Chapter 7. Service procedures for the automatic transaxle are contained in Chapter 7, Part B.

The manual transaxle is a compact, two-piece, lightweight aluminum alloy housing containing both the transmission and differential assemblies.

Because of the complexity, unavailability of replacement parts and special tools necessary, internal repair procedures for the manual transaxle are not recommended for the home mechanic. For readers who wish to tackle a transaxle rebuild, exploded views and a brief *Manual transaxle overhaul - general information* Section are provided. The bulk of information in this Chapter is devoted to removal and installation procedures.

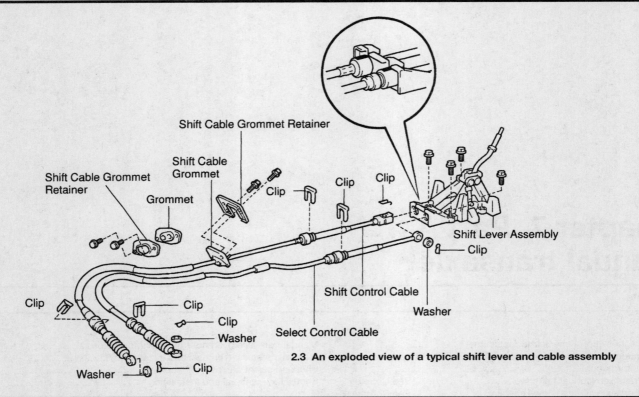

2.3 An exploded view of a typical shift lever and cable assembly

2 Shift and select cables - replacement

Refer to illustration 2.3

1 Remove the center console (see Chapter 11).
2 Raise the vehicle and support it securely on jackstands.
3 Remove the retaining clips and washers from the cable ends at the transaxle **(see illustration)**.
4 Remove the large clips that retain the shift and select cables to the bracket on the transaxle.
5 Remove the screws that attach the cable retainer, if applicable, to the engine side of the firewall. Remove the outer retainer and grommet.
6 Remove the clips and washers that attach the shift and select cables to the shift lever.
7 Remove the large clip-type cable retainers that attach the cables to the bracket at the forward end of the shift lever base.
8 Remove the inner retainer and grommet from the passenger side of the firewall.
9 Pull the cable(s) out through the firewall from the engine side.
10 Installation is the reverse of removal.

3 Shift lever assembly - removal and installation

1 Remove the center console (see Chapter 11).

2 Remove the shift and select cable retainers and disconnect both cables from the shift lever (see Section 2).
3 Remove the retaining bolts and detach the shift lever assembly **(see illustration 2.3)**.
4 Installation is the reverse of removal.

4 Back-up light switch - check and replacement

Refer to illustration 4.2

1 With the ignition key in the On position, place the shift lever in Reverse. The back-up lights should come on.
 a) If the lights don't come on, check the ignition switch, the GAUGE fuse, the bulbs and the wire harness (see Chapter 12).
 b) If the lights remain on all the time, even when the shift lever is not in REVERSE, check the wire harness.
 c) If only one light comes on, but not the other, check the bulb for that light and check the harness.
2 To check the operation of the back-up light switch itself, disconnect the electrical connector and unscrew the switch from the top of the transaxle, then use an ohmmeter to verify that there's continuity when the plunger is depressed, and no continuity when the plunger is released **(see illustration)**. The back-up light switch and harness connector are located on the side of the transaxle near the starter assembly.
3 If the switch doesn't operate as described, replace it.

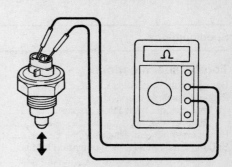

4.2 To check the back-up light switch, verify that there's continuity between the switch connector terminals with the switch plunger depressed, and no continuity with the plunger released

5 Manual transaxle - removal and installation

Removal

Refer to illustrations 5.4, 5.10, 5.11, 5.16, 5.19, 5.25, 5.27, 5.28, 5.32, 5.36a and 5.36b

1 Place protective covers on the fenders and cowl and remove the hood (see Chapter 11).
2 Relieve the fuel system pressure (see Chapter 4).
3 Disconnect the negative cable from the battery. **Caution:** *If the stereo in your vehicle is equipped with an anti-theft system, make sure you have the correct activation code before disconnecting the battery.*

Chapter 7 Part A Manual transaxle

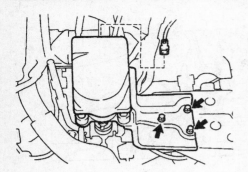

5.4 Location of the cruise control actuator mounting bolts (arrows)

5.10 Location of the ground strap on the transaxle

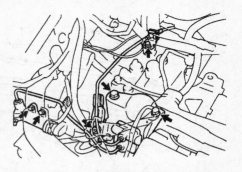

5.11 First remove the clutch release cylinder mounting bolts (left arrows) and then the accumulator mounting bolts (right arrows) (V6 transaxle shown)

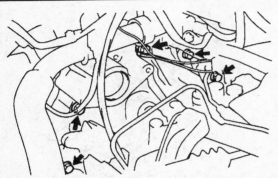

5.16 Remove the transaxle bolts from the upper section working in the engine compartment (V6 model shown, four-cylinder similar)

5.19 Remove the exhaust system from the front section of the exhaust manifold to the catalytic converter

4   If the vehicle is equipped with cruise control, unplug the electrical connector for the actuator and remove the actuator **(see illustration)**.
5   Remove the battery and the battery tray (see Chapter 5).
6   Remove the air intake duct and the air cleaner housing (see Chapter 4).
7   Release the residual fuel pressure in the tank by removing the gas cap, then disconnect the fuel lines connecting the engine to the chassis (see Chapter 4). Plug or cap all open fittings.
8   Remove the fuel filter and fuel filter mounting bracket (see Chapter 1).
9   Remove the starter (see Chapter 5).
10 Disconnect the ground cable from the transaxle **(see illustration)**.
11 Disconnect the clutch release cylinder mounting bolts from the transaxle but do NOT disconnect the hydraulic line **(see illustration)**. Remove the two mounting bolts from the clutch line bracket. Also, on V6 transaxle models (E-153), disconnect the clutch accumulator mounting bolts and set the assembly to the side. **Note:** *The accumulator on the four cylinder transaxle models (S51) does not need to be loosened because it is located away from the transaxle assembly.*
12 Disconnect the throttle linkage and, if equipped, the cruise control cable from the throttle linkage (see Chapter 4).

13 Clearly label, then disconnect all vacuum lines, coolant and emissions hoses, wiring harness connectors (VSS and back-up light switch) and ground straps. Masking tape and/or a touch up paint applicator work well for marking items. Take instant photos or sketch the locations of components and brackets.
14 Remove the cooling fan(s) and shroud(s). Drain the coolant and remove the radiator and all coolant and heater hoses (see Chapter 3).
15 Disconnect the shift and select cables (see Section 2).
16 Remove the upper transaxle mounting bolts **(see illustration**
17 Remove the manifold brace that supports the exhaust manifold and the transaxle.
18 Loosen but do NOT remove the front wheel lug nuts. Raise the vehicle and support it securely on jackstands. Secure the engine using an engine support brace that is installed above the engine compartment. If an engine support brace is not available, install an engine lift and a lifting chain assembly. This will keep the engine stable during the entire transaxle removal procedure. Remove the front wheels.
19 Detach the exhaust pipe(s) from the manifold(s) (see Chapter 4) and from the catalytic converter. Separate the front exhaust pipe from the underside of the vehicle **(see illustration)**.

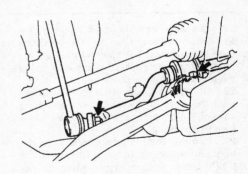

5.25 Disconnect the stabilizer bar bracket and the link (arrows) (left side shown)

20 Remove the splash shields (see Chapter 2A).
21 Drain the engine oil (see Chapter 1).
22 Remove all accessory drivebelts (see Chapter 1).
23 Drain the transaxle fluid (see Chapter 1).
24 Remove the driveaxles (see Chapter 8).
25 On models with power steering, unbolt the power steering gear from the suspension. First, remove the two stabilizer bar mounting nuts and disconnect the stabilizer bar from the links **(see illustration)**.
26 Remove the four set bolts from the stabilizer bar bracket (see Chapter 10).

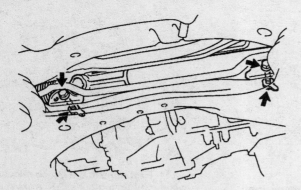

5.27 Steering gear mounting bolts (arrows)

5.28 On V6 transaxles (E-153) remove the shock absorber/coil spring mounting bolts (arrows)

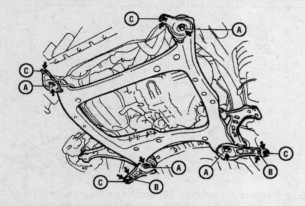

5.32 Location of the subframe mounting bolts and nuts - note the exact locations of the various sizes to insure correct reassembly

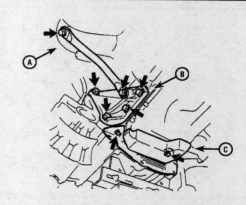

5.36a Remove the bolts from the brace (A), the right side stiffener plate (B) and the rear end plate (C) (four cylinder shown)

27 Use wire to tie the power steering gear to the chassis and remove the mounting bolts **(see illustration)**. The steering gear should remain suspended but out of the way during the transaxle removal procedure.

28 On V6 transaxle models (E153), disconnect the front shock absorber/coil spring assembly mounting bolts from the transaxle **(see illustration)**.

29 Remove the mounting bolts from the front engine mount. Refer to Chapter 2A or 2B for transaxle mount replacement procedures.

30 Remove the left side transaxle mounting insulator and bracket. The mounting insulator is considered a transaxle mount. Refer to Chapter 2A or 2B for additional details.

31 Remove the rear engine mount from the transaxle

32 Remove the subframe mounting bolts from the chassis **(see illustration)**. Be sure to note exactly the size and location of each nut and bolt to insure correct reassembly. **Note:** *The transaxle should be supported by a floor-jack immediately after the subframe is removed from the vehicle. The transaxle will tilt slightly but should remain steady if the engine is properly secured with the engine compartment brace.*

33 Support the transaxle with a floor jack. Place a block of wood on the jack head to prevent damage to the transaxle. Safety chains will help steady the transaxle on the jack.

34 On four cylinder transaxle models (S51), disconnect the steering return pipe from the front suspension cross member (see Chapter 10).

35 Remove the stiffener plate from the left side of the transaxle.

36 Remove the rear end plate **(see illustrations)** and the stiffener plate from the right side of the transaxle.

37 Recheck to be sure nothing is connecting the engine to the vehicle or to the transaxle. Disconnect and label anything still remaining.

38 Slowly lower the transaxle assembly out of the vehicle. Keep the transaxle level as you're separating it from the engine to prevent damage to the input shaft. It may be necessary to pry the mounts away from the frame brackets. **Caution:** *Do not depress the clutch pedal while the transaxle is removed from the vehicle.* **Warning:** *Do not place any part of your body under the transaxle assembly when it's supported only by a hoist or other lifting device.*

39 Move the transaxle assembly away from

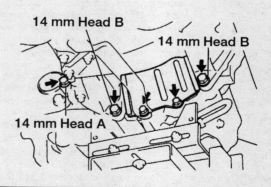

5.36b Locations of the rear end plate bolts (arrows) (V6 engine shown, four-cylinder similar)

the vehicle and carefully place the transaxle assembly on the floor onto wood blocks. Leave enough room for a floor jack underneath the transaxle.

40 The clutch components can now be inspected (see Chapter 8). In most cases, new clutch components should be routinely installed whenever the transaxle is removed.

41 Check the engine and transaxle mounts and the transaxle shock absorber. If any of these components are worn or damaged, replace them.

Installation

42 If removed, install the clutch components (see Chapter 8). Apply a dab of high temperature grease to the pilot bearing.

43 With the transaxle secured to the jack as on removal, raise it into position and then carefully slide it forward, engaging the input shaft with the clutch splines. Do not use excessive force to install the transaxle - if the input shaft does not slide into place, readjust the angle of the transaxle so it is level and/or turn the input shaft so the splines engage properly with the clutch. **Caution:** *Do NOT use transaxle-to-engine bolts to force the engine and transaxle into alignment. Doing so could crack or damage major components. If you experience difficulties, have an assistant help you line up the dowel pins on the block with the transaxle. Some wiggling of the engine and/or the transaxle will probably be necessary to secure proper alignment of the two.*

44 Install the transaxle-to-engine bolts and the engine-to-transaxle bolt. Tighten the bolts to the torque listed in this Chapter's Specifications.

45 Install the right engine mount, the front engine mount, the transaxle shock absorber, the left transaxle mount and the rear engine mount. Install the center bearing support on the rear of the block. Tighten all mounting bolts and nuts securely.

46 Reinstall the remaining components in the reverse order of removal.

47 Remove the jack and hoist and lower the vehicle. Tighten the wheel lug nuts to the torque listed in the Chapter 1 Specifications.

48 Add the specified amounts of coolant, oil and transaxle fluid (see Chapter 1).

49 Connect the negative battery cable. Start the engine and check for proper operation and leaks.

50 Shut off the engine and recheck the fluid levels. Road test the vehicle to check for proper transaxle operation and check for leakage.

6 Manual transaxle overhaul - general information

1 Overhauling a manual transaxle is a difficult job for the do-it-yourselfer. It involves the disassembly and reassembly of many small parts. Numerous clearances must be precisely measured and, if necessary, changed with select fit spacers and snap-rings. As a result, if transaxle problems arise, it can be removed and installed by a competent do-it-yourselfer, but overhaul should be left to a transmission repair shop. Rebuilt transaxles may be available - check with your dealer parts department and auto parts stores. At any rate, the time and money involved in an overhaul is almost sure to exceed the cost of a rebuilt unit.

2 Nevertheless, it's not impossible for an inexperienced mechanic to rebuild a transaxle if the special tools are available and the job is done in a deliberate step-by-step manner so nothing is overlooked.

3 The tools necessary for an overhaul include internal and external snap-ring pliers, a bearing puller, a slide hammer, a set of pin punches, a dial indicator and possibly a hydraulic press. In addition, a large, sturdy workbench and a vise or transaxle stand will be required.

4 During disassembly of the transaxle, make careful notes of how each component was removed, where it fits in relation to other components and what holds it in place.

5 Before taking the transaxle apart for repair, it will help if you have some idea what area of the transaxle is malfunctioning. Certain problems can be closely tied to specific areas in the transaxle, which can make component examination and replacement easier. Refer to the *Troubleshooting* section at the front of this manual for information regarding possible sources of trouble.

Notes

Chapter 7 Part B
Automatic transaxle

Contents

	Section		Section
Automatic transaxle/differential fluid change	See Chapter 1	General information	1
Automatic transaxle/differential lubricant level check	See Chapter 1	Park/Neutral position switch - check, replacement and adjustment	5
Automatic transaxle fluid level check	See Chapter 1	Oil seal replacement	7
Automatic transaxle - removal and installation	8	Shift cable - adjustment and replacement	3
CHECK ENGINE light	See Chapter 6	Shift lock system - description, check and component replacement	6
Diagnosis – general	2	Throttle valve (TV) cable - check and adjustment	4
Electronic control system	9		
Engine mounts – check and replacement	See Chapters 2A or 2B		

Specifications

Throttle valve outer cable end-to-stopper	0.040-inch
Shift lock system	
Shift lock solenoid resistance (TMMK type)	29 to 35 ohms
Key interlock solenoid resistance (TMC and TMMK types)	12.5 to 16.5 ohms

Torque specifications

Ft-lbs (unless otherwise indicated)

Back-up light switch	33
Exhaust manifold brace	
Four-cylinder transaxles	31
V6 transaxles	
California models	25
Except California models	15
Subframe assembly (refer to **illustration 5.32** in Chapter 7A)	
Bolt A	134
Bolt B	24
Nut C	27
Stiffener plate bolts	34
Transaxle-to-engine bolts	
17 mm	48
12 mm	35
Torque converter to driveplate bolts	
Four cylinder engines	20
V6 engines	30
Vehicle speed sensor hold-down bolt	156 in-lbs
Direct clutch speed sensor hold-down bolt	96 in-lbs

1 General information

There are three different type automatic transaxles – one for the four-cylinder engine (A140E) and two for the V6 engine; the U140E used on 1999 and later Lexus ES 300 models and the A541E used on all other models. The transaxles vary in oil capacity, bearing and gear size to compensate for the engine horsepower and torque. Information on the automatic transaxle is included in this Part of Chapter 7. Service procedures for the manual transaxle are contained in Chapter 7, Part A.

Due to the complexity of the automatic transaxles covered in this manual and to the specialized equipment necessary to perform most service operations, this Chapter contains only those procedures related to general diagnosis, routine maintenance, adjustment and removal and installation.

If the transaxle requires major repair work, it should be left to a dealer service department or an automotive or transmission shop. You can, however, remove and install the transaxle yourself and save the expense, even if a transmission shop does the repair work.

2 Diagnosis - general

Automatic transaxle malfunctions may be caused by five general conditions:

a) poor engine performance
b) improper adjustments
c) hydraulic malfunctions
d) mechanical malfunctions
e) malfunctions in the computer or its signal network

Diagnosis of these problems should always begin with a check of the easily repaired items: fluid level and condition (see Chapter 1), shift linkage adjustment and throttle linkage adjustment. Next, perform a road test to determine if the problem has been corrected or if more diagnosis is necessary. If the problem persists after the preliminary tests and corrections are completed, additional diagnosis should be done by a dealer service department or transmission shop. Refer to the *Troubleshooting* section at the front of this manual for information on symptoms of transaxle problems.

Preliminary checks

1 Drive the vehicle to warm the transaxle to normal operating temperature.
2 Check the fluid level as described in Chapter 1:

a) *If the fluid level is unusually low, add enough fluid to bring the level within the designated area of the dipstick, then check for external leaks (see below).*
b) *If the fluid level is abnormally high, drain off the excess, then check the drained fluid for contamination by coolant. The presence of engine coolant in the automatic transaxle fluid indicates that a failure has occurred in the internal radiator walls that separate the coolant from the transaxle fluid (see Chapter 3).*
c) *If the fluid is foaming, drain it and refill the transaxle, then check for coolant in the fluid, or a high fluid level.*

3 Check the engine idle speed. **Note:** *If the engine is malfunctioning, do not proceed with the preliminary checks until it has been repaired and runs normally.*
4 Check the throttle valve cable for freedom of movement. Adjust it if necessary (see Section 4). **Note:** *The throttle cable may function properly when the engine is shut off and cold, but it may malfunction once the engine is hot. Check it cold and at normal engine operating temperature.*
5 Inspect the shift control linkage (see Section 3). Make sure that it's properly adjusted and that the linkage operates smoothly.

Fluid leak diagnosis

6 Most fluid leaks are easy to locate visually. Repair usually consists of replacing a seal or gasket. If a leak is difficult to find, the following procedure may help.
7 Identify the fluid. Make sure it's transmission fluid and not engine oil or brake fluid (automatic transmission fluid is a deep red color).
8 Try to pinpoint the source of the leak. Drive the vehicle several miles, then park it over a large sheet of cardboard. After a minute or two, you should be able to locate the leak by determining the source of the fluid dripping onto the cardboard.
9 Make a careful visual inspection of the suspected component and the area immediately around it. Pay particular attention to gasket mating surfaces. A mirror is often helpful for finding leaks in areas that are hard to see.
10 If the leak still cannot be found, clean the suspected area thoroughly with a degreaser or solvent, then dry it.
11 Drive the vehicle for several miles at normal operating temperature and varying speeds. After driving the vehicle, visually inspect the suspected component again.
12 Once the leak has been located, the cause must be determined before it can be properly repaired. If a gasket is replaced but the sealing flange is bent, the new gasket will not stop the leak. The bent flange must be straightened.
13 Before attempting to repair a leak, check to make sure that the following conditions are corrected or they may cause another leak. **Note:** *Some of the following conditions cannot be fixed without highly specialized tools and expertise. Such problems must be referred to a transmission shop or a dealer service department.*

Gasket leaks

14 Check the pan periodically. Make sure the bolts are tight, no bolts are missing, the gasket is in good condition and the pan is flat (dents in the pan may indicate damage to the valve body inside).
15 If the pan gasket is leaking, the fluid level or the fluid pressure may be too high, the vent may be plugged, the pan bolts may be too tight, the pan sealing flange may be warped, the sealing surface of the transaxle housing may be damaged, the gasket may be damaged or the transaxle casting may be cracked or porous. If sealant instead of gasket material has been used to form a seal between the pan and the transaxle housing, it may be the wrong sealant.

Seal leaks

16 If a transaxle seal is leaking, the fluid level or pressure may be too high, the vent may be plugged, the seal bore may be damaged, the seal itself may be damaged or improperly fitted, the surface of the shaft protruding through the seal may be damaged or a loose bearing may be causing excessive shaft movement.
17 Make sure the dipstick tube seal is in good condition and the tube is properly seated. Periodically check the area around the speedometer gear or sensor for leakage. If fluid is evident, check the O-ring for damage.

Case leaks

18 If the case itself appears to be leaking, the casting is porous and will have to be repaired or replaced.
19 Make sure the oil cooler hose fittings are tight and in good condition.

Fluid comes out vent pipe or fill tube

20 If this condition occurs, the transaxle is overfilled, there is coolant in the fluid, the case is porous, the dipstick is incorrect, the vent is plugged or the drain-back holes are plugged.

3 Shift cable - adjustment and replacement

Adjustment

Refer to illustrations 3.3, 3.4a and 3.4b

1 When the shift lever inside the vehicle is moved from the Neutral position to other positions, it should move smoothly and accurately to each position and the shift indicator should indicate the correct gear position. If the indicator isn't aligned with the correct position, adjust the shift cable as follows:
2 Raise the vehicle and support it securely on jackstands. Remove the splash shields that cover the area between the front of the vehicle and the lower crossmember (see Chapter 11).
3 Loosen the swivel nut on the manual shift lever at the transaxle **(see illustration)**.
4 Place the manual lever in Park, then return it two notches to the Neutral position **(see illustrations)**. **Note:** *On V6 (A541E and U140E) transaxles, push the lever down while on four cylinder (A140E) transaxles, pull the*

Chapter 7 Part B Automatic transaxle

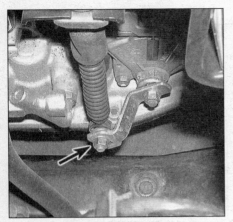

3.3 Before adjusting the shift cable, loosen the swivel nut (arrow) that connects the shift cable to the manual lever on the transaxle

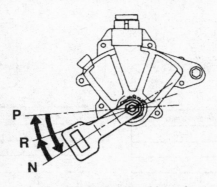

3.4a To adjust the shift cable on four-cylinder models (A140E transaxle), push the manual lever all the way UP, return it two clicks to the Neutral position. Place the shift lever inside the vehicle at the Neutral position and tighten the swivel nut

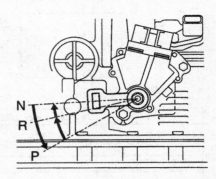

3.4b To adjust the shift cable on the V6 models (A541E and U140E transaxles), push the manual lever all the way DOWN, return it two clicks to the Neutral position, place the shift lever inside the vehicle at the Neutral position and tighten the swivel nut

3.8 To disconnect the shift cable from the transaxle, remove the large C-clip retainer (arrow) from the bracket on the front of the transaxle

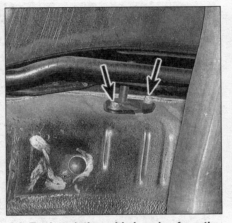

3.9 To detach the cable housing from the firewall, remove the mounting bolts (arrows)

3.11 To detach the shift cable from the shift lever base, pry off this C-clip retainer

lever UP to find the PARK position.
5   Move the shift lever inside the vehicle to the Neutral position.
6   While holding the lever with a slight pressure toward the REVERSE position,

3.12 To disconnect the shift cable from the shift lever, remove the retaining clip and pull out the clevis pin (arrow)

tighten the swivel nut securely.
7   Check the operation of the transaxle in each shift lever position (try to start the engine in each gear - the starter should operate in the Park and Neutral positions only).

Replacement

Refer to illustrations 3.8, 3.9, 3.11 and 3.12

8   Disconnect the cable from the manual lever **(see illustration 3.3)** and remove the large C-clip cable retainer **(see illustration)** from the bracket above the manual lever.
9   Remove the bolts from the cable housing retainer on the firewall **(see illustration)**.
10 Remove the center console (see Chapter 11).
11 Pry off the large C-clip cable retainer **(see illustration)**.
12 Remove the retaining clip **(see illustration)**, pull out the clevis pin and disconnect the cable from the shift lever.
13 Pull the cable through the firewall.
14 Installation is the reverse of removal.
15 Be sure to adjust the cable when you're done.

4 Throttle valve (TV) cable - check and adjustment

Refer to illustration 4.2

1   Have an assistant hold the throttle pedal down while you verify that the throttle valve linkage opens all the way.
2   If the linkage does not open all the way, ask your assistant to hold the pedal down

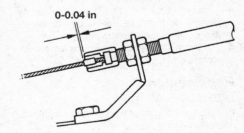

4.2 Throttle valve (TV) cable housing-to-stopper gap details

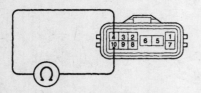

Shift Position	Terminal No. to continuity	Terminal No. to continuity
P	4 – 7	5 – 6
R	4 – 8	–
N	4 – 10	5 – 6
D	4 – 9	–
2	2 – 4	–
L	2 – 3	–

5.4a Park/Neutral position switch terminal guide and continuity table (four-cylinder models)

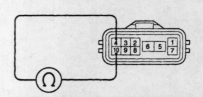

Shift Position	Terminal No. to continuity	Terminal No. to continuity
P	2 – 7	5 – 6
R	2 – 8	–
N	2 – 9	5 – 6
D	2 – 10	–
2	2 – 3	–
L	2 – 4	–

5.4b Park/Neutral position switch terminal guide and continuity table (V6 models)

while you loosen the adjusting nuts and adjust the cable until the mark or stopper is the specified distance from the boot end **(see illustration)**.

3 Tighten the adjusting nuts securely, recheck the clearance and make sure the link opens all the way when the throttle is depressed.

5 Park/Neutral position switch - check, replacement and adjustment

1 The Park/Neutral position switch prevents the engine from starting in any gear other than Park or Neutral. If the engine starts with the shift lever in any position other than Park or Neutral, adjust the switch. The Park/Neutral position switch is also an information sensor for the Electronic Controlled Transaxle (ECT) Electronic Control Unit (ECU). When the shift lever is placed in position, the Park/Neutral position switch sends a voltage signal to the ECU.

Check

Refer to illustrations 5.4a and 5.4b

2 Raise the front of the vehicle and place it securely on jackstands.
3 Disconnect the electrical connector from the Park/Neutral position switch.
4 Using an ohmmeter, check continuity between the indicated terminals for each switch position **(see illustrations)**.
5 If the switch continuity isn't as specified, replace it.

Replacement

Refer to illustration 5.8

6 Raise the front of the vehicle and place it securely on jackstands.
7 Disconnect the electrical connector.
8 Remove the manual lever retaining nut **(see illustration)**.
9 Remove the switch retaining bolts.

5.8 Remove the manual lever retaining nut (left arrow), detach the manual lever and remove the switch retaining bolts (one shown by right arrow)

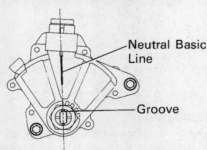

5.12a Park/Neutral position switch alignment details (four-cylinder models)

10 Remove the switch.
11 Installation is the reverse of removal. Be sure to adjust the switch.

Adjustment

Refer to illustrations 5.12a and 5.12b

12 Loosen the switch retaining bolts and rotate the switch until the groove and the neutral basic line are aligned **(see illustrations)**. Hold the switch in this position and tighten the bolts.

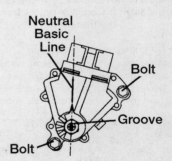

5.12b Park/Neutral position switch alignment details (V6 models)

6 Shift lock system - description, check and component replacement

Description

Refer to illustrations 6.1a and 6.1b

1 The shift lock system prevents the shift lever from being shifted out of Park until the brake pedal is applied. The system consists

Chapter 7 Part B Automatic transaxle

of a stop light switch, a key interlock solenoid, a shift lock override button, a shift lock solenoid, a shift lock control switch and a shift lock control computer (ECU). If the shift lock system doesn't perform as described, check the following components.
Note 1: *Two systems of different design are used on these vehicles, one made by TMC and one by TMMK. Toyota does not provide a way of distinguishing which system is installed on your vehicle, other than by looking at the shift lever assembly with the console cover removed (see illustrations).* **Note 2:** *On TMC type systems, the shift lock solenoid and shift lock control switch are contained within the shift lock control unit assembly. This unit must be replaced as a single component on TMC systems.*

Check

Shift lock control computer

Refer to illustrations 6.3a and 6.3b

2 Remove the console (see Chapter 11). The shift lock system computer is located behind or beside the shift lever **(see illustrations 6.1a and 6.1b)**.

3 Using a digital voltmeter, bridge the indicated terminals of each of the computer connectors; measure the voltage with the ignition switch, brake pedal and/or shift lever in the indicated positions **(see illustrations)**. **Note:** *Because these are running voltages, the connectors must be backprobed, i.e. they must be bridged from the back side of the connector. Do not attempt to push the probes of your voltmeter into the back side of the connector terminals; they're big enough to damage the connector. Instead, use a pair of long*

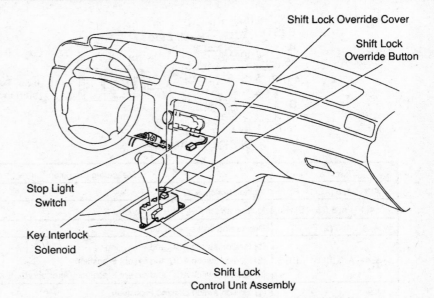

6.1a Shift lock system component locations (systems manufactured by TMC)

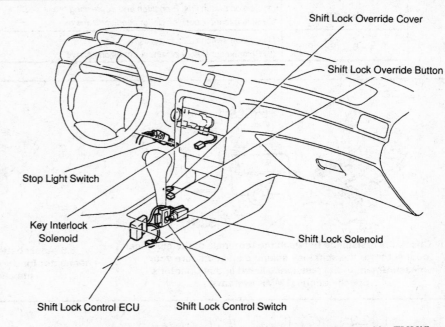

6.1b Shift lock system component locations (systems manufactured by TMMK)

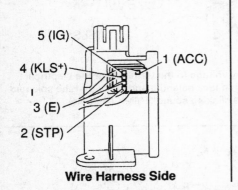

Terminal	Measuring Condition	Voltage (V)
1 – 3 (ACC – E)	Ignition switch ACC	10 – 14
5 – 3 (IG – E)	Ignition switch ON	10 – 14
2 – 3 (STP – E)	Depressing brake pedal	10 – 14
4 – 3 (KLS+ – E)	(1) Ignition switch ACC and P position	0
	(2) Ignition switch ACC and except P position	7.5 – 11
	(3) Ignition switch ACC and except P position (After approx. 1 second)	6 – 9.5

6.3a Terminal guide and voltage table for the shift lock control computer (TMC systems)

7B-6   Chapter 7 Part B   Automatic transaxle

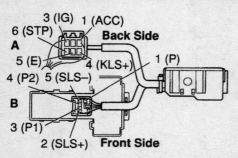

6.3b Terminal guide and voltage table for the shift lock control computer (TMMK systems)

Terminal	Measuring Condition	Voltage (V)
A, 1 – A, 5 (ACC – E)	Ignition switch ACC	10 – 14
A, 3 – A, 5 (IG – E)	Ignition switch ON	10 – 14
A, 6 – A, 5 (STP – E)	Depressing brake pedal	10 – 14
A, 4 – A, 5 (KLS+ – E)	(1) Ignition switch ACC and P position (2) Ignition switch ACC and except P position (3) Ignition switch ACC and except P position (After approx. 1 second)	0 7.5 – 11 6 – 9.5
B, 2 – B, 5 (SLS+ – SLS–)	(1) Ignition switch ON and P position (2) Depress brake pedal (3) Except P position	0 8 – 13.5 0
B, 3 – B, 1 (P1 – P)	(1) Ignition switch ON, P position and depressing brake pedal (2) Shift except P position under conditions above	0 9 – 13.5
B, 4 – B, 1 (P2 – P)	(1) Ignition switch ACC, P position (2) Shift except P position under conditions above	9 – 13.5 0

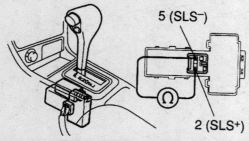

6.5 Check the resistance between the terminals of the electrical connector for the shift lock solenoid and compare your measurement to the resistance listed in this Chapter's Specifications (TMMK systems)

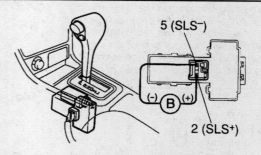

6.6 Apply battery voltage to the terminals of the electrical connector for the shift lock solenoid and verify that the solenoid makes a clicking sound (TMMK systems)

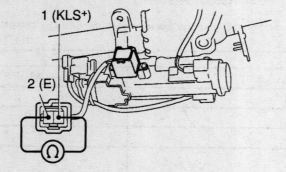

6.8 Check the resistance between the terminals of the electrical connector for the key interlock solenoid and compare your measurement to the resistance listed in this Chapter's Specifications

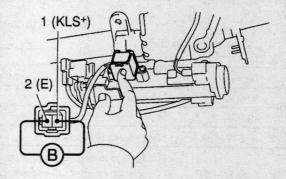

6.9 Apply battery voltage to the terminals of the electrical connector for the key interlock solenoid and verify that the solenoid makes a clicking sound

Chapter 7 Part B Automatic transaxle

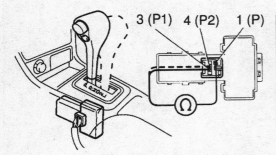

Shift position	Tester connection	Specified value
P position (Release button is not pushed)	1 – 3 (P – P1)	Continuity
P position (Release button is pushed)	1 – 3 (P – P1) 1 – 4 (P – P2)	Continuity
R, N, D, 2, L position	1 – 4 (P – P2)	Continuity

6.11 Terminal guide and continuity table for the shift lock control switch (TMMK)

6.14a The shift lock solenoid (arrow) is located at the front of the shift lever base

6.14b The shift lock control switch is located just ahead of the computer, within the shift lever base

sharp pins, then clip onto them with a pair of alligator clips.
4 If the computer doesn't operate as described, replace it.

Shift lock solenoid (TMMK type)
Refer to illustrations 6.5 and 6.6

5 Disconnect the solenoid electrical connector and, using an ohmmeter, measure the resistance between the two terminals **(see illustration)**. Compare your measurement to the shift lock solenoid resistance listed in this Chapter's Specifications.
6 Apply battery voltage to the connector **(see illustration)** and verify that the solenoid makes a clicking sound.
7 If the shift lock solenoid doesn't perform as described, replace it.

Key interlock solenoid (TMC and TMMK types)
Refer to illustrations 6.8 and 6.9

8 Disconnect the solenoid electrical connector and, using an ohmmeter, measure the resistance between the two terminals **(see illustration)**. Compare your measurement to the key interlock solenoid resistance listed in this Chapter's Specifications.
9 Apply battery voltage to the connector

(see illustration) and verify that the solenoid makes a clicking sound.
10 If the key interlock solenoid doesn't perform as described, replace it.

Shift lock control switch (TMMK)
Refer to illustration 6.11

11 Disconnect the shift lock control switch electrical connector and, using a continuity tester or an ohmmeter, verify that there's continuity between each of the indicated terminals when the shift lever is placed in the indicated positions **(see illustration)**.
12 If the shift lock control switch doesn't perform as described, replace it.

Component replacement
Refer to illustrations 6.14a and 6.14b

13 Remove the center console (see Chapter 11).
14 The shift lock control computer is located to the rear of the shift lever base. The shift lock control switch is located in front of the computer, within the shift lever base, immediately behind and below the shift lever. The shift lock solenoid is located at the front of the shift lever base. To replace one of these units, simply disconnect the electrical connector and unclip or unscrew the device

from its mounting bracket **(see illustrations)**.
15 The key interlock solenoid is located near the ignition switch (see Chapter 12).

7 Oil seal replacement

1 Fluid leaks frequently occur due to wear of the driveaxle oil seals and/or the speedometer drive gear oil seal and O-rings. Replacement of these seals is relatively easy, since the repairs can usually be performed without removing the transaxle from the vehicle.

Driveaxle seals
Refer to illustrations 7.4 and 7.6

2 The driveaxle oil seals are located in either sides of the transaxle, where the driveaxle shaft is splined into the differential. If leakage at the seal is suspected, raise the vehicle and support it securely on jackstands. If the seal is leaking, fluid will be found on the side of the transaxle.
3 Remove the driveaxle assembly (see Chapter 8). If you're replacing the right side driveaxle seal, remove the intermediate shaft and the driveaxle assembly as a single unit.
4 Using a screwdriver or prybar, carefully

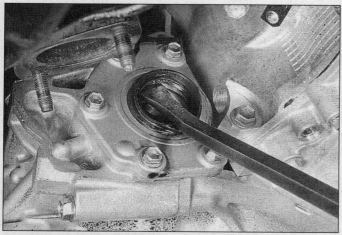

7.4 Carefully pry out the old driveaxle seal with a prybar, screwdriver or a special seal removal tool; make sure you don't gouge or nick the surface of the seal bore

7.6 Drive in the new driveaxle seal with a large socket or a special seal installer

7.9 To remove the vehicle speed sensor from the transaxle, disconnect the electrical connector and remove the sensor hold-down bolt (arrows)

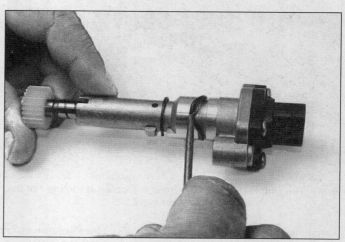

7.10 Remove the O-ring from the speedometer driven gear

pry the oil seal out of the transaxle bore **(see illustration)**.

5 If the oil seal cannot be removed with a screwdriver or prybar, a special oil seal removal tool (available at auto parts stores) will be required.

6 Using a large section of pipe or a large deep socket as a drift, install the new oil seal. Drive it into the bore squarely and make sure that it is completely seated **(see illustration)**. Lubricate the lip of the new seal with multi-purpose grease.

7 Install the driveaxle assembly (see Chapter 8). Be careful not to damage the lip of the new seal.

Speedometer driven gear seal

Refer to illustrations 7.9 and 7.10

8 The vehicle speed sensor and speedometer driven gear housing is located on the differential (rear) part of the transaxle housing. Look for lubricant around the sensor housing to determine if the O-ring is leaking.

9 Disconnect the electrical connector from the vehicle speed sensor and remove the sensor and speedometer driven gear housing from the transaxle **(see illustration)**.

10 Remove the O-ring **(see illustration)**.

11 Install a new O-ring on the driven gear housing and reinstall the speedometer driven gear and vehicle speed sensor housing. Tighten the hold-down bolt securely.

Direct clutch speed sensor seal (V6 models)

12 The direct clutch speed sensor seal is located on the transaxle housing **(see illustrations 9.16b and 9.16c)**. Look for lubricant around the sensor housing to determine if the O-ring is leaking.

13 Disconnect the electrical connector from the direct clutch speed sensor and remove the sensor from the transaxle.

14 Remove the O-ring.

15 Install a new O-ring on the sensor body and reinstall the direct clutch speed sensor. Tighten the hold-down bolt securely.

8 Automatic transaxle - removal and installation

Note: *The following procedure applies to Avalon and Solara models. The manufacturer states that transaxle removal on Camry and Lexus ES300/330 models requires the engine and transaxle to be removed as a unit, then separated once they are out of the vehicle (see Chapter 2, Part C for the engine/transaxle removal procedure).*

Removal

Refer to illustrations 8.5, 8.8, 8.11, 8.29, 8.32a and 8.32b

1 Place protective covers on the fenders and cowl and remove the hood (see Chapter 11).

2 Disconnect the negative cable from the battery (see Chapter 5, Section 1).

3 Remove the battery and the battery tray (see Chapter 5).

Chapter 7 Part B Automatic transaxle   7B-9

8.5 Location of the cruise control actuator

8.8 Disconnect the automatic transaxle fluid cooler lines

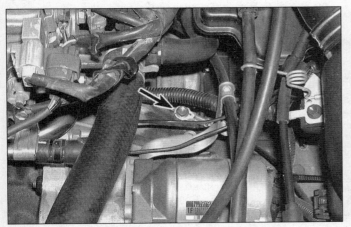

8.11 Location of the ground strap bolt

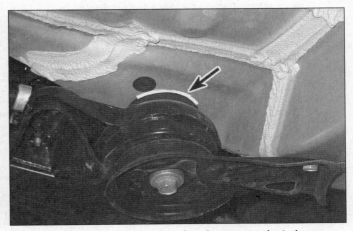

8.29 Paint subframe-to-chassis reference marks to insure correct reassembly

4   On V6 models, remove the engine cover.
5   On models equipped with cruise control, unplug the electrical connector for the actuator and remove the actuator **(see illustration)**.
6   Remove the starter (see Chapter 5).
7   Remove the fluid cooler lines from the transaxle **(see illustration)**. Be sure to position a pan below the line connections to catch any residual fluid.
8   Disconnect the transmission position switch electrical connector (see Section 5).
9   Disconnect the vehicle speed sensor (VSS) connector (see Section 7) and the direct clutch speed sensor (on models so equipped).
10 Disconnect the throttle valve (TV) cable from the throttle linkage (see Section 4).
11 Clearly label, then disconnect any vacuum lines, wiring harnesses, electrical connectors, and ground straps connected to the transaxle **(see illustration)**. Masking tape and/or a touch-up paint applicator works well for marking items. Take instant photos or sketch the locations of components and brackets.

12 Disconnect the shift cable (see Section 3) from the transaxle.
13 Loosen the driveaxle/hub nuts (see Chapter 8).
14 Loosen but do NOT remove the front wheel lug nuts.
15 Remove the transaxle upper mounting bolts.
16 Raise the vehicle and support it securely on jackstands.
17 Secure the engine using an engine support fixture that is fitted above the engine compartment. If an engine support fixture is not available, install an engine hoist and a lifting chain assembly. This will keep the engine stable during the entire transaxle removal procedure. **Warning:** *Be sure the engine/transaxle is securely supported by the fixture or hoist. If it is not securely supported, it could fall during the removal procedure, causing injury or death.* Remove the front wheels.
18 Remove the manifold brace between the exhaust manifold and the transaxle.
19 Remove the under-vehicle splash shields.
20 Detach the exhaust pipe(s) from the

manifold(s) (see Chapter 4). Detach the exhaust pipe from the catalytic converter and separate the pipe from the exhaust system. **Note:** *On some models it will be necessary to remove the front pipe support bracket from the engine (see illustration 8.32a)*.
21 Drain the transaxle fluid (see Chapter 1).
22 Remove the driveaxles (see Chapter 8).
23 Remove the mounting bolts from the front engine mount (see Chapter 2A or 2B).
24 Remove the left side transaxle mount nuts (see Chapter 2A or 2B).
25 Remove the rear engine mount from the transaxle (see Chapter 2A or 2B).
26 Remove the two stabilizer bar mounting nuts and disconnect the stabilizer bar from the links (see Chapter 10).
27 Remove the four set bolts from the stabilizer bar bracket (see Chapter 10).
28 Use wire to tie the power steering gear to a component up above it, then remove the two mounting bolts. The steering gear should remain suspended but out of the way during the transaxle removal procedure.
29 Remove the subframe mounting bolts from the chassis **(see illustration)**. The front section of the subframe is attached with two

8.32a Remove the torque converter cover mounting bolts (A), exhaust pipe flange and brace bolts indicated by B

8.32b Remove the torque converter-to-driveplate bolts - rotate the engine to gain access to the other bolts, also seen clearly here are the lower transaxle-to-engine bolts

bolts and two nuts while the rear section of the subframe is attached with a combination of six nuts and bolts. Be sure to note exactly the size and location of each nut and bolt to insure correct reassembly. **Note:** *The transaxle should be supported by a transmission jack immediately after the subframe is removed from the vehicle. The transaxle will tilt slightly but should remain steady if the engine is properly secured with the engine support fixture or hoist.*

30 Support the transaxle with an approved transaxle jack and safety chains. Floor jacks are often not stable enough to support and lower the transaxle from the vehicle.

31 Remove the stiffener plate from the left side and right side of the transaxle, if equipped.

32 Remove the torque converter mounting bolts **(see illustrations)**. Rotate the engine to gain access to each bolt.

33 Remove the transaxle lower mounting bolts.

34 Recheck to be sure nothing is connecting the transaxle to the engine or to the vehicle. Disconnect and label anything still remaining.

35 Separate the transaxle from the engine, then slowly lower the transaxle assembly out of the vehicle. Keep the transaxle level as you're separating it from the engine to prevent damage to the input shaft. It may be necessary to pry the mounts away from the frame brackets. **Warning:** *Do not place any part of your body under the transaxle assembly or engine when it's supported only by a hoist or other lifting device.*

36 Move the transaxle assembly away from the vehicle and carefully place the transaxle assembly on the floor onto wood blocks. Leave enough room for a floor jack underneath the transaxle.

37 Check the engine and transaxle mounts and the transaxle shock absorber. If any of these components are worn or damaged, replace them.

Installation

38 If removed, install the torque converter on the transaxle input shaft. Make sure the converter hub splines are properly engaged with the splines on the transaxle input shaft.

39 With the transaxle secured to the jack as on removal, and with an assistant holding the torque converter in place, raise the transaxle into position and turn the converter to align the bolt holes in the converter with the bolt holes in the driveplate. Install the converter-to-driveplate bolts. **Note:** *Install all six bolts before tightening any of them.*

40 Install the transaxle-to-engine bolts and the engine-to-transaxle bolt. Tighten the bolts to the torque listed in this Chapter's Specifications. Do not use excessive force to install the transaxle - if something binds and the transaxle won't mate with the engine, alter the angle of the transaxle slightly until it does mate. **Caution:** *Do NOT use transaxle-to-engine bolts to force the engine and transaxle into alignment. Doing so could crack or damage major components. If you experience difficulties, have an assistant help you line up the dowel pins on the block with the transaxle. Some wiggling of the engine and/or the transaxle will probably be necessary to secure proper alignment of the two.*

41 Install the right engine mount, the front engine mount, the transaxle shock absorber, the left transaxle mount and the rear engine mount. Install the center bearing support on the rear of the block. Tighten all mounting bolts and nuts securely.

42 Reinstall the remaining components in the reverse order of removal.

43 Remove all jacks and hoists and lower the vehicle. Tighten the wheel nuts to the torque listed in the Chapter 1 Specifications.

44 Add the specified type and amount of transaxle fluid (see Chapter 1).

45 Connect the negative battery cable. Run the engine, cycle the transaxle through the gears, then check the fluid level (see Chapter 1). check for proper operation and leaks.

46 Road test the vehicle to check for proper transmission operation and check for leakage. Recheck the fluid level.

9 Electronic control system

Trouble codes

1 The electronic control system for the transaxle has some self-diagnostic capabilities. If certain kinds of system malfunctions occur, the PCM stores the appropriate diagnostic trouble code in its memory and the CHECK ENGINE indicator light illuminates to inform the driver. The diagnostic trouble codes can only be extracted from the PCM using a SCAN tool that can be linked to the On Board Diagnostic (OBD II) computer via the 16-pin diagnostic link. Codes are listed below for reference, but can only be extracted with the correct SCAN tool (refer to Chapter 6 for additional information).

Chapter 7 Part B Automatic transaxle

Trouble code/Trouble area

Code	Trouble area
P0500	Defective speedometer gauge (number 1 speed sensor), open in wiring harness or short circuit
P0710	Open or short in SL1 shift solenoid harness or short circuit, blocked or stuck valve body
P0750	Defective Shift Solenoid A (number 1 shift solenoid), valve stuck open or closed
P0753	Defective Shift Solenoid A (number 1 shift solenoid), open in wiring harness or short circuit
P0755	Defective Shift Solenoid B (number 2 shift solenoid), valve stuck open or closed
P0758	Defective Shift Solenoid B (number 2 shift solenoid), open in wiring harness or short circuit
P0765	Defective Shift Solenoid (number S4 shift solenoid), valve stuck open or closed
P0768	Defective Shift Solenoid (number S4 shift solenoid), open in wiring harness or short circuit
P0770	Defective Shift Solenoid E, valve stuck open or closed
P0773	Defective Shift Solenoid E, open in wiring harness or short circuit
P1520	Defective stop light switch, an open or short in the stop light switch circuit or PCM malfunction
P1705	Defective Direct Clutch Speed sensor, an open or short in the direct clutch sensor circuit or PCM malfunction
P1725	Defective input turbine speed sensor, an open or short in the direct clutch sensor circuit or PCM malfunction
P1730	Defective counter gear speed sensor, an open or short in the counter gear speed sensor circuit or PCM malfunction
P1765	Defective Shift Solenoid Valve SLN, an open or short in the shift solenoid valve SLN circuit or PCM malfunction
P1780	Defective Park/Neutral switch, an open or short in the Park/Neutral switch circuit or PCM malfunction

Other Electronic Control System checks

Preliminary checks

2 Check the fluid level and condition. If the fluid smells burned, replace it (see Chapter 1).
3 Check for fluid leaks (see Section 2).
4 Check and, if necessary, adjust the shift cable (see Section 3).
5 Check and, if necessary, adjust the throttle valve (TV) cable (see Section 4).
6 Check and, if necessary, adjust the Park/Neutral position switch (see Section 5).

O/D OFF indicator light check

7 Turn the ignition switch to ON.
8 Verify that the O/D OFF indicator light comes on when the O/D main switch is in the Off (up) position, and goes out when the O/D main switch is pushed to the On position.
9 If the O/D OFF indicator light does not light up, or remains on all the time, have the circuit checked out by a dealer service department.

Manual shifting test

Refer to illustrations 9.11a and 9.11b

10 This test can determine whether a problem lies is in the electronic control system or is a mechanical problem inside the transaxle. This test should only be performed if a SCAN tool capable of resetting trouble codes is

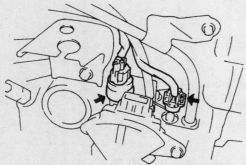

9.11a On four-cylinder models, the solenoid electrical connectors are located right above the Park/Neutral position switch on the front of the transaxle

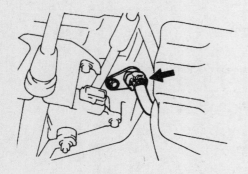

9.11b On V6 models, the solenoid electrical connector is located off to the side of the Park/Neutral switch

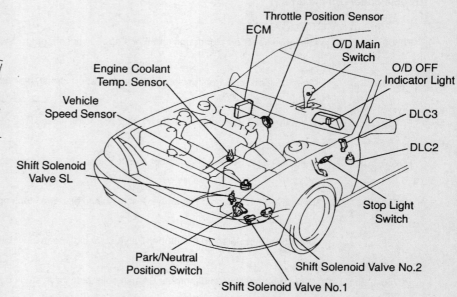

9.16a Electronic control system component locations (four-cylinder models)

available (see Chapter 6).
11 Disconnect the solenoid connector(s) **(see illustrations)**.
12 Drive the vehicle, shifting through the "L," "2" and "D" ranges manually and verify that the gear changes correspond to the shift lever positions.
13 If the transaxle does not perform as described above, the problem is in the transaxle itself and is not an electronic control system problem.
14 Connect the solenoid connectors.
15 Cancel the diagnostic trouble code using a special SCAN tool. Refer to Chapter 6 for additional information.

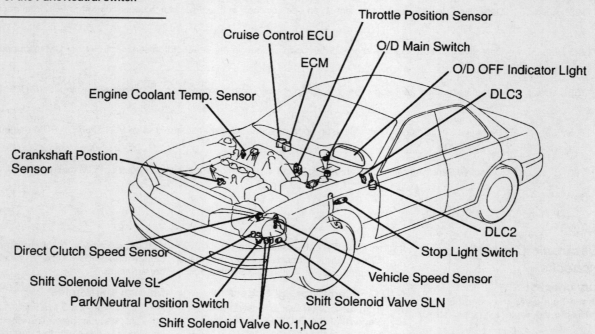

9.16b Electronic control system component locations (V6 models)

Chapter 7 Part B Automatic transaxle

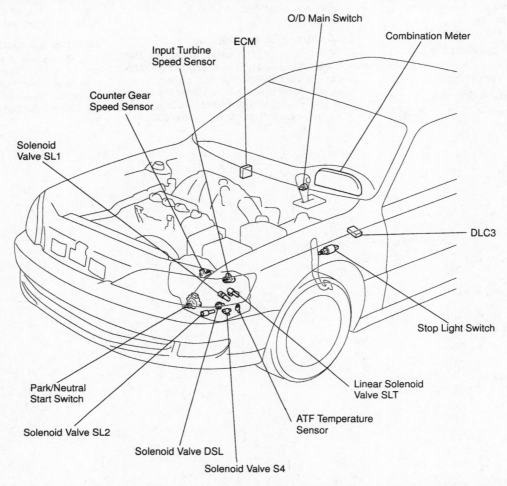

9.16c 1999 and later Lexus ES 300 model U140E transaxle electronic control system component locations

Component check and replacement

Note: *Most electronic control system tests are beyond the scope of this manual. The following procedures are tests you can do at home. Aside from these procedures, diagnosis of the electronic control system should be handled by a dealer service department.*

Solenoids

Refer to illustrations 9.16a, 9.16b, 9.16c and 9.20

16 The accompanying drawings show the locations of the system components **(see illustrations)**.

17 To check a solenoid, disconnect the electrical connector from the solenoid harness at the transaxle **(see illustration 10.11a and 10.11b)**. Measure the resistance on the transaxle-side of the connector with an ohmmeter. The resistance of each solenoid should be between 11 and 15 ohms. If the resistance is less than 8 ohms, there's a short circuit in the solenoid winding; if the resistance is more than 100 k-ohms, there's an open circuit in the solenoid windings. If the resistance of any solenoid is too high or too low, replace it. **Caution:** *Do NOT contact any pins on the PCM side of this electrical connector with the ohmmeter leads. Most ohmmeters use a 9-volt battery which could damage the PCM circuitry.*

18 The solenoids may also be bench tested for proper operation and obstructions, which would restrict fluid flow.

19 Remove the transaxle pan (see Chapter 1). Disconnect the solenoid from the harness and remove the solenoid from the valve body.

20 Apply battery voltage to each solenoid terminal and verify that the solenoids are working **(see illustration)**. When energized, they should make a clicking sound.

21 Apply no more than 65 psi (A541E) or 71 psi (U140E) of compressed air to each solenoid and verify that it doesn't pass air. Now apply battery voltage to each solenoid and verify that the valve opens.

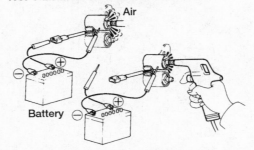

No. 1 and No. 2 Solenoid Valves

9.20 To check the solenoid valves, apply battery voltage to the terminals of each solenoid and verify that it makes a clicking sound; then apply about 65 psi of compressed air to each solenoid and verify that it doesn't pass air; finally, apply battery voltage to each solenoid and verify that the valve opens

22 If a solenoid doesn't operate as described, replace it.
23 Installation is the reverse of removal.

Park/Neutral position switch
24 Refer to Section 5 and check the Park/Neutral position switch.

Throttle Position Sensor
25 Refer to Chapter 6 and check the TPS. A malfunctioning TPS can cause shift-related problems.

Number 2 Speed sensor
26 The number 2 speed sensor is also called the direct clutch speed sensor. This sensor detects the rotation speed of the direct clutch drum and compares the speed signal with the vehicle speed sensor to arrive at the exact shift points for the transmission. Refer to Chapter 6 for the diagnostic and replacement procedures.

Number 1 Speed sensor
27 The number 1 speed sensor is mounted on the side of the transaxle and detects engine rotation from the differential gear. Refer to Chapter 6 for diagnostic and replacement procedures.

Chapter 8
Clutch and driveaxles

Contents

	Section		Section
Clutch components - removal, inspection and installation	3	Clutch release cylinder and accumulator - removal and installation	6
Clutch - description and check	2	Clutch start switch - check and adjustment	8
Clutch fluid level check	See Chapter 1	Driveaxle boot check	See Chapter 1
Clutch hydraulic system - bleeding	7	Driveaxle boot replacement	11
Clutch master cylinder - removal and installation	5	Driveaxle oil seal - replacement	See Chapter 7B
Clutch pedal height and freeplay check and adjustment	See Chapter 1	Driveaxle - removal and installation	10
Clutch release bearing and lever - removal, inspection and installation	4	Driveaxles - general information and inspection	9
		Flywheel - removal and installation	See Chapter 2
		General information	1

Specifications

Clutch
Fluid type	See Chapter 1
Pedal freeplay	See Chapter 1
Pedal height	See Chapter 1

Driveaxle length (standard)

Models with a four cylinder engine
  Solara
    Left side .... 23-63/64 inches (609 mm)
    Right side .... 33-3/16 inches (843 mm)
  Camry
    Left side .... 23-63/64 inches (609 mm)
    Right side .... 34-9/64 inches (867 mm)
Models with a V6 engine
  Manual transaxle
    Left side .... 23-11/16 inches (601.5 mm)
    Right side .... 34-5/16 inches (871.6 mm)
  Automatic transaxle
    Left side .... 23-1/16 inches (586 mm)
    Right side .... 34-45/64 inches (881.6 mm)

Torque specifications
	Ft-lbs (unless otherwise indicated)
Clutch master cylinder mounting nuts	108 in-lbs
Clutch accumulator	
Bracket mounting bolts	15
Hydraulic line threaded fitting	132 in-lbs
Clutch pressure plate-to-flywheel bolts	168 in-lbs
Clutch release cylinder	
Mounting bolts	108 in-lbs
Hydraulic line threaded fitting	132 in-lbs
Driveaxle/hub nut	217
Right driveaxle center bearing lock bolt	24
Wheel lug nuts	See Chapter 1

Chapter 8 Clutch and driveaxles

1 General information

The information in this Chapter deals with the components from the rear of the engine to the front wheels, except for the transaxle, which is dealt with in Chapter 7A and 7B. For the purposes of this Chapter, these components are grouped into two categories: Clutch and driveaxles. Separate Sections within this Chapter offer general descriptions and checking procedures for both groups.

Since nearly all the procedures covered in this Chapter involve working under the vehicle, make sure it's securely supported on sturdy jackstands or a hoist where the vehicle can be easily raised and lowered.

2 Clutch - description and check

Refer to illustration 2.1

1   All vehicles with a manual transaxle use a single dry plate, diaphragm spring type clutch **(see illustration)**. The clutch disc has a splined hub which allows it to slide along the splines of the transaxle input shaft. The clutch and pressure plate are held in contact by spring pressure exerted by the diaphragm in the pressure plate.

2   The clutch release system is operated by hydraulic pressure. The hydraulic release system consists of the clutch pedal, a master cylinder and fluid reservoir, the hydraulic line, an accumulator (on V6 models), a release (or slave) cylinder which actuates the clutch release lever and the clutch release (or throw-out) bearing.

3   When pressure is applied to the clutch pedal to release the clutch, hydraulic pressure is exerted against the outer end of the clutch release lever. As the lever pivots, the shaft fingers push against the release bearing. The bearing pushes against the fingers of the diaphragm spring of the pressure plate assembly, which in turn releases the clutch plate.

4   Terminology can be a problem regarding the clutch components because common names have in some cases changed from that used by the manufacturer. For example, the driven plate is also called the clutch plate or disc, the pressure plate assembly is sometimes referred to as the clutch cover, the clutch release bearing is sometimes called a throw-out bearing, and the release cylinder is sometimes called the operating or slave cylinder.

5   Other than replacing components that have obvious damage, some preliminary checks should be performed to diagnose a clutch system failure.

a) The first check should be of the fluid level in the clutch master cylinder (see Chapter 1). If the fluid level is low, add fluid as necessary and inspect the hydraulic clutch system for leaks. If the master cylinder reservoir has run dry, bleed the system (see Section 7) and re-test the clutch operation.

b) To check "clutch spin-down time," run the engine at normal idle speed with the transaxle in Neutral (clutch pedal up - engaged). Disengage the clutch (pedal down), wait several seconds and shift the transaxle into Reverse. No grinding noise should be heard. A grinding noise would most likely indicate a problem in the pressure plate or the clutch disc.

c) To check for complete clutch release, run the engine (with the parking brake applied to prevent movement) and hold the clutch pedal approximately 1/2-inch from the floor. Shift the transaxle between 1st gear and Reverse several times. If the shift is not smooth, component failure is indicated. Check the release cylinder pushrod travel. With the clutch pedal depressed completely the release cylinder pushrod should extend substantially. If it doesn't, check the fluid level in the clutch master cylinder.

d) Visually inspect the clutch pedal bushing at the top of the clutch pedal to make sure there is no sticking or excessive wear.

e) Under the vehicle, check that the clutch release lever is solidly mounted on the ball stud.

3 Clutch components - removal, inspection and installation

Warning: *Dust produced by clutch wear and deposited on clutch components may contain asbestos, which is hazardous to your health. DO NOT blow it out with compressed air and DO NOT inhale it. DO NOT use gasoline or petroleum based solvents to remove the dust. Brake system cleaner should be used to flush the dust into a drain pan. After the clutch components are wiped clean with a rag, dispose of the contaminated rags and cleaner in a labeled, covered container.*

Removal

Refer to illustration 3.6

1   Access to the clutch components is normally accomplished by removing the transaxle, leaving the engine in the vehicle. If, of course, the engine is being removed for major overhaul, then the opportunity should always be taken to check the clutch for wear and replace worn components as necessary. However, the relatively low cost of the clutch components compared to the time and labor involved in gaining access to them warrants their replacement any time the engine or transaxle is removed, unless they are new or in near-perfect condition. The following procedures assume that the engine will stay in place.

2   Remove the release cylinder (see Section 6). Hang it out of the way with a piece of wire - it's not necessary to disconnect the hose.

3   Remove the transaxle from the vehicle (see Chapter 7A). Support the engine while the transaxle is out. Preferably, an engine hoist should be used to support it from above. However, if a jack is used underneath the engine, make sure a piece of wood is used between the jack and oil pan to spread the load. **Caution:** *The pick-up for the oil pump is very close to the bottom of the oil pan. If the pan is bent or distorted in any way, engine oil starvation could occur.*

4   The release fork and release bearing can remain attached to the transaxle for the time being.

5   To support the clutch disc during removal, install a clutch alignment tool through the clutch disc hub.

6   Carefully inspect the flywheel and pressure plate for indexing marks. The marks are

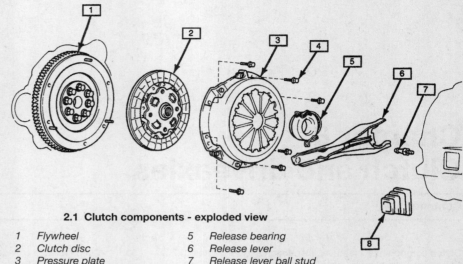

2.1 Clutch components - exploded view

1   Flywheel
2   Clutch disc
3   Pressure plate
4   Pressure plate bolts
5   Release bearing
6   Release lever
7   Release lever ball stud
8   Fork boot

Chapter 8 Clutch and driveaxles

3.6 Mark the relationship of the pressure plate to the flywheel (in case you are going to re-use the same pressure plate)

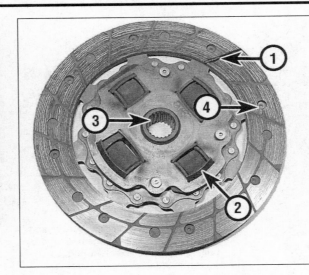

3.10 The clutch disc

1 **Lining** - this will wear down in use
2 **Springs or dampers** - check for cracking and deformation
3 **Splined hub** - the splines must not be worn and should slide smoothly on the transaxle input shaft splines
4 **Rivets** - these secure the lining and will damage the flywheel or pressure plate if allowed to contact the surfaces

NORMAL FINGER WEAR

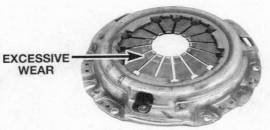
EXCESSIVE FINGER WEAR

BROKEN OR BENT FINGERS

3.12a Replace the pressure plate if any of these conditions are noted

usually an X, an O or a white letter. If they cannot be found, scribe marks yourself so the pressure plate and the flywheel will be in the same alignment during installation **(see illustration)**.

7 Slowly loosen the pressure plate-to-flywheel bolts. Work in a diagonal pattern and loosen each bolt a little at a time until all spring pressure is relieved. Then hold the pressure plate securely and completely remove the bolts, followed by the pressure plate and clutch disc.

Inspection

Refer to illustrations 3.10, 3.12a and 3.12b

8 Ordinarily, when a problem occurs in the clutch, it can be attributed to wear of the clutch driven plate assembly (clutch disc). However, all components should be inspected at this time.
9 Inspect the flywheel for cracks, heat checking, score marks and other damage. If the imperfections are slight, a machine shop can resurface it to make it flat and smooth. Refer to Chapter 2 for the flywheel removal procedure.
10 Inspect the lining on the clutch disc. There should be at least 1/16-inch of lining above the rivet heads. Check for loose rivets, distortion, cracks, broken springs and other

3.12b Examine the pressure plate friction surface for score marks, cracks and evidence of overheating (blue spots)

obvious damage **(see illustration)**. As mentioned above, ordinarily the clutch disc is replaced as a matter of course, so if in doubt about the condition, replace it with a new one.
11 The release bearing should be replaced along with the clutch disc (see Section 4).
12 Check the machined surface and the diaphragm spring fingers of the pressure plate **(see illustrations)**. If the surface is grooved or otherwise damaged, replace the pressure plate assembly. Also check for obvious damage, cracking, etc. Light glazing can be removed with emery cloth or sandpaper. If a new pressure plate is indicated, new or factory rebuilt units are available.

Installation

Refer to illustration 3.14

13 Before installation, carefully wipe the flywheel and pressure plate machined surfaces clean. It's important that no oil or grease is on these surfaces or the lining of the clutch disc. Handle these parts only with clean hands.
14 Position the clutch disc and pressure

8-4   Chapter 8 Clutch and driveaxles

3.14 Center the clutch disc in the pressure plate with a clutch alignment tool

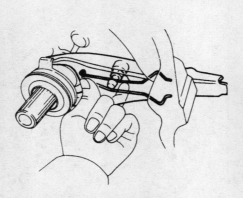

4.3 Reach behind the release lever and disengage the lever from the ball stud by pulling on the retention spring, then remove the lever and bearing

4.4 To check the operation of the bearing, hold it by the outer race and rotate the inner race while applying pressure - the bearing should turn smoothly - if it doesn't, replace it

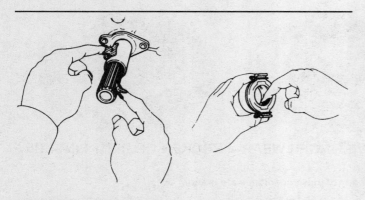

4.5 Apply a light coat of high-temperature grease to the transaxle bearing retainer and also fill the release bearing groove

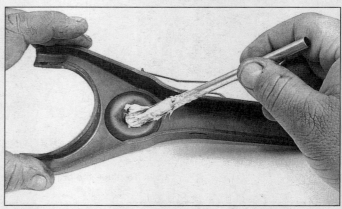

4.6a Using high temperature grease, lubricate the ball stud socket in the back of the release lever . . .

plate with the clutch held in place with an alignment tool **(see illustration)**. Make sure it's installed properly (most replacement clutch plates will be marked "flywheel side" or something similar - if not marked, install the clutch disc with the damper springs or cushion toward the transaxle).

15 Tighten the pressure plate-to-flywheel bolts only finger-tight, working around the pressure plate.

16 Center the clutch disc by ensuring the alignment tool is through the splined hub and into the recess in the crankshaft. Wiggle the tool up, down or side-to-side as needed to bottom the tool. Tighten the pressure plate-to-flywheel bolts a little at a time, working in a crisscross pattern to prevent distortion of the cover. After all of the bolts are snug, tighten them to the torque listed in this Chapter's Specifications. Remove the alignment tool.

17 Using high-temperature grease, lubricate the inner groove of the release bearing (see Section 4). Also place grease on the release lever contact areas and the transaxle input shaft bearing retainer.

18 Install the clutch release bearing (see Section 4).

19 Install the transaxle, release cylinder and all components removed previously, tightening all fasteners to the proper torque specifications.

4 Clutch release bearing and lever - removal, inspection and installation

Warning: *Dust produced by clutch wear and deposited on clutch components may contain asbestos, which is hazardous to your health. DO NOT blow it out with compressed air and DO NOT inhale it. DO NOT use gasoline or petroleum-based solvents to remove the dust. Brake system cleaner should be used to flush it into a drain pan. After the clutch components are wiped clean with a rag, dispose of the contaminated rags and cleaner in a labeled, covered container.*

Removal

Refer to illustration 4.3

1 Disconnect the negative cable from the battery. **Caution:** *If the stereo in your vehicle is equipped with an anti-theft system, make sure you have the correct activation code before disconnecting the battery.*

2 Remove the transaxle (see Chapter 7).

3 Remove the clutch release lever from the ball stud, then remove the bearing from the lever **(see illustration)**.

Inspection

Refer to illustration 4.4

4 Hold the bearing by the outer race and rotate the inner race while applying pressure **(see illustration)**. If the bearing doesn't turn smoothly or if it's noisy, replace the bearing/hub assembly with a new one. Wipe the bearing with a clean rag and inspect it for damage, wear and cracks. Don't immerse the bearing in solvent - it's sealed for life and to do so would ruin it. Also check the release lever for cracks and bends.

Installation

Refer to illustrations 4.5, 4.6a and 4.6b

5 Fill the inner groove of the release bearing with high-temperature grease. Also apply a light coat of the same grease to the

Chapter 8 Clutch and driveaxles

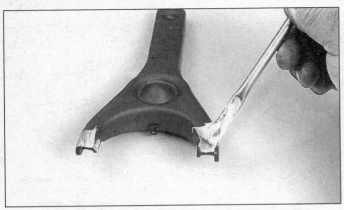

4.6b ... the lever ends and the depression for the cylinder pushrod

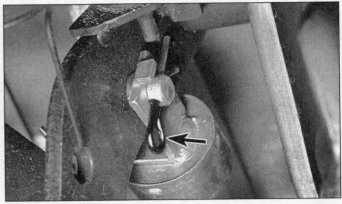

5.3 To release the clutch pushrod from the clutch pedal, remove the clip and clevis pin from the clutch pedal

transaxle input shaft splines and the front bearing retainer (see illustration).
6   Lubricate the release lever ball socket, lever ends and release cylinder pushrod socket with high-temperature grease (see illustrations).
7   Attach the release bearing to the release lever.
8   Slide the release bearing onto the transaxle input shaft front bearing retainer while passing the end of the release lever through the opening in the clutch housing. Push the clutch release lever onto the ball stud until it's firmly seated.
9   Apply a light coat of high-temperature grease to the face of the release bearing where it contacts the pressure plate diaphragm fingers.
10 The remainder of installation is the reverse of the removal procedure.

5   Clutch master cylinder - removal and installation

Removal

Refer to illustration 5.3

1   Disconnect the negative cable from the battery. **Caution:** *If the stereo in your vehicle is equipped with an anti-theft system, make sure you have the correct activation code before disconnecting the battery.*
2   Disconnect the hydraulic line at the clutch master cylinder. If available, use a flare-nut wrench on the fitting, which will prevent the fitting from being rounded off. Have rags handy as some fluid will be lost as the line is removed. **Caution:** *Don't allow brake fluid to come into contact with paint, as it will damage the finish.*
3   Under the dashboard, disconnect the pushrod from the top of the clutch pedal. It's held in place with a clevis pin (see illustration).
4   From under the dash, remove the nuts which secure the master cylinder to the firewall. Remove the master cylinder, again being careful not to spill any of the fluid.

Installation

5   Position the master cylinder on the firewall, installing the mounting nuts finger-tight.
6   Connect the hydraulic line to the master cylinder, moving the cylinder slightly as necessary to thread the fitting properly into the bore. Don't cross-thread the fitting as it's installed.
7   Tighten the mounting nut(s) and the hydraulic line fitting securely.
8   Connect the pushrod to the clutch pedal.
9   Fill the clutch master cylinder reservoir with brake fluid conforming to DOT 3 specifications and bleed the clutch system (see Section 7).
10 Check the clutch pedal height and freeplay and adjust if necessary, following the procedure in Chapter 1.

6   Clutch release cylinder and accumulator - removal and installation

Clutch release cylinder

Removal

Refer to illustration 6.3

1   Disconnect the negative cable from the battery. **Caution:** *If the stereo in your vehicle is equipped with an anti-theft system, make sure you have the correct activation code before disconnecting the battery.*
2   Raise the vehicle and support it securely on jackstands.
3   To disconnect the hydraulic line from the release cylinder, unscrew the threaded fitting. If available, use a flare-nut wrench on the fitting, which will prevent the fitting from being rounded off (see illustration). Have a small can and rags handy, as some fluid will be spilled as the line is removed.
4   Remove the mounting bolts and separate the release cylinder from the transaxle.

Installation

5   Connect the hydraulic line to the release cylinder. Install the release cylinder on the clutch housing, making sure the pushrod is seated in the release fork pocket. Tighten the bolts to the torque listed in this Chapter's Specifications.
6   Tighten the hydraulic line threaded fitting securely.
7   Fill the clutch master cylinder with brake fluid (conforming to DOT 3 specifications).
8   Bleed the system (see Section 7).
9   Lower the vehicle and connect the negative battery cable.

Accumulator

Removal

Refer to illustration 6.15

10 Disconnect the negative cable from the battery. **Caution:** *If the stereo in your vehicle is equipped with an anti-theft system, make sure you have the correct activation code before disconnecting the battery.*
11 Raise the vehicle and support it securely on jackstands.
12 If equipped with cruise control, remove the cruise control actuator mounting bolts and position the assembly off to the side.
13 Remove the battery and the battery tray (see Chapter 5).
14 Remove the starter (see Chapter 5).
15 Unscrew the hydraulic line fittings from

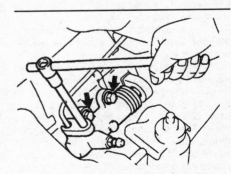

6.3 To remove the release cylinder, disconnect the hydraulic line threaded fitting with a flare-nut wrench and remove the two mounting bolts (arrows)

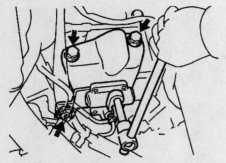

6.15 First remove the hydraulic line threaded fitting and then remove the accumulator mounting bracket bolts

the accumulator, then remove the mounting bolts from the accumulator bracket (see illustration).

Installation

16 Connect the hydraulic lines to the accumulator, but don't tighten them completely yet. Install the accumulator mounting bolts and tighten them to the torque listed in this Chapter's Specifications. Now tighten the hydraulic line fittings securely.
17 The remainder of installation is the reverse of removal.

7 Clutch hydraulic system - bleeding

1 The hydraulic system should be bled of all air whenever any part of the system has been removed or if the fluid level has been allowed to fall so low that air has been drawn into the master cylinder. The procedure is very similar to bleeding a brake system.
2 Fill the master cylinder with new brake fluid conforming to DOT 3 specifications. **Caution:** *Do not re-use any of the fluid coming from the system during the bleeding operation or use fluid which has been inside an open container for an extended period of time.*
3 Raise the vehicle and place it securely on jackstands to gain access to the release cylinder, which is located on the left side of the clutch housing.
4 Remove the dust cap which fits over the bleeder valve and push a length of plastic hose over the valve. Place the other end of the hose into a clear container with about two inches of brake fluid in it. The hose end must be submerged in the fluid.
5 Have an assistant depress the clutch pedal and hold it. Open the bleeder valve on the release cylinder, allowing fluid to flow through the hose. Close the bleeder valve when fluid stops flowing from the hose. Once closed, have your assistant release the pedal.
6 Continue this process until all air is evacuated from the system, indicated by a full, solid stream of fluid being ejected from the bleeder valve each time and no air bubbles in the hose or container. Keep a close watch on the fluid level inside the clutch master cylinder reservoir; if the level drops too low, air will be sucked back into the system and the process will have to be started all over again.
7 Install the dust cap and lower the vehicle. Check carefully for proper operation before placing the vehicle in normal service.

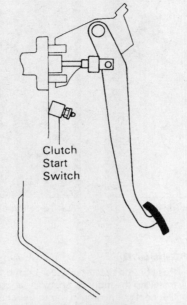

8.4 The clutch start switch is located on a bracket just ahead of the clutch pedal

8 Clutch start switch - check and adjustment

Refer to illustrations 8.4 and 8.5

1 Check the pedal height, pedal freeplay and pushrod play (see Chapter 1).
2 Verify that the engine will not start when the clutch pedal is released.
3 Verify that the engine will start when the clutch pedal is depressed all the way.
4 The clutch start switch is located on a bracket forward of the clutch pedal **(see illustration)**. Using a small flashlight, trace the switch leads from the switch to the electrical connector, unplug the connector and pull it down so that you can see the connector terminals.
5 Verify that there is continuity between the clutch start switch terminals when the switch is ON (pushed) **(see illustration)**.
6 Verify that no continuity exists between the switch terminals when the switch is OFF (released).
7 If the switch fails either of the tests, replace it. This is accomplished by removing the nut nearest the plunger end of the switch and unscrewing the switch. Disconnect the wire harness. Installation is the reverse of removal.
8 To adjust the clutch start switch,

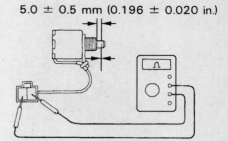

8.5 To check the clutch start switch, verify that there's continuity between the switch connector terminals when the plunger is depressed and no continuity when it's released; to adjust the switch, depress the clutch pedal and turn the switch in or out to achieve the indicated distance

depress the clutch pedal completely and turn the switch in or out to achieve the distance shown in **illustration 8.5**.
9 Verify again that the engine doesn't start when the clutch pedal is released.

9 Driveaxles - general information and inspection

1 Power is transmitted from the transaxle to the wheels through a pair of driveaxles. The inner end of each driveaxle is connected to the transaxle, directly splined to the differential side gears. The outer ends of the driveaxles are splined to the axle hubs and locked in place by a large nut. The left side driveaxle is shorter while the right side driveaxle is longer and equipped with an intermediate shaft that is supported in the middle by a bearing support.
2 The inner ends of the driveaxles are equipped with sliding constant velocity joints, which are capable of both angular and axial motion. Each inner joint assembly consists of either a tripod bearing and a joint tulip (housing) or a ball-and-cage type constant velocity joint in which the joint is free to slide in-and-out as the driveaxle moves up-and-down with the wheel. The joints can be disassembled and cleaned in the event of a boot failure, but if any parts are damaged, the joints must be replaced as a unit (see Section 11).
3 Each outer joint, which consists of ball bearings running between an inner race and an outer race (housing), is capable of angular but not axial movement.
4 The boots should be inspected periodically for damage and leaking lubricant. Torn CV joint boots must be replaced immediately or the joints can be damaged. Boot replacement involves removal of the driveaxle (see Section 10). **Note:** *Some auto parts stores carry "split" type replacement boots, which can be installed without removing the driveaxle from the vehicle. This is a conve-*

Chapter 8 Clutch and driveaxles

10.3 Remove the cotter pin and the nut lock

10.4 You'll need a large breaker bar to loosen the driveaxle/hub nut

10.8 Using a hammer and a brass punch, sharply strike the end of the driveaxle - it should move noticeably (don't push it in too far, though; only until it's loose)

10.11 Pull the steering knuckle out and slide the end of the driveaxle out of the hub. There is a sharp ring around the CV joint just behind the stub axle - wrap a rag around it so you don't cut your hand

nient alternative; however, the driveaxle should be removed and the CV joint disassembled and cleaned to ensure the joint is free from contaminants such as moisture and dirt which will accelerate CV joint wear. The most common symptom of worn or damaged CV joints, besides lubricant leaks, is a clicking noise in turns, a clunk when accelerating after coasting and vibration at highway speeds. To check for wear in the CV joints and driveaxle shafts, grasp each axle (one at a time) and rotate it in both directions while holding the CV joint housings, feeling for play indicating worn splines or sloppy CV joints. Also check the driveaxle shafts for cracks, dents and distortion.

10 Driveaxle - removal and installation

Removal

Refer to illustrations 10.3, 10.4, 10.8, 10.11, 10.12, 10.13a, 10.13b, 10.13c, 10.14 and 10.15

Note: *Not all of the steps in this procedure apply to all models. Read through the procedure carefully and determine which steps apply to the vehicle being worked on before actually beginning any work.*

1 Disconnect the cable from the negative terminal of the battery. **Caution:** *If the stereo in your vehicle is equipped with an anti-theft system, make sure you have the correct activation code before disconnecting the battery.*
2 Set the parking brake.
3 Remove the wheel cover or hub cap. Remove the cotter pin and the bearing nut lock from the driveaxle/hub nut **(see illustration)**.
4 Break loose the driveaxle/hub nut, but don't remove it yet **(see illustration)**.
5 Loosen the front wheel lug nuts, raise the vehicle and support it securely on jackstands. Remove the wheel.
6 Remove any engine splash shields that are in the way (see Chapter 11). Remove the driveaxle/hub nut.
6 **Note:** *Toyota recommends removing the entire right driveaxle assembly as a single unit before attempting to disassemble it because, although you could disassemble it, reattaching the outer driveaxle assembly to the intermediate shaft on the vehicle would be extremely difficult.*

7 Remove the nuts and bolt securing the balljoint to the control arm, then pry the control arm down to separate the components (see Chapter 10).
8 To loosen the driveaxle from the hub splines, tap the end of the driveaxle with a soft-faced hammer or a hammer and a brass punch **(see illustration)**. If the driveaxle is stuck in the hub splines and won't move, it may be necessary to remove the brake disc (see Chapter 9) and push it from the hub with a two-jaw puller.
9 Place a drain pan underneath the transaxle just in case lubricant leaks out.
10 If the transaxle has a case protector (the small plastic cover bolted to the transaxle) over the inner CV joint, remove it.
11 Pull out on the steering knuckle and detach the driveaxle from the hub **(see illustration)**.
12 On right driveaxle assemblies, the intermediate shaft and driveaxle assembly must be removed as a single unit.

10.13a To release the intermediate shaft bearing from the bearing support bracket, remove this lock bolt (arrow)

10.13b To remove the snap-ring from the bearing support bracket, pinch the ends together as shown and pull it out of its groove in the bracket (driveaxle assembly and bearing support bracket removed from the vehicle for clarity)

10.13c To detach the intermediate shaft from the differential side gear, grasp the shaft firmly and pull

10.14 Pry the splined end of the driveaxle from the transaxle using a screwdriver or crowbar

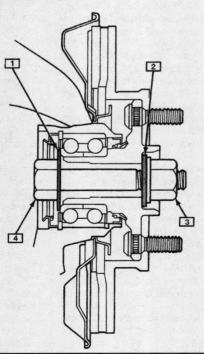

10.15 It isn't a good idea to move the vehicle with a driveaxle removed, but if you must, first install a bolt and a pair of washers through the hub and tighten them securely

1. 2-inch (O.D.) washer
2. 1-3/4 inch (O.D.) washer
3. 9/16-inch nut
4. 9/16-inch bolt

13 If you're removing the right driveaxle on any model, remove the center bearing lock bolt **(see illustration)**, remove the snap-ring **(see illustration)**, grasp the intermediate shaft and pull the splined inner end of the shaft out of the differential side gear **(see illustration)**.

14 If you're removing the left driveaxle, carefully pry the inner CV joint out of the transaxle **(see illustration)**.

15 Should it become necessary to move the vehicle while the driveaxle is out, place a large bolt with two large washers (one on each side of the hub) through the hub and tighten the nut securely **(see illustration)**.

16 Refer to Chapter 7 for the driveaxle seal replacement procedure.

Installation

17 Installation is the reverse of the removal procedure, but with the following additional points:

a) *When installing the left driveaxle or when installing the intermediate shaft on any*

Chapter 8 Clutch and driveaxles

11.3 Cut the old boot clamps off and discard them

11.4 Remove the boot from the inner CV joint and slide the tripod from the joint housing

11.6 Remove the snap-ring with a pair of snap-ring pliers

11.7 Drive the tripod joint from the driveaxle with a brass punch and hammer; be careful not to damage the bearing surfaces or the splines on the shaft

11.9a Clean the outer CV joint thoroughly with solvent and, working the joint through its entire range of motion, inspect the bearing surfaces of the balls; if they're worn or damaged, so are the bearing races

model, push the driveaxle sharply inward to seat the retaining ring on the inner CV joint in the groove in the differential side gear.
b) When installing the right driveaxle/intermediate shaft assembly, be sure to tighten the center bearing lock bolt to the torque listed in this Chapter's Specifications.
c) Install the wheel and lug nuts, lower the vehicle and tighten the lug nuts to the torque listed in the Chapter 1 Specifications.
d) Tighten the driveaxle/hub nut to the torque listed in this Chapter's Specifications, then install the nut lock and a new cotter pin.
e) Check the transaxle lubricant (manual transaxle) or differential lubricant (automatic transaxle) and add, if necessary, to bring it to the proper level (see Chapter 1).

18 Check the intermediate shaft bearing for smooth operation. If it feels rough or sticky it should be replaced. Take it to a dealer service department or other repair shop, as special tools are needed to perform this job.

11 Driveaxle boot replacement

Note: *Complete rebuilt driveaxles are available on an exchange basis, which eliminates much time and work. Check on the cost and availability of parts before disassembling the vehicle.*

1 Remove the driveaxle (see Section 10).
2 Mount the driveaxle in a vise with wood lined jaws (to prevent damage to the axleshaft). Check the CV joint for excessive play in the radial direction, which indicates worn parts. Check for smooth operation throughout the full range of motion for each CV joint. If a boot is torn, the recommended procedure is to disassemble the joint, clean the components and inspect for damage due to loss of lubrication and possible contamination by foreign matter.

Disassembly

Refer to illustrations 11.3, 11.4, 11.6 and 11.7
3 Using diagonal cutters, cut the boot clamps **(see illustration)**, remove the clamps and discard them.

4 Using a screwdriver, carefully pry up on the edge of the outer boot and push it away from the CV joint. Old and worn boots can be cut off. Pull the inner CV joint boot back from the housing and slide the housing from the tripod **(see illustration)**. **Note:** *Right side driveaxles are equipped with an intermediate shaft attached to the inner driveaxle housing.*
5 Mark the tripod and axleshaft to ensure that they are reassembled properly.
6 Remove the tripod joint snap-ring with a pair of snap-ring pliers **(see illustration)**.
7 Use a hammer and a brass punch to drive the tripod joint from the driveaxle **(see illustration)**. **Note:** *The tripod joint must be removed from the driveaxle to be able to slide the inner and outer driveaxle boots over the driveaxle. Do not remove the outer CV joint from the driveaxle.*
8 If you haven't already cut them off, remove both boots.

Check

Refer to illustrations 11.9a and 11.9b
9 Thoroughly clean all components, including the outer CV joint assembly, with sol-

11.9b Check the condition of the center support bearing on the intermediate shaft. Make sure it turns freely, quietly and smoothly; if the bearing is hard to turn, is noisy or feels rough, have it replaced by an automotive machine shop (be sure to have a pair of new dust covers installed too)

11.10a Wrap the splined area of the axleshaft with tape to prevent damage to the boots when removing or installing them

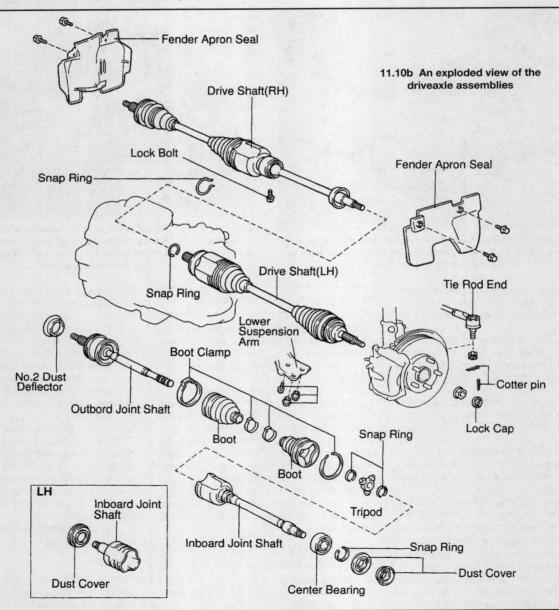

11.10b An exploded view of the driveaxle assemblies

11.10c Install the tripod with the recessed portion of the splines facing the axleshaft, then install a new snap-ring

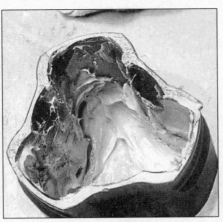

11.10d Place grease at the bottom of the CV joint housing

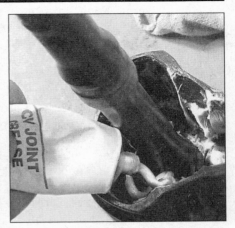

11.10e Install the boot and clamps onto the axleshaft, then insert the tripod into the housing, followed by the rest of the grease

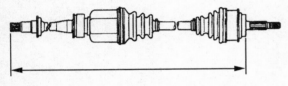

11.12a Measure between the points indicated and move the inner CV joint in or out to set the driveaxle to the length listed in this Chapter's Specifications (right side driveaxle shown)

vent until the old CV joint grease is completely removed. Inspect the bearing surfaces of the inner tripods and housings for cracks, pitting, scoring and other signs of wear. It's not possible to inspect the bearing surfaces of the inner and outer races of the outer CV joint, but you can at least check the surfaces of the ball bearings themselves **(see illustration)**. If they're in good shape, the races probably are, too; if they're not, neither are the races. If the inner CV joint is worn, you can buy a new inner CV joint and install it on the old axleshaft; if the outer CV joint is worn, you'll have to purchase a new outer CV joint *and* axleshaft (they're sold preassembled). Check the condition of the bearing for the intermediate shaft **(see illustration)**. It should turn freely and smoothly. If it's difficult to turn, or makes a grinding noise when rotated, take the intermediate shaft to an automotive machine shop and have a new bearing installed on the shaft.

Reassembly

Refer to illustrations 11.10a, 11.10b, 11.10c, 11.10d, 11.10e, 11.12a, 11.12b and 11.12c

10 Wrap the splines on the inner end the axleshaft with electrical or duct tape to protect the boots from the sharp edges of the splines **(see illustration)**. Slide the clamps and boots onto the axleshaft, outer boot first, then place the tripod on the shaft and install a new snap-ring. Apply grease to the tripod assembly and inside the housing. Insert the tripod into the housing and pack the remainder of the grease around the tripod **(see illustrations)**. If you're repacking the outer joint, be sure to work the entire tube of CV joint grease (included with the boot kit) into the bearing assembly.

11 Slide the boots into place, making sure the ends of both boots seat in their respective grooves in the axleshaft.

12 Adjust the driveaxle to the standard length listed in this Chapter's Specifications, then equalize the pressure in the boot and tighten the boot clamps **(see illustrations)**. The driveaxle is now ready for installation (see Section 10).

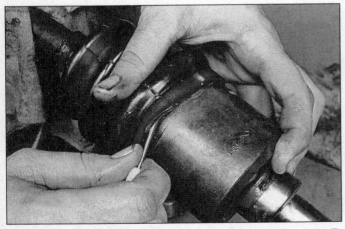

11.12b Equalize the pressure inside the boot by inserting a small, dull screwdriver between the boot and the outer race

11.12c You'll need a special boot clamp installation tool like this one to tighten the new clamps; follow the instructions provided by the tool manufacturer

Notes

Chapter 9 Brakes

Contents

	Section
Anti-lock Brake System (ABS) - general information	2
Brake check	See Chapter 1
Brake disc - inspection, removal and installation	5
Brake fluid level check	See Chapter 1
Brake hoses and lines - inspection and replacement	9
Brake hydraulic system - bleeding	10
Brake light switch - removal, installation and adjustment	15
Brake pedal - adjustment	See Chapter 1
Disc brake caliper - removal and installation	4
Disc brake pads - replacement	3
Drum brake shoes - replacement	6
General information	1
Master cylinder - removal and installation	8
Parking brake - adjustment	13
Parking brake cables - replacement	14
Parking brake shoes (rear disc brakes only) - inspection and replacement	12
Power brake booster - check, removal and installation	11
Wheel cylinder - removal and installation	7

Specifications

General

Brake fluid type	See Chapter 1
Brake pedal specifications	See Chapter 1
Power brake booster pushrod-to-master cylinder piston clearance	0.0 inch
Brake light switch plunger (dimension A)	1/32 to 3/32 inch

Disc brakes

Minimum brake pad thickness	See Chapter 1
Front disc thickness	
Standard	1.102 inches
Minimum*	1.024 inches
Rear disc thickness (models with rear disc brakes)	
Camry and Camry Solara	
Standard	0.394 inch
Minimum*	0.354 inch
Avalon	
Standard	0.354 inch
Minimum*	0.315 inch
Lexus ES 300	
Standard	0.394 inch
Minimum*	0.335 inch
Disc runout limit	
Front	0.0020 inch
Rear	0.0059 inch
Parking brake shoe minimum thickness	1/32 inch

Drum brakes

Brake shoe minimum lining thickness	1/16 inch
Drum inside diameter	
Standard	9.000 inches
Maximum*	9.079 inches

*****Note:*** *If different specifications are cast into the disc or drum, they supersede information printed here.*

Chapter 9 Brakes

Torque specifications — **Ft-lbs** (unless otherwise indicated)

Caliper mounting bolts	
Front caliper	25
Rear caliper	168 in-lbs
Caliper torque plate bolts	
Front torque plate	79
Rear torque plate	34
Brake hose-to-caliper banjo bolt	22
Wheel cylinder mounting bolts	84 in-lbs
Master cylinder-to-brake booster nuts	108 in-lbs
Power brake booster mounting nuts	108 in-lbs
Wheel lug nuts	See Chapter 1

1 General information

The vehicles covered by this manual are equipped with hydraulically operated front and rear brake systems. The front brakes are disc type and the rear brakes are either drum or disc type. Both the front and rear brakes are self adjusting. The disc brakes automatically compensate for pad wear, while the drum brakes incorporate an adjustment mechanism that is activated as the parking brake is applied.

Hydraulic system

The hydraulic system consists of two separate circuits. The master cylinder has separate reservoirs for the two circuits, and, in the event of a leak or failure in one hydraulic circuit, the other circuit will remain operative. A dual proportioning valve on the firewall provides brake balance between the front and rear brakes.

Power brake booster

The power brake booster, utilizing engine manifold vacuum and atmospheric pressure to provide assistance to the hydraulically operated brakes, is mounted on the firewall in the engine compartment.

Parking brake

The parking brake operates the rear brakes only, through cable actuation. It's activated by a lever mounted in the center console (Camry and Solara models) or a pedal mounted on the left side kick panel (Avalon and Lexus ES 300 models).

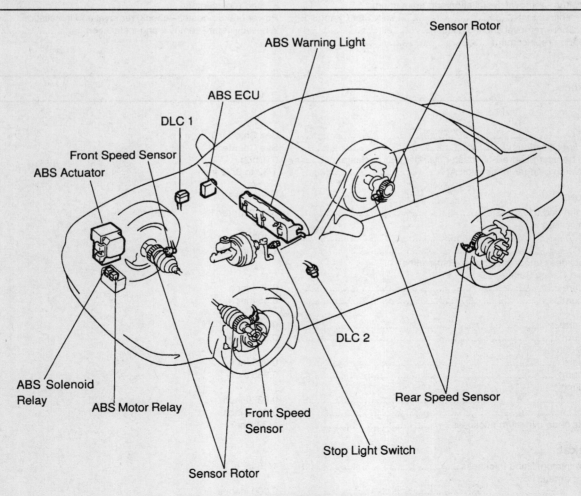

2.1 Anti-lock Brake System (ABS) component location

Chapter 9 Brakes

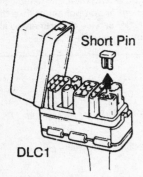

2.10a Remove the short pin from the data link connector (Nippondenso system)

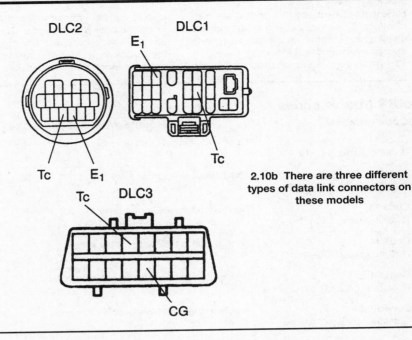

2.10b There are three different types of data link connectors on these models

Service

After completing any operation involving disassembly of any part of the brake system, always test drive the vehicle to check for proper braking performance before resuming normal driving. When testing the brakes, perform the tests on a clean, dry, flat surface. Conditions other than these can lead to inaccurate test results.

Test the brakes at various speeds with both light and heavy pedal pressure. The vehicle should stop evenly without pulling to one side or the other. Avoid locking the brakes, because this slides the tires and diminishes braking efficiency and control of the vehicle.

Tires, vehicle load and wheel alignment are factors which also affect braking performance.

2 Anti-lock Brake System (ABS) - general information

Refer to illustration 2.1

1 The Anti-lock Brake System (ABS) **(see illustration)**, is designed to maintain vehicle steerabilty, directional stability and optimum deceleration under severe braking conditions and on most road surfaces. It does so by monitoring the rotational speed of each wheel and controlling the brake line pressure to each wheel during braking. This prevents the wheel from locking up.

Components

Actuator assembly

2 The actuator assembly consists of the master cylinder, an electric hydraulic pump and four solenoid valves. **Note:** *There are two different manufacturers for the ABS system components; Nippondenso and Bosch. The basic design and component locations are similar, but the Bosch ABS actuator contains the ECU (ABS computer) and solenoid relays, while the Nippondenso ABS system separates these particular components.*

a) *The electric pump provides hydraulic pressure to charge the reservoirs in the actuator, which supplies pressure to the braking system. The pump and reservoirs are housed in the actuator assembly.*
b) *The solenoid valves modulate brake line pressure during ABS operation. The body contains four valves - one for each wheel.*

Speed sensors

3 These sensors are located at each wheel and generate small electrical pulsations when the toothed sensor rings are turning, sending a signal to the electronic controller indicating wheel rotational speed.
4 The front speed sensors are mounted to the front steering knuckle in close relationship to the toothed sensor rings, which are integral with the front driveaxle outer CV joints.
5 The rear wheel sensors are bolted to the axle carriers. The sensor rings are integral with the rear hub assemblies.

ABS computer (ECU)

6 The ABS computer is mounted under the dashboard (Nippondenso systems) or within the ABS actuator (Bosch systems) and is the "brain" for the ABS system. The function of the computer is to accept and process information received from the wheel speed sensors to control the hydraulic line pressure, avoiding wheel lock up. The computer also constantly monitors the system, even under normal driving conditions, to find faults within the system.
7 If a problem develops within the system, an "ABS" light will glow on the dashboard. A diagnostic code will also be stored in the computer and, when retrieved, will indicate the problem area or component.

Diagnostic codes

Refer to illustrations 2.10a and 2.10b

8 The ABS system control unit (computer) has a built-in self-diagnosis system which detects malfunctions in the system sensors and alerts the driver by illuminating an ABS warning light in the instrument panel. The computer stores the failure code until the diagnostic system is cleared or malfunction is repaired.
9 The ABS warning light should come on when the ignition switch is placed in the ON position. When the engine is started, the warning light should go out. If the light remains on, the diagnostic system has detected a malfunction or abnormality in the system.
10 The codes for the ABS can be accessed by turning the ignition key to the OFF position (engine not running). On Nippondenso systems, remove the short pin from the data link connector **(see illustration)**. On Nippondenso and Bosch systems, install a jumper wire or paper clip onto terminals E1 and Tc of the DLC1 or DLC2 or Tc and CG of the DLC3 **(see illustration)** and turn the ignition key ON (engine not running). Observe the codes on the ABS warning light.
11 The diagnostic code is the number of flashes indicated on the ABS light. If any malfunction has been detected, the light will blink the first digit(s) of the code, pause, 1.5 seconds, then blink the second digit of the code. For example, a code 34 (left rear wheel sensor) will first blink three flashes, pause, 1.5 seconds and blink four flashes. If there is more than one code stored in the ECM, the ECM will, pause, 2.5 seconds before flashing the next code. If the system is operating normally (no malfunctions), the warning light will blink once every 0.5 seconds.
12 The accompanying tables explain the

code that will be flashed for each of the malfunctions. The accompanying charts indicate the diagnostic code - in blinks - along with the system, diagnosis and specific areas.

13 After the diagnosis check, clear the trouble codes. First, jump terminals E1 and Tc on the data link connector. Turn the ignition key ON (engine not running) and clear the codes by depressing the brake pedal eight or more times within five seconds. Remove the jumper wire and reinstall the cap on the data link connector. Check the indicated system or component or take the vehicle to a dealer service department to have the malfunction repaired.

ABS trouble codes

Code number	Trouble area	Action to take
Code 11 (1 flash, pause, 1 flash)	Open circuit in solenoid relay circuit	Check the solenoid relay and the relay circuit
Code 12 (1 flash, pause, 2 flashes)	Short circuit in solenoid relay circuit	Check the solenoid relay and the relay circuit
Code 13 (1 flash, pause, 3 flashes)	Open circuit in ABS motor relay circuit	Check the pump motor relay and circuit
Code 14 (1 flash, pause, 4 flashes)	Short circuit in ABS motor relay circuit	Check the solenoid relay and the relay circuit
Code 21 (2 flashes, pause, 1 flash)	Problem in right front wheel solenoid circuit	Check the actuator solenoid and circuit
Code 22 (2 flashes, pause, 2 flashes)	Problem in left front wheel solenoid circuit	Check the actuator solenoid and circuit
Code 23 (2 flashes, pause, 3 flashes)	Problem in right rear wheel solenoid circuit	Check the actuator solenoid and circuit
Code 24 (2 flashes, pause, 4 flashes)	Problem in left rear wheel solenoid circuit	Check the actuator solenoid and circuit
Code 25 (2 flashes, pause, 5 flashes)	SMC1 circuit open or shorted	Check the ABS actuator and circuit
Code 26 (2 flashes, pause, 6 flashes)	SMC2 circuit open or shorted	Check the ABS actuator and circuit
Code 27 (2 flashes, pause, 7 flashes)	SRC1 circuit open or shorted	Check the ABS actuator and circuit
Code 28 (2 flashes, pause, 8 flashes)	SRC2 circuit open or shorted	Check the ABS actuator and circuit
Code 31 (3 flashes, pause, 1 flash)	Sensor signal problem - right front wheel	Check the speed sensor, sensor rotors, wire harness and connector of the speed sensor
Code 32 (3 flashes, pause, 2 flashes)	Sensor signal problem - left front wheel	Check the speed sensor, sensor rotors, wire harness and connector of the speed sensor
Code 33 (3 flashes, pause, 3 flashes)	Sensor signal problem - right rear wheel	Check the speed sensor, sensor rotors, wire harness and connector of the speed sensor
Code 34 (3 flashes, pause, 4 flashes)	Sensor signal problem - left rear wheel	Check the speed sensor, sensor rotors, wire harness and connector of the speed sensor
Code 35 (3 flashes, pause, 5 flashes)	Open circuit - right front speed sensor or circuit	Check the speed sensor, wire harness and electrical connector
Code 36 (3 flashes, pause, 6 flashes)	Open circuit - left front speed sensor or circuit	Check the speed sensor, wire harness and electrical connector
Code 37 (3 flashes, pause, 7 flashes)	Speed sensor rotor has incorrect number of teeth	Check for a damaged sensor rotor
Code 38 (3 flashes, pause, 8 flashes)	Open circuit - right rear speed sensor or circuit	Check the speed sensor, wire harness and electrical connector
Code 39 (3 flashes, pause, 9 flashes)	Open circuit - left rear speed sensor or circuit	Check the speed sensor, wire harness and electrical connector
Code 41 (4 flashes, pause, 1 flash)	Abnormally low battery positive voltage	Check the charging system (alternator, battery and voltage regulator) for any problems (see Chapter 5)

Chapter 9 Brakes

Code 43 (4 flashes, pause, 3 flashes)	ABS control system malfunction	Check all wiring and connections associated with the ABS system
Code 44 (4 flashes, pause, 4 flashes)	NE signal circuit open or shorted	Check wiring harness and connectors between ABS ECU and PCM
Code 45 (4 flashes, pause, 5 flashes)	Deceleration sensor malfunction	Check deceleration sensor and wiring harness connectors
Code 46 (4 flashes, pause, 6 flashes)	Master cylinder sensor malfunction	Check master cylinder sensor and wiring harness connectors
Code 49 (4 flashes, pause, 9 flashes)	Open circuit - brake light switch or circuit	Check the brake light switch or circuit
Code 51 (5 flashes, pause, 1 flash)	Pump motor locked	Check the pump motor, relay and battery for shorts or abnormalities
Code 53 (5 flashes, pause, 3 flashes)	PCM communication circuit malfunction	Check wiring harness and connectors between ABS ECU and PCM
Code 58 (5 flashes, pause, 8 flashes)	Open circuit - brake light switch or circuit	Check the brake light switch or circuit
Code 61 (6 flashes, pause, 1 flash)	Engine control system malfunction	Check for engine control system trouble codes (see Chapter 6)
Code 62 (6 flashes, pause, 2 flashes)	ECU malfunction	ECU problem
Light always ON	ECU malfunction	ECU problem

3 Disc brake pads - replacement

Refer to illustrations 3.5, 3.6a through 3.6t and 3.7a through 3.7l

Warning: *Disc brake pads must be replaced on both front or rear wheels at the same time - never replace the pads on only one wheel. Also, the dust created by the brake system may contain asbestos, which is harmful to your health. Never blow it out with compressed air and don't inhale any of it. An approved filtering mask should be worn when working on the brakes. Do not, under any circumstances, use petroleum-based solvents to clean brake parts. Use brake system cleaner only!*

Note: *This procedure applies to both the front and rear disc brakes.*

1 Remove the cap from the brake fluid reservoir.
2 Loosen the wheel lug nuts, raise the front or rear of the vehicle and support it securely on jackstands. Block the wheels at the opposite end.
3 Remove the wheels. Work on one brake assembly at a time, using the assembled brake for reference if necessary.
4 Inspect the brake disc carefully as outlined in Section 5. If machining is necessary, follow the information in that Section to remove the disc, at which time the pads can be removed as well.
5 Push the piston back into its bore to provide room for the new brake pads. A C-clamp can be used to accomplish this **(see illustration)**. As the piston is depressed to the bottom of the caliper bore, the fluid in the master cylinder will rise. Make sure that it doesn't overflow. If necessary, siphon off some of the fluid.
6 If you're replacing the front brake pads, follow the accompanying photos, beginning with **illustration 3.6a**. Be sure to stay in order and read the caption under each illustration.

3.5 Before removing the caliper, be sure to depress the piston into its bore in the caliper with a large C-clamp to make room for the new pads

3.6a Always wash the brakes with brake cleaner before disassembling anything

Chapter 9 Brakes

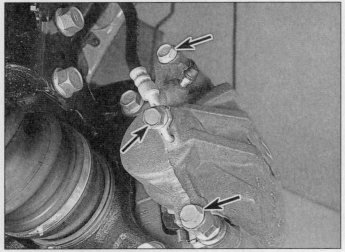

3.6b To remove the caliper, remove the bolts indicated by the upper and lower arrows (the center arrow points to the brake hose banjo bolt, which shouldn't be unscrewed unless the caliper is being completely removed)

3.6c Remove the caliper . . .

3.6d . . . and hang it from the strut coil spring with a piece of wire; do not allow the caliper to hang by the flexible brake hose

3.6e Remove the upper anti-squeal spring (four-cylinder models only) . . .

3.6f . . . and the lower anti-squeal spring (four-cylinder models only)

3.6g Remove the outer shim . . .

3.6h . . . and the inner shim from the outer brake pad

Chapter 9 Brakes

3.6i Remove the outer brake pad

3.6j Remove the outer shim . . .

3.6k . . . and the inner shim from the inner brake pad

3.6l Remove the inner brake pad

3.6m Remove the pad support plates; inspect them for damage and replace as necessary (the plates should "snap" into place in the torque plate; if they're weak or distorted, they should be replaced)

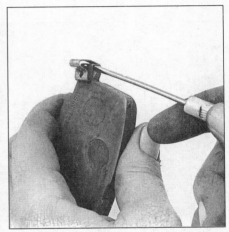
3.6n Pry the wear indicator off the old inner brake pad and transfer it to the new inner pad (if the wear indicator is worn or bent, replace it)

3.6o Install the pad support plates, the new inner brake pad and the shims; make sure the ears on the pad are properly engaged with the pad support plates as shown

3.6p Install the pad support plates, the outer pad and the shims

3.6q Install the upper and lower anti-squeal springs (four-cylinder models only); make sure both springs are properly engaged with the pads

Chapter 9 Brakes

3.6r Pull out the upper and lower sliding pins and clean them off (if either rubber boot is damaged, remove it by levering the flange of the metal bushing that retains the boot) . . .

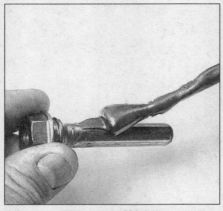

3.6s . . . apply a coat of high-temperature grease to the pins and install them

3.6t Install the caliper and tighten the caliper bolts to the torque listed in this Chapter's Specifications

3.7a Remove the caliper retaining bolt (lower arrow); the upper arrow points to the brake hose banjo fitting bolt, which shouldn't be unscrewed unless the caliper is being removed from the vehicle

3.7b Pivot the caliper up and support it in that position

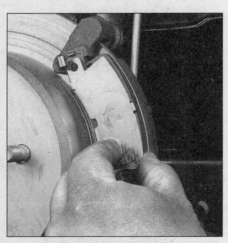

3.7c Remove the outer shim . . .

3.7d . . . and the inner shim from the outer brake pad

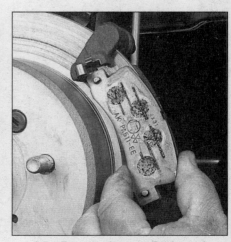

3.7e Remove the outer brake pad

Chapter 9 Brakes

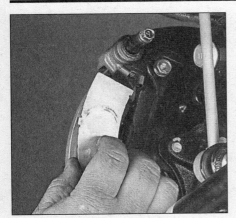

3.7f Remove the outer shim . . .

3.7g . . . and the inner shim from the inner brake pad

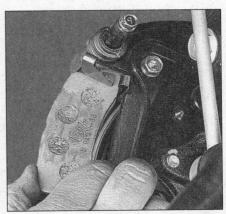

3.7h Remove the inner brake pad

7   If you're replacing the rear brake pads, wash the brake assembly **(see illustration 3.6a)**, then follow the accompanying photos beginning with **illustration 3.7a**. Be sure to stay in order and read the caption under each illustration.

8   When reinstalling the caliper, be sure to tighten the mounting bolts to the torque listed in this Chapter's Specifications. After the job has been completed, firmly depress the brake pedal a few times to bring the pads into contact with the disc. Check the level of the brake fluid, adding some if necessary. Check the operation of the brakes carefully before placing the vehicle into normal service.

4 Disc brake caliper - removal and installation

Warning: *Dust created by the brake system may contain asbestos, which is harmful to your health. Never blow it out with compressed air and don't inhale any of it. An approved filtering mask should be worn when working on the brakes. Do not, under any circumstances, use petroleum-based solvents to clean brake parts. Use brake system cleaner only!*

Note: *Always replace the calipers in pairs - never replace just one of them.*

Removal

Refer to illustrations 4.2, 4.3a, 4.3b and 4.3c

1   Loosen the wheel lug nuts, raise the vehicle and support it securely on jackstands. Remove the wheels.

3.7i Remove the pad support plates; inspect them for wear and replace as necessary (the plates should "snap" into place in the torque plate; if they're weak or distorted, replace them)

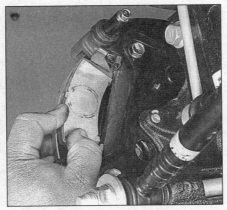

3.7j Install the pad support plates, the inner brake pad and the shims; make sure both shims are properly engaged with the inner brake pad and with each other

3.7k Install the pad support plates, the outer brake pad and the shims; make sure both shims are properly engaged with the outer brake pad and with each other

3.7l Pivot the caliper down over the new pads. Install the caliper retaining bolt and tighten it to the torque listed in this Chapter's Specifications

2 Remove the brake hose banjo bolt and disconnect the hose from the caliper. Plug the hose to keep contaminants out of the brake system and to prevent losing any more brake fluid than is necessary **(see illustration)**.
3 Remove the caliper mounting bolts **(see illustrations)**.
4 Remove the caliper. If necessary, remove the caliper torque plate from the steering knuckle or rear axle carrier **(see illustrations 5.2a and 5.2b)**.

Installation

5 Install the caliper by reversing the removal procedure. Remember to install new sealing washers on either side of the brake hose banjo fitting (they should be included with the rebuild kit). Tighten the caliper mounting bolts (and torque plate bolts, if removed) to the torque listed in this Chapter's Specifications.
6 Bleed the brake system (see Section 10).
7 Install the wheels and lug nuts. Lower the vehicle and tighten the lug nuts to the torque listed in the Chapter 1 Specifications.

4.2 Using a piece of rubber hose of the appropriate size, plug the brake line banjo fitting to prevent brake fluid from leaking out and to prevent dirt and moisture from contaminating the fluid in the hose

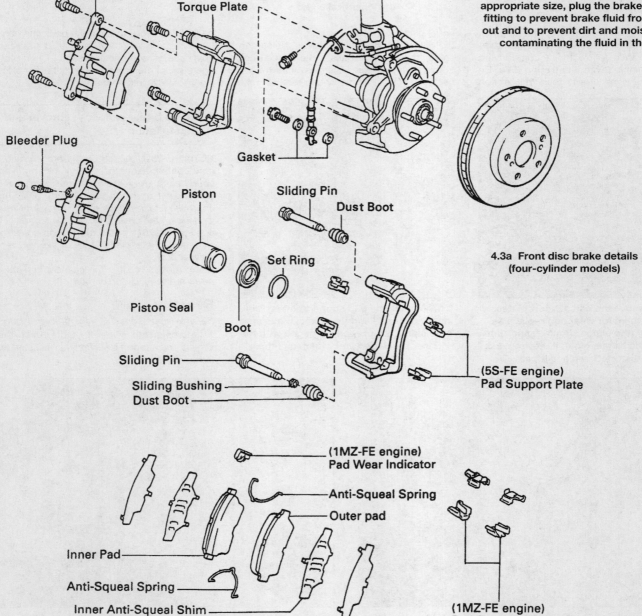

4.3a Front disc brake details (four-cylinder models)

Chapter 9 Brakes 9-11

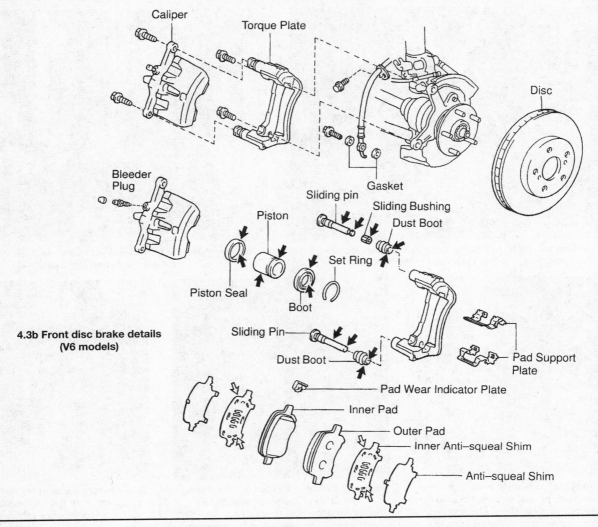

4.3b Front disc brake details (V6 models)

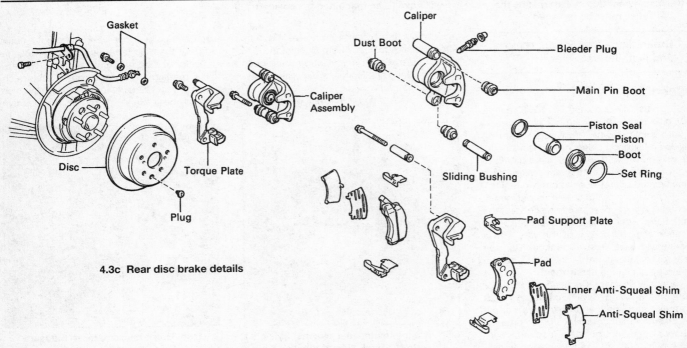

4.3c Rear disc brake details

5.2a To remove the front caliper torque plate from the steering knuckle, remove these two bolts (arrows)

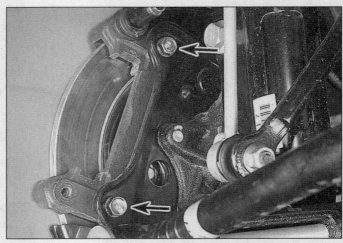

5.2b To remove the rear caliper torque plate from the axle carrier, remove these two bolts (arrows)

5.3 The brake pads on this vehicle were obviously neglected, as they wore down completely and cut deep grooves into the disc - wear this severe means the disc must be replaced

5.4a Use a dial indicator to check disc runout; if the reading exceeds the maximum allowable runout limit, the disc will have to be machined or replaced

5.4b Using a swirling motion, remove the glaze from the disc surface with sandpaper or emery cloth

5 Brake disc - inspection, removal and installation

Inspection

Refer to illustrations 5.2a, 5.2b, 5.3, 5.4a, 5.4b, 5.5a and 5.5b

1   Loosen the wheel lug nuts, raise the vehicle and support it securely on jackstands. Remove the wheel and install the lug nuts to hold the disc in place. If the rear brake disc is being worked on, release the parking brake.

2   Remove the brake caliper as outlined in Section 4. It isn't necessary to disconnect the brake hose. After removing the caliper bolts, suspend the caliper out of the way with a piece of wire **(see illustration 3.6d)**. Remove the torque plate mounting bolts and detach the torque plate **(see illustration)**.

3   Visually inspect the disc surface for score marks and other damage. Light scratches and shallow grooves are normal after use and may not always be detrimental to brake operation, but deep scoring - over 0.039-inch (1.0 mm) - requires disc removal and refinishing by an automotive machine shop. Be sure to check both sides of the disc **(see illustration)**. If pulsating has been noticed during application of the brakes, suspect disc runout.

4   To check disc runout, place a dial indicator at a point about 1/2-inch from the outer edge of the disc **(see illustration)**. Set the indicator to zero and turn the disc. The indicator reading should not exceed the specified allowable runout limit. If it does, the disc should be refinished by an automotive

5.5a The minimum wear dimension is cast into the back side of the disc

5.5b Use a micrometer to measure disc thickness

Chapter 9 Brakes

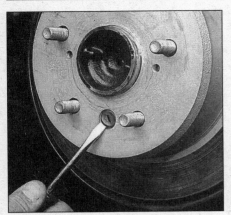

5.6a If the rear disc is difficult to remove, remove this plug . . .

5.6b . . . insert a screwdriver through the hole (the hole must be at the 6 o'clock position, because that's where the adjuster is located) . . .

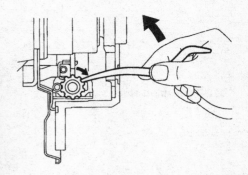

5.6c . . . and rotate the adjuster wheel in a clockwise direction (as seen from the front of the vehicle) to back off the parking brake shoes

machine shop. **Note:** *The discs should be resurfaced regardless of the dial indicator reading, as this will impart a smooth finish and ensure a perfectly flat surface, eliminating any brake pedal pulsation or other undesirable symptoms related to questionable discs. At the very least, if you elect not to have the discs resurfaced, remove the glaze from the surface with sandpaper or emery cloth using a swirling motion* **(see illustration)**.

5 It's absolutely critical that the disc not be machined to a thickness under the specified minimum allowable refinish thickness. The minimum wear (or discard) thickness is cast into the inside of the disc **(see illustration)**. The disc thickness can be checked with a micrometer **(see illustration)**.

Removal

Refer to illustrations 5.6a, 5.6b and 5.6c

6 Remove the lug nuts which were put on to hold the disc in place and remove the disc from the hub. If the rear disc is stuck to the hub and won't come off, remove the plug **(see illustration)** and rotate the adjuster to back the parking brake shoes away from the drum surface within the disc **(see illustrations)**.

Installation

7 Place the disc in position over the threaded studs.
8 Install the torque plate, tightening the bolts to the torque listed in this Chapter's Specifications.
9 Install the caliper, tightening the bolts to the torque listed in this Chapter's Specifications. Bleeding won't be necessary unless the brake hose was disconnected from the caliper.
10 Install the wheel and lug nuts. Lower the vehicle and tighten the lug nuts to the torque listed in the Chapter 1 Specifications. Depress the brake pedal a few times to bring the brake pads into contact with the disc. Check the operation of the brakes carefully before driving the vehicle.

6 Drum brake shoes - replacement

Refer to illustrations 6.4a through 6.4x and 6.5

Warning: *Drum brake shoes must be replaced on both wheels at the same time - never replace the shoes on only one wheel. Also, the dust created by the brake system may contain asbestos, which is harmful to your health. Never blow it out with compressed air and don't inhale any of it. An approved filtering mask should be worn when working on the brakes. Do not, under any circumstances, use petroleum-based solvents to clean brake parts. Use brake system cleaner only!*

Caution: *Whenever the brake shoes are replaced, the return and hold-down springs should also be replaced. Due to the continuous heating/cooling cycle the springs are subjected to, they lose tension over a period of time and may allow the shoes to drag on the drum and wear at a much faster rate than normal.*

1 Loosen the wheel lug nuts, raise the rear of the vehicle and support it securely on jack-

6.4a Mark the relationship of the drum to the hub to insure that the dynamic balance is unaltered

stands. Block the front wheels to keep the vehicle from rolling.
2 Release the parking brake.
3 Remove the wheel. **Note:** *All four rear brake shoes must be replaced at the same time, but to avoid mixing up parts, work on only one brake assembly at a time.*
4 Follow the accompanying illustrations for the brake shoe replacement procedure **(see illustrations 6.4a through 6.4x)**. Be sure to stay in order and read the caption under each illustration. **Note:** *If the brake drum cannot be easily removed, make sure the parking brake is completely released. If the drum still cannot be pulled off, the brake shoes will have to be retracted. This is done by first removing the plug from the backing plate. With the plug removed, push the lever off the adjuster star wheel with a narrow screwdriver while turning the adjuster wheel with another screwdriver, moving the shoes away from the drum* **(see illustration 6.4b)**. *The drum should now come off.*

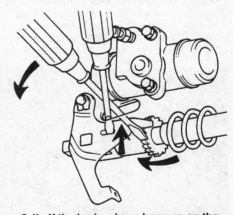

6.4b If the brake shoes hang up on the drum because of excessive wear, insert two screwdrivers through the hole in the backing plate to push the adjuster lever off the star wheel and turn the star wheel to retract the brake shoes

Chapter 9 Brakes

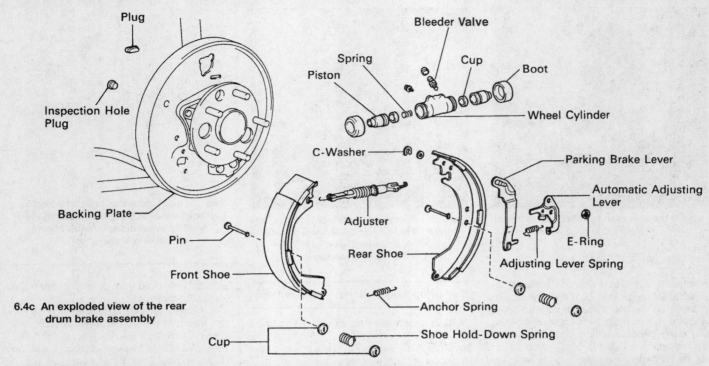

6.4c An exploded view of the rear drum brake assembly

6.4d Before removing anything, clean the brake assembly with brake cleaner and allow it to dry (position a drain pan under the brake to catch the fluid and residue); **DO NOT USE COMPRESSED AIR TO BLOW OFF BRAKE DUST!**

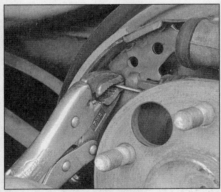

6.4e Unhook the return spring from the front brake shoe. Use a pair of locking pliers to stretch the spring and pull the end out of the hole in the shoe

6.4f Depress the hold-down spring and turn the retainer 90-degrees, then release it; a pair of pliers will work, but a special hold-down spring removal tool makes the job easier (these inexpensive tools are available at most auto parts stores)

6.4g Remove the front shoe from the backing plate and unhook the anchor spring from the end of the shoe

6.4h Remove the hold-down spring from the rear shoe

6.4i Remove the rear shoe and adjuster assembly from the backing plate

Chapter 9 Brakes

6.4j Hold the end of the parking brake cable with a pair of pliers and pull it out of the parking brake lever

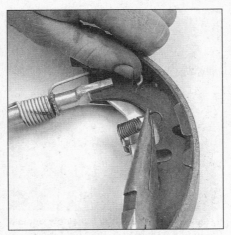

6.4k Remove the adjusting lever spring

6.4l Unhook the return spring from the shoe and slide the adjuster and spring off

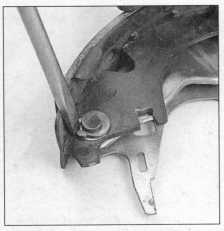

6.4m Pry the C-washer apart and remove it to separate the parking brake lever and adjuster lever from the rear shoe

6.4n Assemble the parking brake lever and adjuster lever to the new rear shoe and crimp the C-washer closed with a pair of pliers (always use a new C-washer)

6.4o Lubricate the moving parts of the adjuster screw with a light coat of high-temperature grease; the screw portion of the adjuster will need to be threaded in further than before to allow the drum to fit over the new shoes

6.4p Install the adjuster assembly on the rear shoe (make sure the end fits properly into the slot in the shoe and hook the spring into the opening in the shoe)

6.4q Install the adjuster lever spring

6.4r Lubricate the brake shoe contact area with high-temperature grease

Chapter 9 Brakes

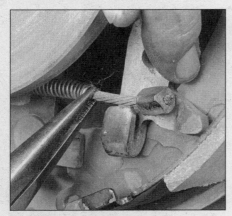

6.4s Pull the parking brake cable spring back and hold it there with a pair of pliers, then place the cable into the hooked end of the parking brake lever

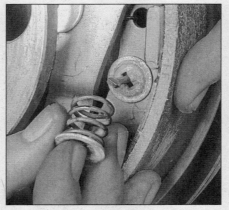

6.4t Place the rear shoe assembly against the backing plate and push the hold-down pin through the shoe. Install the cups (one on each end of the spring) and lock the outer cup to the pin by turning it 90-degrees after the spring has been compressed

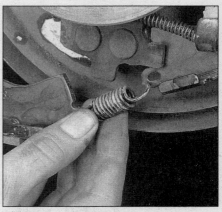

6.4u Connect the anchor spring to the bottom of each shoe and mount the front shoe to the backing plate, then install the hold-down spring and cups

5 Before reinstalling the drum, it should be checked for cracks, score marks, deep scratches and hard spots, which will appear as small discolored areas. If the hard spots cannot be removed with fine emery cloth or if any of the other conditions listed above exist, the drum must be taken to an automotive machine shop to have it resurfaced. **Note:** *Professionals recommend resurfacing the drums each time a brake job is done. Resurfacing will eliminate the possibility of out-of-round drums. If the drums are worn so much that they can't be resurfaced without exceeding the maximum allowable diameter (stamped into the drum), then new ones will be required* **(see illustration)**. *At the very least, if you elect not to have the drums resurfaced, remove the glaze from the surface with emery cloth using a swirling motion.*

6 Install the brake drum on the axle flange.
7 Install the wheel and lug nuts, then lower the vehicle. Tighten the lug nuts to the torque listed in this Chapter's Specifications.
8 Make a number of forward and reverse stops and operate the parking brake to adjust the brakes until satisfactory pedal action is obtained.

9 Check the operation of the brakes carefully before driving the vehicle.

7 Wheel cylinder - removal and installation

Note: *Never replace only one wheel cylinder - always replace both of them at the same time.*

Removal

Refer to illustration 7.4

1 Raise the rear of the vehicle and support it securely on jackstands. Block the front wheels to keep the vehicle from rolling.
2 Remove the brake shoe assembly (see Section 6).
3 Remove all dirt and foreign material from around the wheel cylinder.
4 Disconnect the brake line with a flare-nut wrench, if available **(see illustration)**. Don't pull the brake line away from the wheel cylinder.

6.4v Using a screwdriver, stretch the return spring into its hole in the front shoe

5 Remove the wheel cylinder mounting bolts.
6 Detach the wheel cylinder from the brake backing plate and place it on a clean workbench. Immediately plug the brake line to prevent fluid loss and contamination.

6.4w Pry the parking brake lever forward and verify that the return spring hasn't come unhooked from the rear shoe

6.4x Wiggle the assembly to make sure it's seated properly against the backing plate

6.5 The maximum drum diameter is cast into the inside of the rear drums

Chapter 9 Brakes

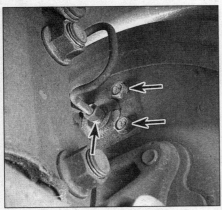

7.4 To remove the wheel cylinder, unscrew the hydraulic line-to-wheel cylinder threaded fitting, then remove the mounting bolts (arrows)

8.1 Unplug the electrical connector for the brake fluid level warning switch

8.4 Loosen the threaded fittings on the brake fluid hydraulic lines

Installation

7 Place the wheel cylinder in position and install the bolts finger tight. Connect the brake line to the cylinder, being careful not to cross-thread the fitting. Tighten the wheel cylinder bolts to the torque listed in this Chapter's Specifications.

8 Tighten the brake line fitting and install the brake shoes and drum.

9 Bleed the brakes (see Section 10). Install the wheel and lug nuts. Lower the vehicle to the ground and tighten the lug nuts to the torque listed in this Chapter's Specifications.

10 Check the operation of the brakes carefully before driving the vehicle.

8 Master cylinder - removal and installation

Removal

Refer to illustrations 8.1, 8.4 and 8.6

1 Unplug the electrical connector for the brake fluid level warning switch **(see illustration)**.

2 Remove as much fluid as possible from the reservoir with a syringe.

3 Place rags under the fittings and prepare caps or plastic bags to cover the ends of the lines once they're disconnected. **Caution:** *Brake fluid will damage paint. Cover all body parts and be careful not to spill fluid during this procedure.*

4 Loosen the fittings at the ends of the brake lines where they enter the master cylinder **(see illustration)**. To prevent rounding off the flats, use a flare-nut wrench, which wraps around the fitting hex.

5 Pull the brake lines away from the master cylinder and plug the ends to prevent contamination.

6 Remove the nuts attaching the master cylinder to the power booster **(see illustration)**. Pull the master cylinder off the studs to remove it. Again, be careful not to spill the fluid as this is done.

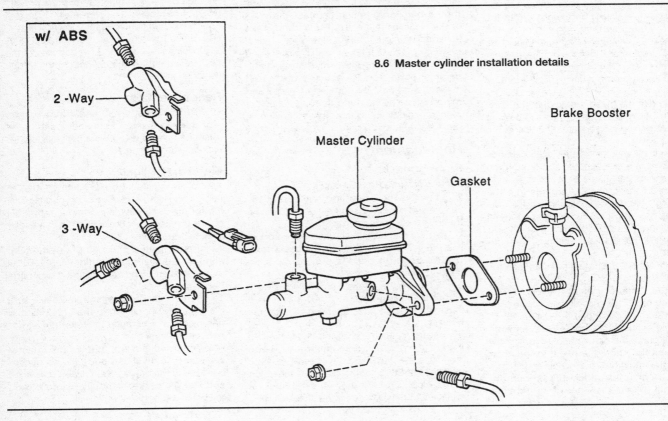

8.6 Master cylinder installation details

8.8 The best way to bleed air from the master cylinder before installing it on the vehicle is with a pair of bleed tubes

9.3 Unscrew the brake line threaded fitting with a flare-nut wrench to protect the fitting corners from being rounded off

9.4 Remove the brake hose-to-brake line U-clip with a pair of pliers

Installation

Refer to illustration 8.8

7   Bench bleed the new master cylinder before installing it. Because it will be necessary to apply pressure to the master cylinder piston and, at the same time, control flow from the brake line outlets, it is recommended that the master cylinder be mounted in a vise, with the jaws of the vise clamping on the mounting flange.

8   Attach a pair of master cylinder bleeder tubes to the outlet ports of the master cylinder **(see illustration)**.

9   Fill the reservoir with brake fluid of the recommended type (see Chapter 1).

10   Slowly push the pistons into the master cylinder (a large Phillips screwdriver can be used for this) - air will be expelled from the pressure chambers and into the reservoir. Because the tubes are submerged in fluid, air can't be drawn back into the master cylinder when you release the pistons.

11   Repeat the procedure until no more air bubbles are present.

12   Remove the bleed tubes, one at a time, and install plugs in the open ports to prevent fluid leakage and air from entering. Install the reservoir cover.

13   Install the master cylinder over the studs on the power brake booster and tighten the attaching nuts only finger tight at this time.

14   Thread the brake line fittings into the master cylinder. Since the master cylinder is still a bit loose, it can be moved slightly in order for the fittings to thread in easily. Do not strip the threads as the fittings are tightened.

15   Fully tighten the mounting nuts, then the brake line fittings.

16   Fill the master cylinder reservoir with fluid, then bleed the master cylinder and the brake system as described in Section 10. To bleed the cylinder on the vehicle, have an assistant pump the brake pedal several times slowly and then hold the pedal to the floor. Loosen the fitting nut to allow air and fluid to escape. Repeat this procedure on both fittings until the fluid is clear of air bubbles.

Caution: *Have plenty of rags on hand to catch the fluid - brake fluid will ruin painted surfaces.*

17   Test the operation of the brake system carefully before placing the vehicle into normal service. **Warning:** *Do not operate the vehicle if you are in doubt about the effectiveness of the brake system.*

9   Brake hoses and lines - inspection and replacement

Inspection

1   About every six months, with the vehicle raised and supported securely on jackstands, the rubber hoses which connect the steel brake lines with the front and rear brake assemblies should be inspected for cracks, chafing of the outer cover, leaks, blisters and other damage. These are important and vulnerable parts of the brake system and inspection should be complete. A light and mirror will be helpful for a thorough check. If a hose exhibits any of the above conditions, replace it with a new one.

Replacement

Front brake hose

Refer to illustrations 9.3 and 9.4

2   Loosen the wheel lug nuts, raise the vehicle and support it securely on jackstands. Remove the wheel.

3   At the frame bracket, unscrew the brake line fitting from the hose **(see illustration)**. Use a flare-nut wrench to prevent rounding off the corners.

4   Remove the U-clip from the female fitting at the bracket with a pair of pliers **(see illustration)**, then pass the hose through the bracket.

5   At the caliper end of the hose, remove the banjo fitting bolt **(see illustration 3.6b)**, then separate the hose from the caliper. Note that there are two copper sealing washers on either side of the fitting - they should be replaced with new ones during installation.

6   Remove the U-clip from the strut bracket, then feed the hose through the bracket.

7   To install the hose, pass the caliper fitting end through the strut bracket, then connect the fitting to the caliper with the banjo bolt and copper washers. Make sure the locating lug on the fitting is engaged with the hole in the caliper, then tighten the bolt to the torque listed in this Chapter's Specifications.

8   Push the metal support into the strut bracket and install the U-clip. Make sure the hose isn't twisted between the caliper and the strut bracket.

9   Route the hose into the frame bracket, again making sure it isn't twisted, then connect the brake line fitting, starting the threads by hand. Install the clip and E-ring, if equipped, then tighten the fitting securely.

10   Bleed the caliper (see Section 10).

11   Install the wheel and lug nuts, lower the vehicle and tighten the lug nuts to the torque specified in Chapter 1.

Rear brake hose

12   Perform Steps 2, 3 and 4 above, then repeat Steps 3 and 4 at the other end of the hose. Be sure to bleed the wheel cylinder (or caliper) (see Section 10).

Metal brake lines

13   When replacing brake lines, be sure to use the correct parts. Don't use copper tubing for any brake system components. Purchase genuine steel brake lines from a dealer or auto parts store.

14   Prefabricated brake line, with the tube ends already flared and fittings installed, is available at auto parts stores and dealer parts departments.

15   When installing the new line, make sure it's securely supported in the brackets and has plenty of clearance between moving or hot components.

16   After installation, check the master cylinder fluid level and add fluid as necessary. Bleed the brake system (see Section 10) and test the brakes carefully before driving the vehicle in traffic.

Chapter 9 Brakes

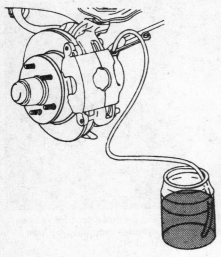

10.8 When bleeding the brakes, a hose is connected to the bleeder valve at the caliper or wheel cylinder and then submerged in brake fluid. Air will be seen as bubbles in the tube and container. All air must be expelled before moving to the next wheel

10 Brake hydraulic system - bleeding

Refer to illustration 10.8

Warning: *Wear eye protection when bleeding the brake system. If the fluid comes in contact with your eyes, immediately rinse them with water and seek medical attention.*

Note: *Bleeding the hydraulic system is necessary to remove any air that manages to find its way into the system when it's been opened during removal and installation of a hose, line, caliper or master cylinder.*

1 You'll probably have to bleed the system at all four brakes if air has entered it due to low fluid level, or if the brake lines have been disconnected at the master cylinder.
2 If a brake line was disconnected only at a wheel, then only that caliper or wheel cylinder must be bled.
3 If a brake line is disconnected at a fitting located between the master cylinder and any of the brakes, that part of the system served by the disconnected line must be bled.
4 Remove any residual vacuum from the brake power booster by applying the brake several times with the engine off.
5 Remove the master cylinder reservoir cover and fill the reservoir with brake fluid. Reinstall the cover. **Note:** *Check the fluid level often during the bleeding operation and add fluid as necessary to prevent the fluid level from falling low enough to allow air bubbles into the master cylinder.*
6 Have an assistant on hand, as well as a supply of new brake fluid, a clear container partially filled with clean brake fluid, a length of tubing to fit over the bleeder valve and a wrench to open and close the bleeder valve.
7 Beginning at the right rear wheel, loosen the bleeder valve slightly, then tighten it to a point where it's snug but can still be loosened quickly and easily.
8 Place one end of the tubing over the bleeder valve and submerge the other end in brake fluid in the container **(see illustration)**.
9 Have the assistant depress the brake pedal slowly, then hold the pedal down firmly.
10 While the pedal is held down, open the bleeder valve just enough to allow a flow of fluid to leave the valve. Watch for air bubbles to exit the submerged end of the tube. When the fluid flow slows after a couple of seconds, close the valve and have your assistant release the pedal.
11 Repeat Steps 9 and 10 until no more air is seen leaving the tube, then tighten the bleeder valve and proceed to the left rear wheel, the right front wheel and the left front wheel, in that order, and perform the same procedure. Be sure to check the fluid in the master cylinder reservoir frequently.
12 Never use old brake fluid. It contains moisture which will deteriorate the brake system components.
13 Refill the master cylinder with fluid at the end of the operation.
14 Check the operation of the brakes. The pedal should feel solid when depressed, with no sponginess. If necessary, repeat the entire process. **Warning:** *Do not operate the vehicle if you're in doubt about the effectiveness of the brake system.*

11 Power brake booster - check, removal and installation

Operating check

1 Depress the brake pedal several times with the engine off and make sure there's no change in the pedal reserve distance.
2 Depress the pedal and start the engine. If the pedal goes down slightly, operation is normal.

Airtightness check

3 Start the engine and turn it off after one or two minutes. Depress the brake pedal slowly several times. If the pedal depresses less each time, the booster is airtight.
4 Depress the brake pedal while the engine is running, then stop the engine with the pedal depressed. If there's no change in the pedal reserve travel after holding the pedal for 30 seconds, the booster is airtight.

Removal

Refer to illustration 11.11

5 Power brake booster units shouldn't be disassembled. They require special tools not normally found in most automotive repair stations or shops. They're fairly complex and, because of their critical relationship to brake performance, should be replaced with a new or rebuilt one.
6 Remove the brake master cylinder (see Section 8).
7 Remove the charcoal canister (see Chapter 6).
8 Remove the air cleaner assembly (see Chapter 4).
9 Disconnect the vacuum hose from the brake booster. Be careful not to damage the hose when removing it from the booster fitting.
10 Inside the vehicle, remove the steering column cover pad (see Chapter 11).
11 Using a flashlight or a drop light, locate the clevis which connects the booster pushrod to the top of the brake pedal. Remove the clevis pin retaining clip with pliers and pull out the pin **(see illustration)**.
12 Remove the four nuts and washers holding the brake booster to the firewall (you may need a light to see them). Slide the booster straight out from the firewall until the studs clear the holes.

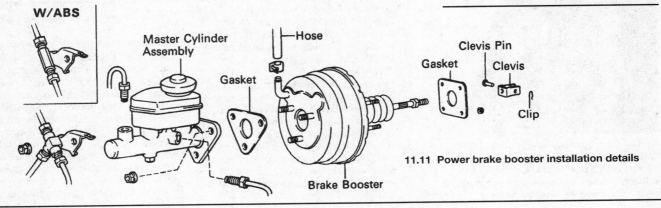

11.11 Power brake booster installation details

9-20   Chapter 9 Brakes

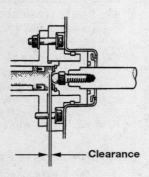

11.14a There should be no clearance between the booster pushrod and the master cylinder pushrod, but no interference either; if there is interference between the two, the brakes may drag; if there is too much clearance, there will be excessive brake pedal travel

11.14b Measure the distance that the pushrod protrudes from the brake booster at the master cylinder mounting surface (including the gasket)

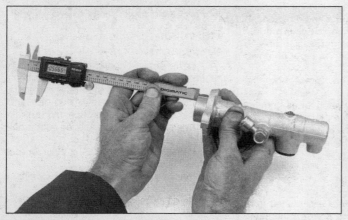

11.14c Measure the distance from the mounting flange to the end of the master cylinder

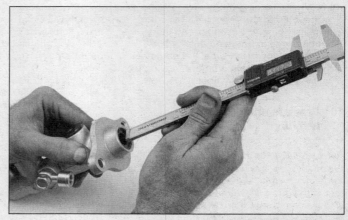

11.14d Measure the distance from the piston pocket to the end of the master cylinder

Installation

Refer to illustrations 11.14a, 11.14b, 11.14c, 11.14d and 11.14e

13   Installation procedures are basically the reverse of removal. Tighten the clevis locknut securely and the booster mounting nuts to the torque listed in this Chapter's Specifications.

11.14e To adjust the length of the booster pushrod, hold the serrated portion of the rod with a pair of pliers and turn the adjusting screw in or out, as necessary, to achieve the desired setting

14   If a new power brake booster unit is being installed, check the pushrod clearance **(see illustration)** as follows:

a) Measure the distance that the pushrod protrudes from the master cylinder mounting surface on the front of the power brake booster, including the gasket. Write down this measurement **(see illustration)**. This is "dimension A."

b) Measure the distance from the mounting flange to the end of the master cylinder **(see illustration)**. Write down this measurement. This is "dimension B."

c) Measure the distance from the end of the master cylinder to the bottom of the pocket in the piston **(see illustration)**. Write down this measurement. This is "dimension C."

d) Subtract measurement B from measurement C, then subtract measurement A from the difference between B and C. This the pushrod clearance.

e) Compare your calculated pushrod clearance to the pushrod clearance listed in this Chapter's Specifications. If necessary, adjust the pushrod length to achieve the correct clearance **(see illustration)**.

15   After the final installation of the master cylinder and brake hoses and lines, the brake pedal height and freeplay must be adjusted and the system must be bled. See the appropriate Sections of this Chapter for the procedures.

12   Parking brake shoes (rear disc brakes only) - inspection and replacement

Refer to illustrations 12.4 and 12.5a through 12.5v

Warning 1: *Dust created by the brake system may contain asbestos, which is hazardous to your health. Never blow it out with compressed air and don't inhale any of it. An approved filtering mask should be worn when working on the brakes. Do not, under any circumstances, use petroleum-based solvents to clean brake parts. Use brake system cleaner only!*

Warning 2: *Parking brake shoes must be replaced on both wheels at the same time - never replace the shoes on only one wheel.*

1   Remove the brake disc (see Section 5).

Chapter 9 Brakes

2 Inspect the thickness of the lining material on the shoes. If the lining has worn down to 1/32-inch or less, the shoes must be replaced.
3 Remove the hub and bearing assembly (see Chapter 10).
4 Wash off the brake parts with brake system cleaner **(see illustration)**.
5 Follow the accompanying illustrations for the brake shoe replacement procedure **(see illustrations 12.5a through 12.5v)**. Be sure to stay in order and read the caption under each illustration.
6 Install the brake disc. Temporarily thread three of the wheel lug nuts onto the studs to hold the disc in place.
7 Remove the hole plug from the brake disc. Adjust the parking brake shoe clearance by turning the adjuster star wheel with a brake adjusting tool or screwdriver until the shoes contact the disc and the disc can't be turned **(see illustrations 5.6a, 5.6b and 5.6c)**. Back-off the adjuster eight notches, then install the hole plug.
8 Install the torque plate **(see illustration 5.2b)** and brake caliper (see Section 4). Be sure to tighten the bolts to the torque listed in this Chapter's Specifications.
9 Install the wheel and tighten the lug nuts to the torque specified in Chapter 1.
10 If the vehicle is equipped with a parking brake lever, pull up on the lever and count the number of clicks that it travels. It should be between five and eight clicks - if it's not, adjust the parking brake as described in the next Section.

12.4 Before disassembling it, be sure to wash the parking brake assembly with brake cleaner

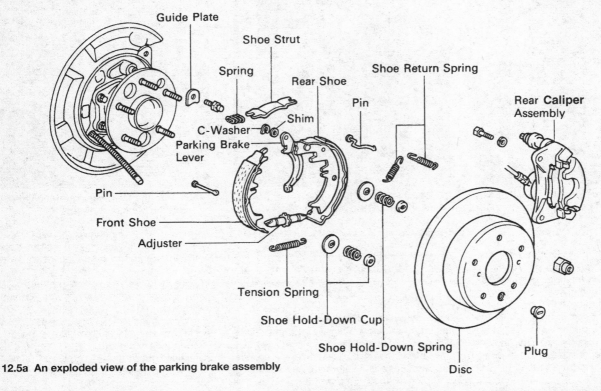

12.5a An exploded view of the parking brake assembly

12.5b Remove the rear parking brake shoe return spring from the anchor pin . . .

12.5c . . . and unhook it from the rear shoe

12.5d Remove the front parking brake shoe return spring from the anchor pin . . .

12.5e ... and unhook it from the front shoe

12.5f Remove the rear shoe hold-down spring and pull out the pin

12.5g Remove the shoe strut from between the shoes

12.5h Remove the front shoe hold-down spring and pull out the pin

12.5i Remove the adjuster and the tension spring (the tension spring, which is not visible in this photo, is behind the adjuster and is attached to both shoes)

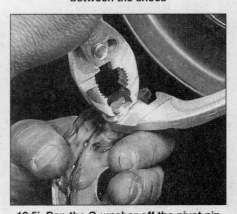

12.5j Pop the C-washer off the pivot pin on the back of the rear shoe ...

11   To bed the shoes to the drum, drive the vehicle at approximately 30 mph on a dry, level road. If the vehicle has a parking brake lever, push in on the parking brake release button and pull up slightly on the lever with about 20 pounds of force; if the vehicle has a pedal-type parking brake system, apply the pedal with about 33 pounds of force. Drive the vehicle with the parking brake applied like this for 1/4-mile. Repeat this procedure two or three times, allowing the brakes to cool between applications.

13 Parking brake - adjustment

Lever type

1   The parking brake lever, when properly adjusted, should travel five to eight clicks when a moderate pulling force is applied. If it travels less than five clicks, there's a chance the parking brake might not be releasing completely and might be dragging on the drum or disc. If the lever can be pulled up more than eight clicks, the parking brake may not hold adequately on an incline, allowing the car to roll.
2   To gain access to the parking brake cable adjuster, remove the center console (see Chapter 11).
3   Loosen the locknut (the upper nut) while holding the adjusting nut (lower nut) with a wrench **(see illustration 13.7)**. Tighten the adjusting nut until the desired travel is attained. Tighten the locknut.
4   Install the console.

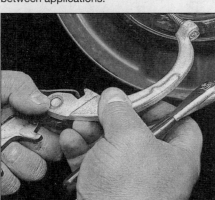

12.5k ... and pull the parking brake lever off the pivot pin

12.5l Apply a thin coat of high-temperature grease to the contact surfaces of the backing plate

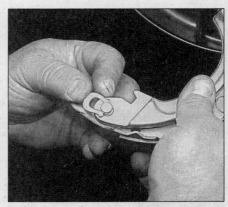

12.5m Slide the parking brake lever onto the pivot pin and install a new C-washer

Chapter 9 Brakes

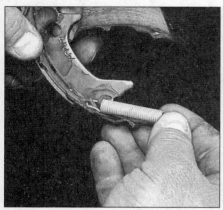

12.5n Attach the tension spring to the back side of the rear shoe . . .

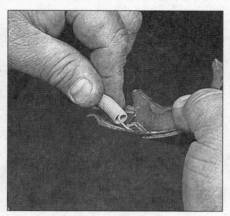

12.5o . . . and to the back side of the front shoe

12.5p Flip the shoes around and install the adjuster; make sure both ends of the adjuster are properly engaged with the shoes as shown

12.5q Place the shoes in position and install the strut and spring as shown; make sure the ends of the strut are properly engaged with the shoes as shown

12.5r Install the front shoe return spring . . .

12.5s . . . and the rear shoe return spring

Pedal type

Refer to illustration 13.7

5   Slowly depress the parking brake pedal all the way and count the number of clicks. It should take about three to six clicks to apply the parking brake. If it travels less than three clicks, there's a chance the parking brake might not be releasing completely and might be dragging on the drum or disc. If it travels more than six clicks, the parking brake may not hold adequately on an incline, allowing the car to roll.

6   Remove the cover plate from inside the center console compartment (see Chapter 11).

7   Loosen the locknut while holding the adjusting nut with a wrench **(see illustration)**. Tighten the adjusting nut until the desired travel is attained. Tighten the locknut.

8   Install the console.

12.5t Install the rear shoe hold-down spring

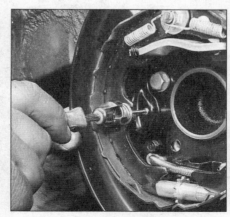

12.5u . . . and the front shoe hold-down spring

12.5v This is how the parking brake assembly should look when you're done!

9-24   Chapter 9 Brakes

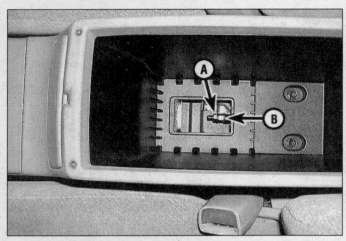

13.7 Parking brake cable locknut A (upper nut) and adjuster nut B (lower nut) for pedal-type parking brake cable systems (same setup is used on models with a parking brake lever)

14.4 To detach either parking brake cable housing from the brake backing plate, remove these two bolts (disc brake backing plate shown, drum brake setup similar)

14 Parking brake cables - replacement

Equalizer-to-parking brake cable

Refer to illustrations 14.4, 14.5, 14.7 and 14.8

1 Loosen the rear wheel lug nuts, raise the rear of the vehicle and support it securely on jackstands. Block the front wheels. Remove the wheel.
2 Make sure the parking brake is completely released, then remove the brake drum or disc.
3 On models with rear drum brakes, remove the brake shoes and disconnect the cable from the parking brake lever (see Section 6); on models with rear disc brakes, remove the parking brake shoes and disconnect the parking brake lever (see Section 12).
4 Unbolt the cable housing from the backing plate **(see illustration)**.
5 Unbolt all cable brackets **(see illustration)**.
6 Remove the center exhaust pipe, or lower it enough to get at the forward part of the cables (see Chapter 4).
7 Trace the cable forward and locate the cable clamp **(see illustration)**. Loosen the clamp bolt and slide the cable housing out of the clamp.
8 Disconnect the cable end from the equalizer by aligning the cable with the slot in the top of the equalizer. Slide the cable end out of the hole **(see illustration)**.
9 Installation is the reverse of removal.
10 Adjust the parking brake (see Section 13).

Intermediate lever-to-equalizer lever cable

11 Remove the center console (see Chapter 11).
12 With the parking brake cable released, remove the locknut and the adjusting nut (see

14.5 This parking brake cable bracket is located near the forward end of the strut rod (but there may be others, depending on the year and model)

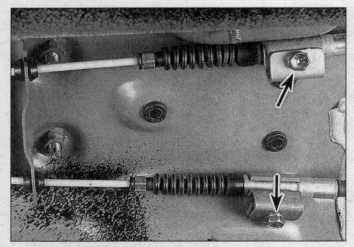

14.7 To release either parking brake cable from its cable clamp, loosen the clamp bolt (arrows) and slide the cable housing out of the clamp

14.8 To detach either parking brake cable from the equalizer, align the cable with the slot in the top of the equalizer and slide the cable end out of the hole

Chapter 9 Brakes

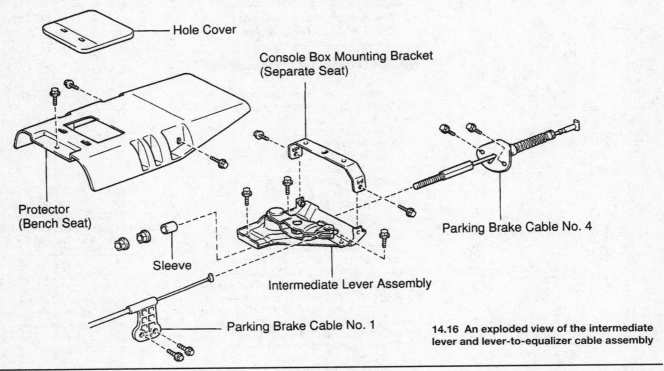

14.16 An exploded view of the intermediate lever and lever-to-equalizer cable assembly

Section 13) and detach the cable.
13 Working under the vehicle, pull the cable to the rear, turn it 90-degrees and pass it through the center of the equalizer (see illustration 14.16). Pull the cable through the hole in the floorpan.
14 Installation is the reverse of the removal procedure. Apply a light coat of grease to the portion of the cable end that contacts the equalizer. Adjust the parking brake cable as outlined previously (see Section 13).

Parking brake pedal-to-intermediate lever cable (Avalon models only)

Refer to illustrations 14.16 and 14.19
15 Remove the front seat (see Chapter 11).
16 Remove the protector (see illustration) on models equipped with a bench seat or remove the center console (see Chapter 11) on models equipped with separate seats.
17 Disconnect the cable from the intermediate lever.
18 Remove the knee bolster (see Chapter 11) to access the parking brake pedal bracket sub-assembly.
19 Disconnect the parking brake switch (see illustration).
20 Remove the clip, pull out the clevis pin and disconnect the parking brake cable from the parking brake pedal assembly.
21 Installation is the reverse of removal.

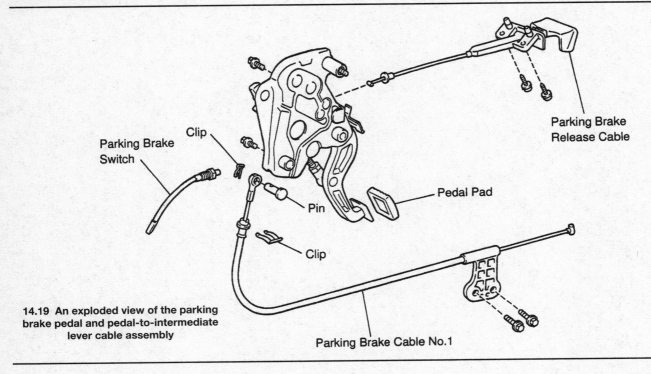

14.19 An exploded view of the parking brake pedal and pedal-to-intermediate lever cable assembly

15 Brake light switch - removal, installation and adjustment

Removal and installation

Refer to illustration 15.1

1   The brake light switch **(see illustration)** is located on a bracket at the top of the brake pedal.
2   Disconnect the wiring harness at the brake light switch.
3   Loosen the locknut and unscrew the switch from the pedal bracket.
4   Installation is the reverse of removal.

Adjustment

Refer to illustration 15.6

5   Check and, if necessary, adjust brake pedal height (see Chapter 1).
6   Loosen the switch locknut, adjust the switch so that the distance the plunger protrudes **(see illustration)** is within the range listed in this Chapter's Specifications. (If you're unable to measure this distance, adjust the plunger so that it lightly contacts the pedal stop.) Tighten the locknut.
7   Plug the electrical connector into the switch and reconnect the battery. Make sure the brake lights come on when the brake pedal is depressed and go off when the pedal is released. If not, repeat the adjustment procedure until the brake lights function properly
8   Check and, if necessary, adjust brake pedal freeplay (see Chapter 1).

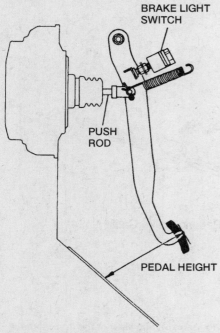

15.1 The brake light switch is located at the top of the brake pedal

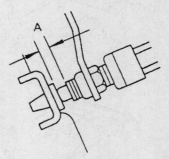

15.6 To adjust the brake light switch, loosen the locknut and rotate the switch until the plunger distance (dimension A) is within the range listed in this Chapter's Specifications, then tighten the locknut

Chapter 10
Suspension and steering systems

Contents

	Section		Section
Balljoints - replacement	6	Steering gear - removal and installation	19
Control arm - removal, inspection and installation	5	Steering knuckle and hub - removal and installation	7
General information	1	Steering system - general information	15
Hub and bearing assembly (front) - removal and installation	8	Steering wheel - removal and installation	16
Hub and bearing assembly (rear) - removal and installation	13	Strut assembly (front) - removal, inspection and installation	3
Power steering fluid level check	See Chapter 1	Strut assembly (rear) - removal, inspection and installation	10
Power Steering Pressure (PSP) switch - check and replacement	21	Strut rod - removal and installation	11
Power steering pump - removal and installation	20	Strut/coil spring assembly - replacement	4
Power steering system - bleeding	22	Suspension arms (rear) - removal and installation	12
Rear axle carrier - removal and installation	14	Tie-rod ends - removal and installation	17
Stabilizer bar and bushings (front) - removal and installation	2	Tire and tire pressure checks	See Chapter 1
Stabilizer bar and bushings (rear) - removal and installation	9	Tire rotation	See Chapter 1
Steering and suspension check	See Chapter 1	Wheel alignment - general information	24
Steering gear boots - replacement	18	Wheels and tires - general information	23

Specifications

Torque specifications

Ft-lbs (unless otherwise indicated)

Front suspension
Balljoint
- Balljoint-to-steering knuckle nut 90
- Balljoint-to-control arm nuts and bolts 94

Control arm
- Front bolts 152
- Rear bolt 152

Stabilizer bar
- Stabilizer bar link-to-control arm bracket nut 29
- Stabilizer bushing/retainer bolts .... 168 in-lbs

Strut
- Strut upper mounting nuts 59
- Strut-to-suspension support (damper shaft) nut
  - Without electronic modulated suspension 36
  - With electronic modulated suspension 26
- Strut-to-steering knuckle bolts/nuts
  - Avalon/Camry/Lexus ES 300 156
  - Solara 162

Torque specifications (continued)

Rear suspension
	Ft-lbs (unless otherwise indicated)
Hub and bearing assembly-to-axle carrier bolts	59

Strut
Strut upper mounting nuts	29
Strut-to-suspension support (damper shaft) nut	
Without electronic modulated suspension	36
With electronic modulated suspension	26
Strut-to-axle carrier nuts/bolts	
New nut	188
Used nut	145

Stabilizer bar
Stabilizer bar link-to-strut assembly	29
Stabilizer bushing/retainer bolts	168 in-lbs

Suspension arms
No. 1/No. 2 arm-to-body through-bolt nut	134
No. 1/No. 2 arm-to-axle carrier through-bolt nut	134
Strut rod-to-body bolt	83
Strut rod-to-axle carrier bolt	83

Steering system
Airbag module retaining screws	63 in-lbs
Steering wheel nut	26
Steering gear mounting bolts/nuts	134
Steering shaft universal joint pinch bolt	19
Tie-rod end-to-steering knuckle nut	36
Power steering pump	
Adjuster and pivot bolts	32
Fluid line banjo bolt	38
Wheel lug nuts	See Chapter 1

1.1 Steering and front suspension components

1   Steering gear assembly
2   Control arm
3   Balljoint fasteners
4   Strut and spring assembly
5   Stabilizer bar link

Chapter 10 Suspension and steering systems

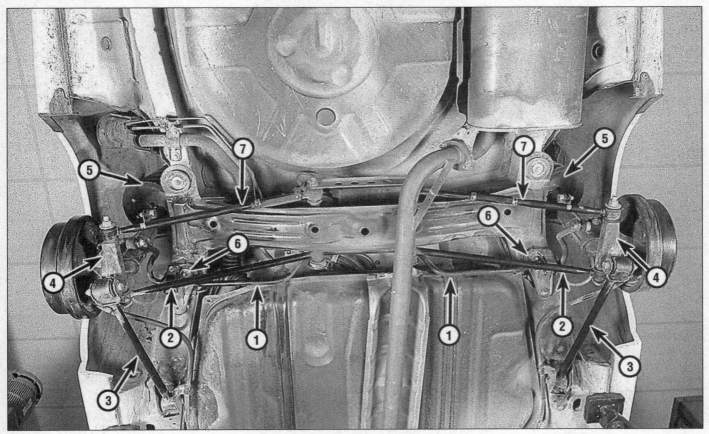

1.2 Rear suspension components

1. Stabilizer bar
2. Suspension arm (no. 1)
3. Strut rod
4. Rear axle carrier
5. Strut assembly
6. Stabilizer bar bushing retainer
7. Suspension arm (no. 2)

1 General information

Refer to illustrations 1.1 and 1.2

The front suspension is a Macpherson strut design. The upper end of each strut is attached to the vehicle's body strut support. The lower end of the strut is connected to the upper end of the steering knuckle. The steering knuckle is attached to a balljoint mounted on the outer end of the suspension control arm **(see illustration)**.

The rear suspension also utilizes strut/coil spring assemblies. The upper end of each strut is attached to the vehicle body by a strut support. The lower end of the strut is attached to an axle carrier. The carrier is located by a pair of suspension arms on each side, and a longitudinally mounted strut rod between the body and each knuckle **(see illustration)**.

Later Lexus ES 300 models are equipped with electronic modulated suspension. Electronic modulated suspension allows the driver to control the adjustment of the shock absorber damping depending on road conditions.

The power-assisted rack-and-pinion steering gear, which is located behind the engine/transaxle assembly, is mounted on the engine cradle. The steering gear actuates the tie-rods, which are attached to the steering knuckles. The steering column is designed to collapse in the event of an accident.

Frequently, when working on the suspension or steering system components, you may come across fasteners which seem impossible to loosen. These fasteners on the underside of the vehicle are continually subjected to water, road grime, mud, etc., and can become rusted or "frozen," making them extremely difficult to remove. In order to unscrew these stubborn fasteners without damaging them (or other components), be sure to use lots of penetrating oil and allow it to soak in for a while. Using a wire brush to clean exposed threads will also ease removal of the nut or bolt and prevent damage to the threads. Sometimes a sharp blow with a hammer and punch will break the bond between a nut and bolt threads, but care must be taken to prevent the punch from slipping off the fastener and ruining the threads. Heating the stuck fastener and surrounding area with a torch sometimes helps too, but isn't recommended because of the obvious dangers associated with fire. Long breaker bars and extension, or "cheater," pipes will increase leverage, but never use an extension pipe on a ratchet - the ratcheting mechanism could be damaged. Sometimes tightening the nut or bolt first will help to break it loose. Fasteners that require drastic measures to remove should always be replaced with new ones.

Since most of the procedures dealt with in this Chapter involve jacking up the vehicle and working underneath it, a good pair of jackstands will be needed. A hydraulic floor jack is the preferred type of jack to lift the vehicle, and it can also be used to support certain components during various operations. **Warning:** *Never, under any circumstances, rely on a jack to support the vehicle while working on it. Whenever any of the suspension or steering fasteners are loosened or removed they must be inspected and, if necessary, replaced with new ones of the same part number or of original equipment quality and design. Torque specifications must be followed for proper reassembly and component retention. Never attempt to heat or straighten any suspension or steering components. Instead, replace any bent or damaged part with a new one.*

2.4 To detach the stabilizer bar link from the bar, remove the nut (upper arrow); if you're removing the control arm, remove the lower nut

2.5 To detach the stabilizer bar from the engine cradle, remove these bushing retainer bolts (arrows) and remove the retainer

2 Stabilizer bar and bushings (front) - removal and installation

Removal

Refer to illustrations 2.4, 2.5 and 2.9

1 Loosen the front wheel lug nuts. Raise the front of the vehicle and support it securely on jackstands. Apply the parking brake and block the rear wheels to keep the vehicle from rolling off the stands. Remove the front wheels.

2 Remove the left and right fender apron seals. Each seal is retained by three bolts.

3 Disconnect the left and right tie-rod ends from the steering knuckles (see Section 17).

4 Disconnect the stabilizer bar links from the bar **(see illustration)**. If the ballstud turns with the nut, use an Allen wrench to hold the stud.

5 Detach both stabilizer bar bushing retainers from the engine cradle **(see illustration)**.

6 Remove the front exhaust pipe (see Chapter 4).

7 Remove the steering gear mounting bolts (see Section 19).

8 Lift the steering gear assembly and remove the stabilizer bar by working it out through the left wheel housing.

9 While the stabilizer bar is off the vehicle, slide off the retainer bushings and inspect them. If they're cracked, worn or deteriorated, replace them. It's also a good idea to inspect the stabilizer bar link. To check it, flip the balljoint stud side-to-side five or six times as shown **(see illustration)**, then install the nut. Using an inch-pound torque wrench, turn the nut continuously one turn every two to four seconds and note the torque reading on the fifth turn. It should be about 0.4 to 8.7 in-lbs. If it isn't, replace the link assembly.

10 Clean the bushing area of the stabilizer bar with a stiff wire brush to remove any rust or dirt.

Installation

11 Lubricate the inside and outside of the new bushings with vegetable oil (used in cooking) to simplify reassembly. **Caution:** *Don't use petroleum or mineral-based lubricants or brake fluid - they will lead to deterioration of the bushings.*

12 Installation is the reverse of removal.

3 Strut assembly (front) - removal, inspection and installation

Removal

Refer to illustrations 3.2, 3.4, 3.6a and 3.6b

1 Loosen the wheel lug nuts, raise the vehicle and support it securely on jackstands. Remove the wheel. Support the control arm with a floor jack.

2 Remove the brake hose bracket from the strut **(see illustration)**.

3 If the vehicle is equipped with ABS, detach the speed sensor wiring harness from the strut by removing the clamp bracket bolt. On models equipped with electronic modulated suspension, remove the shock absorber cap and disconnect the shock absorber wiring harness. If the strut is to be disassembled, loosen, but do not remove, the damper shaft (center) nut (this will require a special socket if the vehicle is equipped with electronic modulated suspension; check with your local auto parts store or tool dealer).

4 Remove the strut-to-knuckle nuts **(see illustration)** and knock the bolts out with a

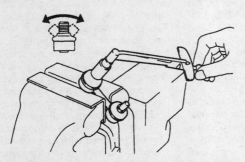

2.9 To check the balljoint in the stabilizer bar link, flip the balljoint stud side-to-side five or six times as shown, install the nut and, using an inch-pound torque wrench, turn the nut continuously one turn every two to four seconds, then note the torque reading on the fifth turn. It should be about 0.4 to 8.7 in-lbs; if it isn't, replace the link

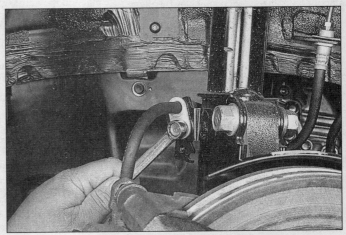

3.2 Detach the brake hose bracket from the strut

Chapter 10 Suspension and steering systems

3.4 To disconnect the lower end of the strut from the steering knuckle, remove these two nuts (arrows) and knock out the bolts with a hammer and punch

3.6a To disconnect the upper end of the strut from the vehicle body, remove these three nuts (arrows)

hammer and punch.

5 Separate the strut from the steering knuckle. Be careful not to overextend the inner CV joint. Also, don't let the steering knuckle fall outward, as the brake hose could be damaged.

6 Support the strut and spring assembly with one hand and remove the three strut-to-shock tower nuts (this is only if the strut is going to be disassembled) **(see illustration)**. On models equipped with electronic modulated suspension, loosen but do not remove, the nut in the center of the strut (this is only necessary if the strut is going to be disassembled) **(see illustration)**. Remove the assembly out from the fenderwell.

Inspection

7 Check the strut body for leaking fluid, dents, cracks and other obvious damage which would warrant repair or replacement.

8 Check the coil spring for chips or cracks in the spring coating (this can cause premature spring failure due to corrosion). Inspect the spring seat for cuts, hardness and general deterioration.

9 If any undesirable conditions exist, proceed to the strut disassembly procedure (see Section 4).

Installation

Refer to illustration 3.14

10 Guide the strut assembly up into the fenderwell and insert the three upper mounting studs through the holes in the shock tower. Once the three studs protrude from the shock tower, install the nuts so the strut won't fall back through. This is most easily accomplished with the help of an assistant, as the strut is quite heavy and awkward.

11 Slide the steering knuckle into the strut flange and insert the two bolts. Install the nuts and tighten them to the torque listed in this Chapter's Specifications.

12 Reattach the brake hose bracket to the strut. If the vehicle is equipped with ABS, install the speed sensor wiring harness bracket.

13 Install the wheel and lug nuts, then lower the vehicle and tighten the lug nuts to the torque listed in the Chapter 1 Specifications.

14 Tighten the three upper mounting nuts to the torque listed in this Chapter's Specifications.

15 If you're working on a model equipped with electronic modulated suspension and the strut had been disassembled, tighten the nut in the center of the strut to the torque listed in this Chapter's Specifications, using a tool like the one shown (check with your local auto parts store, dealer parts department or tool dealer regarding tool availability) **(see illustration)**. Connect the electrical connector and install cap.

4 Strut/coil spring assembly - replacement

1 If the struts or coil springs exhibit the telltale signs of wear (leaking fluid, loss of damping capability, chipped, sagging or cracked coil springs) explore all options before beginning any work. The strut/shock absorber assemblies are not serviceable and must be replaced if a problem develops. However, strut assemblies complete with springs may be available on an exchange basis, which eliminates much time and work. Whichever route you choose to take, check on the cost and availability of parts before disassembling your vehicle. **Warning:** *Disassembling a strut is potentially dangerous and utmost attention must be directed to the job, or serious injury may result. Use only a high-quality spring compressor and carefully follow the manufacturer's instructions furnished with the tool. After removing the coil spring from the strut assembly, set it aside in a safe, isolated area.*

Disassembly

Refer to illustrations 4.3, 4.4, 4.5a, 4.5b, 4.6 and 4.7

2 Remove the strut and spring assembly (see Section 3 [front] or Section 10 [rear]). Mount the strut assembly in a vise. Line the vise jaws with wood or rags to prevent damage to the unit and don't tighten the vise excessively.

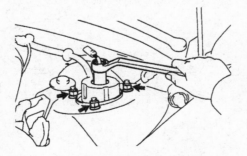

3.6b On electronic modulated suspension equipped models, loosen, but do not remove the center nut before removing the strut assembly (only do this if the strut is to be disassembled)

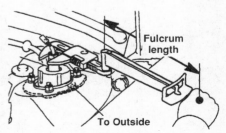

3.14 On models with electronic modulated suspension, a tool like this one, and a torque wrench with a fulcrum length of approximately 13-9/16 inches is necessary to properly tighten the strut center nut

10-6   Chapter 10   Suspension and steering systems

4.3 Install the spring compressor in accordance with the tool manufacturer's instructions and compress the spring until all pressure is relieved from the upper spring seat

4.4 Remove the damper shaft nut

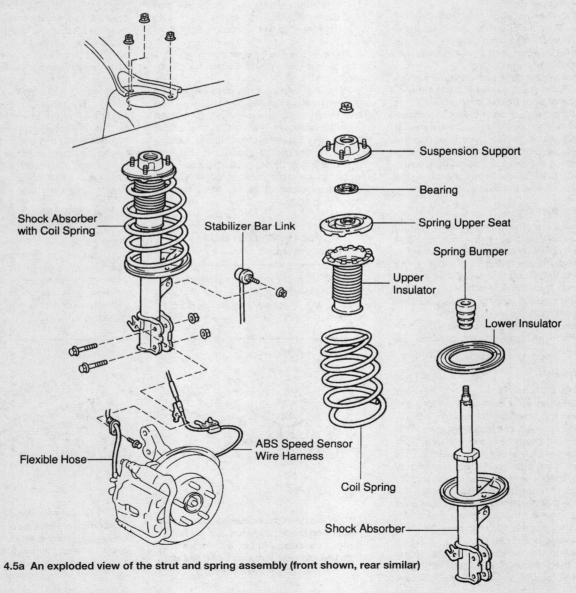

4.5a An exploded view of the strut and spring assembly (front shown, rear similar)

Chapter 10 Suspension and steering systems

4.5b Lift the suspension support off the damper shaft

4.6 Remove the spring seat from the damper shaft

4.7 Remove the compressed spring assembly - keep the ends of the spring pointed away from your body

4.11 When installing the spring, make sure the end fits into the recessed portion of the lower seat (arrow)

4.12 The flats on the damper shaft (arrow) must match up with the flats in the spring seat

3   Following the tool manufacturer's instructions, install the spring compressor (which can be obtained at most auto parts stores or equipment yards on a daily rental basis) on the spring and compress it sufficiently to relieve all pressure from the upper spring seat **(see illustration)**. This can be verified by wiggling the spring.

4   Loosen the damper shaft nut **(see illustration)**.

5   Remove the nut and suspension support **(see illustrations)**. Inspect the bearing in the suspension support for smooth operation. If it doesn't turn smoothly, replace the suspension support. Check the rubber portion of the suspension support for cracking and general deterioration. If there is any separation of the rubber, replace it.

6   Remove the upper spring seat from the damper shaft **(see illustration)**. Check the spring seat for cracking and hardness; replace it if necessary. Remove the upper insulator from the damper shaft.

7   Carefully lift the compressed spring from the assembly **(see illustration)** and set it in a safe place. **Warning:** *Never place your head near the end of the spring!*

8   Slide the rubber bumper off the damper shaft.

9   Check the lower insulator for wear, cracking and hardness and replace it if necessary.

Reassembly

Refer to illustrations 4.11, 4.12, 4.13, 4.14a and 4.14b

10   If the lower insulator is being replaced, set it into position with the dropped portion seated in the lowest part of the seat. Extend the damper rod to its full length and install the rubber bumper.

11   Carefully place the coil spring onto the lower insulator, with the end of the spring resting in the lowest part of the insulator **(see illustration)**.

12   Install the upper insulator and spring seat, making sure that the flats in the hole in the seat match up with the flats on the damper shaft **(see illustration)**.

13   Align the OUT mark of the spring upper seat with the mark of the upper insulator **(see illustration)**.

14   If you're working on a front strut, make sure the arrow on the spring seat faces toward the lower bracket, where the steering

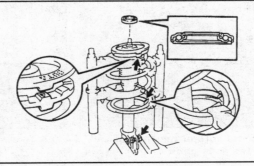

4.13 Align the OUT mark on the upper spring seat with the mark on the upper insulator

Chapter 10 Suspension and steering systems

knuckle fits **(see illustration)**. If you're working on a rear strut, make sure the suspension support is positioned with the pointed part of the support facing out, with the hole centered on the strut bracket **(see illustration)**.

15 Install the dust seal and suspension support to the damper shaft.

16 Install the nut and, if you're working on a model without electronic modulated suspension, tighten it to the torque listed in this Chapter's Specifications (on models with electronic modulated suspension, final tightening of this nut is carried out when the strut is installed on the vehicle). Remove the spring compressor tool.

17 Install the strut/spring assembly (see Section 3 [front] or 10 [rear]). If you're working on a model with electronic modulated suspension, tighten the damper shaft nut **(see illustration 3.14)**.

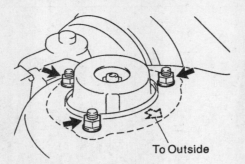

4.14a Make sure the arrow on the spring seat faces toward the lower bracket, where the steering knuckle fits

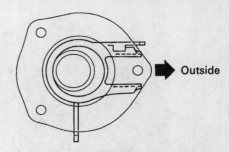

4.14b Proper rear suspension support-to-strut relationship

5 Control arm - removal, inspection and installation

Removal

Refer to illustrations 5.3a, 5.3b, 5.4 and 5.5

1 Loosen the wheel lug nuts on the side to be dismantled, raise the front of the vehicle, support it securely on jackstands and remove the wheel.

2 Disconnect the stabilizer bar link from the control arm (see Section 2).

3 Remove the balljoint retaining bolt and nuts **(see illustration)**. Use a prybar to disconnect the control arm from the steering knuckle **(see illustration)**.

4 Remove the two bolts that attach the front of the control arm to the engine cradle **(see illustration)**.

5 Remove the bolt and nut that attach the rear of the control arm to the engine cradle **(see illustration)**.

6 Remove the control arm.

Inspection

7 Make sure the control arm is straight. If it's bent, replace it. Do not attempt to straighten a bent control arm.

8 Inspect the bushings. If they're cracked, torn or worn out, replace the control arm.

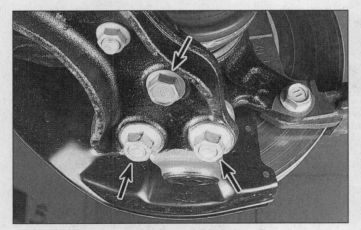

5.3a Remove these nuts and this bolt to disconnect the control arm from the balljoint

5.3b Separate the control arm from the steering knuckle with a prybar

5.4 To detach the front end of the control arm from the engine cradle, remove these two bolts (arrows)

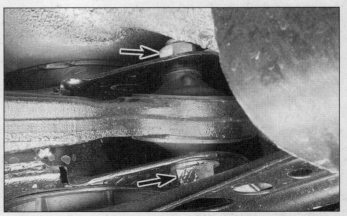

5.5 To detach the rear end of the control arm from the engine cradle, remove this nut and bolt (arrows)

Chapter 10 Suspension and steering systems

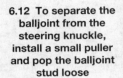

6.12 To separate the balljoint from the steering knuckle, install a small puller and pop the balljoint stud loose

Installation

9 Installation is the reverse of removal. Be sure to tighten all fasteners to the torque listed in this Chapter's Specifications.
10 Install the wheel and lug nuts, lower the vehicle and tighten the lug nuts to the torque listed in the Chapter 1 Specifications.
11 It's a good idea to have the front wheel alignment checked, and if necessary, adjusted after this job has been performed.

6 Balljoints - replacement

1 Loosen the wheel lug nuts, raise the vehicle and support it securely on jackstands. Remove the wheel. **Note:** *If you're going to remove the balljoint using a puller (as described in Steps 9 through 13), loosen the driveaxle/hub nut before removing the wheel (see Chapter 8).*

Picklefork method

Caution: *The following procedure is the quickest way to detach a balljoint from the steering knuckle, but it will very likely damage the balljoint boot. If you want to save the boot, proceed to Step 9.*

2 Remove the cotter pin from the balljoint stud and loosen the nut (but don't remove it yet).
3 Separate the balljoint from the steering knuckle with a picklefork-type balljoint separator. Lubricate the rubber boot with grease and work carefully so as not to tear the boot. Remove the balljoint stud nut.
4 Remove the bolt and nuts securing the balljoint to the control arm **(see illustration 5.3a)**. Separate the balljoint from the control arm with a prybar **(see illustration 5.3b)**.
5 To install the balljoint, position it on the steering knuckle and install the nut, but don't tighten it yet.
6 Attach the balljoint to the control arm and install the bolt and nuts, tightening them to the torque listed in this Chapter's Specifications.
7 Tighten the balljoint stud nut to the torque listed in this Chapter's Specifications and install a new cotter pin. If the cotter pin hole doesn't line up with the slots on the nut, tighten the nut additionally until it does line up - don't loosen the nut to insert the cotter pin.
8 Install the wheel and lug nuts. Lower the vehicle and tighten the lug nuts to the torque listed in the Chapter 1 Specifications.

Puller method

Refer to illustration 6.12

9 Separate the control arm from the balljoint (see Section 5).
10 Pull the outer end of the driveaxle from the steering knuckle (see Chapter 8) and suspend the driveaxle with a piece of wire.
11 Remove the cotter pin from the balljoint stud and loosen the nut (but don't remove it yet).
12 Install a small puller **(see illustration)** and pop the balljoint stud from the steering knuckle.
13 Remove the nut and remove the balljoint.
14 Install the new balljoint into the steering knuckle and tighten the nut to the torque listed in this Chapter's Specifications. Install a new cotter pin. If the cotter pin hole doesn't line up with the slots on the nut, tighten the nut additionally until it does line up - don't loosen the nut to insert the cotter pin.
15 Insert the outer end of the driveaxle through the steering knuckle and install the nut. Tighten it securely, but don't attempt to tighten it completely yet.
16 Connect the balljoint to the lower arm and tighten the fasteners to the torque listed in this Chapter's Specifications.
17 Install the wheel and lug nuts, lower the vehicle and tighten the lug nuts to the torque listed in the Chapter 1 Specifications.
18 Tighten the driveaxle/hub nut to the torque listed in the Chapter 8 Specifications. Install the lock washer and cotter pin.

7 Steering knuckle and hub - removal and installation

Warning: *Dust created by the brake system may contain asbestos, which is harmful to your health. Never blow it out with compressed air and don't inhale any of it. Do not, under any circumstances, use petroleum-based solvents to clean brake parts. Use brake system cleaner only.*

Removal

1 Loosen the hub nut (see Chapter 8). Loosen the wheel lug nuts, raise the vehicle and support it securely on jackstands. Remove the wheel.
2 Remove the brake caliper and support it with a piece of wire as described in Chapter 9. Remove the caliper torque plate and separate the brake disc from the hub.
3 Loosen, but do not remove the strut-to-steering knuckle bolts **(see illustration 3.4)**.
4 Separate the tie-rod from the steering knuckle arm (see Section 17).
5 Remove the balljoint-to-lower arm bolt and nuts **(see illustration 5.3a and 5.3b)**. The strut-to-knuckle bolts can now be removed.
6 Push the driveaxle from the hub as described in Chapter 8. Support the end of the driveaxle with a piece of wire.
7 Separate the steering knuckle from the strut. If necessary, detach the balljoint from the steering knuckle.

Installation

8 Guide the knuckle and hub assembly into position, inserting the driveaxle into the hub.
9 Push the knuckle into the strut flange and install the bolts and nuts, but don't tighten them yet.
10 Connect the balljoint to the control arm and install the bolt and nuts (don't tighten them yet).
11 Attach the tie-rod to the steering knuckle arm (see Section 17). Tighten the strut bolt nuts, the balljoint-to-control arm bolt and nuts and the tie-rod nut to the torque values listed in this Chapter's Specifications.
12 Place the brake disc on the hub and install the caliper as outlined in Chapter 9.
13 Install the hub nut and tighten it securely (final tightening will be carried out when the vehicle is lowered).
14 Install the wheel and lug nuts.
15 Lower the vehicle and tighten the lug nuts to the torque listed in the Chapter 1 Specifications. Tighten the driveaxle/hub nut and tighten it to the torque listed in the Chapter 8 Specifications)

8 Hub and bearing assembly (front) - removal and installation

Due to the special tools and expertise required to press the hub and bearing from the steering knuckle, this job should be left to a professional mechanic. However, the steering knuckle and hub may be removed and the assembly taken to a dealer service department or other repair shop. See Section 7 for the steering knuckle and hub removal procedure.

Chapter 10 Suspension and steering systems

9.3 Remove the nut (arrow) and detach the link from the stabilizer bar (do this on each side)

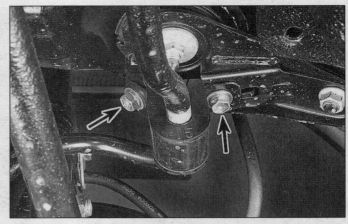

9.4 Remove the bushing retainer bolts (arrows) from both the left and right retainers (left retainer shown)

9 Stabilizer bar and bushings (rear) - removal and installation

Refer to illustrations 9.3 and 9.4

1 Loosen the rear wheel lug nuts. Raise the rear of the vehicle and place it securely on jackstands. Remove the rear wheels.
2 Remove the heat insulator from the exhaust system.
3 Detach the stabilizer bar links from the bar **(see illustration)**. If the ballstud turns with the nut, use a hex wrench to hold the stud.
4 Unbolt the stabilizer bar bushing retainers from the body **(see illustration)**.
5 The stabilizer bar can now be removed from the vehicle. Pull the retainers off the stabilizer bar (if they haven't fallen off already) using a rocking motion.
6 Check the bushings for wear, hardness, distortion, cracking and other signs of deterioration, replacing them if necessary. Check the stabilizer bar links as described in Section 2, Step 9.
7 Using a wire brush, clean the areas of the bar where the bushings ride. Installation is the reverse of the removal procedure. If necessary, use a light coat of vegetable oil to ease bushing and U-bracket installation (don't use petroleum-based products or brake fluid, as these will damage the rubber).
8 Installation is the reverse of removal.

10 Strut assembly (rear) - removal, inspection and installation

Removal

Refer to illustrations 10.7 and 10.8

1 Remove the rear seat and package tray trim panel (see Chapter 11).
2 Loosen the rear wheel lug nuts, raise the rear of the vehicle and support it securely on jackstands. Remove the wheel.
3 On models equipped with ABS, remove the flexible hose and ABS speed sensor from the shock absorber.

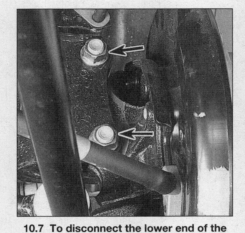

10.7 To disconnect the lower end of the strut from the rear axle carrier, remove these nuts (arrows) and knock out the bolts (but don't remove the bolts until after the upper strut nuts are removed and the axle carrier is supported by a floor jack - the strut assembly is heavy, so it's a good idea to have an assistant standing by to lend a hand if necessary)

4 Detach the brake hose from the strut **(see illustration 3.2)**. If the vehicle is equipped with ABS, detach the ABS sensor wire from the strut. On models equipped with electronic modulated suspension, remove clip and clamp and disconnect the shock absorber wiring harness.
5 Disconnect the stabilizer bar link from the strut **(see illustration 9.3)**.
6 Support the axle carrier with a floor jack.
7 Loosen the strut-to-axle carrier bolt nuts **(see illustration)**.
8 Remove the three upper strut-to-body mounting nuts **(see illustration)**. On models equipped with electronic modulated suspension, loosen but do not remove the damper shaft (center) nut (this is only necessary if the strut is going to be disassembled), and will require a special socket (check with your local auto parts store or tool dealer) **(see illustration 3.6b)**. Remove the assembly out from the fenderwell.

10.8 To disconnect the upper end of the strut from the vehicle, remove these three nuts (arrows)

9 Lower the axle carrier with the jack and remove the two strut-to-axle carrier bolts.
10 Remove the strut assembly.

Inspection

11 Follow the inspection procedures described in Section 3. If you determine that the strut assembly must be disassembled for replacement of the strut or the coil spring, refer to Section 4.

Installation

12 Maneuver the assembly up into the fenderwell and insert the mounting studs through the holes in the body. Install the nuts, but don't tighten them yet.
13 Push the axle carrier into the strut lower bracket and install the bolts and nuts, tightening them to the torque listed in this Chapter's Specifications.
14 Connect the stabilizer bar link to the strut bracket.
15 Attach the brake hose bracket to the strut. If the vehicle is equipped with ABS, attach the ABS wire to the strut.
16 Install the wheel and lug nuts, lower the vehicle and tighten the lug nuts to the torque

Chapter 10 Suspension and steering systems   10-11

11.2 To disconnect the strut rod from the rear axle carrier, remove the bolt indicated by the lower arrow; to disconnect the suspension arms from the carrier, remove the nut and long through-bolt

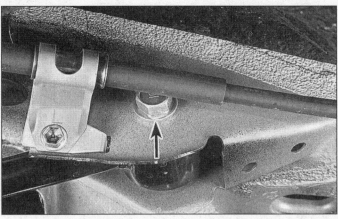

11.3 To disconnect the forward end of the strut rod from the vehicle body, remove this bolt (arrow)

listed in the Chapter 1 Specifications.
17 Tighten the three strut upper mounting nuts to the torque listed in this Chapter's Specifications.
18 If you're working on a model equipped with electronic modulated suspension and the strut has been disassembled, tighten the strut center nut to the torque listed in this Chapter's Specifications, using a setup like the one in **illustration 3.14**. Connect the electrical connector.
19 Install the package tray trim and seat.

11 Strut rod - removal and installation

Refer to illustrations 11.2 and 11.3

1 Loosen the wheel lug nuts, raise the vehicle and support it securely on jackstands. Remove the wheel.
2 Remove the strut rod-to-axle carrier bolt **(see illustration)**.
3 Remove the strut rod-to-body bracket bolt **(see illustration)** and detach the rod from the vehicle.
4 Installation is the reverse of the removal procedure. Be sure to tighten the bolts to the torque listed in this Chapter's Specifications.

12 Suspension arms (rear) - removal and installation

Removal

Refer to illustration 12.5

1 Raise the rear of the vehicle and support it securely on jackstands. Block the front wheels.
2 Disconnect the strut rod from the axle carrier **(see illustration 11.2)**.
3 Remove the exhaust center section (see Chapter 4).
4 Remove the suspension arm-to-rear axle carrier bolt and nut **(see illustration 11.2)**.
5 Remove the suspension arm-to-suspen-

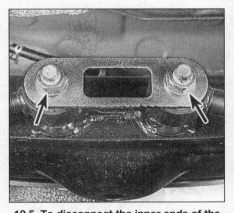

12.5 To disconnect the inner ends of the suspension arms from the suspension member lower support, remove these nuts (arrows) and drive out the bolts

sion member lower support nut and bolt **(see illustration)**.
6 Remove the No. 2 (rear) suspension arm.
7 Remove the left and right stabilizer bar bushing retainers **(see illustration 9.4)**.
8 Remove the exhaust center pipe and tail pipe (see Chapter 4).
9 Remove the No. 1 (front) suspension arm.

Installation

10 Installation is the reverse of removal. Be sure to tighten all fasteners to the torque listed in this Chapter's Specifications.
11 Install the wheel and lug nuts, then lower the vehicle to the ground. Tighten the wheel lug nuts to the torque listed in the Chapter 1 Specifications.
12 Have the rear wheel alignment checked by a dealer service department or an alignment shop.

13 Hub and bearing assembly (rear) - removal and installation

Warning: Dust created by the brake system

13.3 To remove the four bolts which attach the hub and bearing assembly to the rear axle carrier, rotate the hub flange and align one of the holes in the flange with each of the bolts

may contain asbestos, which is harmful to your health. Never blow it out with compressed air and don't inhale any of it. Do not, under any circumstances, use petroleum-based solvents to clean brake parts. Use brake system cleaner only.
Note: The rear hub and bearing assembly is not serviceable. If found to be defective, it must be replaced as a unit.

Removal

Refer to illustrations 13.3 and 13.5

1 Loosen the wheel lug nuts, raise the vehicle and support it securely on jackstands. Remove the wheel.
2 Remove the brake drum or disc from the hub (see Chapter 9).
3 Remove the four hub-to-axle carrier bolts, accessible by turning the hub flange so that the large circular cutout exposes each bolt **(see illustration)**.
4 Remove the hub and bearing assembly from its seat, maneuvering it out through the brake assembly.

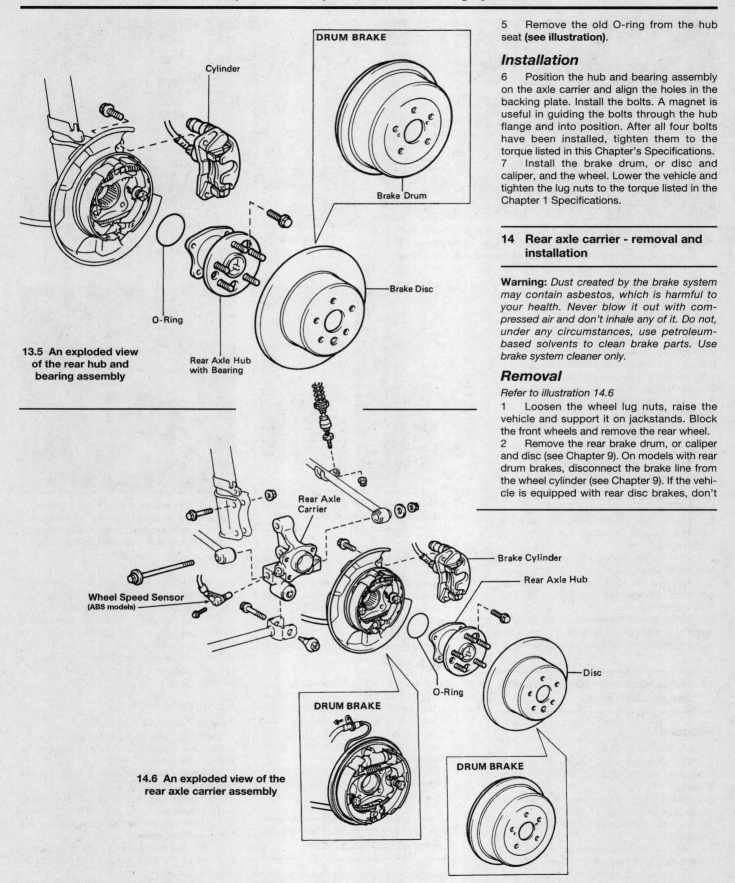

5 Remove the old O-ring from the hub seat **(see illustration)**.

Installation

6 Position the hub and bearing assembly on the axle carrier and align the holes in the backing plate. Install the bolts. A magnet is useful in guiding the bolts through the hub flange and into position. After all four bolts have been installed, tighten them to the torque listed in this Chapter's Specifications.

7 Install the brake drum, or disc and caliper, and the wheel. Lower the vehicle and tighten the lug nuts to the torque listed in the Chapter 1 Specifications.

14 Rear axle carrier - removal and installation

Warning: *Dust created by the brake system may contain asbestos, which is harmful to your health. Never blow it out with compressed air and don't inhale any of it. Do not, under any circumstances, use petroleum-based solvents to clean brake parts. Use brake system cleaner only.*

Removal

Refer to illustration 14.6

1 Loosen the wheel lug nuts, raise the vehicle and support it on jackstands. Block the front wheels and remove the rear wheel.

2 Remove the rear brake drum, or caliper and disc (see Chapter 9). On models with rear drum brakes, disconnect the brake line from the wheel cylinder (see Chapter 9). If the vehicle is equipped with rear disc brakes, don't

13.5 An exploded view of the rear hub and bearing assembly

14.6 An exploded view of the rear axle carrier assembly

Chapter 10 Suspension and steering systems

detach the brake hose from the caliper.
3   On drum brake models, loosen and remove the bolt that retains the flexible hose to the strut.
4   Remove the rear hub and bearing assembly (see Section 13).
5   It isn't necessary to disassemble the brake shoe assembly or disconnect the parking brake cable from the backing plate. On models with rear drum brakes, detach the backing plate and rear brake assembly from the axle carrier; on models with rear disc brakes, detach the backing plate and rear parking brake assembly from the axle carrier. Suspend the backing plate and brake assembly from the coil spring with a piece of wire.
6   On models with ABS, remove the wheel speed sensor from the axle carrier **(see illustration)**.
7   Loosen, but don't remove the strut-to-axle carrier bolts **(see illustration 10.7)**.
8   Remove the suspension arm-to-axle carrier bolt, nut and washers and remove the rear strut rod-to-axle carrier bolt **(see illustration 11.2)**.
9   Remove the loosened strut-to-axle carrier bolts while supporting the carrier so it doesn't fall and detach the axle carrier from the strut bracket.

Installation

10 Inspect the carrier bushing for cracks, deformation and signs of wear. If it is worn out, take the carrier to a dealer service department or other repair shop to have the old one pressed out and a new one pressed in.
11 Push the axle carrier into the strut bracket, aligning the two bolt holes. Insert the two strut-to-carrier bolts and tighten them finger tight.
12 Install the suspension arm-to-axle carrier bolt (from the front), washers and nut. Tighten the nut by hand.
13 Place a jack under the carrier and raise it to simulate normal ride height.
14 Tighten the strut-to-carrier bolts to the torque listed in this Chapter's Specifications.
15 Connect the strut rod and tighten the bolt to the torque listed in this Chapter's Specifications.
16 Tighten the suspension arm bolt/nut to the torque listed in this Chapter's Specifications.
17 On drum brake models, attach the flexible hose to the strut.

18 On models with ABS, reattach the wheel speed sensor to the axle carrier.
19 Attach the brake backing plate to the axle carrier, install the hub and tighten the four bolts to the torque listed in this Chapter's Specifications.
20 On models with rear drum brakes, connect the brake line to the wheel cylinder (see Chapter 9). Be careful not to damage the line when bending it back into place.
21 Install the rear brake drum or disc and caliper (see Chapter 9).
22 Install the wheel and lug nuts.
23 Bleed the wheel cylinder (see Chapter 9).
24 Lower the vehicle and tighten the lug nuts to the torque listed in the Chapter 1 Specifications.

15 Steering system - general information

All models are equipped with rack-and-pinion steering. The steering gear is bolted to the engine cradle and operates the steering knuckles via tie-rods. The inner ends of the tie-rods are protected by rubber boots which should be inspected periodically for secure attachment, tears and leaking lubricant.

The power assist system consists of a belt-driven pump and the associated lines and hoses. The fluid level in the power steering pump reservoir should be checked periodically (see Chapter 1).

The steering wheel operates the steering shaft, which actuates the steering gear through universal joints. Looseness in the steering can be caused by wear in the steering shaft universal joints, the steering gear, the tie-rod ends and loose retaining bolts.

16 Steering wheel - removal and installation

Warning: *The models covered by this manual are equipped with Supplemental Restraint Systems (SRS), more commonly known as airbags. Always disable the airbag system before working in the vicinity of any airbag system components to avoid the possibility of accidental deployment of the airbag(s), which could cause personal injury (see Chapter 12).*

Removal

Refer to illustrations 16.2, 16.3a, 16.3b, 16.3c, 16.4 and 16.6

1   Turn the ignition key to Off, then disconnect the cable from the negative terminal of the battery. Also disconnect the positive cable and wait at least two minutes before proceeding. **Caution:** *If the stereo in your vehicle is equipped with an anti-theft system, make sure you have the correct activation code before disconnecting the battery.*
2   Turn the steering wheel so that the wheels are pointing straight ahead, then pry off the small covers on either side of the steering wheel and loosen the Torx screws that attach the airbag module to the steering wheel **(see illustration)**. Loosen each screw until the groove in the circumference of the screw catches on the screw case.
3   Pull the airbag module off the steering wheel **(see illustration)** and disconnect the module electrical connector **(see illustration)**. **Warning:** *Carry the airbag module with the trim side facing away from you and set it down in an isolated area with the trim side facing up.*

16.3a Remove the airbag module . . .

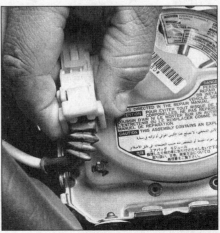

16.3b . . . flip up the lock for the module electrical connector . . .

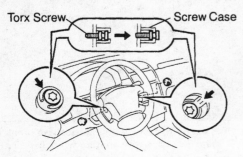

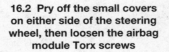

16.2 Pry off the small covers on either side of the steering wheel, then loosen the airbag module Torx screws

Chapter 10 Suspension and steering systems

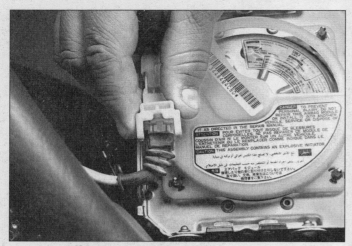

16.3c ... and unplug the electrical connector

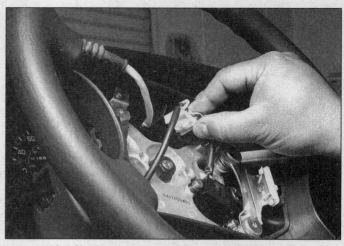

16.4 Unplug the electrical connector for the horn and cruise control

16.6 Use a steering wheel puller to remove the steering wheel

4   Unplug the electrical connector for the horn and cruise control system **(see illustration)**.
5   Remove the steering wheel retaining nut, then mark the relationship of the steering shaft to the hub (if marks don't already exist or don't line up) to simplify installation and ensure steering wheel alignment.
6   Use a puller to disconnect the steering wheel from the shaft **(see illustration)**. **Caution:** *Don't hammer on the shaft in an attempt to remove the wheel.*

Installation

Refer to illustration 16.7

7   Make sure that the front wheels are facing straight ahead. Turn the spiral cable counterclockwise by hand until it becomes harder to turn the cable. Rotate the cable clockwise about three turns and align the two red pointers **(see illustration)**.
8   To install the wheel, align the mark on the steering wheel hub with the mark on the shaft and slip the wheel onto the shaft. Install the nut and tighten it to the torque listed in this Chapter's Specifications.
9   Plug in the horn and cruise control connector.
10 Plug in the electrical connector for the airbag module.
11 Install the airbag module and tighten the Torx retaining screws to the torque listed in this Chapter's Specifications.
12 Connect the battery cables (positive first, negative last).

17 Tie-rod ends - removal and installation

Removal

Refer to illustrations 17.2a, 17.2b and 17.4

1   Loosen the wheel lug nuts. Raise the front of the vehicle, support it securely on jackstands, block the rear wheels and set the

16.7 To align the spiral cable, turn the cable counterclockwise until it's harder to turn, rotate the cable clockwise three turns and align the two red marks (the cable should be able to rotate about three turns in either direction when it's properly centered)

17.2a Hold the tie-rod end with a wrench and break the jam nut loose with another wrench

Chapter 10 Suspension and steering systems

17.2b Back off the jam nut and mark the exposed threads to ensure that the new tie-rod end is threaded on the same number of turns

17.4 Install a small puller as shown to separate the tie-rod from the steering knuckle

parking brake. Remove the front wheel.
2 Hold the tie-rod with a pair of locking pliers or wrench and loosen the jam nut enough to mark the position of the tie-rod end in relation to the threads **(see illustrations)**.
3 Remove the cotter pin and loosen the nut on the tie-rod end stud.
4 Disconnect the tie-rod from the steering knuckle arm with a puller **(see illustration)**. Remove the nut and separate the tie-rod.
5 Unscrew the tie-rod end from the tie-rod.

Installation

6 Thread the tie-rod end on to the marked position and insert the tie-rod stud into the steering knuckle arm. Tighten the jam nut securely.
7 Install the castle nut on the stud and tighten it to the torque listed in this Chapter's Specifications. Install a new cotter pin.
8 Install the wheel and lug nuts. Lower the vehicle and tighten the lug nuts to the torque listed in the Chapter 1 Specifications.

9 Have the alignment checked by a dealer service department or an alignment shop.

18 Steering gear boots - replacement

Refer to illustration 18.3

1 Loosen the lug nuts, raise the vehicle and support it securely on jackstands. Remove the wheel.
2 Remove the tie-rod end and jam nut (see Section 17).
3 Remove the steering gear boot clamps **(see illustration)** and slide off the boot.
4 Before installing the new boot, wrap the threads and serrations on the end of the steering rod with a layer of tape so the small end of the new boot isn't damaged.
5 Slide the new boot into position on the steering gear until it seats in the groove in the steering rod and install new clamps.
6 Remove the tape and install the tie-rod end (see Section 17).
7 Install the wheel and lug nuts. Lower the

vehicle and tighten the lug nuts to the torque listed in the Chapter 1 Specifications.

19 Steering gear - removal and installation

Warning: *Make sure the steering shaft is not turned while the steering gear is removed or you could damage the spiral cable for the airbag system. To prevent the shaft from turning, place the ignition key in the lock position or thread the seat belt through the steering wheel and clip it into place.*

Removal

Refer to illustrations 19.2, 19.3 and 19.6

1 Park the vehicle with the front wheels pointing straight ahead. Loosen the front wheel lug nuts, raise the front of the vehicle and support it securely on jackstands. Apply the parking brake and remove the wheels. Remove the engine under-covers on models so equipped.
2 Place a drain pan under the steering

18.3 Remove the outer clamp (arrow) from the steering gear boot with a pair of pliers; the inner clamp (not visible in this photo) must be cut or pried off

19.2 Disconnect the power steering fluid line fittings (arrows) from the steering gear

10-16   Chapter 10 Suspension and steering systems

19.3 Remove the U-joint pinch bolt and nut (arrows)

gear. Detach the power steering pressure and return lines **(see illustration)** and cap the ends to prevent excessive fluid loss and contamination. Detach the bracket from the top of the steering gear assembly.
3   Mark the relationship of the lower universal joint to the steering gear input shaft. Remove the lower intermediate shaft pinch bolt **(see illustration)**.
4   Separate the tie-rod ends from the steering knuckle arms (see Section 17).
5   Remove the stabilizer bar bushing retainer bolts (see Section 2). (You can't remove the steering gear mounting bolts until the stabilizer is lifted up out of the way.)
6   Remove the steering gear mounting nuts **(see illustration)**, lift up the stabilizer bar and knock out the steering gear mounting bolts.
7   Separate the intermediate shaft from the steering gear input shaft and pull the steering gear assembly out from the right side.

8   Check the steering gear mounting grommets for excessive wear or deterioration, replacing them if necessary.

Installation
9   Raise the steering gear into position and connect the U-joint, aligning the marks.
10 Install the mounting bolts and nuts and tighten them to the torque listed in this Chapter's Specifications.
11 Connect the tie-rod ends to the steering knuckle arms (see Section 17).
12 Install the U-joint pinch bolt and tighten it to the torque listed in this Chapter's Specifications.
13 Connect the power steering pressure and return hoses to the steering gear and fill the power steering pump reservoir with the recommended fluid (see Chapter 1). Reattach the bracket to the top of the steering gear assembly.

14 Install the stabilizer bar bushing retainer bolts and tighten them to the torque listed in this Chapter's Specifications.
15 Lower the vehicle and bleed the steering system (see Section 21).

20 Power steering pump - removal and installation

Removal
Refer to illustrations 20.4a and 20.4b
1   Disconnect the cable from the negative battery terminal. **Caution:** *If the stereo in your vehicle is equipped with an anti-theft system, make sure you have the correct activation code before disconnecting the battery.*
2   Using a large syringe or suction gun, suck as much fluid out of the power steering fluid reservoir as possible. Place a drain pan under the vehicle to catch any fluid that spills out when the hoses are disconnected.
3   Loosen the right front wheel lug nuts, raise the vehicle and support it securely on jackstands. Remove the right front wheel.
4   Remove the right front fender apron seal **(see illustrations)**.
5   Loosen the clamp and disconnect the fluid return hose from the pump.
6   Remove the pressure line-to-pump union bolt and separate the line from the pump. Remove the sealing washers on each side of the fitting - these should be replaced when installing the pump.
7   Loosen the adjuster and pivot bolts and remove the drivebelt.
8   Remove the pivot, adjuster and any other mounting bolts, and remove the pump.

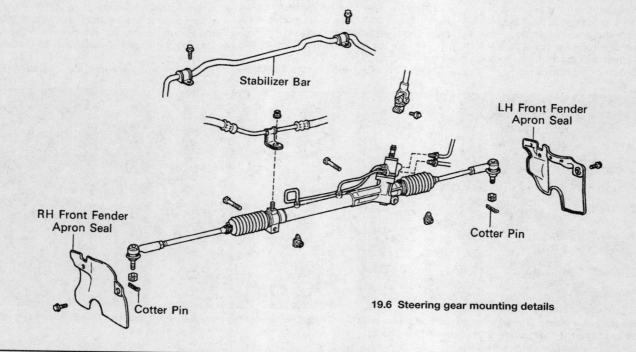

19.6 Steering gear mounting details

Chapter 10 Suspension and steering systems

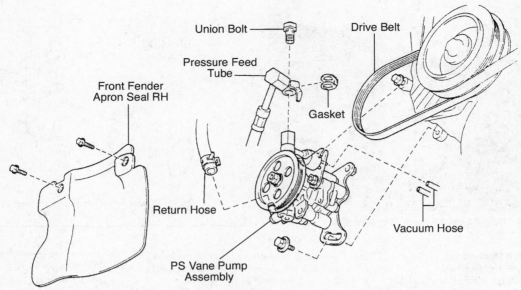

20.4a Installation details of the power steering pump assembly (four-cylinder engine)

Installation

9 Installation is the reverse of removal. Be sure to tighten the fluid line banjo bolt and the adjuster and pivot bolts to the torque listed in this Chapter's Specifications. Adjust the drivebelt tension (see Chapter 1).

10 Top up the fluid level in the reservoir (see Chapter 1) and bleed the system (see Section 21).

21 Power Steering Pressure (PSP) switch - check and replacement

Check

Refer to illustration 21.4

1 The power steering pressure (PSP) switch is located at the high pressure line outlet fitting of the power steering pump.

2 When steering system pressure reaches a high-pressure setpoint, the PSP switch closes and sends a signal to the PCM that the PCM uses to maintain engine idle speed during parking maneuvers.

3 Check the operation of the PSP switch if the engine stalls during parking or if the engine idles continuously at high rpm.

4 Disconnect the PSP switch connector and connect an ohmmeter to the terminals on

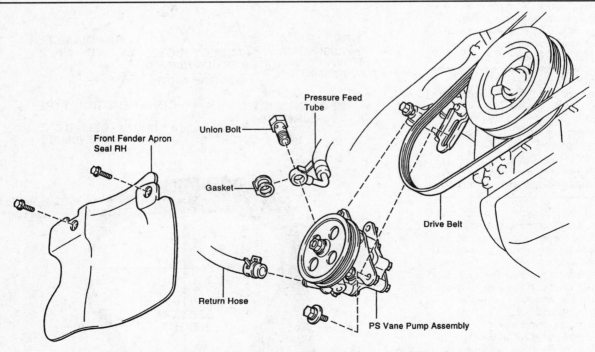

20.4b Installation details of the power steering pump assembly (V6 engine)

Chapter 10 Suspension and steering systems

21.4 Location of the Power Steering Pressure (PSP) switch

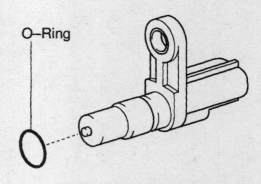

21.11 Power Steering Pressure switch details

the switch body **(see illustration)**.

5   Start the engine and let it idle.

6   Turn the steering wheel to point the front wheels straight ahead and read the ohmmeter. It should indicate no continuity (infinite resistance).

7   Turn the steering wheel to either side and watch the ohmmeter. The PSP switch should close as the wheel nears the steering stop on either side, and the meter should indicate continuity (zero ohms).

8   If the switch fails either test, replace it.

Replacement

Refer to illustration 21.11

9   Raise the vehicle and support it securely on jackstands.

10 Disconnect the cable from the negative battery terminal. **Caution:** *If the stereo in your vehicle is equipped with an anti-theft system, make sure you have the correct activation code before disconnecting the battery.*

11 Disconnect the electrical connector from the switch and unscrew the switch from the fitting on the steering gear.

12 Install and connect the new switch and lower the vehicle to the ground.

13 Refer to Section 22 and bleed air from the power steering system. Add fluid as required (see Chapter 1).

22 Power steering system - bleeding

1   Following any operation in which the power steering fluid lines have been disconnected, the power steering system must be bled to remove all air and obtain proper steering performance.

2   With the front wheels in the straight ahead position, check the power steering fluid level and, if low, add fluid until it reaches the Cold mark on the dipstick.

3   Start the engine and allow it to run at fast idle. Recheck the fluid level and add more if necessary to reach the Cold mark on the dipstick.

4   Bleed the system by turning the wheels from side to side, without hitting the stops. This will work the air out of the system. Keep the reservoir full of fluid as this is done.

5   When the air is worked out of the system, return the wheels to the straight ahead position and leave the vehicle running for several more minutes before shutting it off.

6   Road test the vehicle to be sure the steering system is functioning normally and noise free.

7   Recheck the fluid level to be sure it is up to the Hot mark on the dipstick while the engine is at normal operating temperature. Add fluid if necessary (see Chapter 1).

23 Wheels and tires - general information

Refer to illustration 23.1

1   All vehicles covered by this manual are equipped with metric-sized fiberglass or steel belted radial tires **(see illustration)**. Use of other size or type of tires may affect the ride and handling of the vehicle. Don't mix different types of tires, such as radials and bias belted, on the same vehicle as handling may be seriously affected. It's recommended that tires be replaced in pairs on the same axle,

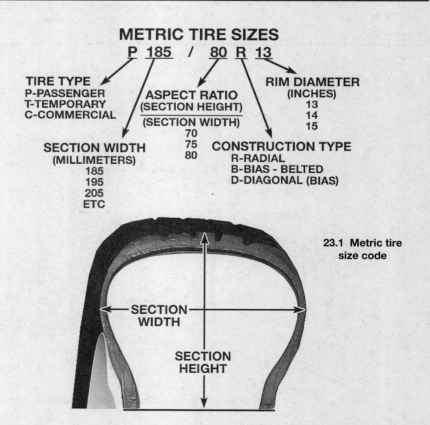

23.1 Metric tire size code

but if only one tire is being replaced, be sure it's the same size, structure and tread design as the other.

2   Because tire pressure has a substantial effect on handling and wear, the pressure on all tires should be checked at least once a month or before any extended trips (see Chapter 1).

3   Wheels must be replaced if they are bent, dented, leak air, have elongated bolt holes, are heavily rusted, out of vertical symmetry or if the lug nuts won't stay tight. Wheel repairs that use welding or peening are not recommended.

4   Tire and wheel balance is important in the overall handling, braking and performance of the vehicle. Unbalanced wheels can adversely affect handling and ride characteristics as well as tire life. Whenever a tire is installed on a wheel, the tire and wheel should be balanced by a shop with the proper equipment.

24 Wheel alignment - general information

Refer to illustration 24.1

A wheel alignment refers to the adjustments made to the wheels so they are in proper angular relationship to the suspension and the ground. Wheels that are out of proper alignment not only affect vehicle control, but also increase tire wear. The alignment angles normally measured are camber, caster and toe-in **(see illustration)**. Toe-in is the only adjustable angle on the front or the rear. The other angles should be measured to check for bent or worn suspension parts.

Getting the proper wheel alignment is a very exacting process, one in which complicated and expensive machines are necessary to perform the job properly. Because of this, you should have a technician with the proper equipment perform these tasks. We will, however, use this space to give you a basic idea of what is involved with a wheel alignment so you can better understand the process and deal intelligently with the shop that does the work.

Toe-in is the turning in of the wheels. The purpose of a toe specification is to

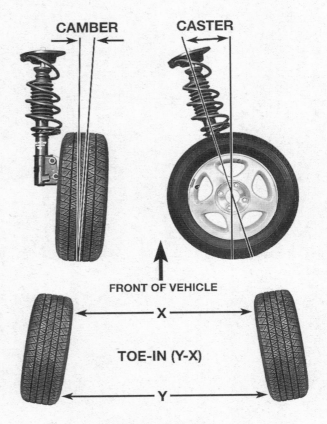

24.1 Camber, caster and toe-in angles

ensure parallel rolling of the wheels. In a vehicle with zero toe-in, the distance between the front edges of the wheels will be the same as the distance between the rear edges of the wheels. The actual amount of toe-in is normally only a fraction of an inch. On the front end, toe-in is controlled by the tie-rod end position on the tie-rod. On the rear end, it's controlled by a threaded adjuster on the rear (number two) suspension arm. Incorrect toe-in will cause the tires to wear improperly by making them scrub against the road surface.

Camber is the tilting of the wheels from vertical when viewed from one end of the vehicle. When the wheels tilt out at the top, the camber is said to be positive (+). When the wheels tilt in at the top the camber is negative (-). The amount of tilt is measured in degrees from vertical and this measurement is called the camber angle. This angle affects the amount of tire tread which contacts the road and compensates for changes in the suspension geometry when the vehicle is cornering or traveling over an undulating surface.

Caster is the tilting of the front steering axis from the vertical. A tilt toward the rear is positive caster and a tilt toward the front is negative caster.

Notes

Chapter 11 Body

Contents

	Section		Section
Body - maintenance	2	Hinges and locks - maintenance	7
Body repair - major damage	6	Hood latch and release cable - removal and installation	10
Body repair - minor damage	5	Hood - removal, installation and adjustment	9
Bumper covers - removal and installation	11	Instrument panel - removal and installation	25
Center console - removal and installation	22	Mirrors - removal and installation	21
Cowl cover - removal and installation	26	Rear package shelf - removal and installation	28
Dashboard trim panels - removal and installation	23	Seats - removal and installation	27
Door latch, lock cylinder and handle - removal and installation	18	Steering column covers - removal and installation	24
		Trunk lid latch and lock cylinder - removal and installation	14
Door - removal, installation and adjustment	17	Trunk lid - removal, installation and adjustment	13
Door trim panels - removal and installation	16	Trunk release and fuel door cable - removal and installation	15
Door window glass - removal and installation	19	Upholstery and carpets - maintenance	4
Door window glass regulator - removal and installation	20	Vinyl trim - maintenance	3
Front fender - removal and installation	12	Windshield and fixed glass - replacement	8
General information	1		

1 General information

Warning: *The front seat belts on some models are equipped with pre-tensioners, which are pyrotechnic (explosive) devices designed to retract the seat belts in the event of a collision. On models equipped with pre-tensioners, do not remove the front seat belt retractor assemblies, and do not disconnect the electrical connectors leading to the assemblies. Problems with the pre-tensioners will turn on the SRS (airbag) warning light on the dash. If any pre-tensioner problems are suspected, take the vehicle to a dealer service department.*

These models feature a "unibody" layout, using a floor pan with front and rear frame side rails which support the body components, front and rear suspension systems and other mechanical components.

Certain components are particularly vulnerable to accident damage and can be unbolted and repaired or replaced. Among these parts are the body moldings, bumpers, front fenders, the hood and trunk lid, doors and all glass.

Only general body maintenance practices and body panel repair procedures within the scope of the do-it-yourselfer are included in this Chapter.

2 Body - maintenance

1 The condition of your vehicle's body is very important, because the resale value depends a great deal on it. It's much more difficult to repair a neglected or damaged body than it is to repair mechanical components. The hidden areas of the body, such as the wheel wells, the frame and the engine compartment, are equally important, although they don't require as frequent attention as the rest of the body.
2 Once a year, or every 12,000 miles, it's a good idea to have the underside of the body steam-cleaned. All traces of dirt and oil will be removed and the area can then be inspected carefully for rust, damaged brake lines, frayed electrical wires, damaged cables and other problems.
3 At the same time, clean the engine and the engine compartment with a steam cleaner or water-soluble degreaser.
4 The wheel wells should be given close attention, since undercoating can peel away and stones and dirt thrown up by the tires can cause the paint to chip and flake, allowing rust to set in. If rust is found, clean down to the bare metal and apply an anti-rust paint.
5 The body should be washed about once a week. Wet the vehicle thoroughly to soften the dirt, then wash it down with a soft sponge and plenty of clean soapy water. If the surplus dirt is not washed off very carefully, it can wear down the paint.
6 Spots of tar or asphalt thrown up from the road should be removed with a cloth soaked in kerosene. Scented lamp oil is available in most hardware stores and the smell is easier to work with than straight kerosene.
7 Once every six months, wax the body and chrome trim. If a chrome cleaner is used to remove rust from any of the vehicle's plated parts, remember that the cleaner also removes part of the chrome, so use it sparingly. On any plated parts where chrome cleaner is used, use a good paste wax over the plating for extra protection.

3 Vinyl trim - maintenance

Don't clean vinyl trim with detergents, caustic soap or petroleum-based cleaners. Plain soap and water works just fine, with a soft brush to clean dirt that may be ingrained. Wash the vinyl as frequently as the rest of the vehicle.

After cleaning, application of a high quality rubber and vinyl protectant will help prevent oxidation and cracks. The protectant can also be applied to weather stripping, vacuum lines and rubber hoses, which often fail as a result of chemical degradation, and to the tires.

4 Upholstery and carpets - maintenance

1 Every three months remove the floormats and clean the interior of the vehicle (more frequently if necessary). Use a stiff whisk broom to brush the carpeting and loosen dirt and dust, then vacuum the upholstery and carpets thoroughly, especially along seams and crevices.
2 Dirt and stains can be removed from carpeting with basic household or automotive carpet shampoos available in spray cans. Follow the directions and vacuum again, then use a stiff brush to bring back the "nap" of the carpet.
3 Most interiors have cloth or vinyl upholstery, either of which can be cleaned and maintained with a number of material-specific cleaners or shampoos available in auto supply stores. Follow the directions on the product for usage, and always spot-test any upholstery cleaner on an inconspicuous area (bottom edge of a backseat cushion) to ensure that it doesn't cause a color shift in the material.
4 After cleaning, vinyl upholstery should be treated with a protectant. **Note:** *Make sure the protectant container indicates the product can be used on seats - some products may make a seat too slippery.* **Caution:** *Do not use protectant on steering wheels.*
5 Leather upholstery requires special care. It should be cleaned regularly with saddlesoap or leather cleaner. Never use alcohol, gasoline, nail polish remover or thinner to clean leather upholstery.
6 After cleaning, regularly treat leather upholstery with a leather conditioner, rubbed in with a soft cotton cloth. Never use car wax on leather upholstery.
7 In areas where the interior of the vehicle is subject to bright sunlight, cover leather seating areas of the seats with a sheet if the vehicle is to be left out for any length of time.

5 Body repair - minor damage

Flexible plastic body panels (front and rear bumper fascia)

The following repair procedures are for minor scratches and gouges. Repair of more serious damage should be left to a dealer service department or qualified auto body shop. Below is a list of the equipment and materials necessary to perform the following repair procedures on plastic body panels. Although a specific brand of material may be mentioned, it should be noted that equivalent products from other manufacturers may be used instead.

> Wax, grease and silicone removing solvent
> Cloth-backed body tape
> Sanding discs
> Drill motor with three-inch disc holder
> Hand sanding block
> Rubber squeegees
> Sandpaper
> Non-porous mixing palette
> Wood paddle or putty knife
> Curved-tooth body file
> Flexible parts repair material

1 Remove the damaged panel, if necessary or desirable. In most cases, repairs can be carried out with the panel installed.
2 Clean the area(s) to be repaired with a wax, grease and silicone removing solvent applied with a water-dampened cloth.
3 If the damage is structural, that is, if it extends through the panel, clean the backside of the panel area to be repaired as well. Wipe dry.
4 Sand the rear surface about 1-1/2 inches beyond the break.
5 Cut two pieces of fiberglass cloth large enough to overlap the break by about 1-1/2 inches. Cut only to the required length.
6 Mix the adhesive from the repair kit according to the instructions included with the kit, and apply a layer of the mixture approximately 1/8-inch thick on the backside of the panel. Overlap the break by at least 1-1/2 inches.
7 Apply one piece of fiberglass cloth to the adhesive and cover the cloth with additional adhesive. Apply a second piece of fiberglass cloth to the adhesive and immediately cover the cloth with additional adhesive in sufficient quantity to fill the weave.
8 Allow the repair to cure for 20 to 30 minutes at 60-degrees to 80-degrees F.
9 If necessary, trim the excess repair material at the edge.
10 Remove all of the paint film over and around the area(s) to be repaired. The repair material should not overlap the painted surface.
11 With a drill motor and a sanding disc (or a rotary file), cut a "V" along the break line approximately 1/2-inch wide. Remove all dust and loose particles from the repair area.
12 Mix and apply the repair material. Apply a light coat first over the damaged area; then continue applying material until it reaches a level slightly higher than the surrounding finish.
13 Cure the mixture for 20 to 30 minutes at 60-degrees to 80-degrees F.
14 Roughly establish the contour of the area being repaired with a body file. If low areas or pits remain, mix and apply additional adhesive.
15 Block sand the damaged area with sandpaper to establish the actual contour of the surrounding surface.
16 If desired, the repaired area can be temporarily protected with several light coats of primer. Because of the special paints and techniques required for flexible body panels, it is recommended that the vehicle be taken to a paint shop for completion of the body repair.

Steel body panels

See photo sequence

Repair of minor scratches

17 If the scratch is superficial and does not penetrate to the metal of the body, repair is very simple. Lightly rub the scratched area with a fine rubbing compound to remove loose paint and built up wax. Rinse the area with clean water.
18 Apply touch-up paint to the scratch, using a small brush. Continue to apply thin layers of paint until the surface of the paint in the scratch is level with the surrounding paint. Allow the new paint at least two weeks to harden, then blend it into the surrounding paint by rubbing with a very fine rubbing compound. Finally, apply a coat of wax to the scratch area.
19 If the scratch has penetrated the paint and exposed the metal of the body, causing the metal to rust, a different repair technique is required. Remove all loose rust from the bottom of the scratch with a pocket knife, then apply rust inhibiting paint to prevent the formation of rust in the future. Using a rubber or nylon applicator, coat the scratched area with glaze-type filler. If required, the filler can be mixed with thinner to provide a very thin paste, which is ideal for filling narrow scratches. Before the glaze filler in the scratch hardens, wrap a piece of smooth cotton cloth around the tip of a finger. Dip the cloth in thinner and then quickly wipe it along the surface of the scratch. This will ensure that the surface of the filler is slightly hollow. The scratch can now be painted over as described earlier in this Section.

Repair of dents

20 When repairing dents, the first job is to pull the dent out until the affected area is as close as possible to its original shape. There is no point in trying to restore the original shape completely as the metal in the damaged area will have stretched on impact and cannot be restored to its original contours. It is better to bring the level of the dent up to a point which is about 1/8-inch below the level of the surrounding metal. In cases where the dent is very shallow, it is not worth trying to pull it out at all.
21 If the back side of the dent is accessible, it can be hammered out gently from behind using a soft-face hammer. While doing this, hold a block of wood firmly against the oppo-

site side of the metal to absorb the hammer blows and prevent the metal from being stretched.

22 If the dent is in a section of the body which has double layers, or some other factor makes it inaccessible from behind, a different technique is required. Drill several small holes through the metal inside the damaged area, particularly in the deeper sections. Screw long, self tapping screws into the holes just enough for them to get a good grip in the metal. Now the dent can be pulled out by pulling on the protruding heads of the screws with locking pliers.

23 The next stage of repair is the removal of paint from the damaged area and from an inch or so of the surrounding metal. This is easily done with a wire brush or sanding disk in a drill motor, although it can be done just as effectively by hand with sandpaper. To complete the preparation for filling, score the surface of the bare metal with a screwdriver or the tang of a file or drill small holes in the affected area. This will provide a good grip for the filler material. To complete the repair, see the Section on filling and painting.

Repair of rust holes or gashes

24 Remove all paint from the affected area and from an inch or so of the surrounding metal using a sanding disk or wire brush mounted in a drill motor. If these are not available, a few sheets of sandpaper will do the job just as effectively.

25 With the paint removed, you will be able to determine the severity of the corrosion and decide whether to replace the whole panel, if possible, or repair the affected area. New body panels are not as expensive as most people think and it is often quicker to install a new panel than to repair large areas of rust.

26 Remove all trim pieces from the affected area except those which will act as a guide to the original shape of the damaged body, such as headlight shells, etc. Using metal snips or a hacksaw blade, remove all loose metal and any other metal that is badly affected by rust. Hammer the edges of the hole in to create a slight depression for the filler material.

27 Wire brush the affected area to remove the powdery rust from the surface of the metal. If the back of the rusted area is accessible, treat it with rust inhibiting paint.

28 Before filling is done, block the hole in some way. This can be done with sheet metal riveted or screwed into place, or by stuffing the hole with wire mesh.

29 Once the hole is blocked off, the affected area can be filled and painted. See the following subsection on filling and painting.

Filling and painting

30 Many types of body fillers are available, but generally speaking, body repair kits which contain filler paste and a tube of resin hardener are best for this type of repair work. A wide, flexible plastic or nylon applicator will be necessary for imparting a smooth and contoured finish to the surface of the filler material. Mix up a small amount of filler on a clean piece of wood or cardboard (use the hardener sparingly). Follow the manufacturer's instructions on the package, otherwise the filler will set incorrectly.

31 Using the applicator, apply the filler paste to the prepared area. Draw the applicator across the surface of the filler to achieve the desired contour and to level the filler surface. As soon as a contour that approximates the original one is achieved, stop working the paste. If you continue, the paste will begin to stick to the applicator. Continue to add thin layers of paste at 20-minute intervals until the level of the filler is just above the surrounding metal.

32 Once the filler has hardened, the excess can be removed with a body file. From then on, progressively finer grades of sandpaper should be used, starting with a 180-grit paper and finishing with 600-grit wet-or-dry paper. Always wrap the sandpaper around a flat rubber or wooden block, otherwise the surface of the filler will not be completely flat. During the sanding of the filler surface, the wet-or-dry paper should be periodically rinsed in water. This will ensure that a very smooth finish is produced in the final stage.

33 At this point, the repair area should be surrounded by a ring of bare metal, which in turn should be encircled by the finely feathered edge of good paint. Rinse the repair area with clean water until all of the dust produced by the sanding operation is gone.

34 Spray the entire area with a light coat of primer. This will reveal any imperfections in the surface of the filler. Repair the imperfections with fresh filler paste or glaze filler and once more smooth the surface with sandpaper. Repeat this spray-and-repair procedure until you are satisfied that the surface of the filler and the feathered edge of the paint are perfect. Rinse the area with clean water and allow it to dry completely.

35 The repair area is now ready for painting. Spray painting must be carried out in a warm, dry, windless and dust free atmosphere. These conditions can be created if you have access to a large indoor work area, but if you are forced to work in the open, you will have to pick the day very carefully. If you are working indoors, dousing the floor in the work area with water will help settle the dust which would otherwise be in the air. If the repair area is confined to one body panel, mask off the surrounding panels. This will help minimize the effects of a slight mismatch in paint color. Trim pieces such as chrome strips, door handles, etc., will also need to be masked off or removed. Use masking tape and several thickness of newspaper for the masking operations.

36 Before spraying, shake the paint can thoroughly, then spray a test area until the spray painting technique is mastered. Cover the repair area with a thick coat of primer. The thickness should be built up using several thin layers of primer rather than one thick one. Using 600-grit wet-or-dry sandpaper, rub down the surface of the primer until it is very smooth. While doing this, the work area should be thoroughly rinsed with water and the wet-or-dry sandpaper periodically rinsed as well. Allow the primer to dry before spraying additional coats.

37 Spray on the top coat, again building up the thickness by using several thin layers of paint. Begin spraying in the center of the repair area and then, using a circular motion, work out until the whole repair area and about two inches of the surrounding original paint is covered. Remove all masking material 10 to 15 minutes after spraying on the final coat of paint. Allow the new paint at least two weeks to harden, then use a very fine rubbing compound to blend the edges of the new paint into the existing paint. Finally, apply a coat of wax.

6 Body repair - major damage

1 Major damage must be repaired by an auto body shop specifically equipped to perform unibody repairs. These shops have the specialized equipment required to do the job properly.

2 If the damage is extensive, the body must be checked for proper alignment or the vehicle's handling characteristics may be adversely affected and other components may wear at an accelerated rate.

3 Due to the fact that some of the major body components (hood, fenders, doors, etc.) are separate and replaceable units, any seriously damaged components should be replaced rather than repaired. Sometimes the components can be found in a wrecking yard that specializes in used vehicle components, often at considerable savings over the cost of new parts.

7 Hinges and locks - maintenance

Once every 3000 miles, or every three months, the hinges and latch assemblies on the doors, hood and trunk (or liftgate) should be given a few drops of light oil or lock lubricant. The door latch strikers should also be lubricated with a thin coat of grease to reduce wear and ensure free movement. Lubricate the door and trunk (or liftgate) locks with spray-on graphite lubricant.

8 Windshield and fixed glass - replacement

Replacement of the windshield and fixed glass requires the use of special fast-setting adhesive/caulk materials and some specialized tools and techniques. These operations should be left to a dealer service department or a shop specializing in glass work.

These photos illustrate a method of repairing simple dents. They are intended to supplement *Body repair - minor damage* in this Chapter and should not be used as the sole instructions for body repair on these vehicles.

1 If you can't access the backside of the body panel to hammer out the dent, pull it out with a slide-hammer-type dent puller. In the deepest portion of the dent or along the crease line, drill or punch hole(s) at least one inch apart . . .

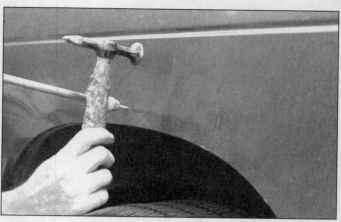

2 . . . then screw the slide-hammer into the hole and operate it. Tap with a hammer near the edge of the dent to help 'pop' the metal back to its original shape. When you're finished, the dent area should be close to its original contour and about 1/8-inch below the surface of the surrounding metal

3 Using coarse-grit sandpaper, remove the paint down to the bare metal. Hand sanding works fine, but the disc sander shown here makes the job faster. Use finer (about 320-grit) sandpaper to feather-edge the paint at least one inch around the dent area

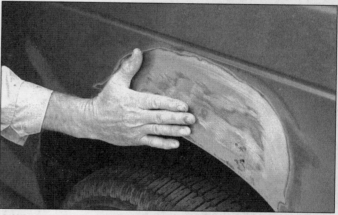

4 When the paint is removed, touch will probably be more helpful than sight for telling if the metal is straight. Hammer down the high spots or raise the low spots as necessary. Clean the repair area with wax/silicone remover

5 Following label instructions, mix up a batch of plastic filler and hardener. The ratio of filler to hardener is critical, and, if you mix it incorrectly, it will either not cure properly or cure too quickly (you won't have time to file and sand it into shape)

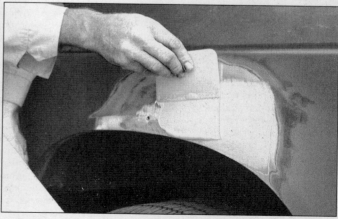

6 Working quickly so the filler doesn't harden, use a plastic applicator to press the body filler firmly into the metal, assuring it bonds completely. Work the filler until it matches the original contour and is slightly above the surrounding metal

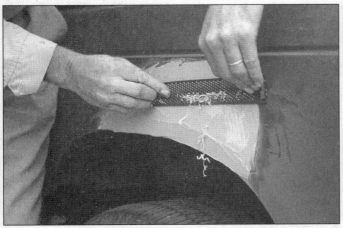

7 Let the filler harden until you can just dent it with your fingernail. Use a body file or Surform tool (shown here) to rough-shape the filler

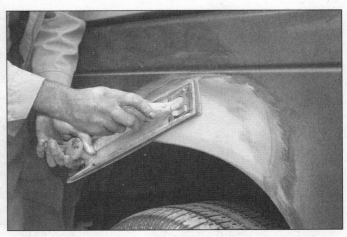

8 Use coarse-grit sandpaper and a sanding board or block to work the filler down until it's smooth and even. Work down to finer grits of sandpaper - always using a board or block - ending up with 360 or 400 grit

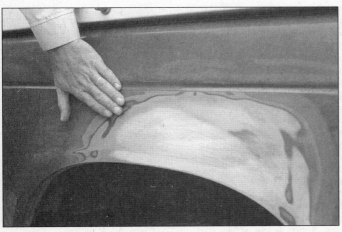

9 You shouldn't be able to feel any ridge at the transition from the filler to the bare metal or from the bare metal to the old paint. As soon as the repair is flat and uniform, remove the dust and mask off the adjacent panels or trim pieces

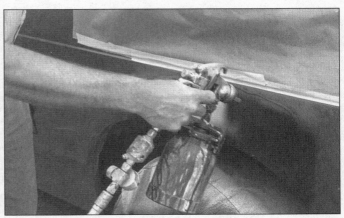

10 Apply several layers of primer to the area. Don't spray the primer on too heavy, so it sags or runs, and make sure each coat is dry before you spray on the next one. A professional-type spray gun is being used here, but aerosol spray primer is available inexpensively from auto parts stores

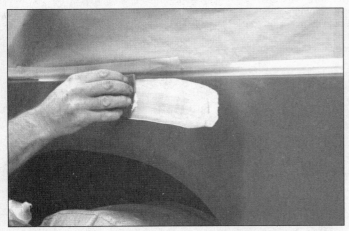

11 The primer will help reveal imperfections or scratches. Fill these with glazing compound. Follow the label instructions and sand it with 360 or 400-grit sandpaper until it's smooth. Repeat the glazing, sanding and respraying until the primer reveals a perfectly smooth surface

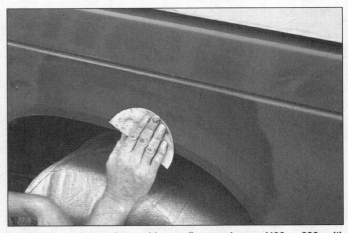

12 Finish sand the primer with very fine sandpaper (400 or 600-grit) to remove the primer overspray. Clean the area with water and allow it to dry. Use a tack rag to remove any dust, then apply the finish coat. Don't attempt to rub out or wax the repair area until the paint has dried completely (at least two weeks)

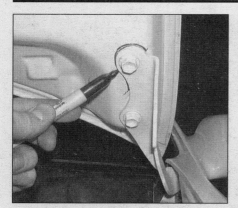

9.3 Draw alignment marks around the hood hinges to ensure proper alignment of the hood when it's reinstalled

9.4 On Avalon models, detach the hood support struts by prying out the clip (arrow) and pulling the strut from the stud

9.10 To adjust the hood latch horizontally or vertically, loosen these bolts (arrows)

9 Hood - removal, installation and adjustment

Note: *The hood is somewhat awkward to remove and install, at least two people should perform this procedure.*

Removal and installation

Refer to illustrations 9.3 and 9.4

1 Open the hood, then place blankets or pads over the fenders and cowl area of the body. This will protect the body and paint as the hood is lifted off.

2 Disconnect any cables or wires that will interfere with removal. Disconnect the windshield washer tubing from the nozzles on the hood.

3 Make marks around the hood hinge to ensure proper alignment during installation **(see illustration)**.

4 Have an assistant support the weight of the hood and, on Avalon models, detach the support struts by prying out the clips at the top **(see illustration)**.

5 Remove the hinge-to-hood bolts and lift off the hood.

6 Installation is the reverse of removal. Align the hinge bolts with the marks made in Step 3.

Adjustment

Refer to illustrations 9.10 and 9.11

7 Fore-and-aft and side-to-side adjustment of the hood is done by moving the hinge plate slot after loosening the bolts or nuts. **Note:** *The factory bolts are "centering" type that will not allow adjustment. To adjust the hood in relation to the hinges, these bolts must be replaced with standard bolts with flat washers and lock washers.*

8 Mark around the entire hinge plate so you can determine the amount of movement.

9 Loosen the bolts and move the hood into correct alignment. Move it only a little at a time. Tighten the hinge bolts and carefully lower the hood to check the position.

10 If necessary after installation, the entire hood latch assembly can be adjusted up-and-down as well as from side-to-side on the radiator support so the hood closes securely and flush with the fenders. Scribe a line or mark around the hood latch mounting bolts to provide a reference point, then loosen them and reposition the latch assembly, as necessary **(see illustration)**. Following adjustment, retighten the mounting bolts.

11 Finally, adjust the hood bumpers on the radiator support so the hood, when closed, is flush with the fenders **(see illustration)**.

12 The hood latch assembly, as well as the hinges, should be periodically lubricated with white, lithium-base grease to prevent binding and wear.

10 Hood latch and release cable - removal and installation

Warning: *The models covered by this manual are equipped with Supplemental Restraint systems (SRS), more commonly known as airbags. Always disconnect the negative battery cable, then the positive battery cable and wait two minutes before working in the vicinity of any airbag system component to avoid the possibility of accidental deployment of the airbag, which could cause personal injury (see Chapter 12). The yellow wiring harnesses and connectors are for this system. Do not use electrical test equipment on any of the airbag system wiring or tamper with it in any way.*

Caution: *If the stereo in your vehicle is equipped with an anti-theft system, make sure you have the correct activation code before disconnecting the battery.*

Latch

Refer to illustration 10.2

1 Scribe a line around the latch to aid

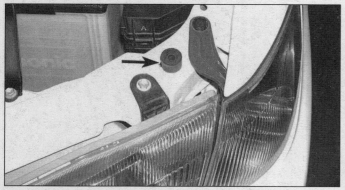

9.11 To adjust the vertical height of the leading edge of the hood so that it's flush with the fenders, turn each edge cushion (arrow indicates one) clockwise (to lower the hood) or counterclockwise (to raise the hood)

10.2 Pry out the cable retainer from the backside of the hood latch assembly, then disengage the cable

Chapter 11 Body

10.8 Remove the two screws (arrows) retaining the hood release lever to the body

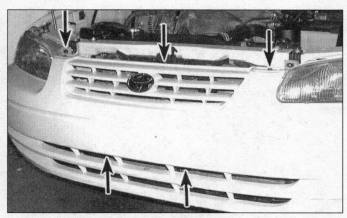

11.3a On Camry models, remove the bolts and plastic pins (arrows) retaining the front bumper cover - the grille is part of the bumper cover - two pins are accessed from below

alignment when installing, then remove the plastic cover over the latch, retained by plastic pins. Remove the retaining bolts securing the hood latch to the radiator support **(see illustration 9.10)**. Remove the latch.

2   Disconnect the hood release cable by disengaging the cable from the latch assembly **(see illustration)**.

3   Installation is the reverse of removal. **Note:** *Adjust the latch so the hood engages securely when closed and the hood bumpers are slightly compressed.*

Cable

Refer to illustration 10.8

4   Disconnect the hood release cable from the latch assembly as described in step 1.

5   Attach a piece of thin wire or string to the end of the cable and unclip all remaining cable retaining clips. at the radiator support.

6   Refer to Section 12 and remove the plastic inner fenderwell.

7   Working in the passenger compartment, remove the driver's side door sill cover and kick panel.

8   Remove the two hood release lever mounting bolts and detach the hood release lever **(see illustration)**.

9   Pull the cable and grommet rearward into the passenger compartment until you can see the wire or string. Ensure that the new cable has a grommet attached then remove the wire or string from the old cable and fasten it to the new cable.

10   With the new cable attached to the wire or string, pull the wire or string back through the firewall until the new cable reaches the latch assembly.

11   Working in the passenger compartment, reinstall the new cable into the hood release lever, making sure the cable housing fits snugly into the notch in the handle bracket.

12   The remainder of the installation is the reverse of removal. **Note:** *Push on the grommet with your fingers from the passenger compartment to seat the grommet in the firewall correctly.*

11 Bumper covers - removal and installation

Warning: *The models covered by this manual are equipped with Supplemental Restraint systems (SRS), more commonly known as airbags. Always disconnect the negative battery cable, then the positive battery cable and wait two minutes before working in the vicinity of any airbag system component to avoid the possibility of accidental deployment of the airbag, which could cause personal injury* (see Chapter 12). *The yellow wiring harnesses and connectors are for this system. Do not use electrical test equipment on any of the airbag system wiring or tamper with it in any way.*
Caution: *If the stereo in your vehicle is equipped with an anti-theft system, make sure you have the correct activation code before disconnecting the battery.*

Front bumper

Refer to illustrations 11.3a, 11.3b, 11.4a, 11.4b, 11.4c and 11.4d

1   Apply the parking brake, raise the vehicle and support it securely on jackstands.

2   Disconnect the negative battery cable, then the positive battery cable and wait two minutes before proceeding any further.

3   On Camry models, detach the screws securing the top, bottom and sides of the bumper cover **(see illustrations)**. **Note:** *Use a small screwdriver to pop the center button up on the plastic fasteners, but do not try to remove the center buttons. They stay in the ferrules.*

4   On Avalon models, remove the headlights and parking lights (see Chapter 12) and

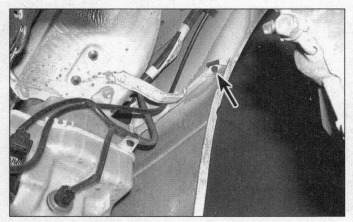

11.3b Remove the bolt (arrow) on each side where the bumper cover attaches to the fender (Camry models)

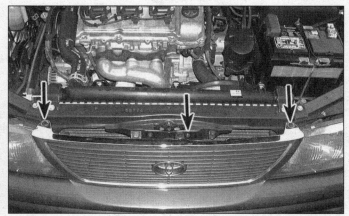

11.4a Remove the two screws and one pin (arrows) holding the grille in place on Avalon models

11.4b On Avalon models, disconnect the fog light connectors, if equipped, then remove the upper bolts (arrows) retaining the front bumper cover

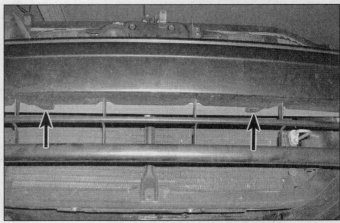

11.4c Remove two pins (arrows) from the bottom (Avalon) . . .

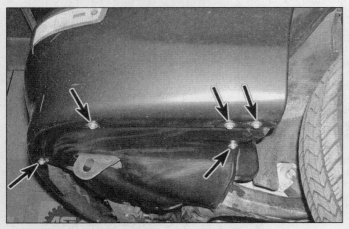

11.4d . . . then remove the screws (arrows) holding the splash panels to the bottom of the Avalon front bumper cover, providing access to the remaining bolt holding the bumper cover to the bottom of the fender

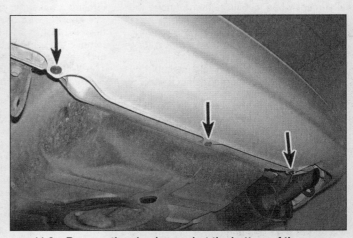

11.6a Remove the pins (arrows) at the bottom of the rear bumper cover - Camry models

grille **(see illustrations)**. Remove the cover.

5 Installation is the reverse of removal. Make sure the tabs on the back of the bumper cover fit into the corresponding clips on the body before attaching the bolts and screws. An assistant would be helpful at this point.

Rear bumper

Refer to illustrations 11.6a, 11.6b 11.7 and 11.8

6 Working under the vehicle, detach the plastic clips and screws securing the lower edge of the bumper cover **(see illustrations)**.

7 Remove the nuts securing the bumper cover in each corner of the trunk interior **(see illustration)**.
8 Open the trunk lid and remove the clips

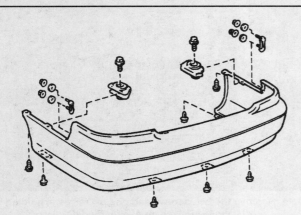

11.6b Rear bumper cover fasteners - Avalon models

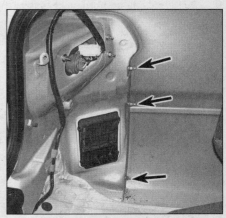

11.7 Inside the trunk with the liner pulled back, remove the three nuts (arrows) holding the ends of the rear bumper cover

Chapter 11 Body

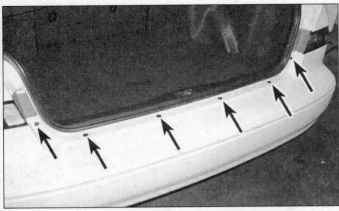

11.8 Remove the pins (arrows) securing the top edge of the rear bumper cover to the trunk opening

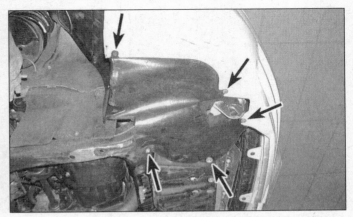

12.3a Remove the bolts (arrows) at the front lower portion of the inner fenderwell

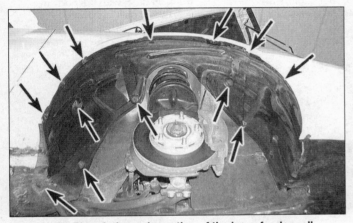

12.3b Detach the main portion of the inner fenderwell, secured by bolts, screws and plastic clips (arrows)

12.4 Remove the upper fender bolt (arrow) with the door open

or pins securing the upper edge of the bumper cover **(see illustration)**. Pull the bumper cover out and away from the vehicle. **Note:** *Use a small screwdriver to pop the center button up on the plastic fasteners, but do not try to remove the center buttons. They stay in the ferrules. Where studs/bolts stick through the body, use a plastic hammer to tap them out of the body.*

9 Installation is the reverse of removal.

12 Front fender - removal and installation

Refer to illustrations 12.3a, 12.3b, 12.4, 12.5, 12.6a, 12.6b and 12.6c

1 Raise the vehicle, support it securely on jackstands and remove the front wheel.
2 Remove the front parking/turn signal light (see Chapter 12).
3 Detach the inner fenderwell screws and clips, then remove the inner fenderwell and mud shield **(see illustrations)**.
4 Open the front door, and remove the upper fender-to-body bolt **(see illustration)**.
5 Remove the lower fender-to-body bolt **(see illustration)**.
6 Remove the remaining fender mounting bolts **(see illustrations)**.
7 Lift off the fender. It's a good idea to

12.5 Remove the bottom bolt (arrow) retaining the fender to the rocker panel area

12.6a Remove the bumper cover bolt (A) at the lower front of the fender, then the fender-to-brace bolt (B)

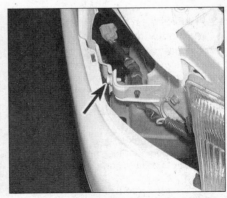

12.6b With the parking/turn signal light housing removed (see Chapter 12), remove the front fender-to-brace bolt (arrow)

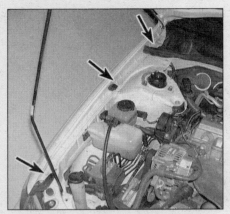

12.6c Remove the three bolts (arrows) along the top of the fender

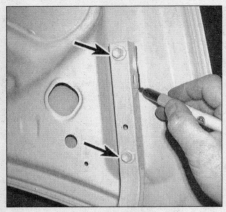

13.3 Draw around the trunk hinges with a marking pen before loosening the bolts (arrows) to ensure proper alignment of the trunk lid when it's reinstalled

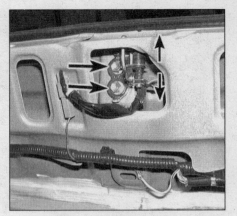

13.7 Loosen the bolts (arrows) and move the striker as necessary to adjust the trunk lid flush with the body in the closed position

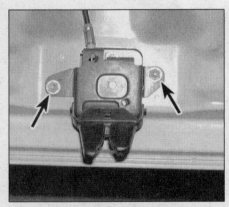

14.1 Scribe around the trunk latch on the trunk lid, then remove the two latch mounting bolts (arrows)

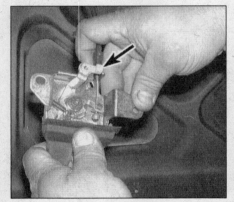

14.3 Pry the release cable (arrow) out of the latch assembly

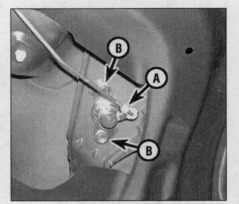

14.5 Pry the rod from its clip (A) and remove the lock cylinder mounting bolt (B) - Avalon shown, Camry similar

have an assistant support the fender while it's being moved away from the vehicle to prevent damage to the surrounding body panels.

8   Installation is the reverse of removal. Check the alignment of the fender to the hood and front edge of the door before final tightening of the fender fasteners.

13 Trunk lid - removal, installation and adjustment

Note: *The trunk lid is heavy and somewhat awkward to remove and install - at least two people should perform this procedure.*

Removal and installation

Refer to illustration 13.3

1   Open the trunk lid and cover the edges of the trunk compartment with pads or cloths to protect the painted surfaces when the lid is removed.
2   Unplug the electrical connectors for the license plate, tail and brake lights (see Chapter 12), trunk lock actuator, and remove the wire harness and actuator cable from the trunk lid (see Section 14). Tie string or wire to the cables before withdrawing them from the trunk lid so they can be pulled back into the trunk lid when its reinstalled. **Note:** *The trunk lid inner liner must be removed to perform these procedures* (see Chapter 12).
3   Scribe or draw alignment marks around the trunk hinges **(see illustration)**.
4   Remove the hinge-to-trunk lid bolts from both sides and lift off the trunk lid.
5   Installation is the reverse of removal. Be sure to align the hinge flanges with the marks made on the trunk lid during removal.

Adjustment

Refer to illustration 13.7

6   After installation, close the lid and see if it's in proper alignment with the adjacent body surfaces. Fore-and-aft and side-to-side adjustments of the lid are controlled by the position of the hinge bolts in the slots. To adjust it, loosen the hinge bolts, reposition the lid and retighten the bolts.
7   The height of the rear of the lid in relation to the surrounding body panels when closed can be adjusted by loosening the lock striker bolts, moving the striker up/down or left/right, then re-tightening the bolts **(see illustration)**.
Note: *Make a reference mark around the striker before making adjustments.*

14 Trunk lid latch and lock cylinder - removal and installation

Trunk lid latch

Refer to illustrations 14.1 and 14.3

1   Scribe a line around the trunk lid latch assembly for a reference point to aid the installation procedure **(see illustration)**.
2   Detach the two retaining bolts and remove the latch.
3   Remove the end of the latch release cable from the latch **(see illustration)**.
4   Installation is the reverse of removal.

Trunk lock cylinder

Refer to illustration 14.5

5   Open the trunk and remove the screws and the trunk lid trim panel. Remove the lock cylinder rod from its clip and remove the lock cylinder mounting bolt **(see illustration)**. On models so equipped, disconnect the electrical connector from the lock.
6   Twist the lock about 45-degrees and remove it from the trunk.
7   Installation is the reverse of removal.

Chapter 11 Body

11-11

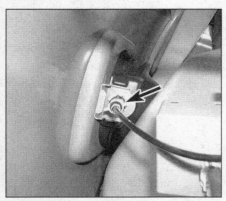

15.2a Twist the fuel door cable end (arrow) one quarter-turn to remove it from the fuel door housing (Camry models)

15.2b On Avalon models, the fuel door cable can be pulled out of its bracket (arrow) and lifted from the mechanism

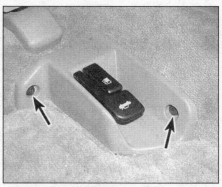

15.6 Remove the two screws (arrows) holding the fuel door/trunklid release assembly to the floor

15 Trunk release and fuel door cable - removal and installation

Refer to illustrations 15.2a, 15.2b, 15.6 and 15.7

1 Refer to Section 14 for removal of the trunk latch and disengagement of the cable from the latch.
2 Inside the trunk, remove the floor mat and left-side panel for access to the fuel door cable end. Twist the fuel door latch cable end (white plastic part) one quarter-turn counter-clockwise and pull it from the fuel door housing **(see illustrations)**.
3 Remove the rear seat bottom as described in Section 27. On coupe models, both seat back and bottom must be removed.
4 Pry up the driver's side door sill covers (only one cover on coupe models). On four-door models, remove the center door pillar's lower trim panel.
5 Peel back the carpeting to access the release cable. Open all of the clips holding the cable to the body.
6 Remove the two screws holding the release handle assembly to the floor **(see illustration)**.
7 Turn the housing over and remove the cable housings from the clips and the eyes from the levers **(see illustration)**.

8 Attach a piece of thin wire to the end of the cable.
9 Working in the trunk compartment, pull the cable towards the rear of the vehicle until you can see the wire.
10 Attach the wire to the front of the new cable and fish it back through the body until it can be attached to the lever. The remainder of the installation is the reverse of removal.

16 Door trim panels - removal and installation

Refer to illustrations 16.1, 16.3a, 16.3b, 16.5a, 16.5b, 16.5c, 16.5d and 16.7

Warning: *The models covered by this manual are equipped with Supplemental Restraint systems (SRS), more commonly known as airbags. Always disconnect the negative battery cable, then the positive battery cable and wait two minutes before working in the vicinity of any airbag system component to avoid the possibility of accidental deployment of the airbag, which could cause personal injury (see Chapter 12). The yellow wiring harnesses and connectors are for this system. Do not use electrical test equipment on any of the airbag system wiring or tamper with it in any way.*

15.7 Remove the cables from the clips (arrows) and the cable eyes from the slots in the levers

Caution: *Wear gloves when working inside the door openings to protect against cuts from sharp metal edges.*

Removal

1 On manual window models, remove the window crank handle **(see illustration)**.
2 Remove the door lock knob.
3 On power window models, pry up the window switch plate to access one panel mounting screw underneath **(see illustrations)**.
4 Pry out the outside mirror cover.

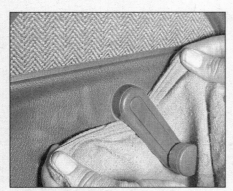

16.1 On models equipped with manual windows, remove the window crank handle by working a cloth behind it to release the clip

16.3a Pry up the power window switch plate, disconnect the electrical connectors . . .

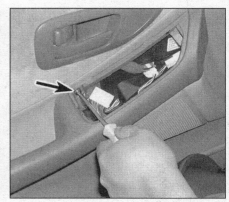

16.3b . . . and remove the one screw (arrow) through the door panel

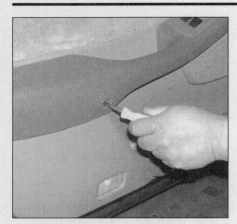

16.5a Pry off the cover, then remove the mounting screw

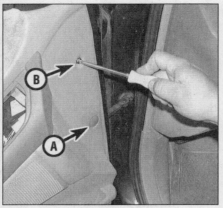

16.5b At the front of the door, remove the plastic covers (A) and the screws (B)

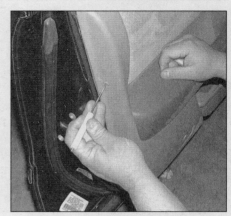

16.5c Pry off the protective cap and remove the screw at the rear edge of the door

5 Remove the door trim panel retaining screws and the screw in the door arm rest, then carefully pry the panel out until the clips disengage **(see illustrations)**. Work slowly and carefully around the outer edge of the trim panel until it's free. **Note:** *Most of the screws on the door panel are covered by small plastic discs, which must be pried out before removing the screws.*

6 Once all of the clips are disengaged, pull the trim panel up from the door, unplug any wiring harness connectors and remove the panel.

7 For access to the door outside handle or the door window regulator inside the door, raise the window fully, remove the power window control unit (if equipped), the door panel bracket and the speaker assembly (see Chapter 12), then carefully peel back the plastic watershield **(see illustration)**.

Installation

8 Prior to installation of the door trim panel, be sure to reinstall any clips in the panel which may have come out when you removed the panel.

9 Plug in the wire harness connectors for the power door lock switch and the power window switch, if equipped, and place the panel in position in the door. Press the door panel into place until the clips are seated. Install the inner door handle and its screw. Install the power door lock switch assembly, if equipped, or the pull-pocket and its screw. Install the window regulator crank handle or power window switch assembly.

17 Door - removal, installation and adjustment

Warning: *The models covered by this manual are equipped with Supplemental Restraint systems (SRS), more commonly known as airbags. Always disconnect the negative battery cable, then the positive battery cable and wait two minutes before working in the vicinity of any airbag system component to avoid the possibility of accidental deployment of the airbag, which could cause personal injury* (see Chapter 12). *The yellow wiring harnesses and connectors are for this system. Do not use electrical test equipment on any of the airbag system wiring or tamper with it in any way.*

Caution: *Wear gloves when working inside the door openings to protect against cuts from sharp metal edges.*

Note: *The door is heavy and somewhat awkward to remove and install - at least two people should perform this procedure.*

Removal and installation

Refer to illustrations 17.6, 17.8a and 17.8b

1 Lower the window completely in the door and then disconnect the negative cable from the battery.

2 Open the door all the way and support it from the ground on jacks or blocks covered with rags to prevent damaging the paint.

3 Remove the door trim panel and watershield as described in Section 16.

4 Disconnect all electrical connections, ground wires and harness retaining clips from the door. **Note:** *It is a good idea to label all connections to aid the reassembly process.*

5 From the door side, detach the rubber conduit between the body and the door. Then pull the wiring harness through the conduit hole and remove it from the door.

6 Remove the door stop strut bolt **(see illustration)**.

7 Mark around the door hinges with a pen or a scribe to facilitate realignment during reassembly.

8 With an assistant holding the door, remove the hinge-to-door bolts **(see illustrations)** and lift the door off. **Note:** *Draw a reference line around the hinges before removing the bolts.*

9 Installation is the reverse of removal.

16.5d Carefully use a trim panel tool to pry all around the panel and release the clips

16.7 Carefully peel back the plastic watershield for access to the inner door

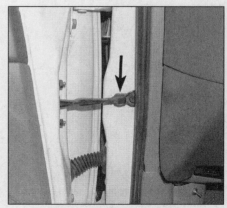

17.6 Remove the bolt retaining the door stop strut (arrow)

Chapter 11 Body

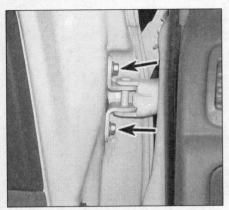

17.8a Remove the door hinge bolts with the door supported (arrows indicate bolts for the top hinge, bottom hinge similar)

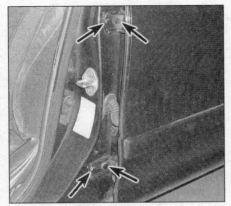

17.8b Open the front door to access the rear door hinge-to-body bolts (arrows)

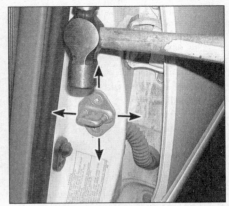

17.13 Adjust the door lock striker by loosening the mounting screws and gently tapping the striker in the desired direction (arrows)

Adjustment

Refer to illustration 17.13

10 Having proper door-to-body alignment is a critical part of a well-functioning door assembly. First check the door hinge pins for excessive play. Fully open the door and lift up and down on the door without lifting the body. If a door has 1/16-inch or more excessive play, the hinges should be replaced.

11 Door-to-body alignment adjustments are made by loosening the hinge-to-body bolts or hinge-to-door bolts and moving the door. Proper body alignment is achieved when the top of the doors are parallel with the roof section, the front door is flush with the fender, the rear door is flush with the rear quarter panel and the bottom of the doors are aligned with the lower rocker panel. If these goals can't be reached by adjusting the hinge-to-body or hinge-to-door bolts, body alignment shims may have to be purchased and inserted behind the hinges to achieve correct alignment.

12 To adjust the door-closed position, scribe a line or mark around the striker plate to provide a reference point, then check that the door latch is contacting the center of the latch striker. If not adjust the up and down position first.

13 Finally adjust the latch striker sideways position, so that the door panel is flush with the center pillar or rear quarter panel and provides positive engagement with the latch mechanism **(see illustration)**.

18 Door latch, lock cylinder and handle - removal and installation

Caution: *Wear gloves when working inside the door openings to protect against cuts from sharp metal edges.*

Door latch

Refer to illustration 18.2

1 Raise the window, then remove the door trim panel and watershield (see Section 16).
2 Working through the large access hole, disengage the outside door handle-to-latch rod, outside door lock-to-latch rod, and the inside handle-to-latch rod **(see illustration)**.
3 All door lock rods are attached by plastic clips. The plastic clips can be removed by unsnapping the portion engaging the connecting rod and then pulling the rod out of its locating hole. On models with power door locks, disconnect the electrical connectors at the latch.

4 Remove the screws securing the latch to the door **(see illustration 18.2)**. Remove the latch assembly through the door opening.
5 Installation is the reverse of removal.

Outside handle and door lock cylinder

Refer to illustrations 18.7 and 18.8

6 To remove the outside handle and lock cylinder assembly, raise the window and remove the door trim panel and watershield (see Section 16). **Caution:** *Take care not to scratch the paint on the outside of the door. Wide masking tape applied around the handle opening before beginning the procedure can help avoid scratches.*

7 Working through the access hole, disengage the plastic clips that secure the outside door lock-to-latch rod, and the bolts holding the lock and handle assembly **(see illustration)**. **Note:** *The handle can be unbolted and pulled toward the outside of the door to make disconnecting the lock rod easier.*

8 Remove the handle and lock cylinder assembly from the vehicle. The lock cylinder can be removed from the handle assembly by

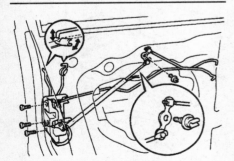

18.2 Detach the plastic clips on the actuating rods leading to the latch - the latch is secured to the end of the door with three screws

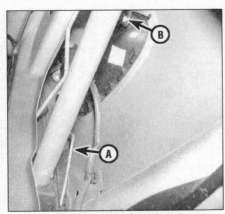

18.7 From inside the door opening, detach the actuating rod (A) - B indicates one of the two bolts retaining the outer handle

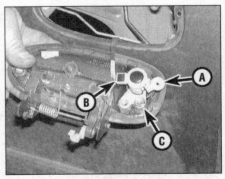

18.8 If not already done from inside, detach the rod at the lock (A), then pull the clip (non-power lock) or disconnect the electrical connector (B) and the lock cylinder bolts (C indicates one, the other is also a handle retaining bolt, already removed)

removing a clip, or the electrical connector can be unbolted and the cylinder left on the handle assembly **(see illustration)**.
9 Installation is the reverse of removal.

19 Door window glass - removal and installation

Caution: *Wear gloves when working inside the door openings to protect against cuts from sharp metal edges.*

Front door glass

Refer to illustrations 19.3a, 19.3b and 19.3c

1 Remove the door trim panel and the plastic watershield (see Section 16).
2 Lower the window glass all the way down into the door.
3 Raise the window just enough to access the window retaining bolts through the holes in the door frame **(see illustrations)**.
4 Place a rag over the glass to help prevent scratching the glass and remove the two glass mounting bolts.
5 Remove the glass by pulling it up and out.
6 Installation is the reverse of removal.

Rear door glass (four-door models)

Refer to illustrations 19.8a and 19.8b

7 Remove the door panel and watershield (see Section 16).
8 Lower the window enough to see the bolts on the glass guide bar. Remove the bolts while holding up the glass **(see illustrations)**. Push the glass from the guide bar and remove the glass through the window opening.
9 Installation is the reverse of the removal procedure.

Rear quarter glass (Solara models)

10 Replacement of the fixed quarter glass requires the use of special fast-setting adhesive/caulk materials and some specialized tools and techniques. These operations should be left to a dealer service department or a shop specializing in glass work.

20 Door window glass regulator - removal and installation

Caution: *Wear gloves when working inside the door openings to protect against cuts from sharp metal edges.*

Front

Refer to illustrations 20.4 and 20.5

1 Remove the door trim panel and the plastic watershield (see Section 16).
2 Remove the window glass assembly (see Section 19).

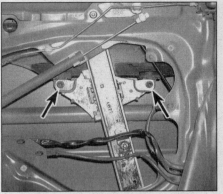

19.3a Raise the window to access the glass retaining bolts (arrows) through the holes in the door frame - Avalon model shown

3 On power-operated windows, disconnect the electrical connector from the window regulator motor.
4 On Avalon and Solara models, remove the regulator/motor assembly mounting bolts **(see illustration)**.
5 On Camry models, the regulator assembly is a scissors-type. Before removing the bolts, mark the position of the rear bolt of the upper roller guide. Remove the bolts for the lower roller guide (equalizer), then the six screws holding the regulator/motor (crank assembly on non-power window models) to the door **(see illustration)**.

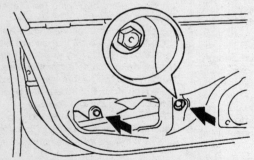

19.3b On Camry models, the two access holes are near the bottom of the door

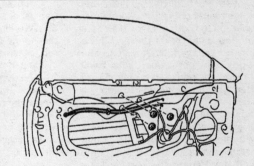

19.3c On Solara and Lexus models, raise the glass until the three nuts can be removed

19.8a On Avalon models, lower the glass until the two bolts (arrows) can be removed - when reinstalling glass, don't tighten the bolts until measurements A and B are equal

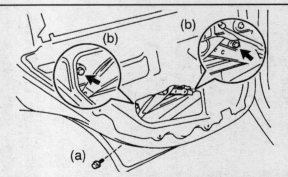

19.8b On Camry models, remove the rear run channel bolt (A), and the two glass retaining bolts (B)

Chapter 11 Body

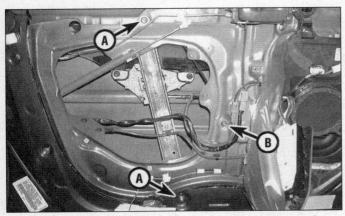

20.4 Remove the regulator nuts (A) and the motor mounting nuts (B indicates one of three) - Lexus, Avalon and Solara models

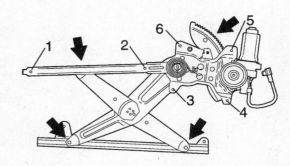

20.5 On Camry models, slide the equalizer bar from the bottom rollers, then remove the 6 mounting screws to remove the regulator/motor from the door - arrows indicate lubrication points

6   Pull the equalizer arm and regulator assemblies through the service hole in the door frame to remove it.
7   Installation is the reverse of removal. Lubricate the rollers and wear points on the regulator with white grease before installation.

Rear

8   Remove the door trim panel and the plastic watershield (see Section 16).
9   Remove the window glass assembly (see Section 19).
10 On power operated windows, disconnect the electrical connector from the window regulator motor.
11 Remove the regulator/motor assembly mounting bolts. On all models, the rear regulator/motor assembly is similarly mounted to that model's assembly for the front doors (see above).
12 Remove the equalizer bar and raise the regulator assembly through the service hole in the door frame to remove it.
13 Installation is the reverse of removal. Lubricate the rollers and wear points on the regulator with white grease before installation.

21 Mirrors - removal and installation

Outside mirrors

Refer to illustration 21.4
1   On power mirror equipped models, remove the door trim panel and the plastic watershield (see Section 16).
2   Pry off the mirror trim cover on the inside of the door.
3   Disconnect the electrical connector from the mirror (if equipped).
4   Remove the three mirror retaining bolts and detach the mirror from the vehicle (see illustration).
5   Installation is the reverse of removal.

Inside mirror

Refer to illustration 21.6
6   Pry between the mirror mount and the notch in the base of the mirror stalk with a screwdriver tip covered with tape (see illustration). There is a hairpin-type spring holding the mirror stalk in the base. Push the screwdriver in about 3/4-inch to release the spring.
7   To install the mirror, reinsert the spring if it was removed earlier. Insert the mirror stalk's lug into the mount, pushing downward until the mirror is secured.
8   If the mount plate itself has come off the windshield, adhesive kits are available at auto parts stores to resecure it. Follow the instructions included with the kit.

22 Center console - removal and installation

Refer to illustration 22.3
Warning: *The models covered by this manual are equipped with Supplemental Restraint systems (SRS), more commonly known as airbags. Always disconnect the negative battery cable, then the positive battery cable and wait two minutes before working in the vicinity of any airbag system component to avoid the possibility of accidental deployment of the airbag, which could cause personal injury (see Chapter 12). The yellow wiring harnesses and connectors are for this system. Do not use electrical test equipment on any of the airbag system wiring or tamper with it in any way.*
1   Disconnect the battery cables (negative first, then positive). **Caution:** *If the stereo in your vehicle is equipped with an anti-theft system, make sure you have the correct activation code before disconnecting the battery.*
2   Using a screwdriver with the tip taped to prevent scratching the panels, pry the center instrument cluster finish panel off.
3   Pry out the center console upper panel from the console (see illustration on following page). On manual transaxle models, refer to Chapter 7 and remove the shift knob.
4   Lift up the armrest and the cover plate at the floor of the armrest compartment to access the screws for the rear console box.
5   Remove the screws at the front of the rear console box and remove the box.
6   Remove the front console screws and lift the console up and over the shift lever. Disconnect any electrical connections and remove the console from the vehicle.
7   Installation is the reverse of removal.

21.4 Remove the three mirror mounting bolts (arrows) - on electric mirrors, trace the electrical harness down into the door and disconnect the connector

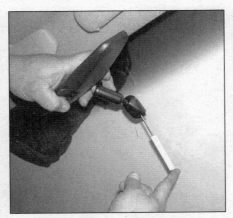

21.6 While pulling up on the inside mirror, push a screwdriver into the slot in the bottom of the base to release it

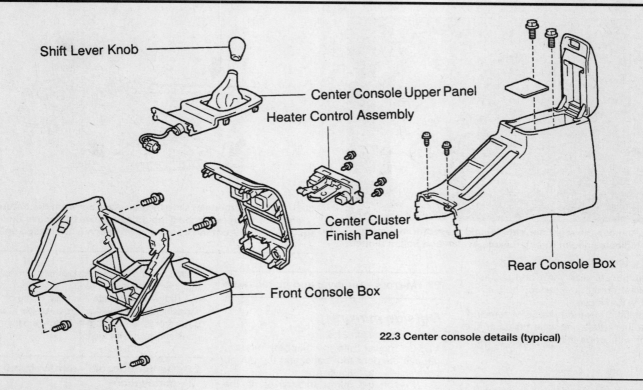

22.3 Center console details (typical)

23 Dashboard trim panels

Warning: *The models covered by this manual are equipped with Supplemental Restraint systems (SRS), more commonly known as airbags. Always disconnect the negative battery cable, then the positive battery cable and wait two minutes before working in the vicinity of any airbag system component to avoid the possibility of accidental deployment of the airbag, which could cause personal injury (see Chapter 12). The yellow wiring harnesses and connectors are for this system. Do not use electrical test equipment on any of the airbag system wiring or tamper with it in any way.*

1 Disconnect the negative battery cable, then the positive cable. **Caution:** *If the stereo in your vehicle is equipped with an anti-theft system, make sure you have the correct activation code before disconnecting the battery.*

Instrument cluster bezel

Refer to illustration 23.3

2 If equipped with a tilt steering column, tilt the column all the way down.
3 Remove the two screws at the top of the instrument cluster bezel **(see illustration)**.
4 Using a screwdriver with the tip taped with masking tape, carefully pry the lower portion of the bezel away from the instrument panel until the clips are released. Take care not to scratch the surrounding trim on the instrument panel.
5 Installation is the reverse of the removal procedure. Make sure the clips are engaged properly before pushing the bezel firmly into place.

Knee bolster

Refer to illustrations 23.8 and 23.9

6 Remove the screws holding the hood-release lever to the knee bolster (some models).
7 Squeeze the side of the pull-out coin box (if equipped) to release it from the bolster.
8 Remove the two bolts and pull the knee bolster out to disengage the clips behind it **(see illustration)**. On Avalon models, remove the cover over the driver's side fuse panel.

23.3 Detach the retaining screws (arrows) at the top of the bezel, then pull the bezel outward to disengage the clips at the bottom - Camry shown (on Avalon models there are four screws)

23.8 Remove the two knee boltster bolts (arrows) - Avalon shown, Camry similar

Chapter 11 Body

11-17

23.9 The two bolts removed for the knee bolster are also two of the bolts retaining the bolster reinforcement panel - remove these two additional bolts (arrows) and remove the reinforcement panel

23.11 Pry the center bezel away from the clips - on Avalon models (shown) pull the trim panel forward enough to disconnect the electrical connector at the back of the hazard warning switch (arrow) between the center dash vents

9 Remove the retaining bolts securing the knee bolster reinforcement, if needed for access to components under the dashboard **(see illustration)**. Pull outward on the lower edge of the knee bolster reinforcement panel and detach it from the vehicle.
10 Installation is the reverse of removal.

Center trim panel
Refer to illustration 23.11

11 Pry the center instrument panel trim bezel carefully from the dashboard with a taped screwdriver **(see illustration)**.
12 Installation is the reverse of removal.

Passenger's side lower panel and glove box
Refer to illustration 23.15

13 The passenger's side lower panel is held in place under the glove box area by clips. Pry and pull it down to release it.
14 On Camry models open the glove box and squeeze the sides in to allow the compartment to come back and down, then remove the mounting screws. On Solara models, remove the two lower screws, then open the glove box and remove the upper screws. The glove box comes out with the panel around it.
15 On Avalon models, remove the three nuts below the glove box door and remove the door **(see illustration)**.

24 Steering column covers - removal and installation

Refer to illustration 24.2
Warning: *The models covered by this manual are equipped with Supplemental Restraint systems (SRS), more commonly known as airbags. Always disconnect the negative battery cable, then the positive battery cable and wait two minutes before working in the vicinity of any airbag system component to avoid the possibility of accidental deployment of the airbag, which could cause personal injury (see Chapter 12). The yellow wiring harnesses and connectors are for this system. Do not use electrical test equipment on any of the airbag system wiring or tamper with it in any way.*
Caution: *If the stereo in your vehicle is equipped with an anti-theft system, make sure you have the correct activation code before disconnecting the battery.*

1 On tilt steering columns, move the column to the lowest position. Refer to Chapter 10 and remove the steering wheel.
2 Remove the screws, then separate the halves and remove the upper and lower steering column covers **(see illustration)**.
3 Installation is the reverse of the removal procedure.

25 Instrument panel - removal and installation

Refer to illustrations 25.7, 25.8a, 25.8b and 25.9
Warning: *The models covered by this manual are equipped with Supplemental Restraint systems (SRS), more commonly known as airbags. Always disconnect the negative battery cable, then the positive battery cable and wait two minutes before working in the vicinity of any airbag system component to avoid the possibility of accidental deployment of the airbag, which could cause personal injury*

23.15 Remove the lower glove box mounting screws (arrows) - Avalon model shown

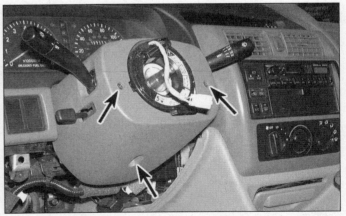

24.2 Remove the screws (arrows), then remove upper and lower covers - Avalon shown; on Camry models, one screw is on the left side behind a plastic access panel

11-18   Chapter 11 Body

25.7 Remove the bolts (arrows) and carefully lower the steering column

25.8a Label and disconnect the main instrument panel electrical connectors at the left . . .

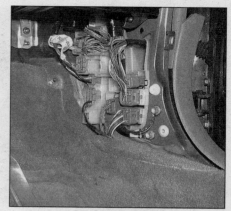

25.8b . . . and the right sides, below the instrument panel

(see Chapter 12). *The yellow wiring harnesses and connectors are for this system. Do not use electrical test equipment on any of the airbag system wiring or tamper with it in any way.*
Caution: *If the stereo in your vehicle is equipped with an anti-theft system, make sure you have the correct activation code before disconnecting the battery.*

1 Disconnect the negative battery cable, then the positive battery cable.
2 Remove the dashboard trim panels (see Section 23) and the center floor console (see Section 22).
3 Remove the instrument cluster (see Chapter 12) and the glove box (see Section 23).
4 Disconnect the passenger's side air bag

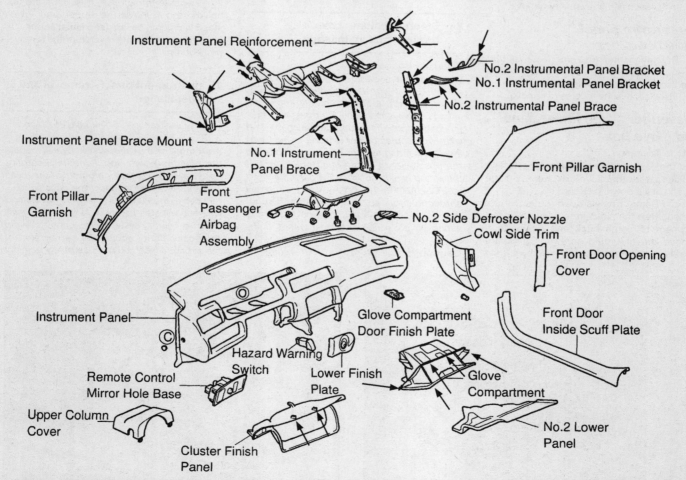

25.9 Remove the fasteners holding the instrument panel to the body - four are at the left end, five at the steering column area (four are up inside the steering column cavity), three are at the console, and two are inside the glove box cavity

Chapter 11 Body

11-19

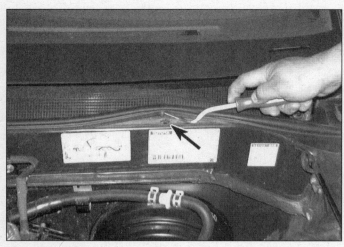

26.2 Pry up the plastic pins (arrow indicates one) to remove the cowl rubber and cowl cover

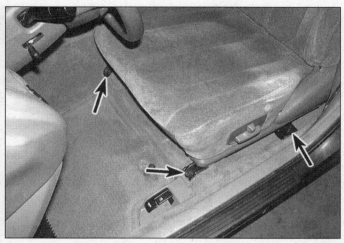

27.2 Typical front seat track retaining bolts (arrows indicate three of the four; the track covers have been removed)

(if equipped) and remove it (see Chapter 12).

5   Remove the audio unit from the center of the dashboard (see Chapter 12).

6   Remove the air conditioning control panel (see Chapter 3).

7   Remove the driver's knee bolster and reinforcement panel (see Section 23). Then detach the bolts securing the steering column and lower it away from the instrument panel **(see illustration)**.

8   A number of electrical connectors must be disconnected in order to remove the instrument panel. Most are designed so that they will only fit on the matching connector (male or female), but if there is any doubt, mark the connectors with masking tape and a marking pen before disconnecting them **(see illustrations)**.

9   Remove all of the fasteners (bolts, screws and nuts) holding the instrument panel to the body **(see illustration)**. Once all are removed, lift the panel then pull it away from the windshield and take it out through the driver's door opening. **Note:** *This is a two-person job, and you may also have to remove the bolts to the instrument panel reinforcement tube for complete removal of the*

instrument panel.

10   Installation is the reverse of removal.

26 Cowl cover - removal and installation

Refer to illustration 26.2

1   Remove the wiper arms.

2   Remove the two screws, one at each end near the fender. Carefully pry up the cowl rubber seal with a trim tool, the plastic pins holding the cowl cover are part of the rubber seal **(see illustration)**. **Note:** *Slide a thin tool along the rubber-to-cowl seam to find where the pins are located.*

3   Installation is the reverse of removal.

27 Seats - removal and installation

Front seat

Refer to illustration 27.2

Warning 1: *The front seat belts on some models are equipped with pre-tensioners,*

which are pyrotechnic (explosive) devices designed to retract the seat belts in the event of a collision. On models equipped with pre-tensioners, do not remove the front seat belt retractor assemblies, and do not disconnect the electrical connectors leading to the assemblies. Problems with the pre-tensioners will turn on the SRS (airbag) warning light on the dash. If any pre-tensioner problems are suspected, take the vehicle to a dealer service department.

Warning 2: *On models with side-impact airbags, be sure to disarm the airbag system before beginning this procedure* (see Chapter 12).

1   Pry out the plastic covers to access the seat tracks and their mounting bolts.

2   Remove the retaining bolts **(see illustration)**.

3   Tilt the seat upward to access the underside, then disconnect any electrical connectors and lift the seat from the vehicle.

4   Installation is the reverse of removal.

Rear seat

Refer to illustration 27.6

5   Lift up on the front edge of the rear seat bottom cushion, remove the rear mounting bolt and remove the cushion from the vehicle. **Note:** *Work the seat belts out of the slits at the back of the cushion.*

6   On Camry and Solara models, pull the seat back locks up and flip the two seat backs down one at a time and remove the mounting bolts. On Avalon models, the four seat back mounting bolts are revealed when the lower cushion is removed **(see illustration)**.

7   Installation is the reverse of removal.

28 Rear package shelf - removal and installation

Refer to illustration 28.3

1   To remove the package shelf, first

27.6 Remove the bolts (arrows) retaining the seat back - Avalon model shown (on Camry and Solara, flip the seat backs down to access the bolts)

remove the high-mounted brake light assembly (see Chapter 12).

2   Unlock and flip forward the seat backs (Camry and Solara) or remove the seat back on Avalon models (see Section 27).

3   Remove the plastic pins retaining the front of the package shelf to the body **(see illustration)**.

4   To fully remove the package shelf from the vehicle, the bolts must be removed from the bottom ends of the shoulder harness belts, and the belts fed through the holes in the package shelf.

5   Installation is the reverse of the removal procedure. If removed, torque the lower seat belt bolts to 31 ft-lbs.

28.3 With the seat back removed (or folded down), remove the plastic pins (arrows)

Chapter 12
Chassis electrical system

Contents

	Section		Section
Airbag - general information	27	Horn - check and replacement	20
Antenna - check and repair	15	Ignition switch and lock cylinder - check and replacement	9
Bulb replacement	21	Instrument cluster - removal and installation	12
Circuit breakers - general information	5	Instrument panel gauges - check	11
Cruise control system - description and check	23	Instrument panel switches - check and replacement	10
Daytime Running Lights (DRL) - general information	26	Power door lock system - description and check	25
Electric side view mirrors - description and check	22	Power window system - description and check	24
Electrical troubleshooting - general information	2	Radio and speakers - removal and installation	14
Fuses - general information	3	Rear window defogger - check and repair	16
Fusible links - general information	4	Relays - general information and testing	6
General information	1	Steering column switches - check and replacement	8
Headlight housing - removal and installation	19	Turn signal and hazard flasher - check and replacement	7
Headlights - adjustment	18	Wiper motor - check and replacement	13
Headlight bulbs - removal and installation	17	Wiring diagrams - general information	28

1 General information

The electrical system is a 12-volt, negative ground type. Power for the lights and all electrical accessories is supplied by a lead/acid-type battery which is charged by the alternator.

This Chapter covers repair and service procedures for the various electrical components not associated with the engine. Information on the battery, alternator, distributor and starter motor can be found in Chapter 5.

It should be noted that when portions of the electrical system are serviced, the cable should be disconnected from the negative battery terminal to prevent electrical shorts and/or fires. **Caution:** *If the stereo in your vehicle is equipped with an anti-theft system, make sure you have the correct activation code before disconnecting the battery.*

2 Electrical troubleshooting - general information

Refer to illustration 2.15

A typical electrical circuit consists of an electrical component, any switches, relays, motors, fuses, fusible links or circuit breakers related to that component and the wiring and electrical connectors that link the component to both the battery and the chassis. To help you pinpoint an electrical circuit problem, wiring diagrams are included at the end of this Chapter.

Before tackling any troublesome electrical circuit, first study the appropriate wiring diagrams to get a complete understanding of what makes up that individual circuit. Trouble spots, for instance, can often be narrowed down by noting if other components related to the circuit are operating properly. If several components or circuits fail at one time, chances are the problem is in a fuse or ground connection, because several circuits are often routed through the same fuse and ground connections.

Electrical problems usually stem from simple causes, such as loose or corroded connections, a blown fuse, a melted fusible link or a bad relay. Visually inspect the condition of all fuses, wires and connections in a problem circuit before troubleshooting it.

If testing instruments are going to be utilized, use the diagrams to plan ahead of time where you will make the necessary connections in order to accurately pinpoint the trouble spot.

The basic tools needed for electrical troubleshooting include a circuit tester or voltmeter (a 12-volt bulb with a set of test leads can also be used), a continuity tester, which includes a bulb, battery and set of test

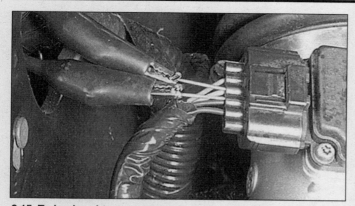

2.15 To backprobe a connector, insert a small, sharp probe (such as a straight-pin) into the back of the connector alongside the desired wire until it contacts the metal terminal inside; connect your meter leads to the probes - this allows you to test a circuit in operation

3.1a The interior fuse box is located on the dashboard, to the left of the steering column (Avalon model shown)

leads, and a jumper wire, preferably with a circuit breaker incorporated, which can be used to bypass electrical components. Before attempting to locate a problem with test instruments, use the wiring diagram(s) to decide where to make the connections.

Voltage checks

Voltage checks should be performed if a circuit is not functioning properly. Connect one lead of a circuit tester to either the negative battery terminal or a known good ground. Connect the other lead to a electrical connector in the circuit being tested, preferably nearest to the battery or fuse. If the bulb of the tester lights, voltage is present, which means that the part of the circuit between the electrical connector and the battery is problem free. Continue checking the rest of the circuit in the same fashion. When you reach a point at which no voltage is present, the problem lies between that point and the last test point with voltage. Most of the time the problem can be traced to a loose connection. **Note:** *Keep in mind that some circuits receive voltage only when the ignition key is in the Accessory or Run position.*

Finding a short

One method of finding shorts in a circuit is to remove the fuse and connect a test light or voltmeter in its place. There should be no voltage present in the circuit. Move the wiring harness from side to side while watching the test light. If the bulb goes on, there is a short to ground somewhere in that area, probably where the insulation has rubbed through. The same test can be performed on each component in the circuit, even a switch.

Ground check

Perform a ground test to check whether a component is properly grounded. Disconnect the battery and connect one lead of a self-powered test light, known as a continuity tester, to a known good ground. Connect the other lead to the wire or ground connection being tested. If the bulb goes on, the ground is good. If the bulb does not go on, the ground is not good. **Caution:** *If the stereo in your vehicle is equipped with an anti-theft system, make sure you have the correct activation code before disconnecting the battery.*

Continuity check

A continuity check is done to determine if there are any breaks in a circuit - if it is passing electricity properly. With the circuit off (no power in the circuit), a self-powered continuity tester can be used to check the circuit. Connect the test leads to both ends of the circuit (or to the "power" end and a good ground), and if the test light comes on the circuit is passing current properly. If the light doesn't come on, there is a break somewhere in the circuit. The same procedure can be used to test a switch, by connecting the continuity tester to the power in and power out sides of the switch. With the switch turned On, the test light should come on.

Finding an open circuit

When diagnosing for possible open circuits, it is often difficult to locate them by sight because oxidation or terminal misalignment are hidden by the electrical connectors. Merely wiggling an electrical connector on a sensor or in the wiring harness may correct the open circuit condition. Remember this when an open circuit is indicated when troubleshooting a circuit. Intermittent problems may also be caused by oxidized or loose connections.

Electrical troubleshooting is simple if you keep in mind that all electrical circuits are basically electricity running from the battery, through the wires, switches, relays, fuses and fusible links to each electrical component (light bulb, motor, etc.) and to ground, from which it is passed back to the battery. Any electrical problem is an interruption in the flow of electricity to and from the battery.

Connectors

Most electrical connections on these vehicles are made with multiwire plastic connectors. The mating halves of many connectors are secured with locking clips molded into the plastic connector shells. The mating halves of large connectors, such as some of those under the instrument panel, are held together by a bolt through the center of the connector.

To separate a connector with locking clips, use a small screwdriver to pry the clips apart carefully, then separate the connector halves. Pull only on the shell, never pull on the wiring harness as you may damage the individual wires and terminals inside the connectors. Look at the connector closely before trying to separate the halves. Often the locking clips are engaged in a way that is not immediately clear. Additionally, many connectors have more than one set of clips.

Each pair of connector terminals has a male half and a female half. When you look at the end view of a connector in a diagram, be sure to understand whether the view shows the harness side or the component side of the connector. Connector halves are mirror images of each other, and a terminal shown on the right side end view of one half will be on the left side end view of the other half.

It is often necessary to take circuit voltage measurements with a connector connected. Whenever possible, carefully insert the test probes of your meter into the rear of the connector shell to contact the terminal inside. This kind of connection is called "backprobing" **(see illustration).** When inserting a test probe into a male terminal, be careful not to distort the terminal opening. Doing so can lead to a poor connection and corrosion at that terminal later.

3   Fuses - general information

Refer to illustrations 3.1a, 3.1b 3.1c and 3.3

The electrical circuits of the vehicle are protected by a combination of fuses, circuit breakers and fusible links. Fuse blocks are located under the instrument panel and in the engine compartment depending on the

Chapter 12 Chassis electrical system

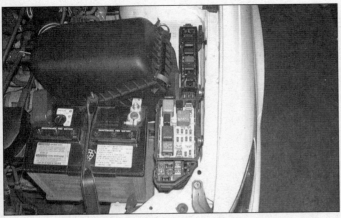

3.1b The main engine compartment fuse/relay box is one-piece on Camry and Solara models (shown with cover removed) - there are two separate boxes next to each other on Avalon models

3.1c Several relays are in this box (arrow) near the driver's side shock tower

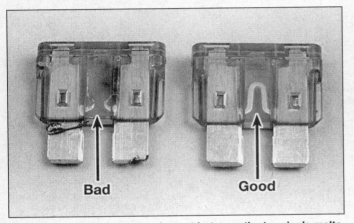

3.3 When a fuse blows, the element between the terminals melts

4.2 Cartridge-type fusible links (arrows) for high-current applications are found in this box to the rear of the main engine compartment fuse/relay box

model year of the vehicle (see illustrations).

Each of the fuses is designed to protect a specific circuit, and the various circuits are identified on the fuse panel cover.

Miniaturized fuses are employed in the fuse blocks. These compact fuses, with blade terminal design, allow fingertip removal and replacement. If an electrical component fails, always check the fuse first. The best way to check a fuse is with a test light. Check for power at the exposed terminal tips of each fuse. If power is present on one side of the fuse but not the other, the fuse is blown. A blown fuse can also be confirmed by visually inspecting it (see illustration).

Be sure to replace blown fuses with the correct type. Fuses of different ratings are physically interchangeable, but only fuses of the proper rating should be used. Replacing a fuse with one of a higher or lower value than specified is not recommended. Each electrical circuit needs a specific amount of protection. The amperage value of each fuse is molded into the fuse body.

If the replacement fuse immediately fails, don't replace it again until the cause of the problem is isolated and corrected. In most cases, this will be a short circuit in the wiring caused by a broken or deteriorated wire.

4 Fusible links - general information

Refer to illustration 4.2

Some circuits are protected by fusible links. The links are used in circuits which are not ordinarily fused, such as the ignition circuit, or which carry high current.

Cartridge type fusible links are located in the engine compartment fuse/relay block and are similar to a large fuse (see illustration) and, after disconnecting the negative battery cable, are simply unplugged and replaced by a unit of the same amperage. **Caution:** *If the stereo in your vehicle is equipped with an anti-theft system, make sure you have the correct activation code before disconnecting the battery.*

On models with Traction Control, a second fusible link box is located on the right side of the radiator support. Some fusible links are held in place by a screw which must be loosened before removing the link.

5 Circuit breakers - general information

Circuit breakers protect components such as power windows, power door locks and headlights.

On some models the circuit breaker resets itself automatically, so an electrical overload in a circuit-breaker-protected system will cause the circuit to fail momentarily, then come back on. If the circuit doesn't come back on, check it immediately. Once the condition is corrected, the circuit breaker will resume its normal function. Some circuit breakers must be reset manually.

6 Relays - general information and testing

General information

1 Several electrical accessories in the vehicle, such as the fuel injection system, horns, starter, air conditioning/heating system and fog lights use relays to transmit the electrical signal to the component. Relays

use a low-current circuit (the control circuit) to open and close a high-current circuit (the power circuit). If the relay is defective, that component will not operate properly. The various relays are mounted in engine compartment fuse/relay boxes **(see illustrations 3.1a, 3.1b and 3.1c)**. If a faulty relay is suspected, it can be removed and tested using the procedure below or by a dealer service department or a repair shop. Defective relays must be replaced as a unit. **Note:** *Relay identifications are often found on the cover of the relay, and always on a decal on the inside of the fuse/relay box cover.*

Testing

Refer to illustrations 6.4a, 6.4b and 6.4c

2 It's best to refer to the wiring diagram for the circuit to determine the proper hook-ups for the relay you're testing. However, if you're not able to determine the correct hook-up from the wiring diagrams, you may be able to determine the test hook-ups from the information that follows.

3 On most relays, two of the terminals are the relay's control circuit (they connect to the relay coil which, when energized, closes the large contacts to complete the circuit). The other terminals are the power circuit (they are connected together within the relay when the control-circuit coil is energized).

4 The relays are usually marked as an aid to help you determine which terminals are the control circuit and which are the power circuit **(see illustrations)**.

5 Connect a fused jumper wire between one of the two control circuit terminals and the positive battery terminal. Connect another jumper wire between the other control circuit terminal and ground. When the connections are made, the relay should click. On some relays, polarity may be critical, so, if the relay doesn't click, try swapping the jumper wires on the control circuit terminals.

6 With the jumper wires connected, check for continuity between the power circuit terminals as indicated by the markings on the relay.

7 If the relay fails any of the above tests, replace it.

7 Turn signal and hazard flasher - check and replacement

Refer to illustration 7.1

1 The turn signal/hazard flasher are contained in a small, canister-shaped unit located in the relay center under the dash near the steering column **(see illustration)**.

2 When the flasher unit is functioning properly, an audible click can be heard during its operation. If the turn signals fail on one side or the other and the flasher unit does not make its characteristic clicking sound, a faulty turn signal bulb is indicated. **Note:** *When the flasher is operating properly, the clicking is steady. Normal driving will give you the pace of the clicks. If the flasher pace sud-

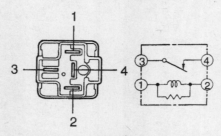

6.4a Diagram and pin identification for a typical four-terminal relay

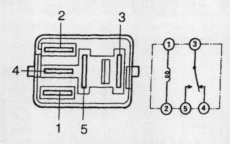

6.1c Diagram and pin identification for typical five-terminal relay

denly seems slow than usual, or does not flash at all, one of the turn signal bulbs is probably burned out.*

3 If both turn signals fail to blink, the problem may be due to a blown fuse, a faulty flasher unit, a broken switch or a loose or open connection. If a quick check of the fuse box indicates that the turn signal fuse has blown, check the wiring for a short before installing a new fuse.

4 To replace the flasher, simply pull it out of its holder and unplug it.

5 Make sure that the replacement unit is identical to the original. Compare the old one to the new one before installing it.

6 Installation is the reverse of removal.

8 Steering column switches - check and replacement

Warning: *The models covered by this manual are equipped with Supplemental Restraint Systems (SRS), more commonly known as airbags. Always disable the airbag system before working in the vicinity of any airbag system components to avoid the possibility of accidental deployment of the airbag(s), which could cause personal injury (see Section 27).*

Check

Refer to illustrations 8.3a and 8.3b

1 There are two column-mounted switches. On the left side is the turn signal/headlight switch, while on the right side is the wiper/washer switch.

2 Trace the wire from the combination

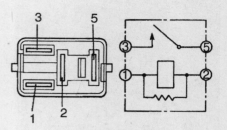

6.4b Some four-terminal relays are marked with 1, 2, 3 and 5, but they are four-terminal types tested the same as ones marked 1 through 4

7.1 Turn signal/hazard flasher location (arrow) in a plastic holder behind the interior fuse panel, on the firewall

switch to the main wiring harness connector and disconnect the connectors.

3 Using an ohmmeter or self-powered test light and the accompanying diagrams, check for continuity between the indicated switch terminals with the switch in each of the indicated positions **(see illustrations)**. If the continuity isn't as specified, replace the switch.

Replacement (either switch)

Refer to illustration 8.8

4 Disconnect the cable from the negative terminal of the battery. **Caution:** *If the stereo in your vehicle is equipped with an anti-theft system, make sure you have the correct activation code before disconnecting the battery.*

5 Remove the steering wheel (see Chapter 10).

6 Remove the steering column covers (see Chapter 11).

7 Unplug the electrical connectors from the combination switch.

8 Remove the retaining screws and pull the combination switch from the column **(see illustration)**.

9 Individual switches can now be removed from the switch body and replaced as necessary. Detach the retaining screws or pins and remove the defective switch.

10 Installation is the reverse of removal.

Chapter 12 Chassis electrical system

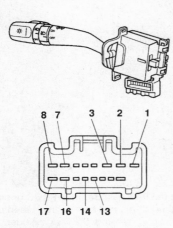

8.3a Combination switch connector terminal guide and continuity chart

Combination switch position	Continuity between terminals
Headlight switch Off	None
Headlight switch - taillight	14 and 16
Headlight switch On	13, 14 and 16
Headlight dimmer: low beam	16 and 17
Headlight dimmer: high beam	7 and 16
Flash-to-pass	7, 8 and 16
Turn signal, Left	1 and 2
Turn signal, Neutral	None
Turn signal, Right	2 and 3

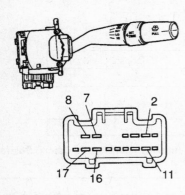

8.3b Washer/wiper switch connector terminal guide and continuity chart

Washer/wiper switch position	Continuity between terminals
Off	7 and 16
Int	7 and 16
Lo	7 and 17
Hi	8 and 17
Washer On	2 and 11

8.8 Combination switch retaining screws (arrows) - turn signal/headlight switch shown, wiper/washer switch similar

9 Ignition switch and lock cylinder - check and replacement

Warning: *The models covered by this manual are equipped with Supplemental Restraint Systems (SRS), more commonly known as airbags. Always disable the airbag system before working in the vicinity of any airbag system components to avoid the possibility of accidental deployment of the airbag(s), which could cause personal injury (see Section 27).*

Check

Refer to illustration 9.2

1 Trace the wire from the ignition switch to the main wiring harness connector and disconnect the connector.
2 Using an ohmmeter or self-powered test light and the accompanying diagrams, check for continuity between the indicated switch terminals with the switch in each of the indicated positions **(see illustrations)**. If the continuity isn't as specified, replace the switch.

Replacement

Refer to illustrations 9.9 and 9.12

3 Disconnect the cable from the negative terminal of the battery. **Caution:** *If the stereo in your vehicle is equipped with an anti-theft system, make sure you have the correct activation code before disconnecting the battery.*
4 Place the ignition key in the ACC position.
5 Remove the steering wheel (see Chapter 10).
6 Remove the steering column covers (see Chapter 11).
7 Remove the steering column switches (see Section 8).

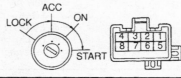

9.2 Ignition switch connector terminal guide and continuity chart

Switch position	Tester connection	Specified condition
LOCK	–	No continuity
ACC	2 – 3	Continuity
ON	2 – 3 – 4 6 – 7	Continuity
START	1 – 2 – 4 6 – 7 – 8	Continuity

9.9 Detach the retaining screws (arrows) to remove the ignition switch from the steering column housing

9.12 With the lock cylinder in the ACC position, depress the retaining pin with a small screwdriver, then pull straight out to remove the lock cylinder

Ignition switch

8 Unplug the ignition switch wiring harness connectors.
9 Detach the switch retaining screws **(see illustration)** remove the switch assembly from the steering column.
10 Installation is reverse of the removal.

Lock cylinder

11 Use a small screwdriver or punch to depress the lock cylinder retaining pin.
12 Pull straight out on the lock cylinder assembly to remove it from the vehicle **(see illustration)**.
13 To install the lock cylinder, depress the retaining pin and guide the lock cylinder into the steering column housing until the retaining pin extends itself back into the locating hole in the steering column housing.
14 The remainder of the installation is the reverse of removal.

10 Instrument panel switches - check and replacement

Warning: *The models covered by this manual are equipped with Supplemental Restraint Systems (SRS), more commonly known as airbags. Always disable the airbag system before working in the vicinity of any airbag system components to avoid the possibility of accidental deployment of the airbag(s), which could cause personal injury (see Section 27).*
Caution: *If the stereo in your vehicle is equipped with an anti-theft system, make sure you have the correct activation code before disconnecting the battery.*

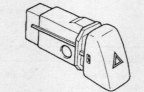

Switch position	Tester connection	Specified condition
Switch OFF	7 – 10	Continuity
Switch ON	5 – 6 – 9	Continuity
	7 – 8	
Illumination circuit	2 – 3	Continuity

10.2a Hazard warning switch terminal guide and continuity chart for the Camry, Solara and 1999 and earlier Avalon models

Hazard warning switch

Check
Refer to illustrations 10.2a and 10.2b

1 To check the switch it must first be removed from the center dashboard trim bezel (see below).
2 Test for continuity between the terminals of the switch **(see illustrations)**. If the switch fails, replace it.

Replacement
Refer to illustration 10.4

3 Carefully pry the center bezel from the dashboard (see Chapter 11).
4 Pull the bezel out until you can disconnect the 10-pin electrical connector at the rear of the switch. Depress the tab and remove the switch from the bezel **(see illustration)**.

Power mirror control switch

Check
Refer to illustrations 10.6a, 10.6b, 10.6c and 10.6d

5 The power mirror control switch is mounted at the left end of the instrument panel. On Camry and Avalon models, it is mounted below the left dashboard vent, and to the right of the vent on Solara models. Remove the switch for testing (see Step 7).
6 Test for continuity between the terminals of the switch **(see illustrations)**. If the continuity isn't as specified, replace the switch.

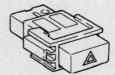

Switch position	Tester connection	Specified condition
Switch OFF	5 – 7	Continuity
Switch ON	1 – 2 – 3 – 4	Continuity
	5 – 6	
Illumination circuit	8 – 9	Continuity

10.2b Hazard warning switch terminal guide and continuity chart for the Lexus ES 300 and 1999 and later Avalon models

Chapter 12 Chassis electrical system

12-7

10.4 Carefully squeeze the clips (arrow) retaining the hazard switch to the center dash bezel

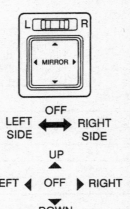

Switch position	Tester connection	Specified condition
OFF	–	No continuity
UP	1 – 9, 6 – 10	Continuity
DOWN	1 – 10, 6 – 9	Continuity
LEFT	5 – 9, 6 – 10	Continuity
RIGHT	5 – 10, 6 – 9	Continuity

Switch position	Tester connection	Specified condition
OFF	–	No continuity
UP	6 – 10, 7 – 9	Continuity
DOWN	6 – 9, 7 – 10	Continuity
LEFT	6 – 10, 8 – 9	Continuity
RIGHT	6 – 9, 8 – 10	Continuity

10.6a Power mirror switch terminal guide and continuity charts - Camry and Solara models

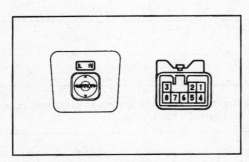

10.6b Power mirror switch terminal identification and continuity charts - 1999 and earlier Avalon models

Left side

Switch position	Tester connection	Specified condition
OFF		No continuity
UP	2 – 6, 1 – 3 – 8	Continuity
DOWN	3 – 6, 1 – 2 – 8	Continuity
LEFT	1 – 3 – 6, 2 – 8	Continuity
RIGHT	1 – 2 – 6, 3 – 8	Continuity

Right side

Switch position	Tester connection	Specified condition
OFF		No continuity
UP	2 – 5, 1 – 3 – 7	Continuity
DOWN	3 – 5, 1 – 2 – 7	Continuity
LEFT	1 – 3 – 5, 2 – 7	Continuity
RIGHT	3 – 7, 1 – 2 – 5	Continuity

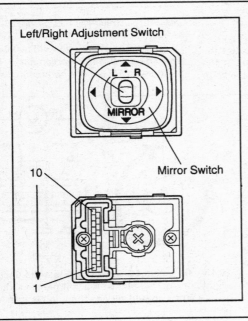

Left side

Switch position	Tester connection	Specified condition
OFF	–	No continuity
UP	3 – 4, 7 – 8	Continuity
DOWN	3 – 8, 4 – 7	Continuity
LEFT	4 – 9, 7 – 8	Continuity
RIGHT	4 – 7, 8 – 9	Continuity

Right side

Switch position	Tester connection	Specified condition
OFF	–	No continuity
UP	2 – 4, 7 – 8	Continuity
DOWN	2 – 8, 4 – 7	Continuity
LEFT	4 – 10, 7 – 8	Continuity
RIGHT	4 – 7, 8 – 10	Continuity

10.6c Power mirror switch terminal identification and continuity charts - 2000 and later Avalon models

Replacement

7 Carefully pry the switch from the panel, using tape on the screwdriver tip to prevent scratching the instrument panel.
8 Disconnect the electrical connector and remove the switch.
9 Installation is the reverse of the removal procedure.

Traction control switch

Check

10 The traction control switch works in conjunction with the ABS braking system to provide better traction in some driving conditions. Remove the switch (see Step 12) to test it for proper operation. It is mounted in a panel to the left of the steering wheel.
11 Test for continuity between the two terminals at either end of the switch. There should be continuity with the switch pressed in, and no continuity with the switch released. If the continuity is not as described, replace the switch.

Replacement

12 Carefully pry the switch from the panel, using tape on the screwdriver tip to prevent scratching the instrument panel.
13 Disconnect the electrical connector and remove the switch.
14 Installation is the reverse of the removal procedure.

Instrument panel light control switch

Check

Refer to illustrations 10.16a and 10.16b

15 On some models, the instrument panel light control is a thumbwheel at the top left of the instrument panel. On others, it is a knob in the switch panel below the left dashboard vent. In all cases, the functional unit is a rheostat that controls the voltage to the instrument panel lights.
16 Remove the switch (see Step 17). Test the rheostat on Toyota models by applying battery voltage to terminal 1 and a ground jumper wire to terminal 3 **(see illustration)**. Attach voltmeter leads (positive to terminal 2 and negative to terminal 3), and rotate the knob to see if voltage varies. On Lexus models, apply battery voltage to terminal 5 and a ground jumper wire to terminal 1 **(see illustration)**. Attach voltmeter leads (positive to terminal 5 and negative to terminal 3), and rotate the knob to see if voltage varies. Replace the rheostat if the voltage doesn't vary.

Replacement

17 On models with the rheostat below the left vent, pry the switch panel out of the dashboard with a screwdriver (tape the tip to prevent scratches). On models with a thumbwheel, remove the instrument cluster to access the switch (see Section 12).

11 Instrument panel fuel and temperature gauges - check

Refer to illustrations 11.3a, 11.3b, 11.3c, 11.3d and 11.3e

Warning: *The models covered by this manual are equipped with Supplemental Restraint Systems (SRS), more commonly known as*

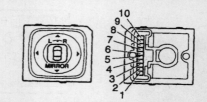

Left side

Switch position	Tester connection		Specified condition
OFF	—		No continuity
UP	3 – 4	7 – 8	Continuity
DOWN	3 – 8	4 – 7	Continuity
LEFT	4 – 9	7 – 8	Continuity
RIGHT	4 – 7	8 – 9	Continuity

Right side

Switch position	Tester connection	Specified condition
OFF	—	No continuity
UP	2 – 4 1 – 7 – 8	Continuity
DOWN	4 – 7 1 – 2 – 8	Continuity
LEFT	4 – 10 1 – 7 – 8	Continuity
RIGHT	4 – 7 1 – 8 – 10	Continuity

10.6d Power mirror switch terminal identification and continuity charts - Lexus ES 300 models

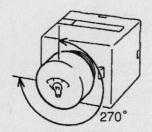

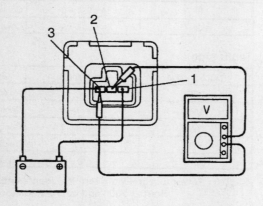

10.16a Test the dashboard light control rheostat with a voltmeter - voltage should vary as the knob is turned (Toyota models)

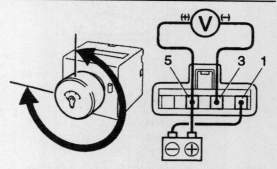

10.16b On Lexus ES 300 models test the dashboard light control rheostat with a voltmeter as shown - voltage should vary as the knob is turned

Chapter 12 Chassis electrical system

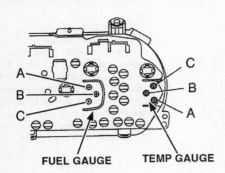

11.3a Camry, Solara and 1999 and earlier Avalon test gauge resistance: fuel gauge, A-B = 126 ohms, A-C = 280 ohms, and B-C = 154 ohms; temperature gauge, A-B = 175 ohms, A-C = 54 ohms, and B-C = 229 ohms

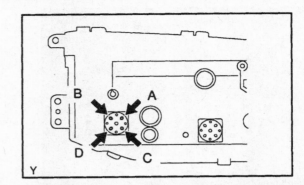

Tester connection	Resistance (Ω)
A – B	250
C – D	250

11.3b 2000 and later model Avalon fuel gauge terminal guide and resistance chart

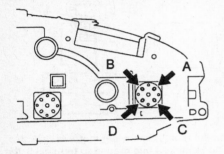

11.3c 2000 and later model Avalon engine coolant gauge test resistance test: A-B = 250 ohms, C-D = 250 ohms

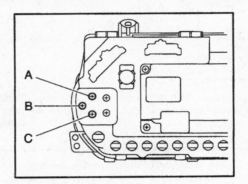

Tester connection	Resistance (Ω)
A – B	Approx. 270.8
A – C	Approx. 91.3
B – C	Approx. 179.5

11.3d Lexus ES 300 fuel gauge terminal guide and resistance chart

airbags. Always disable the airbag system before working in the vicinity of any airbag system components to avoid the possibility of accidental deployment of the airbag(s), which could cause personal injury (see Section 27).

1 Turn the ignition switch to the ON position.
2 If the gauge pointer does not move from the empty, low or cold positions, check the fuse. If the fuse is OK, locate the particular sending unit for the circuit you're working on (see Chapter 4 for fuel sending unit location or Chapter 3 for the temperature sending unit location). Connect the sending unit connector to ground. If the pointer goes to the full, high or hot position, replace the sending unit. If the pointer stays in same position, use a jumper wire to ground the sending unit terminal on the back of the gauge (if necessary, refer to the wiring diagrams at the end of this Chapter). If the pointer moves, the problem lies in the wire between the gauge and the sending unit. If the pointer does not move with the sending unit terminal on the back of the gauge grounded, check for voltage at the other terminal of the gauge. If voltage is present, replace the gauge.
3 To check the gauge units alone, remove the instrument cluster (see Section 12) and test the resistance on the terminals for the fuel and temperature gauges **(see illustrations)**.

12 Instrument cluster - removal and installation

Refer to illustration 12.3

Warning: *The models covered by this manual are equipped with Supplemental Restraint Systems (SRS), more commonly known as airbags. Always disable the airbag system before working in the vicinity of any airbag system components to avoid the possibility of accidental deployment of the airbag(s), which could cause personal injury (see Section 27).*

1 Disconnect the negative battery cable.
Caution: *If the stereo in your vehicle is*

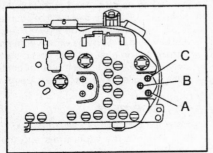

Tester connection	Resistance (Ω) *
A – B	Approx. 175.7
A – C	Approx. 54.0
B – C	Approx. 229.7

11.3e Lexus ES 300 fuel gauge terminal guide and resistance chart

Chapter 12 Chassis electrical system

12.3 Remove the four screws (arrows) and pull out the instrument cluster

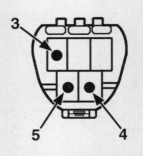

13.2a Toyota model wiper motor test: With a battery-power jumper on terminal 5 and a ground on terminal 4, the motor should operate at Low speed; with the power applied to terminal 3 and ground to terminal 4, the motor should operate at High speed

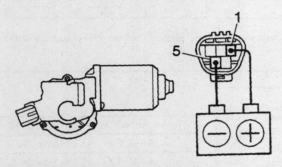

13.2b Lexus ES 300 model Low Speed wiper motor test: With a battery-power jumper on terminal 1 and a ground on terminal 5, the motor should operate at Low speed

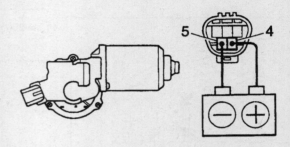

13.2c Lexus ES 300 model High Speed wiper motor test: With a battery-power jumper to terminal 4 and ground to terminal 5, the motor should operate at High speed

equipped with an anti-theft system, make sure you have the correct activation code before disconnecting the battery.
2 Remove the instrument cluster bezel (see Chapter 11).
3 Remove the cluster mounting screws **(see illustrations)** and pull the instrument cluster towards the steering wheel.
4 Disconnect any electrical connectors that would interfere with removal.
5 Cover the steering column with a cloth to protect the trim covers, then remove the instrument cluster from the vehicle.
6 Installation is the reverse of removal.

13 Wiper motor - check and replacement

Check

Refer to illustrations 13.2a, 13.2b, 13.2c
Note: *Refer to the wiring diagrams for wire colors and locations in the following checks. Keep in mind that power wires are generally larger in diameter and brighter colors, and ground wires are usually smaller in diameter and darker colors. When checking for voltage, probe a grounded 12-volt test light to each terminal at a connector until it lights; this verifies voltage (power) at the terminal.*
1 **If the wipers work slowly**, make sure the battery is in good condition and has a strong charge (see Chapter 1). If the battery is in good condition, remove the wiper motor (see below) and operate the wiper arms by hand. Check for binding linkage and pivots. Lubricate or repair the linkage or pivots as necessary. Reinstall the wiper motor. If the wipers still operate slowly, check for loose or corroded connections, especially the ground connection. If all connections look OK, replace the motor.
2 **If the wipers fail to operate when activated**, check the fuse. If the fuse is OK, connect a jumper wire between the wiper motor and ground, then retest. If the motor works now, repair the ground connection. If the motor still doesn't work, turn on the wipers and check for voltage at the motor. If there's voltage at the motor, remove the motor and check it off the vehicle with fused jumper wires from the battery **(see illustrations)**. If the motor now works, check for binding linkage (see Step 1 above). If the motor still doesn't work, replace it. If there's no voltage at the motor, check for voltage at the switch. If there's no voltage at the switch, check the wiring between the switch and fuse panel for continuity. If the wiring is OK, the switch is probably bad.
3 **If the wipers only work on one speed**, check the continuity of the wires between the switch and motor, and the continuity of the terminals on the motor. If the wires are OK, replace the switch.
4 **If the interval (delay) function is inoperative**, check the continuity of all the wiring between the switch and motor. If the wiring is OK, replace the interval module.
5 **If the wipers stop at the position they're in when the switch is turned off (fail to park)**, check for voltage at the wiper motor when the wiper switch is OFF but the ignition is ON. If voltage is present, the limit switch in the motor is malfunctioning. Replace the wiper motor. If no voltage is present, trace and repair the limit switch wiring between the fuse panel and wiper motor.
6 **If the wipers won't shut off unless the ignition is OFF**, disconnect the wiring from the wiper control switch. If the wipers stop, replace the switch. If the wipers keep running, there's a defective limit switch in the motor; replace the motor.
7 **If the wipers won't retract below the hoodline**, check for mechanical obstructions in the wiper linkage or on the vehicle's body which would prevent the wipers from parking. If there are no obstructions, check the wiring between the switch and motor for continuity. If the wiring is OK, replace the wiper motor.

Replacement

Refer to illustrations 13.9a, 13.9b and 13.9c
8 Remove the wiper arms and the cowl cover.

Chapter 12 Chassis electrical system

13.9a Front wiper motor and linkage retaining bolts (A) and electrical connector (B) - Camry, Solara, 2000 and later Avalon and Lexus ES 300 models

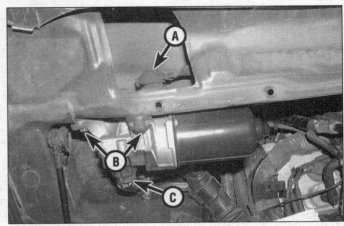

13.9b On 1999 and earlier Avalon models, remove the clip (A) from the linkage, remove the mounting bolts (B indicates two of the three) and disconnect the electrical connector (C)

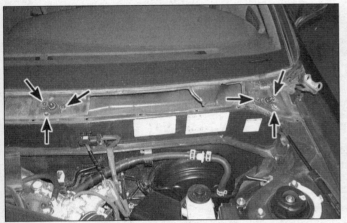

13.9c If the 1999 and earlier Avalon wiper linkage must be removed, remove these bolts (arrows) and slide the linkage over until it can be removed from the cowl

14.3a Remove the radio mounting screws (arrows)

9 Disconnect the electrical connector at the motor. Remove the wiper linkage retaining bolts and remove the assembly with the motor **(see illustrations)**.
10 Remove the bolts and remove the motor from the linkage assembly.
11 Installation is the reverse of removal.

14 Radio and speakers - removal and installation

Warning: *The models covered by this manual are equipped with Supplemental Restraint Systems (SRS), more commonly known as airbags. Always disable the airbag system before working in the vicinity of any airbag system components to avoid the possibility of accidental deployment of the airbag(s), which could cause personal injury (see Section 27).*

1 Disconnect the negative battery cable.
Caution: *If the stereo in your vehicle is equipped with an anti-theft system, make sure you have the correct activation code before disconnecting the battery.*

14.3b Pull the radio forward, then disconnect the antenna lead (A) and the electrical connector (B)

Radio

Refer to illustrations 14.3a and 14.3b
2 Remove the center dashboard trim bezel (see Chapter 11).
3 Remove the retaining screws and pull the radio outward to access the backside, then disconnect the electrical connectors and the antenna lead **(see illustrations)**.
4 Installation is the reverse of removal.

Door speakers

Refer to illustration 14.6
5 Remove the door trim panel (see Chapter 11). All models have front door-mounted speakers, while Avalon models have speakers in both front and rear doors (rear door speakers mounted similarly to the front door speakers).
6 Remove the speaker retaining screws.

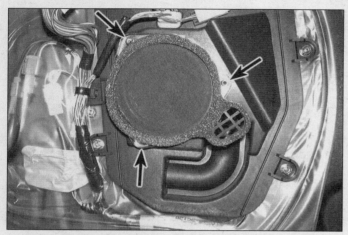

14.6 Remove the speaker retaining screws (arrows) and disconnect the electrical connector to remove the speaker from the vehicle (Avalon front door shown)

16.4 When measuring the voltage at the rear window defogger grid, wrap a piece of aluminum foil around the positive probe of the voltmeter and press the foil against the wire with your finger

Disconnect the electrical connector and remove the speaker from the vehicle **(see illustration)**.

7   Installation is the reverse of removal.

Rear package shelf speakers (some models)

8   Refer to Chapter 11 and remove the clips holding the rear package shelf in place, to access the speaker screws. **Note:** *Some Camry and Solara models have a speaker at each side of the package shelf, while Avalon models have one larger woofer in the center of the package shelf. Others models have speakers in the rear doors only and none on the package shelf.*
9   Remove the speaker retaining screws. From inside the trunk, disconnect the electrical connector and remove the speaker from the vehicle.
10   Installation is the reverse of removal.

Tweeters

11   In addition to the standard door and rear speakers, some models are equipped with tweeters for improved high-range sound. Mounted in the front doors just ahead of the window glass, they can be removed by simply prying under the edge of the tweeter with a screwdriver tip covered with tape. Pull the tweeter out and disconnect the electrical connector.
12   Installation is the reverse of removal.

15   Antenna - removal and installation

1   The vehicles covered by this manual are equipped with a wire grid-type antenna attached to the rear window glass.
2   If there is a problem with radio reception, examine the antenna wires by checking for continuity with an ohmmeter.

3   Wrap small pieces of aluminum foil around the tips of your ohmmeter probes and touch them to the wires of the antenna grid. With one probe at one of the antenna terminals, move the other probe along the wire, checking for continuity until a break is found.
4   If a break is found, the wire can be repaired in the same manner as a rear window defogger wire (see Section 16).

16   Rear window defogger - check and repair

1   The rear window defogger consists of a number of horizontal elements baked onto the glass surface.
2   Small breaks in the element can be repaired without removing the rear window.

Check

Refer to illustrations 16.4, 16.5 and 16.7
3   Turn the ignition switch and defogger system switches to the ON position. Using a voltmeter, place the positive probe against the defogger grid positive terminal and the negative probe against the ground terminal. If battery voltage is not indicated, check the fuse, defogger switch and related wiring. If voltage is indicated, but all or part of the defogger doesn't heat, proceed with the following tests.
4   When measuring voltage during the next two tests, wrap a piece of aluminum foil around the tip of the voltmeter positive probe and press the foil against the heating element with your finger **(see illustration)**. Place the negative probe on the defogger grid ground terminal.
5   Check the voltage at the center of each heating element **(see illustration)**. If the voltage is 5 or 6-volts, the element is okay (there is no break). If the voltage is 0-volts, the element is broken between the center of the element and the positive end. If the voltage is 10

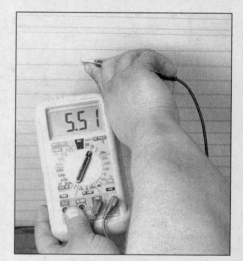

16.5 To determine if a heating element has broken, check the voltage at the center of each element - if the voltage is 5 or 6-volts, the element is unbroken - if the voltage is 10 or 12-volts, the element is broken between the center and the ground side - if there is no voltage, the element is broken between the center and the positive side

to 12-volts the element is broken between the center of the element and ground. Check each heating element.
6   Connect the negative lead to a good body ground. The reading should stay the same. If it doesn't, the ground connection is bad.
7   To find the break, place the voltmeter negative probe against the defogger ground terminal. Place the voltmeter positive probe with the foil strip against the heating element at the positive terminal end and slide it toward the negative terminal end. The point at which the voltmeter deflects from several volts to zero is the point at which the heating element is broken **(see illustration)**.

Chapter 12 Chassis electrical system

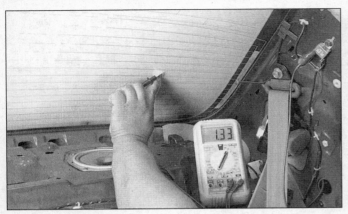

16.7 To find the break, place the voltmeter negative lead against the defogger ground terminal, place the voltmeter positive lead with the foil strip against the heating element at the positive terminal end and slide it toward the negative terminal end - the point at which the voltmeter reading changes abruptly is the point at which the element is broken

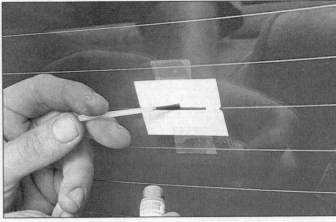

16.13 To use a defogger repair kit, apply masking tape to the inside of the window at the damaged area, then brush on the special conductive coating

17.2 Rotate the bulb holder (arrow) out of the headlight housing - remove the bulb from the holder to replace it

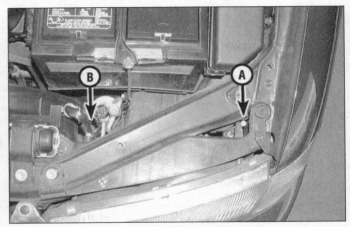

18.1 Headlight adjustment screws

A   Vertical adjustment   B   Horizontal adjustment

Repair

Refer to illustration 16.13

8 Repair the break in the element using a repair kit specifically recommended for this purpose, available at most auto parts stores. Included in this kit is plastic conductive epoxy.
9 Prior to repairing a break, turn off the system and allow it to cool off for a few minutes.
10 Lightly buff the element area with fine steel wool, then clean it thoroughly with rubbing alcohol.
11 Use masking tape to mask off the area being repaired.
12 Thoroughly mix the epoxy, following the instructions provided with the repair kit.
13 Apply the epoxy material to the slit in the masking tape, overlapping the undamaged area about 3/4-inch on either end **(see illustration)**.
14 Allow the repair to cure for 24 hours before removing the tape and using the system.

17 Headlight bulbs - removal and installation

Refer to illustration 17.2

Warning: *Halogen gas filled bulbs are under pressure and may shatter if the surface is scratched or the bulb is dropped. Wear eye protection and handle the bulbs carefully, grasping only the base whenever possible. Do not touch the surface of the bulb with your fingers because the oil from your skin could cause it to overheat and fail prematurely. If you do touch the bulb surface, clean it with rubbing alcohol.*

1 Reach behind the headlight assembly and unplug the electrical connector.
2 Grasp the bulb holder securely and rotate it counterclockwise to remove it from the housing **(see illustration)**.
3 Pull straight out on the bulb to remove it from the bulb holder. Without touching the glass with your bare fingers, insert the new bulb assembly into the headlight housing.
4 Plug in the electrical connector.

18 Headlights - adjustment

Refer to illustrations 18.1 and 18.3

Note: *The headlights must be aimed correctly. If adjusted incorrectly they could blind the driver of an oncoming vehicle and cause a serious accident or seriously reduce your ability to see the road. The headlights should be checked for proper aim every 12 months and any time a new headlight is installed or front end body work is performed. It should be emphasized that the following procedure is only an interim step which will provide temporary adjustment until the headlights can be adjusted by a properly equipped shop.*

1 Halogen bulb type headlights have two adjustment screws located on the top of each headlight housing. Adjustments are made by turning two screws on the headlight housings **(see illustration)**.
2 There are several methods of adjusting the headlights. The simplest method requires masking tape, a blank wall and a level floor.
3 Position masking tape vertically on the

Chapter 12 Chassis electrical system

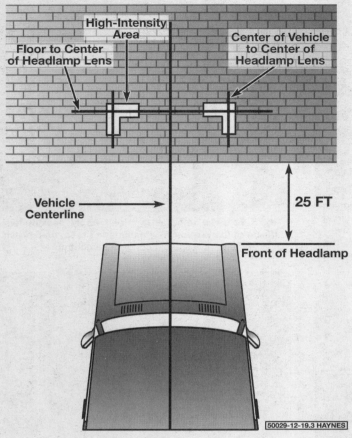

18.3 Headlight adjustment details

19.3 Headlight housing retaining bolts (arrows)

20.3 Check for power at the horn terminal (right arrow) with the horn button depressed - left arrow is the mounting bolt

wall in reference to the vehicle centerline and the centerlines of both headlights **(see illustration)**.

4 Position a horizontal tape line in reference to the centerline of all the headlights.
Note: *It may be easier to position the tape on the wall with the vehicle parked only a few inches away.*

5 Adjustment should be made with the vehicle parked 25 feet from the wall, sitting level, the gas tank half-full and no unusually heavy load in the vehicle.

6 Starting with the low beam adjustment, position the high intensity zone so it is two inches below the horizontal line and two inches to the side of the headlight vertical line, away from oncoming traffic. Adjustment is made by turning the vertical adjusting screw to raise or lower the beam. The horizontal adjusting screw should be used in the same manner to move the beam left or right.

7 With the high beams on, the high intensity zone should be vertically centered with the exact center just below the horizontal line.
Note: *It may not be possible to position the headlight aim exactly for both high and low beams. If a compromise must be made, keep in mind that the low beams are the most used and have the greatest effect on driver safety.*

8 Have the headlights adjusted by a dealer service department or service station at the earliest opportunity.

19 Headlight housing - removal and installation

Refer to illustration 19.3

Warning: *These vehicles are equipped with halogen gas-filled headlight bulbs which are under pressure and may shatter if the surface is damaged or the bulb is dropped. Wear eye protection and handle the bulbs carefully, grasping only the base whenever possible. Do not touch the surface of the bulb with your fingers because the oil from your skin could cause it to overheat and fail prematurely. If you do touch the bulb surface, clean it with rubbing alcohol.*

1 Remove the headlight bulb (see Section 17).
2 Remove the side marker light (see Section 21).
3 Remove the retaining bolts, detach the housing and withdraw it from the vehicle **(see illustration)**.
4 Installation is the reverse of removal. Be sure to check headlight adjustment (see Section 18).

20 Horn - check and replacement

Check

Refer to illustration 20.3

Note: *Check the fuses before beginning electrical diagnosis.*

1 Disconnect the electrical connector from the horn.
2 To test the horn, connect battery voltage to the terminal with a jumper wire. If the horn doesn't sound, replace it.
3 If the horn does sound, check for voltage at the terminal when the horn button is depressed **(see illustration)**. If there's voltage at the terminal, check for a bad ground at the horn.
4 If there's no voltage at the horn, check the relay (see Section 6). Note that most horn relays are either the four-terminal or externally grounded, three-terminal type.
5 If the relay is OK, check for voltage to the relay power and control circuits. If either of the circuits is not receiving voltage, inspect the wiring between the relay and the fuse panel.
6 If both relay circuits are receiving voltage, depress the horn button and check the circuit from the relay to the horn button for continuity to ground. If there's no continuity,

Chapter 12 Chassis electrical system

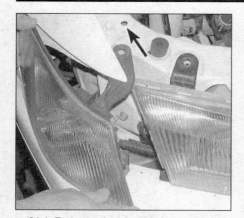

21.1 Remove the plastic pin or screw (arrow) on top of the park/turn housing, then rotate the housing outward to access the bulb - rotate the bulb holder to pull it out for bulb replacement

21.3a Remove the two screws (arrows) and push the tail light housing toward the side of the vehicle to release it from the body (Avalon models)

21.3b Pull back the trunk liner to access the tail light bulbs (arrows) on Camry and Solara models

21.4 With the tail light housing removed (Avalon models), there is access to replace the bulbs for the turn signal, the brake/tail light and the rear side marker bulb

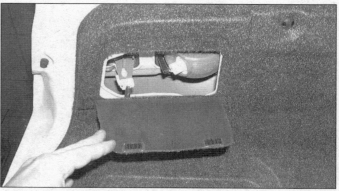

21.5 Inspection covers in the trunk lid liner allow easy replacement of the bulbs

check the circuit for an open. If there's no open circuit, replace the horn button.
7 If there's continuity to ground through the horn button, check for an open or short in the circuit from the relay to the horn.

Replacement

8 To replace the horn(s), disconnect the electrical connector and remove the bracket bolt **(see illustration 20.3)**.
9 Installation is the reverse of removal.

21 Bulb replacement

Front park/turn signal lights

Refer to illustration 21.1

1 Remove the plastic pin (a screw on some models) that secures the top of the park/turn signal light housing. Then pull the side light housing outward to access the bulbs located on the backside. Rotate the bulb holder counterclockwise and pull the bulb out **(see illustration)**.
2 Installation is the reverse of removal.

Rear tail light/brake light/turn signal

Refer to illustrations 21.3a, 21.3b, 21.4 and 21.5

3 These vehicles have tail/brake/turn lights on both the rear corners of the body and on the trunk lid. To remove the body-mounted lights for bulb replacement on Avalon models, remove the two screws and detach the tail light lens **(see illustration)**. To replace the bulbs on Camry and Solara models, pull back the trunk lining material to expose the three bulbs, which are on the inside of the body **(see illustration)**.
4 Rotate the bulb holders counterclockwise and pull the bulbs out to remove them **(see illustration)**.
5 To access the bulbs in the trunk-lid part of the rear lighting group, pull down the Velcro-latched inspection covers **(see illustration)**.
6 Installation is the reverse of removal.

License plate light

Refer to illustrations 21.8a and 21.8b

7 With the trunk lid liner removed **(see illustration)**, remove the license plate bulb

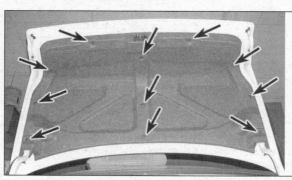

21.7a Remove the plastic clips (arrows) and the trunk lid liner to access the license plate light bulbs

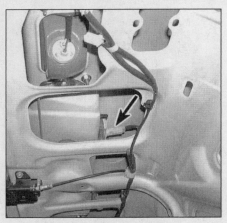

21.7b Twist the license plate light bulb socket (arrow) to remove it

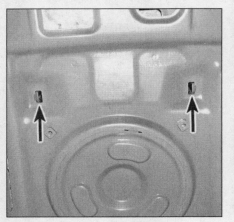

21.9 From inside the trunk, release the two clips (arrows) holding the high-mounted brake light housing to the package shelf

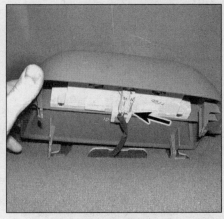

21.10 With the high-mounted brake light housing pulled up, the bulb holder (arrow) can be removed and the bulb replaced

sockets from the trunk lid **(see illustration)**. Replace the bulbs.
8   Installation is the reverse of removal.

High-mounted brake light

Refer to illustrations 21.10 and 21.11

9   Open the trunk to release two plastic clips for the high-mounted brake light, under the rear seat package shelf, then pull the housing up from inside the vehicle **(see illustration)**. **Note:** *Some Solara models are equipped with a trunk lid-mounted spoiler, which incorporates the high-mount brake light. On these models, remove the two screws at the lens and pull the light out of the spoiler to change the bulb.*
10   Twist the bulb holder counterclockwise to remove it, then pull the bulb straight out of the holder **(see illustration)**.
11   Installation is the reverse of removal.

Instrument cluster lights

Refer to illustration 21.14

12   To gain access to the instrument cluster illumination bulbs, the instrument cluster will have to be removed (see Section 12). The bulbs can then be removed and replaced from the rear of the cluster.
13   Rotate the bulb counterclockwise to remove it **(see illustration)**.
14   Installation is the reverse of removal.

Interior light

15   Pry the interior lens off the interior light housing.
16   Detach the bulb from the terminals. It may be necessary to pry the bulb out - if this is the case, pry only on the ends of the bulb (otherwise the glass may shatter).
17   Installation is the reverse of removal.

22 Electric side view mirrors - description and check

1   Most electric rear view mirrors use two motors to move the glass; one for up and down adjustments and one for left-right adjustments.
2   The control switch has a selector portion which sends voltage to the left or right side mirror. With the ignition ON but the engine OFF, roll down the windows and operate the mirror control switch through all functions (left-right and up-down) for both the left and right side mirrors.
3   Listen carefully for the sound of the electric motors running in the mirrors.
4   If the motors can be heard but the mirror glass doesn't move, there's a problem with the drive mechanism inside the mirror. Remove and disassemble the mirror to locate the problem.
5   If the mirrors do not operate and no sound comes from the mirrors, check the fuse (see Chapter 1).
6   If the fuse is OK, remove the mirror control switch from the dashboard. Check the continuity of the switch (see Section 10).
7   Re-connect the switch. Locate the wire going from the switch to ground. Leaving the switch connected, connect a jumper wire between this wire and ground. If the mirror works normally with this wire in place, repair the faulty ground connection.

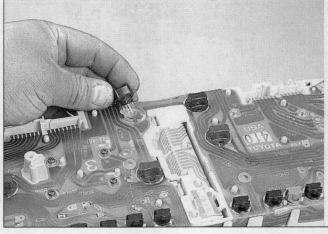

21.13 Remove the instrument cluster bulbs by rotating them 1/4-turn counterclockwise and pulling straight out

8   If the mirror still doesn't work, remove the mirror and check the wires at the mirror for voltage. Check with ignition ON and the mirror selector switch on the appropriate side. Operate the mirror switch in all its positions. There should be voltage at one of the switch-to-mirror wires in each switch position (except the neutral "off" position).
9   If there's not voltage in each switch position, check the circuit between the mirror and control switch for opens and shorts.
10   If there's voltage, remove the mirror and test it off the vehicle with jumper wires. Replace the mirror if it fails this test.

23 Cruise control system - description and check

1   The cruise control system maintains vehicle speed with a electrically-operated motor located in the engine compartment, which is connected to the accelerator pedal by a cable. The system consists of the cruise control unit, brake switch, control switches and vehicle speed sensor. Some features of the system require special testers and diagnostic procedures which are beyond the

Chapter 12 Chassis electrical system

24 Power window system - description and check

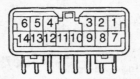

Switch	Position	Continuity between terminals
Left front	Up	3, 8 and 9 4, 5 and 6
Left front	Off	3, 4 and 5 4, 5 and 6
Left front	Down	6, 8 and 9 3, 4 and 5
Right front	Up	8, 9 and 11 4, 5 and 13*
Right front	Off	4, 5 and 11 4, 5 and 13 11 and 13**
Right front	Down	8, 9 and 13 4, 5 and 11*
Left rear	Up	8, 9 and 10 4, 5 and 12*
Left rear	Off	4, 5 and 10 4, 5 and 12 10 and 12**
Left rear	Down	8, 9 and 12 4, 5 and 10 8, 9 and 12**
Right rear	Up	7, 8 and 9 4, 5 and 14*
Right rear	Off	4, 5 and 7 4, 5 and 14 7 and 14**
Right rear	Down	8, 9 and 14 4, 5 and 7*

* only with switch Unlocked, ** only with switch Locked

24.10a Power window master control switch terminal guide and continuity chart (Toyota models)

Refer to illustrations 24.10a through 24.10j

1 The power window system operates electric motors, mounted in the doors, which lower and raise the windows. The system consists of the control switches, relays, the motors, regulators, glass mechanisms and associated wiring.

2 The power windows can be lowered and raised from the master control switch by the driver or by remote switches located at the individual windows. Each window has a separate motor that is reversible. The position of the control switch determines the polarity and therefore the direction of operation.

3 The circuit is protected by a fuse and a circuit breaker. Each motor is also equipped with an internal circuit breaker; this prevents one stuck window from disabling the whole system.

4 The power window system will only operate when the ignition switch is ON. In addition, many models have a window lock-out switch at the master control switch that, when activated, disables the switches at the rear windows and, sometimes, the switch at the passenger's window also. Always check these items before troubleshooting a window problem.

5 These procedures are general in nature, so if you can't find the problem using them, take the vehicle to a dealer service department or other properly equipped repair facility.

6 If the power windows won't operate, always check the fuse and circuit breaker first.

7 If only the rear windows are inoperative, or if the windows only operate from the master control switch, check the rear window lockout switch for continuity in the unlocked position. Replace it if it doesn't have continuity.

8 Check the wiring between the switches and fuse panel for continuity. Repair the wiring, if necessary.

9 If only one window is inoperative from the master control switch, try the other control switch at the window. **Note:** *This doesn't apply to the drivers door window.*

10 If the same window works from one switch, but not the other, check the switch for continuity **(see illustrations on the following page)**. If the continuity is not as specified replace the switch.

11 If the switch tests OK, check for a short or open in the circuit between the affected switch and the window motor.

12 If one window is inoperative from both switches, remove the trim panel from the affected door and check for voltage at the switch and at the motor while the switch is operated.

13 If voltage is reaching the motor, disconnect the glass from the regulator (see Chapter 11). Move the window up and down by hand while checking for binding and damage.

---

scope of this manual. Listed below are some general procedures that may be used to locate common problems.

2 Locate and check the fuse (see Section 3).

3 Have an assistant operate the brake lights while you check their operation (voltage from the brake light switch deactivates the cruise control).

4 If the brake lights don't come on or stay on all the time, correct the problem and retest the cruise control.

5 Visually inspect the control cable between cruise control motor and the throttle linkage for free movement, replace if necessary.

6 The cruise control system uses a speed-sensing device. The speed sensor is located in the speedometer. Test the speed sensor (see Chapter 6).

7 Test drive the vehicle to determine if the cruise control is now working. If it isn't, take it to a dealer service department or an automotive electrical specialist for further diagnosis.

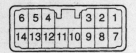

Left Front

Window unlock:

Switch position	Tester connection	Specified condition
UP	1 – 13 6 – 7	Continuity
OFF	1 – 6 – 13	Continuity
DOWN	1 – 6 7 – 13	Continuity

Window lock:

Switch position	Tester connection	Specified condition
UP	1 – 13 6 – 7	Continuity
OFF	1 – 6 – 13	Continuity
DOWN	1 – 6 7 – 13	Continuity

Right Front

Window unlock:

Switch position	Tester connection	Specified condition
UP	1 – 5 7 – 12	Continuity
OFF	1 – 5 – 12	Continuity
DOWN	1 – 12 5 – 7	Continuity

Window lock:

Switch position	Tester connection	Specified condition
UP	7 – 12	Continuity
OFF	5 – 12	Continuity
DOWN	5 – 7	Continuity

Left Rear

Window unlock:

Switch position	Tester connection	Specified condition
UP	1 – 9 7 – 10	Continuity
OFF	1 – 9 – 10	Continuity
DOWN	1 – 10 7 – 9	Continuity

Window lock:

Switch position	Tester connection	Specified condition
UP	7 – 10	Continuity
OFF	9 – 10	Continuity
DOWN	7 – 9	Continuity

Right Rear

Window unlock:

Switch position	Tester connection	Specified condition
UP	1 – 14 7 – 11	Continuity
OFF	1 – 11 – 14	Continuity
DOWN	1 – 11 7 – 14	Continuity

Window lock:

Switch position	Tester connection	Specified condition
UP	7 – 11	Continuity
OFF	11 – 14	Continuity
DOWN	7 – 14	Continuity

24.10b Power window master control switch terminal guide and continuity chart (1998 and earlier Lexus ES 300)

Also check for binding and damage to the regulator. If the regulator is not damaged and the window moves up and down smoothly, replace the motor. If there's binding or damage, lubricate, repair or replace parts, as necessary.

14 If voltage isn't reaching the motor, check the wiring in the circuit for continuity between the switches and motors. You'll need to consult the wiring diagram for the vehicle. If the circuit is equipped with a relay, check that the relay is grounded properly and receiving voltage.

15 Test the windows after you are done to confirm proper repairs.

Left Front

Window unlock:

Switch position	Tester connection	Specified condition
UP AUTO	4 – 9 – 11	Continuity
UP MANUAL	4 – 9	Continuity
OFF	–	No Continuity
DOWN MANUAL	2 – 4	Continuity
DOWN AUTO	2 – 4 – 11	Continuity

Right Front

Window unlock:

Switch position	Tester connection	Specified condition
UP AUTO	4 – 13 – 15	Continuity
UP MANUAL	4 – 13	Continuity
OFF	–	No Continuity
DOWN MANUAL	4 – 5	Continuity
DOWN AUTO	4 – 5 – 15	Continuity

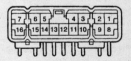

Left Rear

Window unlock:

Switch position	Tester connection	Specified condition
UP	8 – 12 7 – 10	Continuity
OFF	7 – 8 – 10	Continuity
DOWN	10 – 12 7 – 8	Continuity

Right Rear

Window unlock:

Switch position	Tester connection	Specified condition
UP	12 – 14 7 – 16	Continuity
OFF	7 – 14 – 16	Continuity
DOWN	12 – 16 7 – 14	Continuity

24.10c Power window master control switch terminal guide and continuity chart (1999 and later Lexus ES 300)

Chapter 12 Chassis electrical system

Front driver's switch (Window unlock):

Switch position	Tester connection	Specified condition
UP	1 – 4, 3 – 9	Continuity
OFF	1 – 3 – 4	Continuity
DOWN	1 – 3, 4 – 9	Continuity

Front driver's switch (Window lock):

Switch position	Tester connection	Specified condition
UP	1 – 4, 3 – 9	Continuity
OFF	1 – 3 – 4	Continuity
DOWN	1 – 3, 4 – 9	Continuity

Front passenger's switch (Window unlock):

Switch position	Tester connection	Specified condition
UP	1 – 10, 8 – 9	Continuity
OFF	1 – 8 – 10	Continuity
DOWN	1 – 8, 9 – 10	Continuity

Front passenger's switch (Window lock):

Switch position	Tester connection	Specified condition
UP	8 – 9	Continuity
OFF	8 – 10	Continuity
DOWN	9 – 10	Continuity

24.10d Power window master control switch terminal guide and continuity chart (2000 and later Solara)

Switch position	Tester connection	Specified condition
UP	1 – 2, 3 – 4	Continuity
OFF	1 – 2, 3 – 5	Continuity
DOWN	1 – 4, 3 – 5	Continuity

24.10e Passenger front power window switch terminal guide and continuity chart (Toyota)

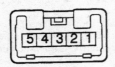

Switch position	Tester connection	Specified condition
LOCK	2 – 3	Continuity
OFF	–	No continuity
UNLOCK	1 – 2	Continuity

24.10f Passenger front power window switch terminal guide and continuity chart (1998 and earlier Lexus ES 300)

Switch position	Tester connection	Specified condition
UP	1 – 3, 4 – 5	Continuity
OFF	1 – 2, 4 – 5	Continuity
DOWN	1 – 2, 3 – 5	Continuity

24.10g Passenger front power window switch terminal guide and continuity chart (1999 and later Lexus ES 300)

Switch position	Tester connection		Specified condition
UP MANUAL	3 – 6		Continuity
UP AUTO	3 – 6	3 – 5	Continuity
OFF	–		No Continuity
DOWN AUTO	3 – 4	3 – 5	Continuity
DOWN MANUAL	3 – 4		Continuity

24.10h Rear door power window switch terminal guide and continuity chart (Camry, Solar, 1998 and earlier Lexus ES 300 and 1999 and earlier Avalon)

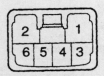

Switch position	Tester connection		Specified condition
UP	1 – 3	2 – 5	Continuity
OFF	1 – 3	2 – 4	Continuity
DOWN	2 – 4	3 – 5	Continuity

24.10i Rear door power window switch terminal guide and continuity chart (1999 and later Lexus ES 300)

Switch position	Tester connection	Specified condition
UP	5 – 4, 2 – 6	Continuity
OFF	4 – 6, 2 – 6	No continuity
DOWN	2 – 5, 4 – 6	Continuity

24.10j Rear door power window switch terminal guide and continuity chart (2000 and later Avalon)

25 Power door lock system - description and check

Refer to illustrations 25.6a, 25.6b, 25.6c, 25.6e, 25.f and 25.9

1 A power door lock system operates the door lock actuators mounted in each door. The system consists of the switches, actuators, a control unit and associated wiring. Diagnosis can usually be limited to simple checks of the wiring connections and actuators for minor faults that can be easily repaired. Since this system uses an electronic control unit in-depth diagnosis should be left to a dealership service department. The door lock control unit is located behind the instrument panel, to the right of the fuse box.

2 Power door lock systems are operated by bi-directional solenoids located in the doors. The lock switches have two operating positions: Lock and Unlock. When activated, the switch sends a ground signal to the door lock control unit to lock or unlock the doors. Depending on which way the switch is activated, the control unit reverses polarity to the solenoids, allowing the two sides of the circuit to be used alternately as the feed (positive) and ground side.

3 Some vehicles may have an anti-theft systems incorporated into the power locks. If you are unable to locate the trouble using the following general Steps, consult a dealer service department or other qualified repair shop.

4 Always check the circuit protection first. Some vehicles use a combination of circuit breakers and fuses.

5 Operate the door lock switches in both directions (Lock and Unlock) with the engine off. Listen for the click of the solenoids operating.

6 Test the switches for continuity **(see illustrations)**. Replace the switch if there's not continuity in both switch positions.

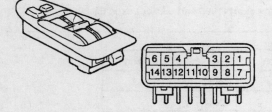

Switch position	Tester connection	Specified condition
LOCK	2 – 4	Continuity
OFF	–	No continuity
UNLOCK	2 – 3	Continuity

25.6a Driver's power door lock switch terminal guide and continuity chart (Toyota Camry, Solara and 1999 and earlier Avalon)

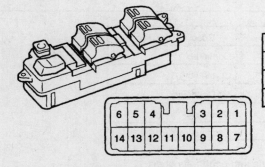

Switch position	Tester connection	Specified condition
LOCK	1 – 4	Continuity
OFF	–	No continuity
UNLOCK	1 – 3	Continuity

25.6b Driver's power door lock switch terminal guide and continuity chart (1998 and earlier Lexus ES 300)

Chapter 12 Chassis electrical system

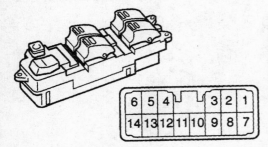

Switch position	Tester connection	Specified condition
LOCK	1 – 4	Continuity
OFF	–	No continuity
UNLOCK	1 – 3	Continuity

25.6c Driver's power door lock switch terminal guide and continuity chart (1999 and later Lexus ES 300)

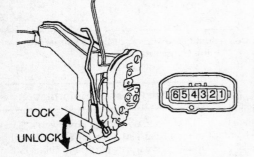

Switch position	Tester connection	Specified condition
LOCK	3 – 5	Continuity
OFF	–	No continuity
UNLOCK	3 – 6	Continuity

25.6d Driver's power door lock switch terminal guide and continuity chart (2000 and later Avalon)

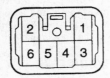

Switch position	Tester connection	Specified condition
LOCK	3 – 6	Continuity
OFF	–	No continuity
UNLOCK	3 – 5	Continuity

25.6e Passenger's power door lock switch terminal guide and continuity chart (Camry, Solara and 1999 and earlier Avalon)

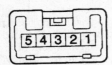

Switch position	Tester connection	Specified condition
LOCK	2 – 3	Continuity
OFF	–	No continuity
UNLOCK	1 – 2	Continuity

25.6f Passenger's power door lock switch terminal guide and continuity chart (Lexus ES 300)

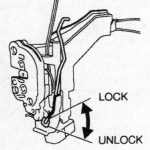

Switch position	Tester connection	Specified condition
LOCK	2 – 4	Continuity
OFF	–	No continuity
UNLOCK	1 – 4	Continuity

25.6g Passenger's power door lock switch terminal guide and continuity chart (2000 and later Avalon)

Chapter 12 Chassis electrical system

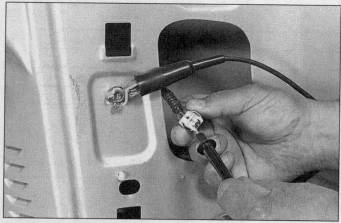

25.9 Check for voltage at the lock solenoid while the lock switch is operated

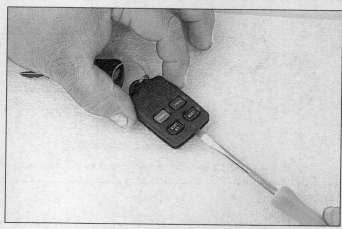

25.14 Use a small screwdriver or coin to separate the transmitter halves

7   Check the wiring between the switches, control unit and solenoids for continuity. Repair the wiring if there's no continuity.
8   Check for a bad ground at the switches or the control unit.
9   If all but one lock solenoids operate, remove the trim panel from the affected door (see Chapter 11) and check for voltage at the solenoid while the lock switch is operated. One of the wires should have voltage in the Lock position; the other should have voltage in the Unlock position **(see illustration)**.
10   If the inoperative solenoid is receiving voltage, replace the solenoid.
11   If the inoperative solenoid isn't receiving voltage, check the relay for an open or short in the wire between the lock solenoid and the control unit. **Note:** *It's common for wires to break in the portion of the harness between the body and door (opening and closing the door fatigues and eventually breaks the wires).*

Keyless entry system

Refer to illustrations 25.14 and 25.15

12   The keyless entry system consists of a remote control transmitter that sends a coded infrared signal to a receiver which then operates the door lock system. On models so equipped, the transmitter may also engage the alarm system and provide a "panic" button which flashes the lights and blows the horn for emergencies.
13   Replace the transmitter batteries when the red LED light on the case doesn't light when the button is pushed. As the batteries deteriorate with age, the distance at which the remote transmitter operates will diminish.
14   Use a coin or small screwdriver to carefully separate the case halves for battery replacement **(see illustration)**.
15   Replace the two lithium batteries with the same type as originally installed, observing the polarity diagram on the case **(see illustration)**.
16   Snap the case halves together.

26   Daytime Running Lights (DRL) - general information

The Daytime Running Lights (DRL) system used on Canadian models illuminates the headlights whenever the engine is running. The only exception is with the engine running and the parking brake engaged. Once the parking brake is released, the lights will remain on as long as the ignition switch is on, even if the parking brake is later applied.

The DRL system supplies reduced power to the headlights so they won't be too bright for daytime use, while prolonging headlight life.

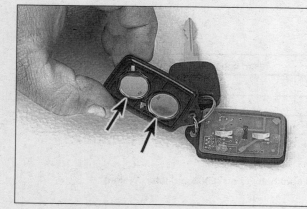

25.15 Replace the lithium batteries with the same type as used originally

27   Airbag - general information

All models are equipped with a Supplemental Restraint System (SRS), more commonly known as an airbag. This system is designed to protect the driver, and the front seat passenger, from serious injury in the event of a head-on or frontal collision. It consists of an airbag module in the center of the steering wheel and the right side of the instrument panel and a sensing/diagnostic module mounted in the center of the vehicle, ahead of the floor console.

Airbag module

Driver's side

The airbag inflator module contains a housing incorporating the cushion (airbag) and inflator unit, mounted in the center of the steering wheel The inflator assembly is mounted on the back of the housing over a hole through which gas is expelled, inflating the bag almost instantaneously when an electrical signal is sent from the system. A coil assembly on the steering column under the module carries this signal to the module.

This coil assembly can transmit an electrical signal regardless of steering wheel position. The igniter in the air bag converts the electrical signal to heat and ignites the sodium azide/copper oxide powder, producing nitrogen gas, which inflates the bag.

Passenger's side

The airbag is mounted above the glove compartment and designated by the letters SRS (Supplemental Restraint System). It consists of an inflator containing an igniter, a bag assembly, a reaction housing and a trim cover.

The airbag is considerably larger that the steering wheel-mounted unit and is supported by the steel reaction housing. The trim cover is textured and painted to match the instrument panel and has a molded seam which splits when the bag inflates. As with

Chapter 12 Chassis electrical system

the steering-wheel-mounted air bag, the igniter electrical signal converts to heat, converting sodium azide/iron oxide powder to nitrogen gas, inflating the bag.

Side-impact airbags

Extra protection is provided on some later models with the addition of side-impact airbags. These are smaller devices located in the seat backs on the side toward the exterior of the vehicle. The impact sensors for the side-impact airbags are located at the bottom of the door pillars in the body.

Sensing and diagnostic module

The sensing and diagnostic module supplies the current to the airbag system in the event of the collision, even if battery power is cut off. It checks this system every time the vehicle is started, causing the "AIR BAG" light to go on then off, if the system is operating properly. If there is a fault in the system, the light will go on and stay on, flash, or the dash will make a beeping sound. If this happens, the vehicle should be taken to your dealer immediately for service.

Precautions

Warning: *Failure to follow these precautions could result in accidental deployment of the airbag and personal injury.*
Caution: *If the stereo in your vehicle is equipped with an anti-theft system, make sure you have the correct activation code before disconnecting the battery.*

Whenever working in the vicinity of the steering wheel, steering column or any of the other SRS system components, the system must be disarmed. To disarm the system:

a) *Point the wheels straight ahead and turn the key to the Lock position.*
b) *Disconnect the cable from the negative battery terminal, then the positive cable.*
c) *Wait at least two minutes for the back-up power supply to be depleted.*

Whenever handling an airbag module, always keep the airbag opening (the trim side) pointed away from your body. Never place the airbag module on a bench or other surface with the airbag opening facing the surface. Always place the airbag module in a safe location with the airbag opening facing up.

Never measure the resistance of any SRS component. An ohmmeter has a built-in battery supply that could accidentally deploy the airbag.

Never use electrical welding equipment on a vehicle equipped with an airbag without first disconnecting the yellow airbag connector, located under the steering column near the combination switch connector (driver's airbag) and behind the glove box (passenger's airbag).

Never dispose of a live airbag module. Return it to a dealer service department or other qualified repair shop for safe deployment and disposal.

Wire colors are indicated by an alphabetical code.

B	= Black	L	= Blue	R	= Red
BR	= Brown	LG	= Light Green	V	= Violet
G	= Green	O	= Orange	W	= White
GR	= Gray	P	= Pink	Y	= Yellow

The first letter indicates the basic wire color and the second letter indicates the color of the stripe.

Example: L—Y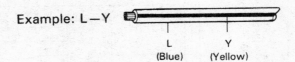

L   Y
(Blue) (Yellow)

Wiring diagram color code chart

28 Wiring diagrams - general information

Since it isn't possible to include all wiring diagrams for every year covered by this manual, the following diagrams are those that are typical and most commonly needed.

Prior to troubleshooting any circuits, check the fuse and circuit breakers (if equipped) to make sure they're in good condition. Make sure the battery is properly charged and check the cable connections (see Chapter 1).

When checking a circuit, make sure that all connectors are clean, with no broken or loose terminals. When unplugging a connector, do not pull on the wires. Pull only on the connector housings themselves.

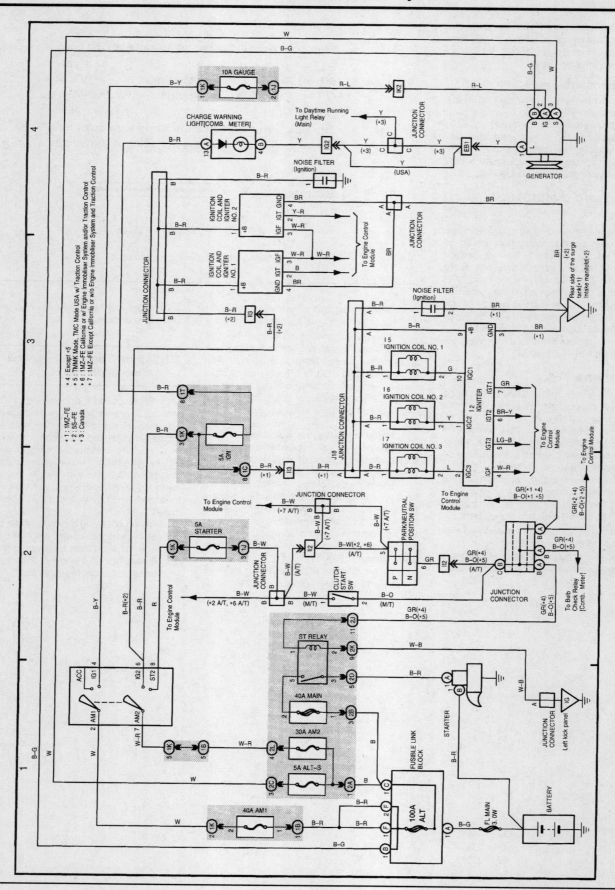

Chapter 12 Chassis electrical system

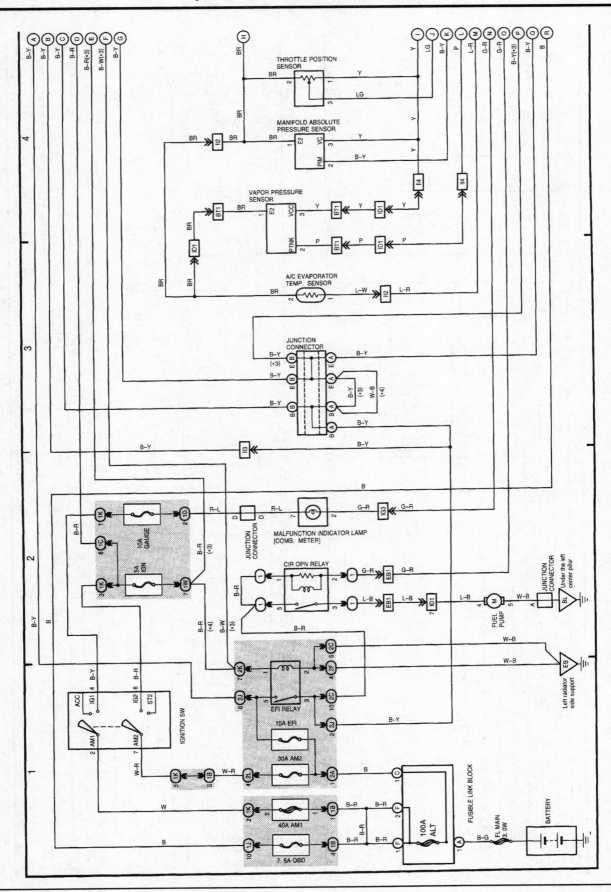

Toyota engine control system - four-cylinder models (1 of 3)

Chapter 12 Chassis electrical system

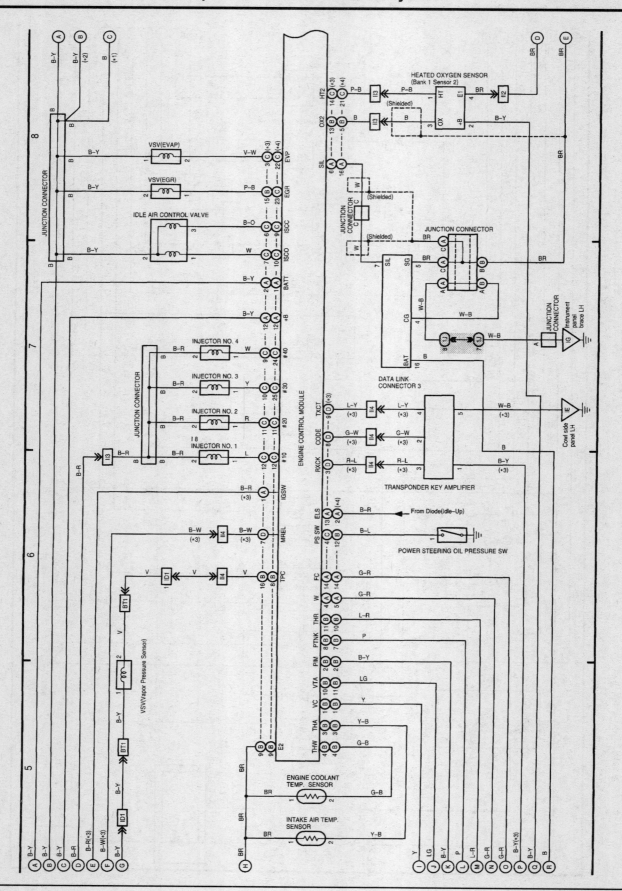

Toyota engine control system - four-cylinder models (2 of 3)

Chapter 12 Chassis electrical system

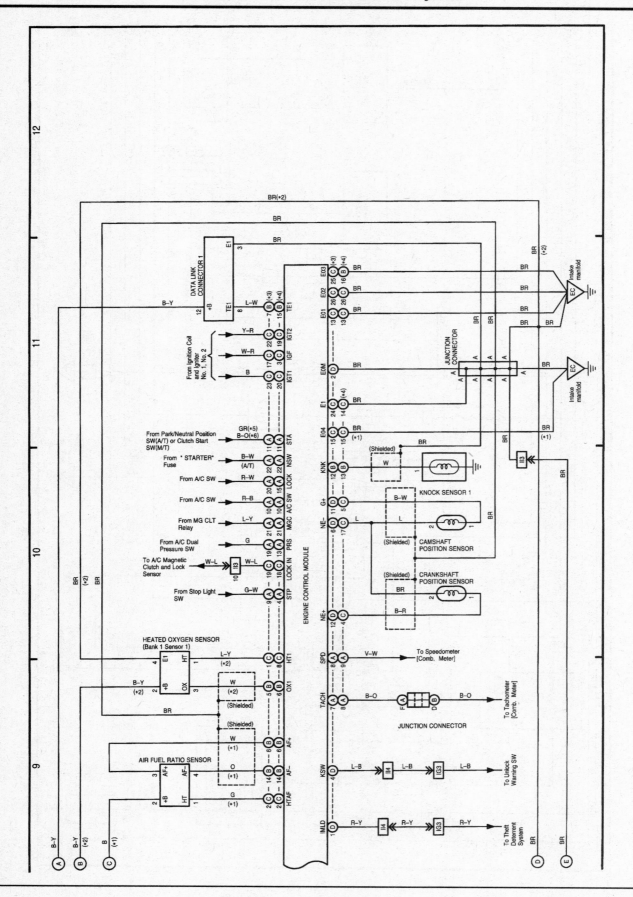

Toyota engine control system - four-cylinder models (3 of 3)

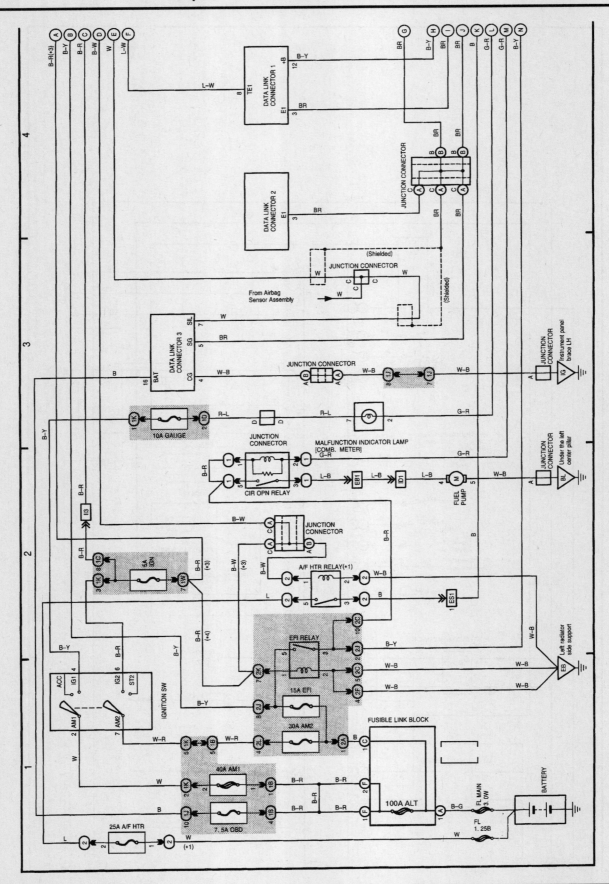

Chapter 12 Chassis electrical system

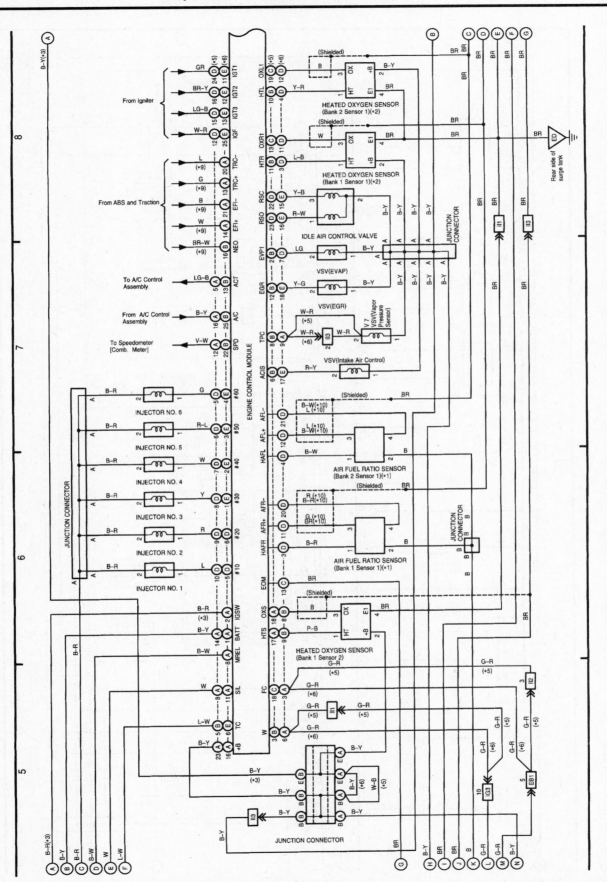

Toyota engine control system - V6 models (2 of 3)

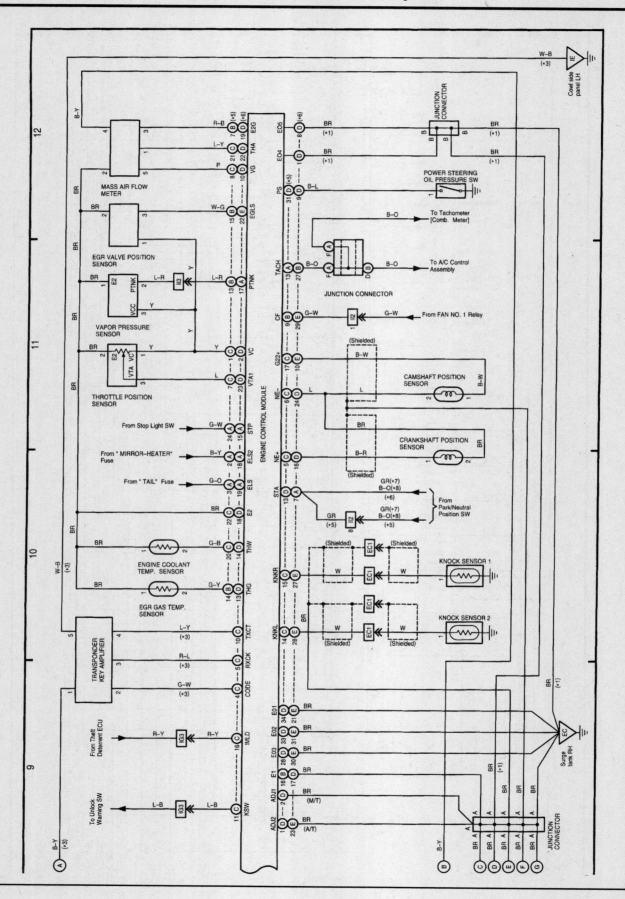

Chapter 12 Chassis electrical system

Toyota engine control system - V6 models (3 of 3)

Chapter 12 Chassis electrical system

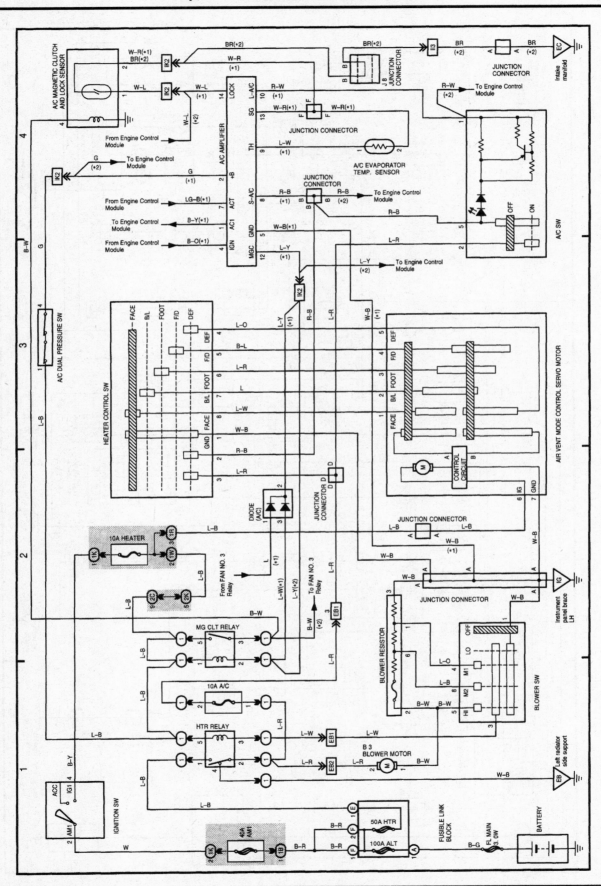

Typical manual heating/air conditioning system (Toyota)

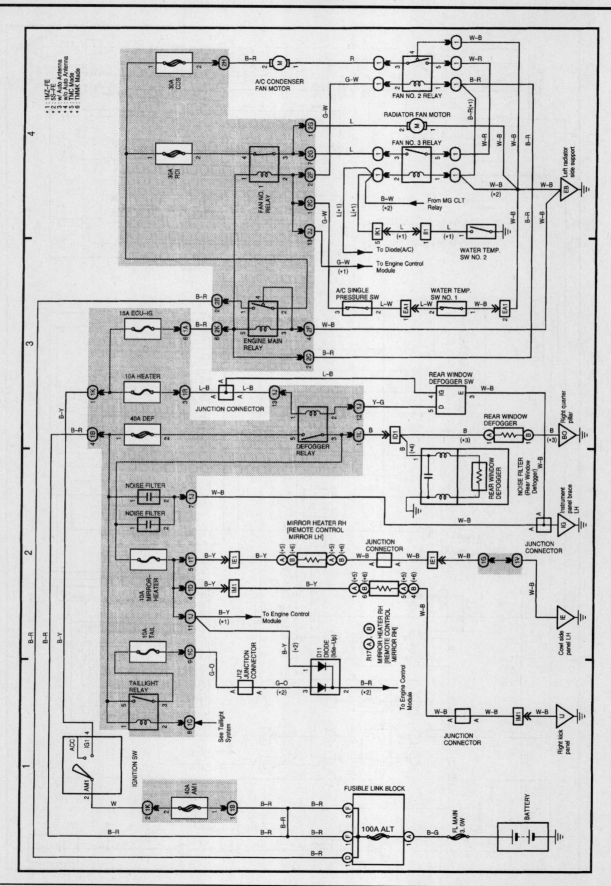

Typical engine cooling fan system; rear window defogger and mirror heater systems (Toyota)

Chapter 12 Chassis electrical system

12-33

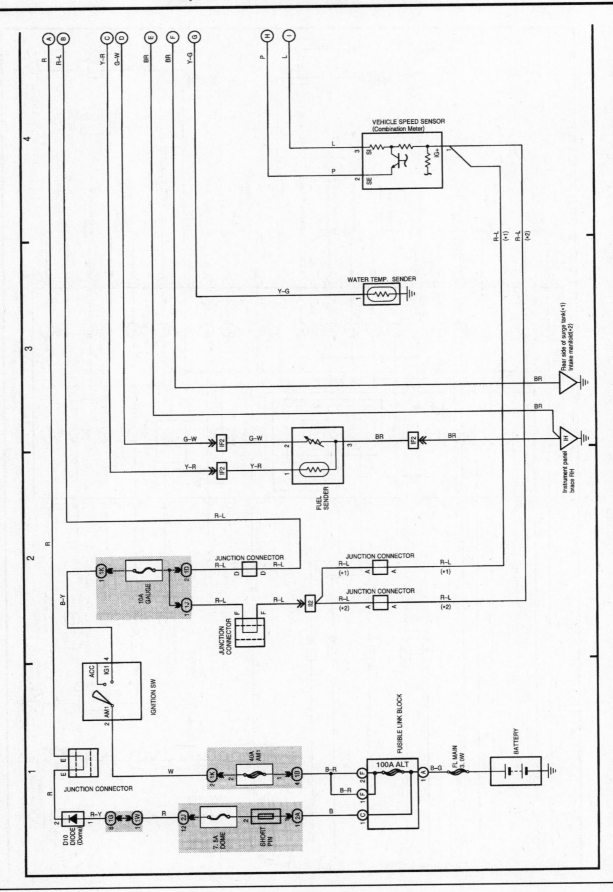

Typical Toyota instrument panel gauges and warning light system (1 of 2)

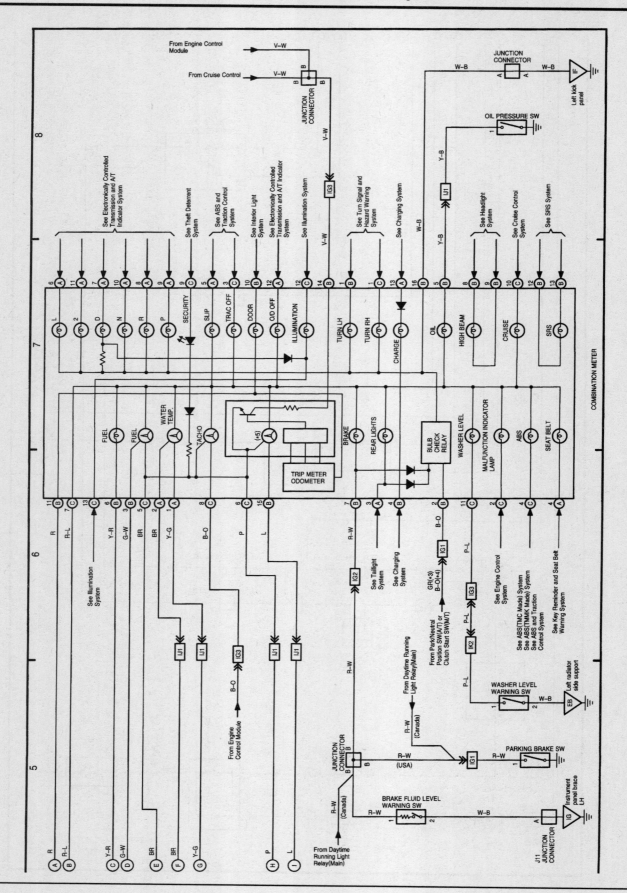

Typical Toyota instrument panel gauges and warning light system (2 of 2)

Chapter 12 Chassis electrical system

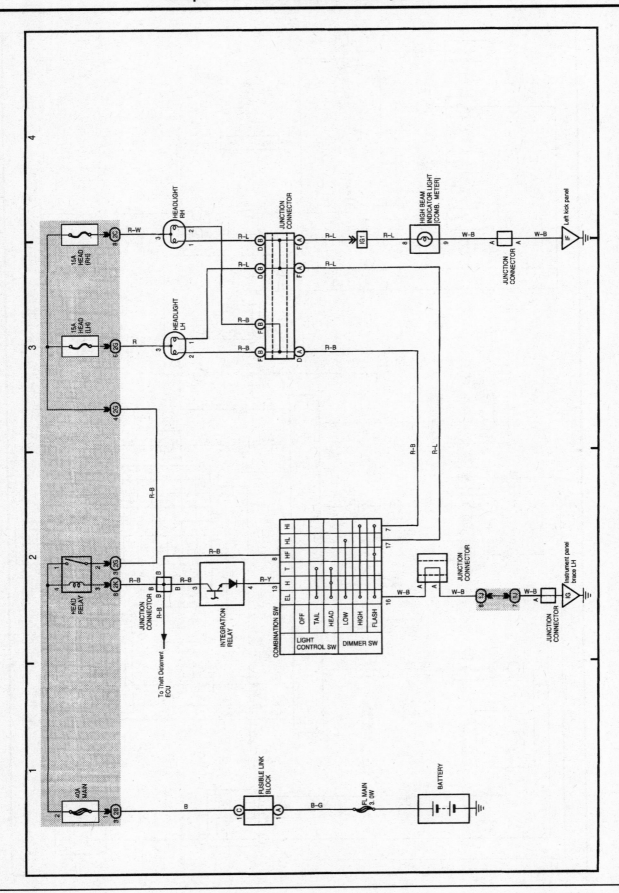

Typical Toyota headlight system (USA models)

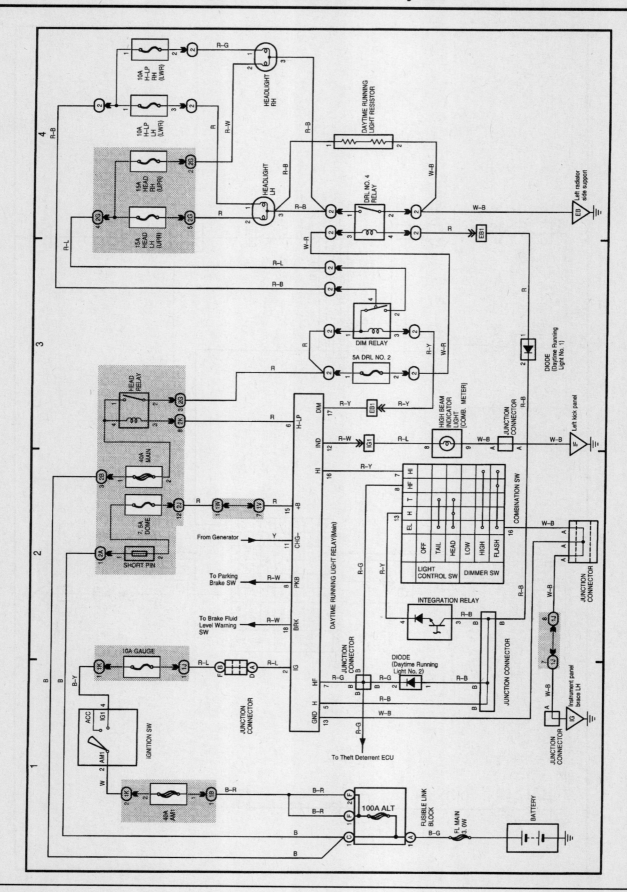

Chapter 12 Chassis electrical system

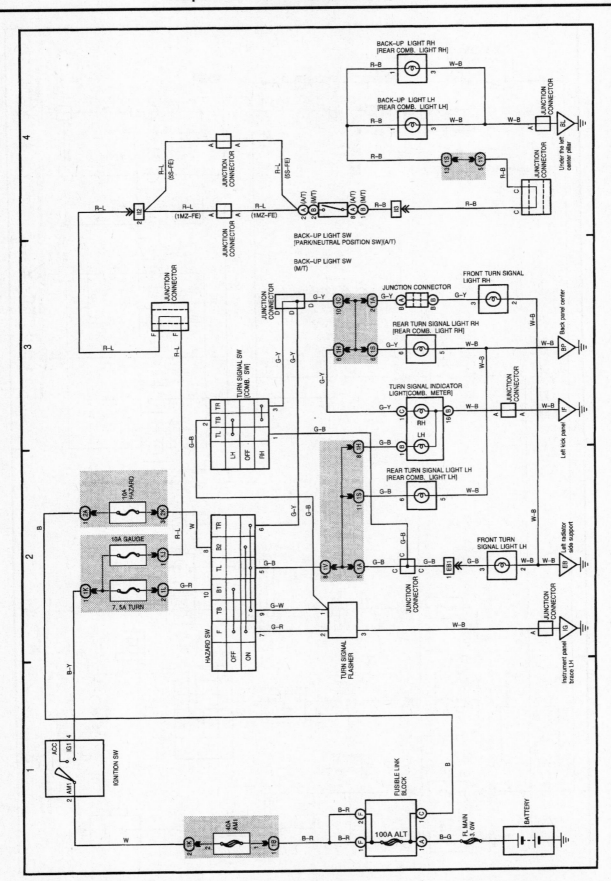

Typical turn signal, hazard warning light and backup light systems (Toyota)

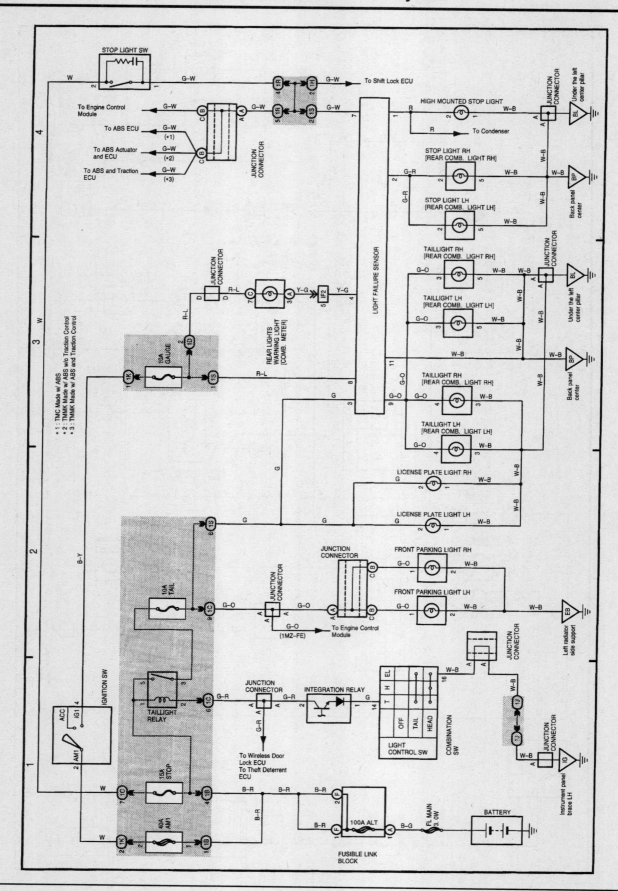

Chapter 12 Chassis electrical system

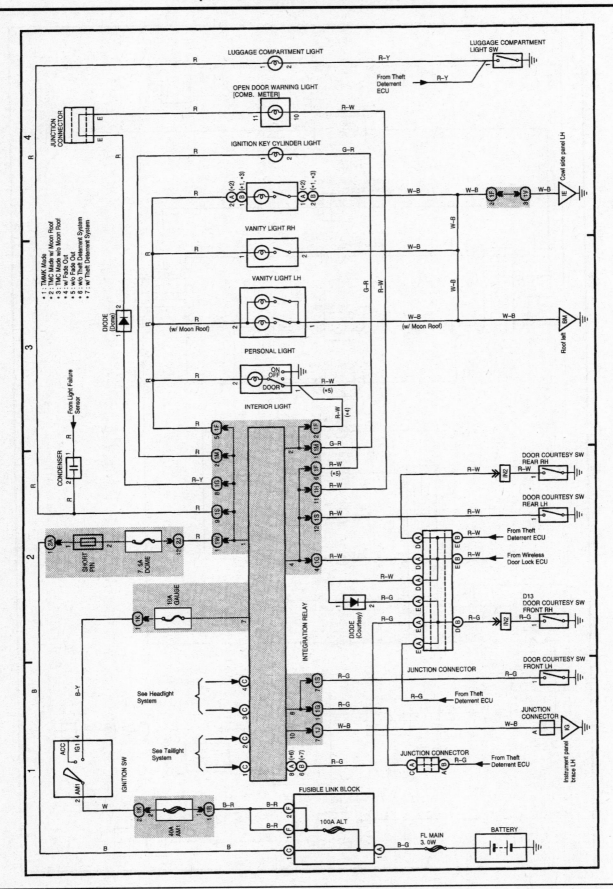

Typical Toyota interior lighting system (1 of 2)

Chapter 12 Chassis electrical system

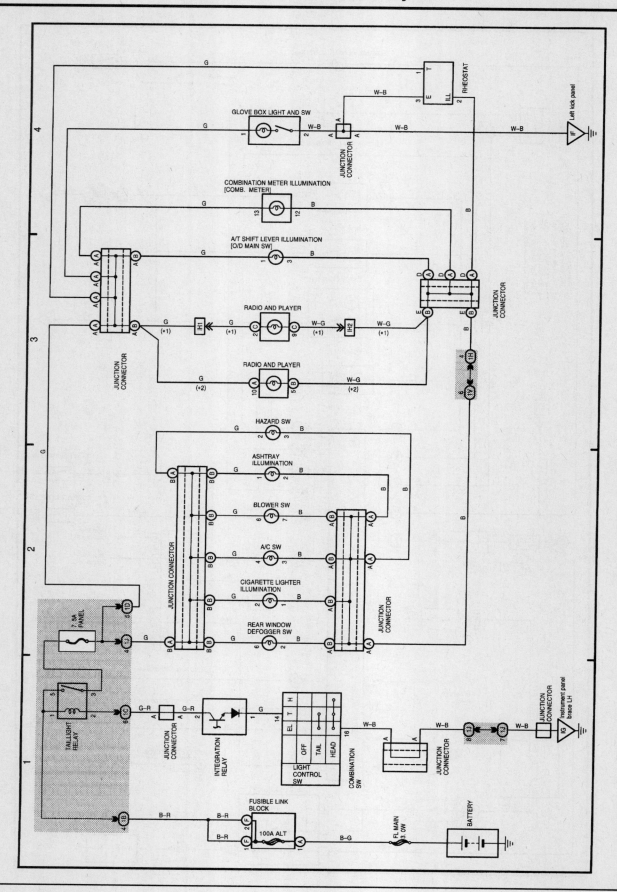

Typical Toyota interior lighting system (2 of 2)

Chapter 12 Chassis electrical system

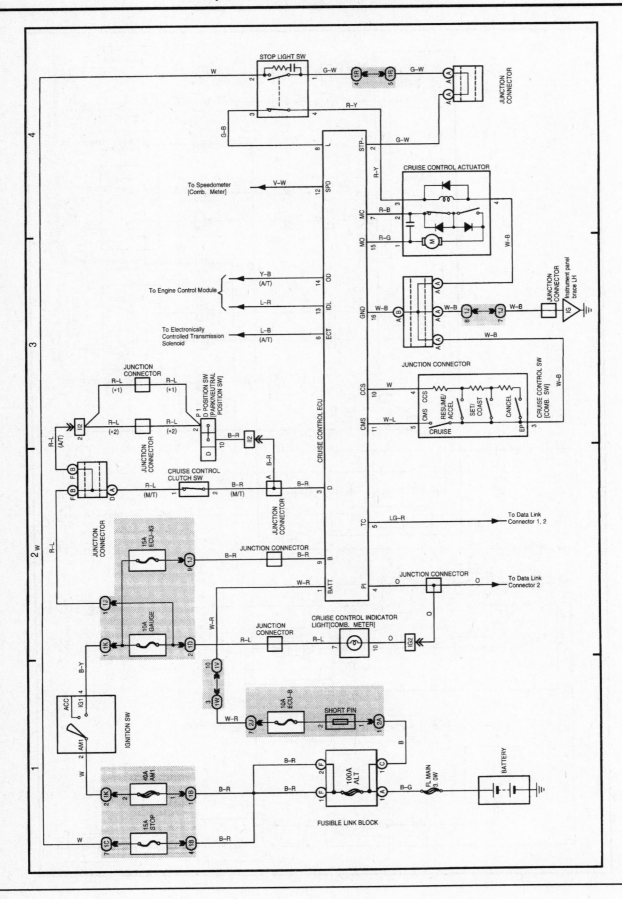

Typical cruise control system (Toyota)

12-42 Chapter 12 Chassis electrical system

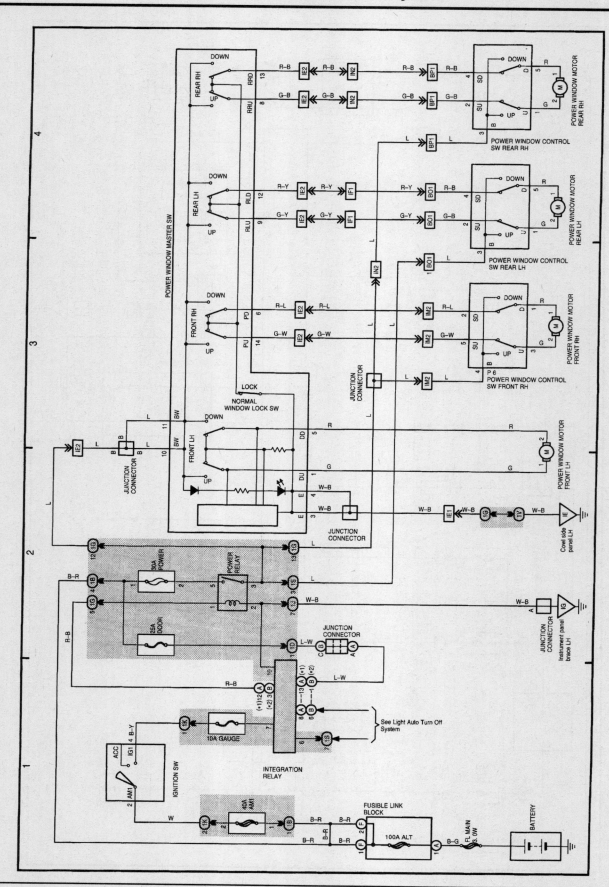

Typical power window system (Toyota)

Chapter 12 Chassis electrical system

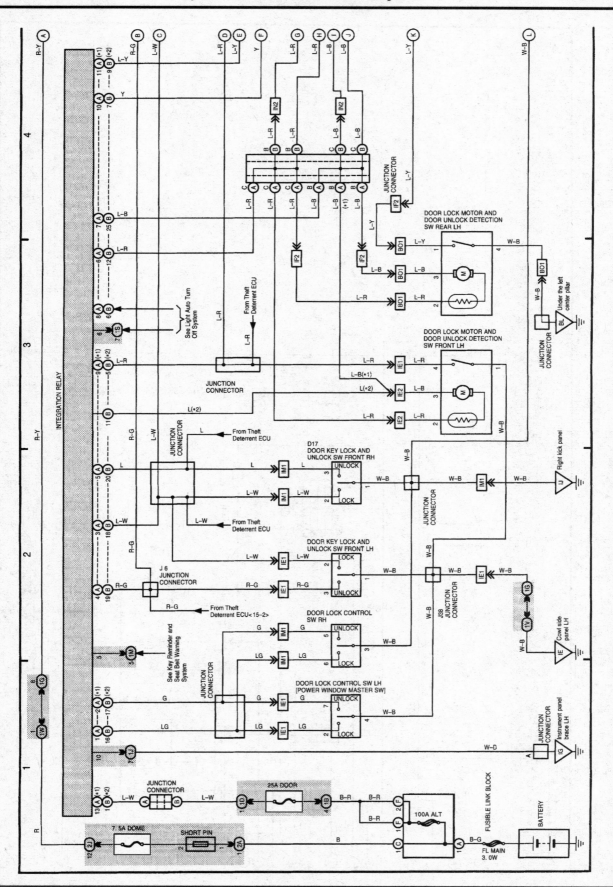

Typical Toyota power door lock system (1 of 2)

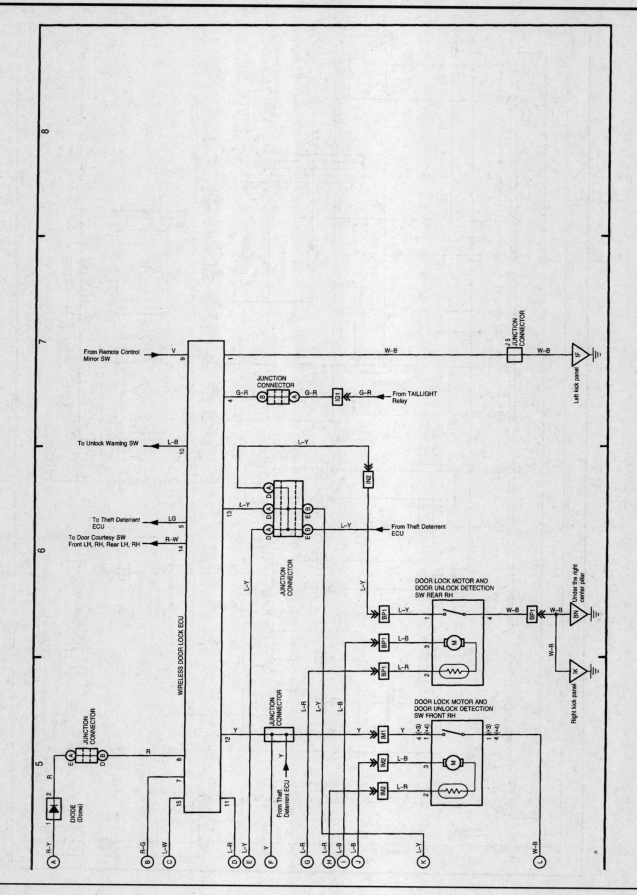

Typical Toyota power door lock system (2 of 2)

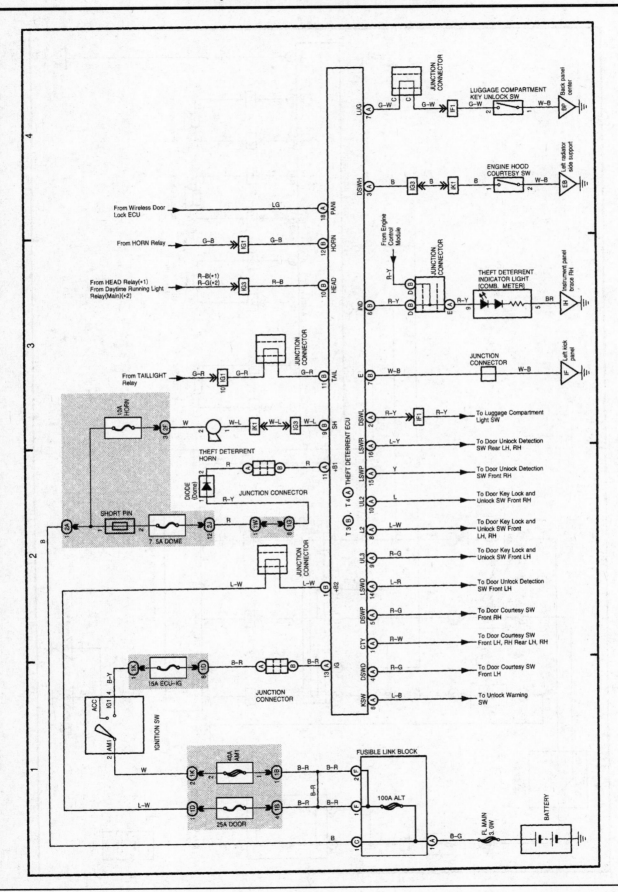

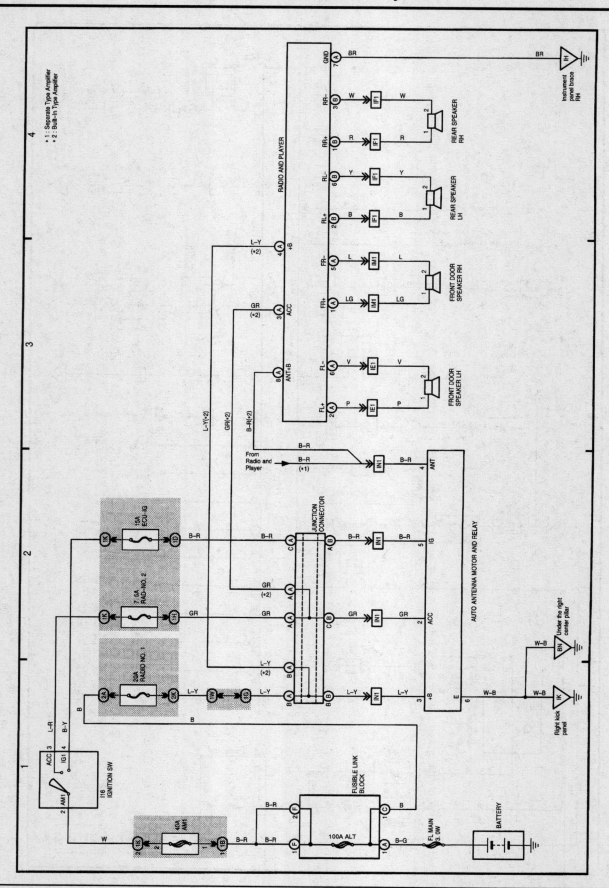

Typical Toyota audio system (with built-in amplifier)

Chapter 12 Chassis electrical system

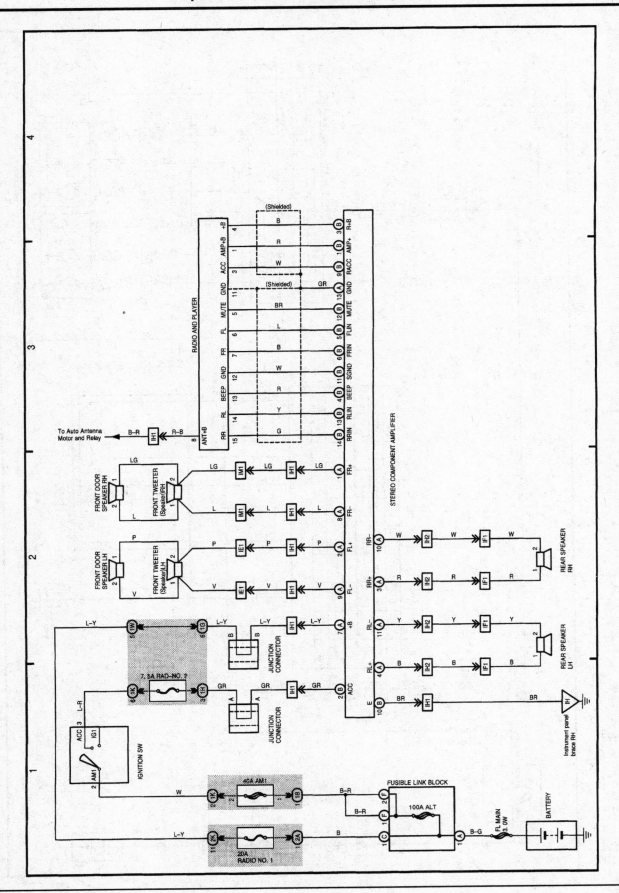

Typical Toyota audio system (with separate amplifier)

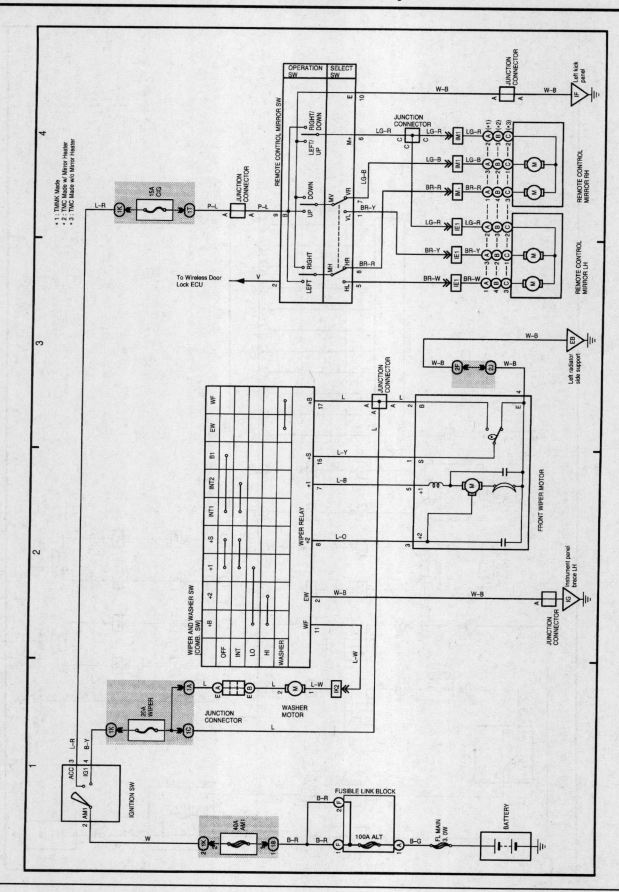

Typical windshield wiper and washer system; power mirror system (Toyota)

Chapter 12 Chassis electrical system

12-49

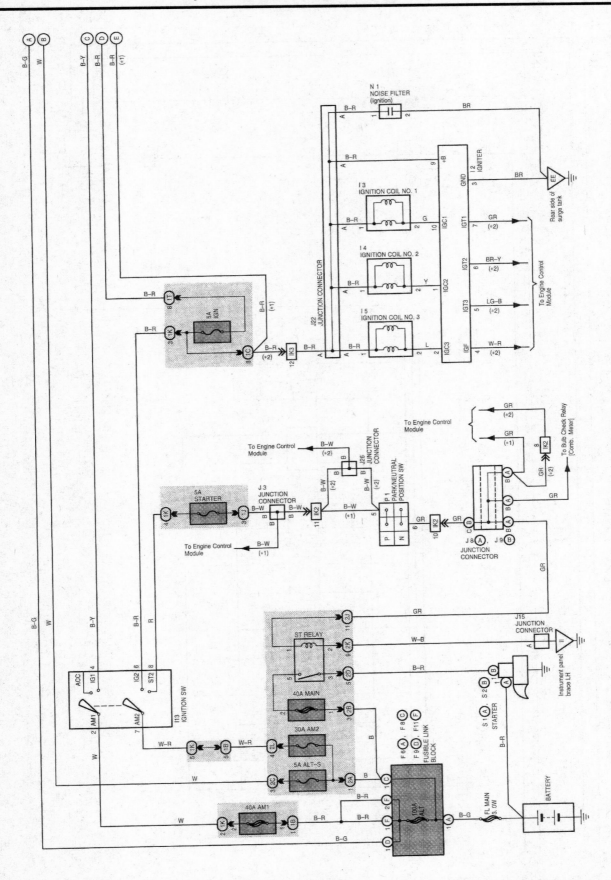

Typical Lexus ES 300 starting, charging and ignition systems (1 of 2)

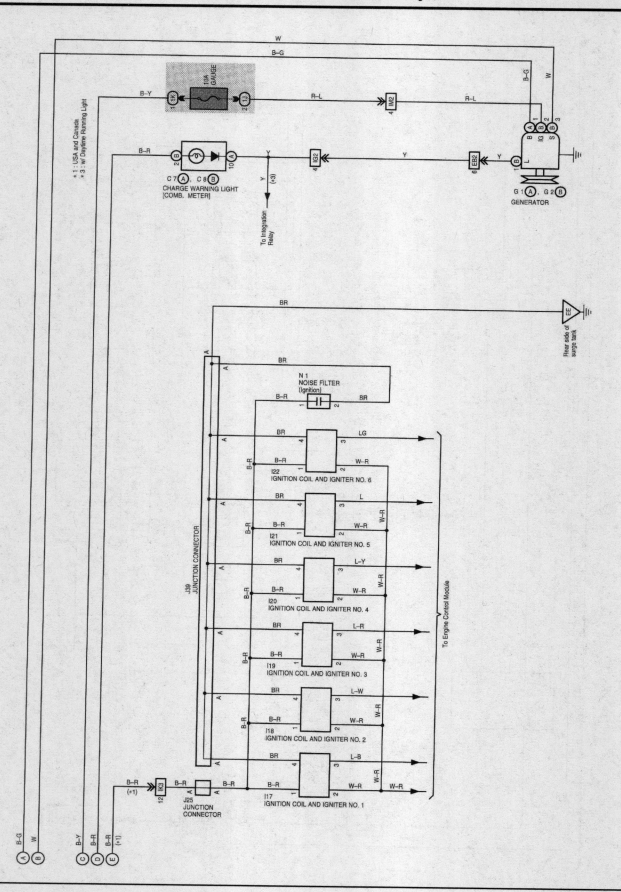

Typical Lexus ES 300 starting, charging and ignition systems (2 of 2)

Chapter 12 Chassis electrical system

12-51

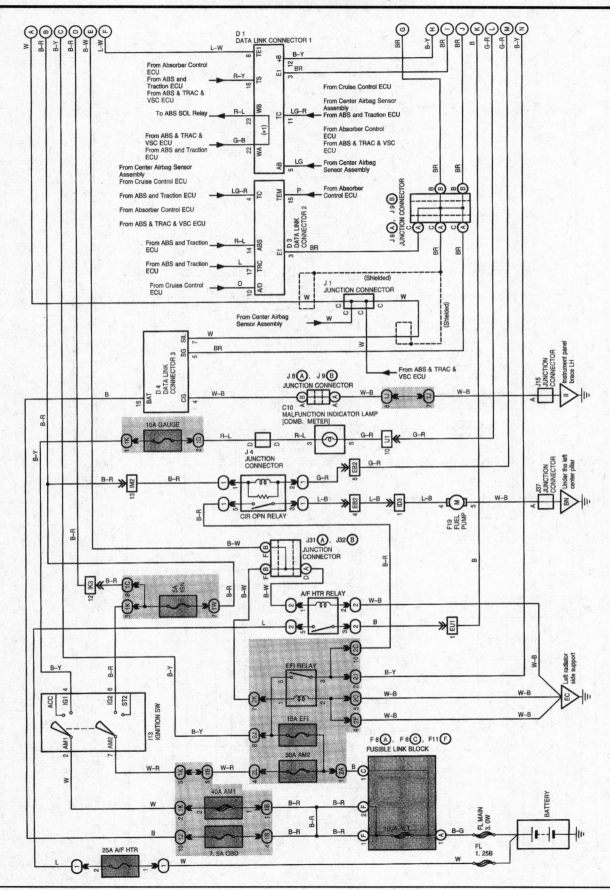

Typical Lexus ES 300 engine control and immobiliser system (1 of 3)

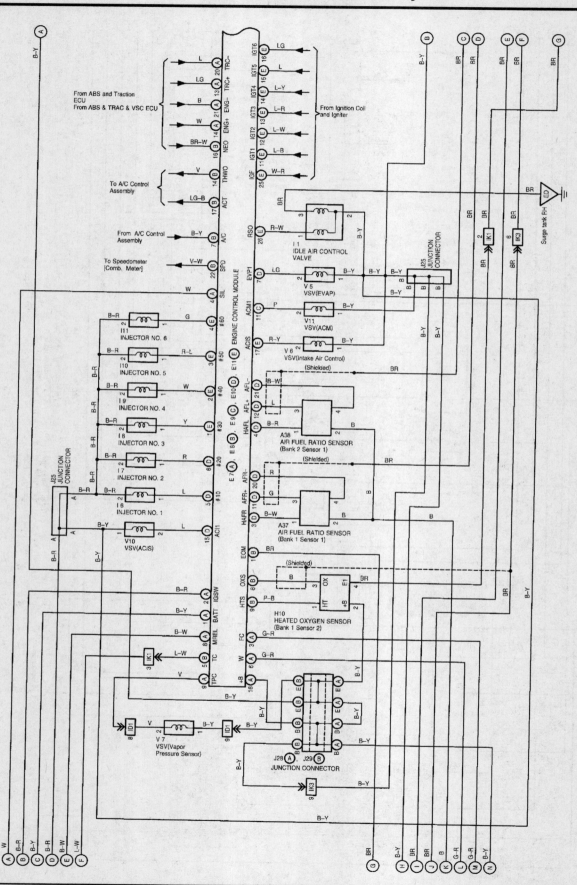

Typical Lexus ES 300 engine control and immobiliser system (2 of 3)

Chapter 12 Chassis electrical system

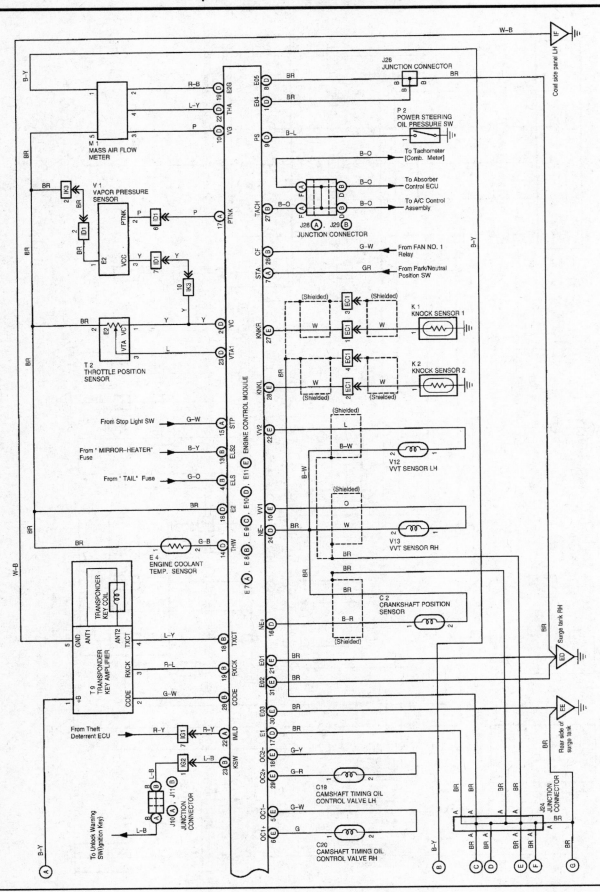

Typical Lexus ES 300 engine control and immobiliser system (3 of 3)

12-54 Chapter 12 Chassis electrical system

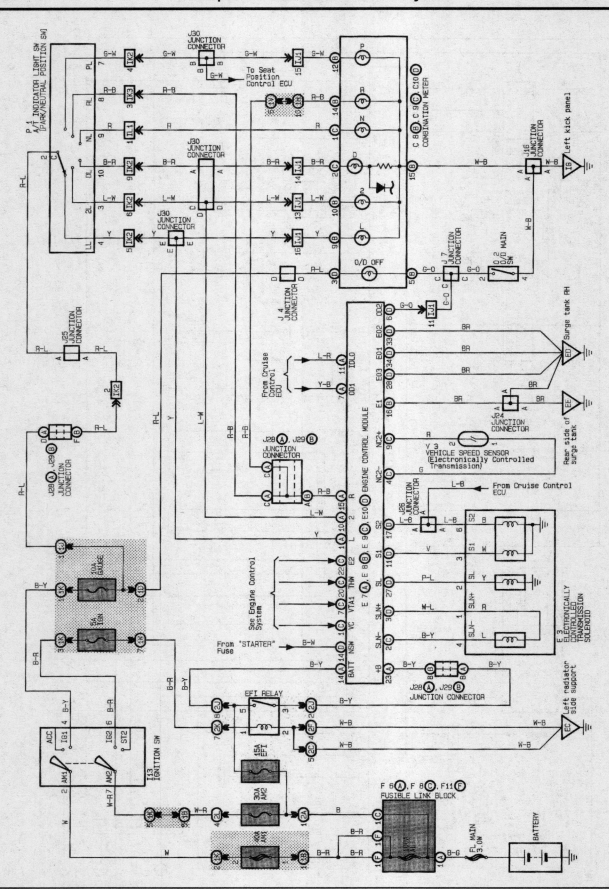

Typical Lexus ES 300 A541E electronically controlled transaxle system

Chapter 12 Chassis electrical system

12-55

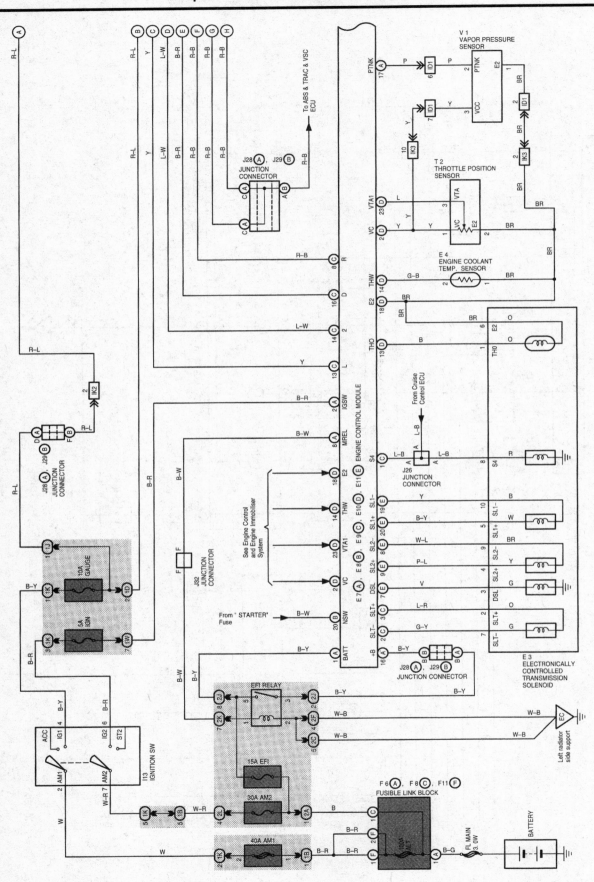

Typical Lexus ES 300 U140E electronically controlled transaxle system (1 of 2)

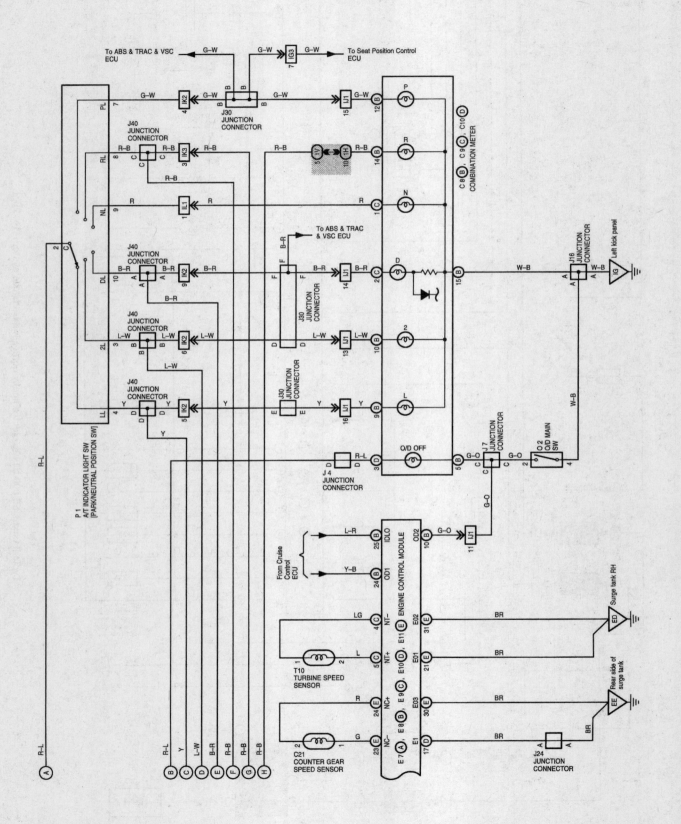

Typical Lexus ES 300 U140E electronically controlled transaxle system (2 of 2)

Chapter 12 Chassis electrical system

12-57

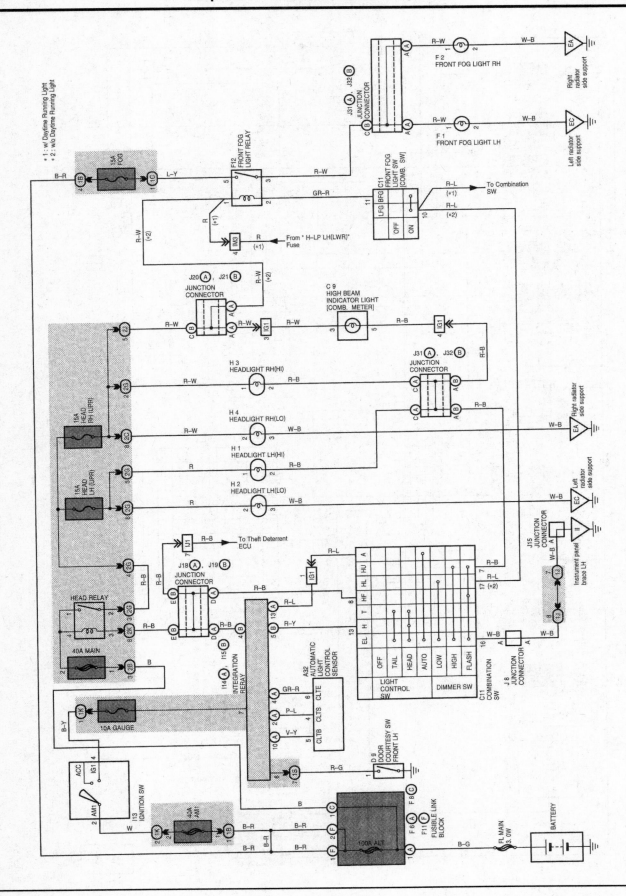

Typical Lexus ES 300 headlight system (without daytime running lights)

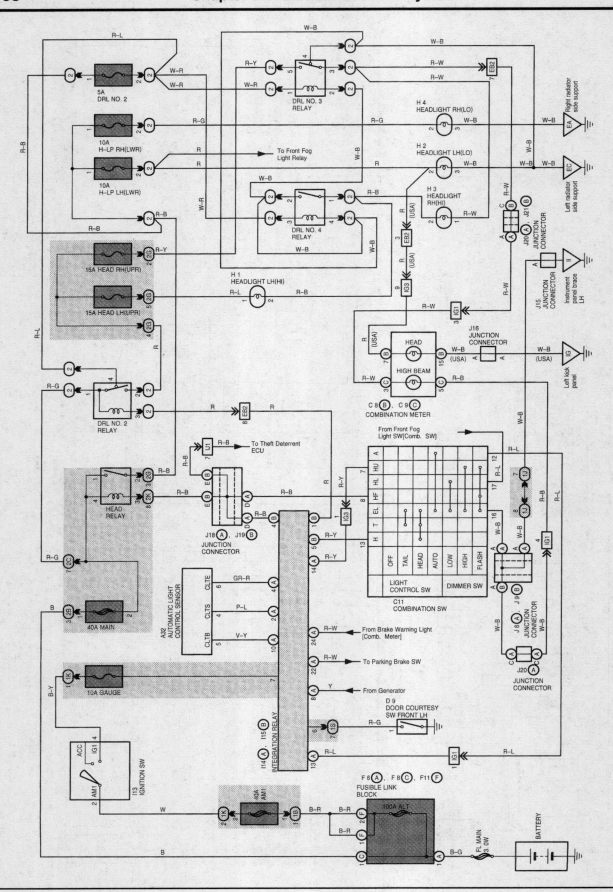

Typical Lexus ES 300 headlight system (with daytime running lights)

Chapter 12 Chassis electrical system

12-59

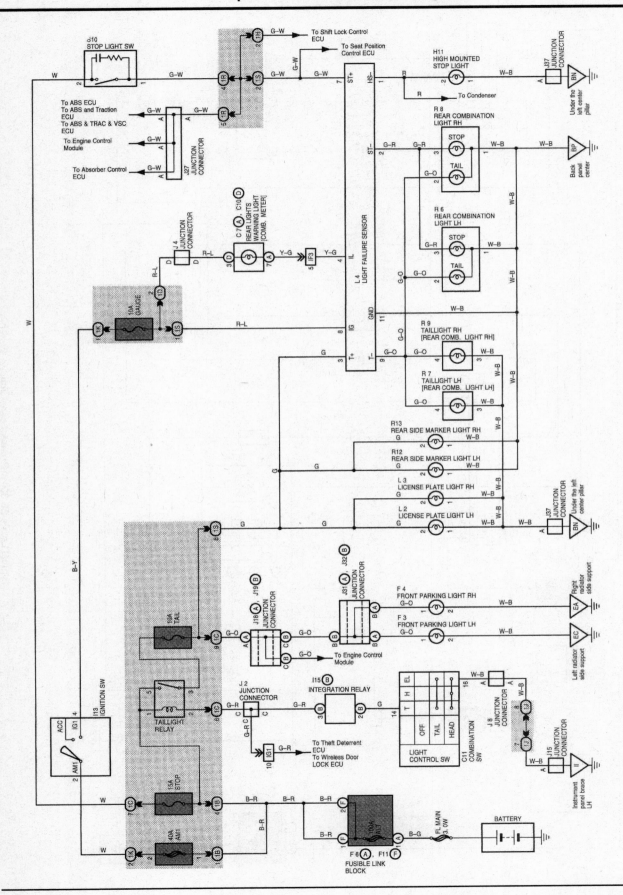

Typical Lexus ES 300 tail light and brake light systems

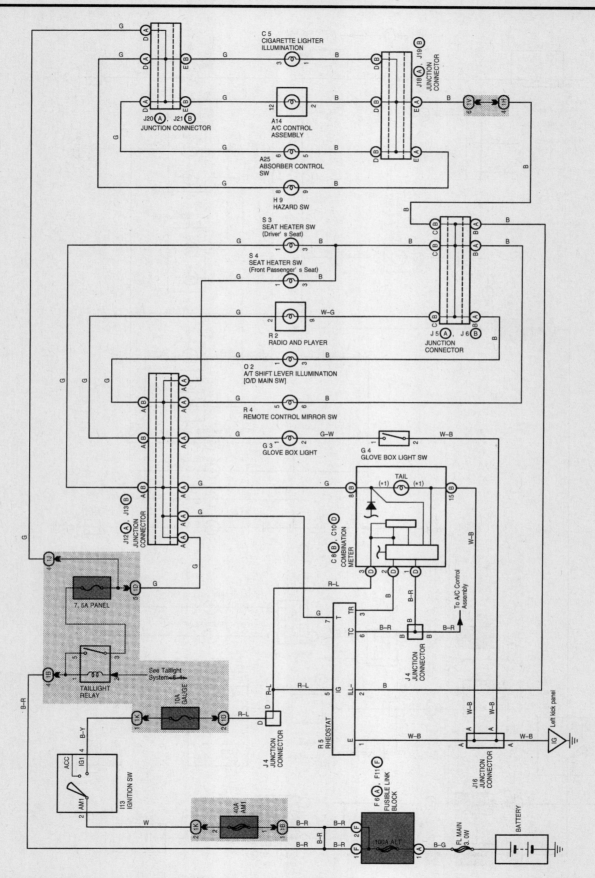

Typical Lexus ES 300 instrument panel lighting systems

Chapter 12 Chassis electrical system

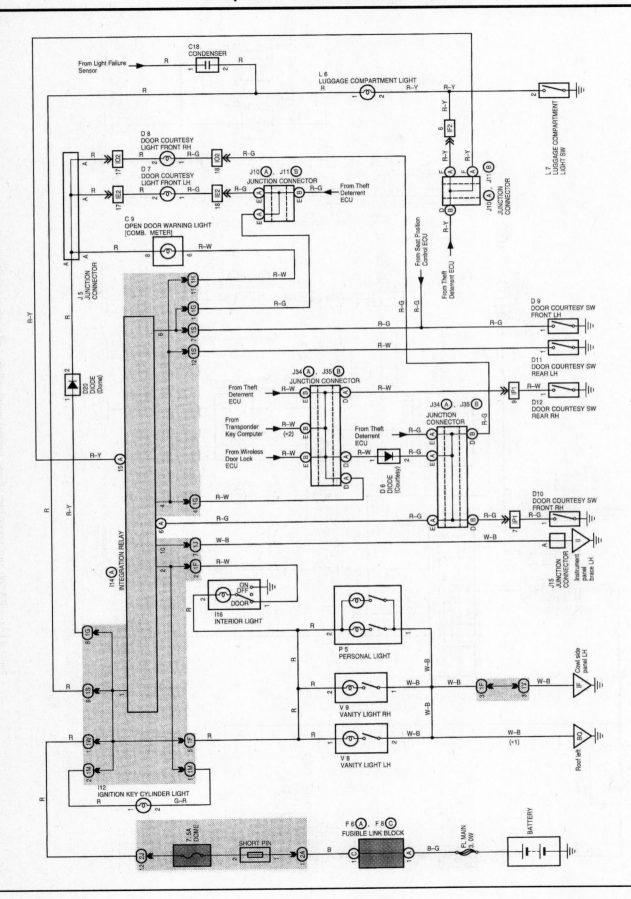

Typical Lexus ES 300 interior lighting systems

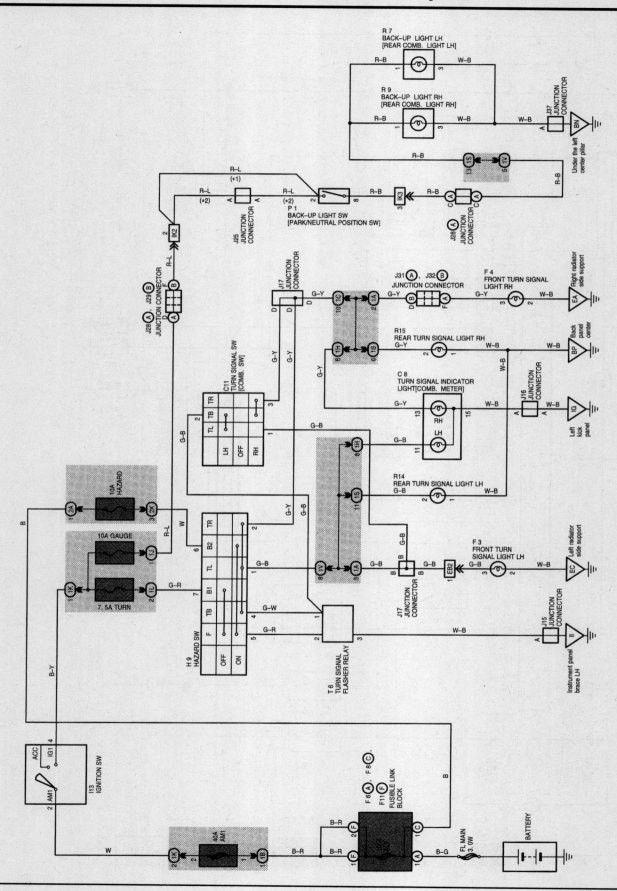

Typical Lexus ES 300 turn signal turn signal, hazard warning light and backup light systems

Chapter 12 Chassis electrical system

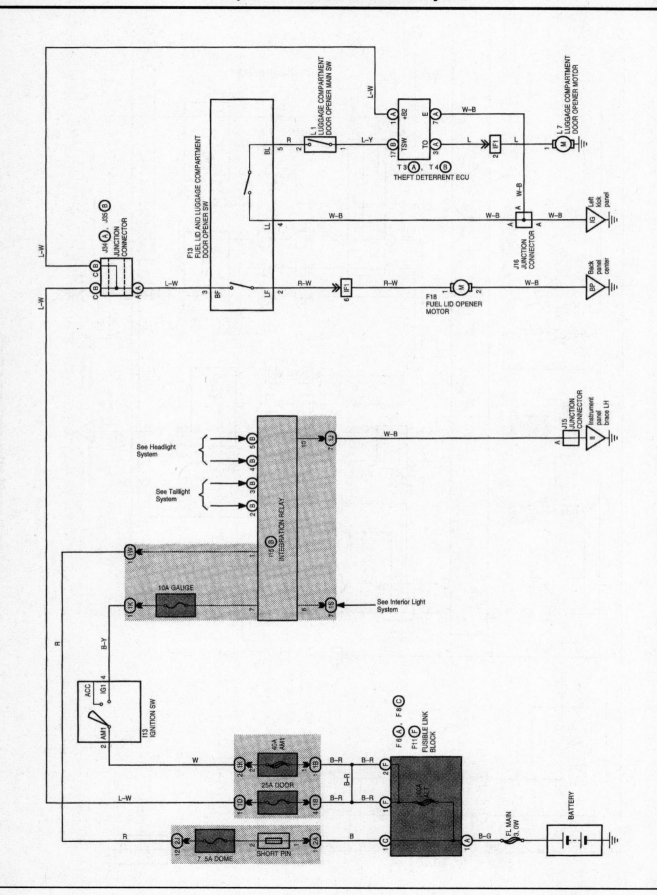

Typical Lexus ES 300 rear lighting system

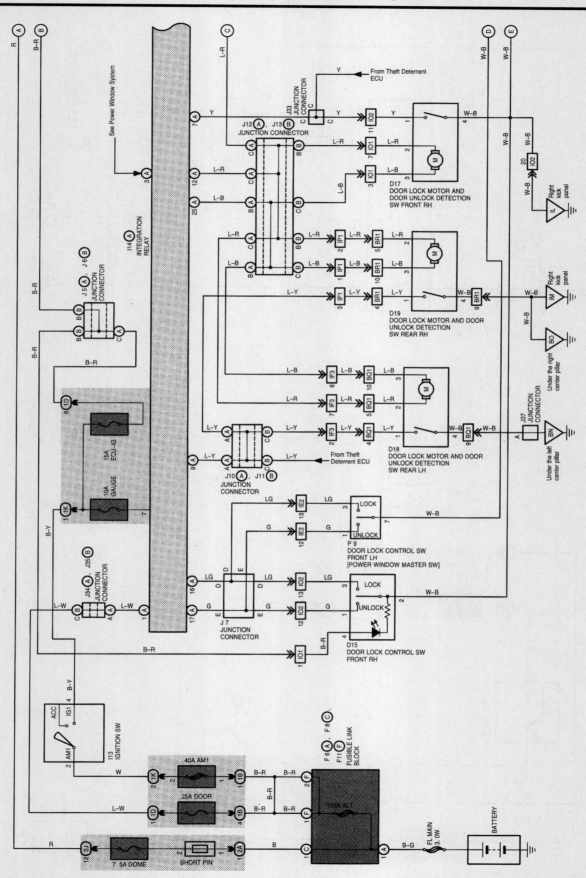

Typical Lexus ES 300 door lock system (1 of 2)

Chapter 12 Chassis electrical system

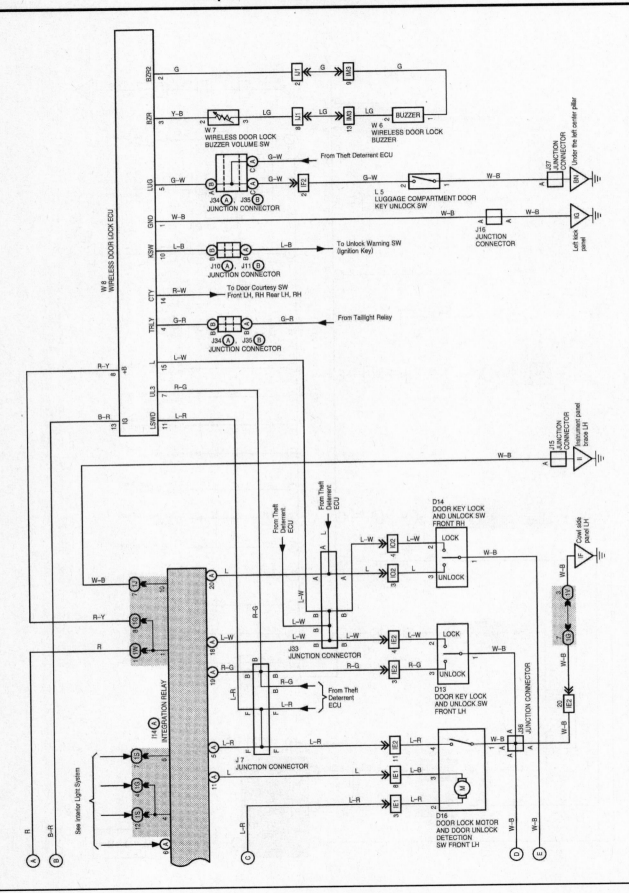

Typical Lexus ES 300 door lock system (2 of 2)

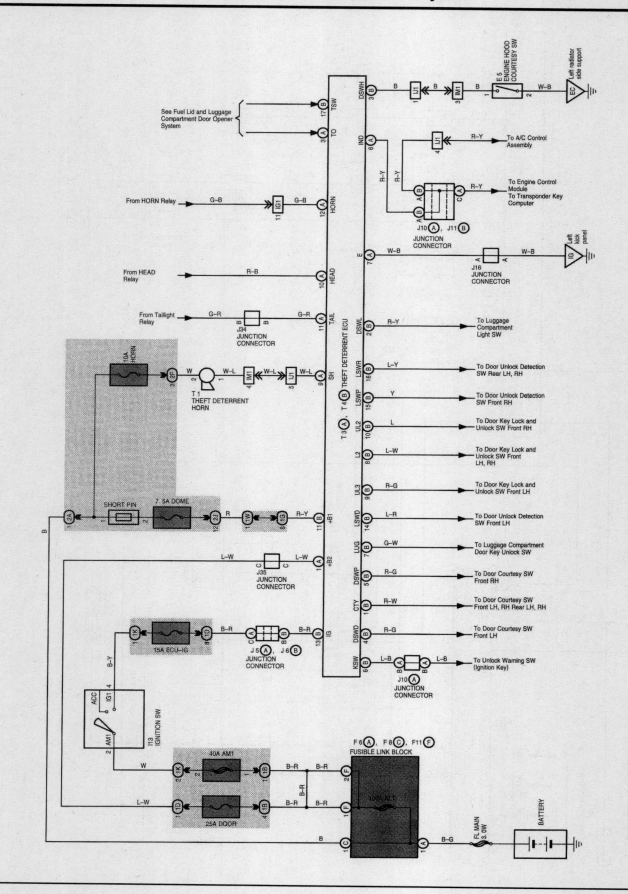

Typical Lexus ES 300 theft deterrent system

Chapter 12 Chassis electrical system

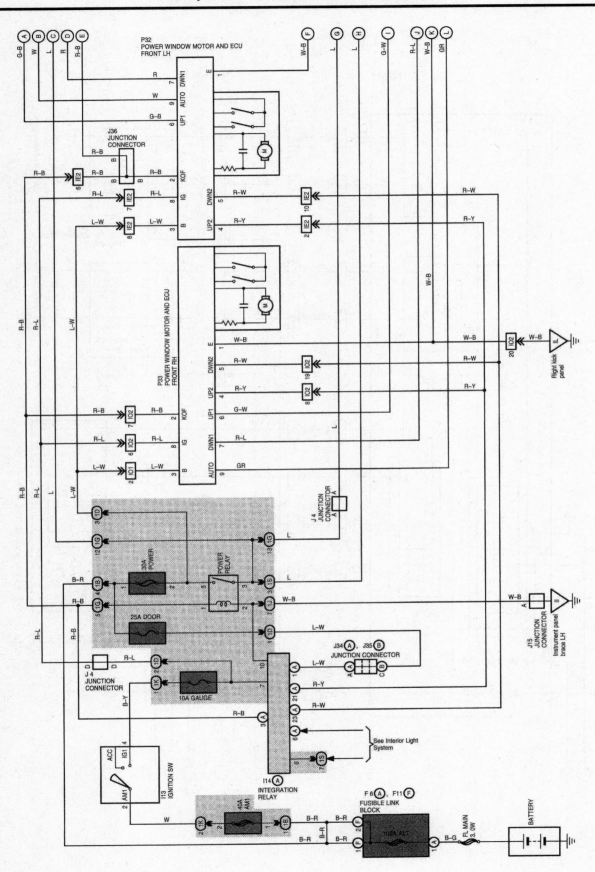

Typical Lexus ES 300 power window system (1 of 2)

12-68 Chapter 12 Chassis electrical system

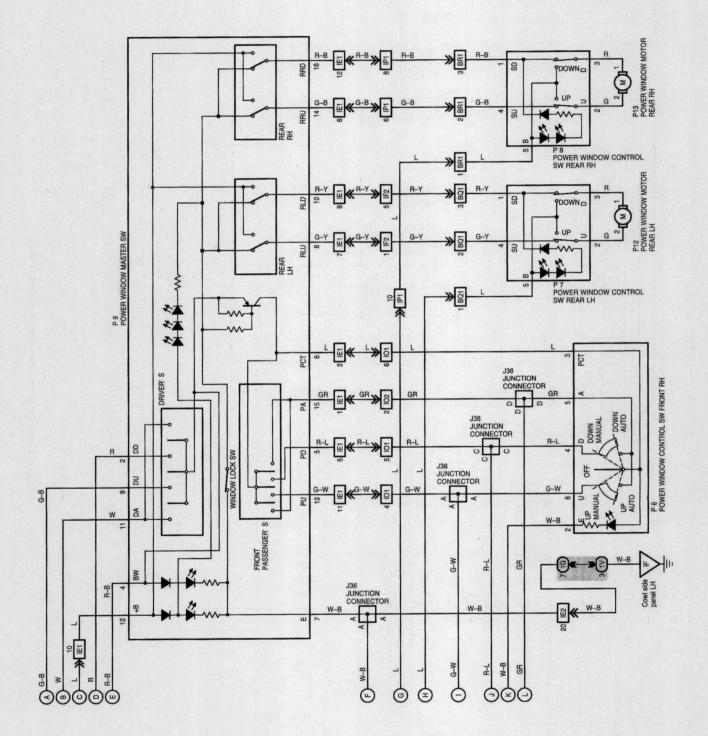

Typical Lexus ES 300 power window system (2 of 2)

Chapter 12 Chassis electrical system

12-69

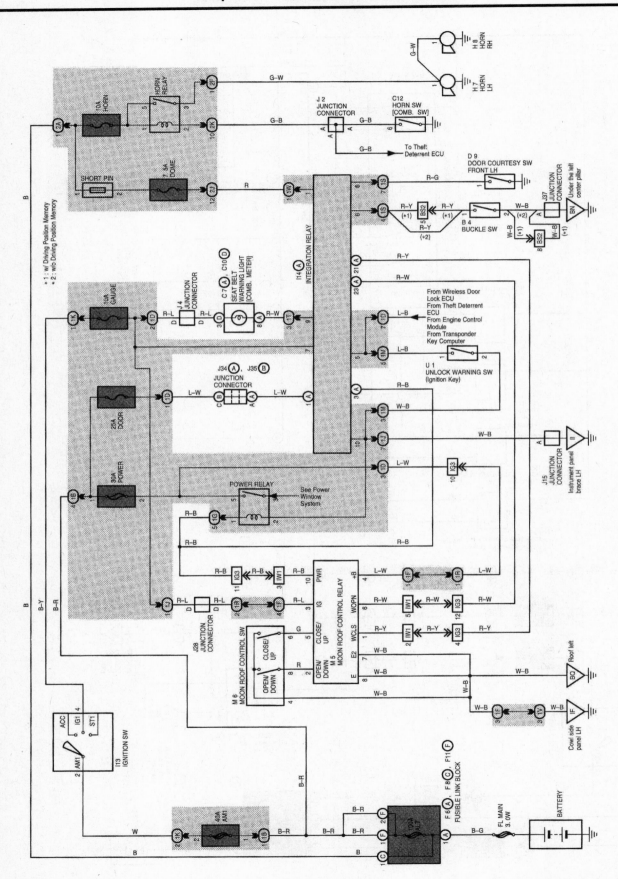

Typical Lexus ES 300 moon roof and seat belt warning system

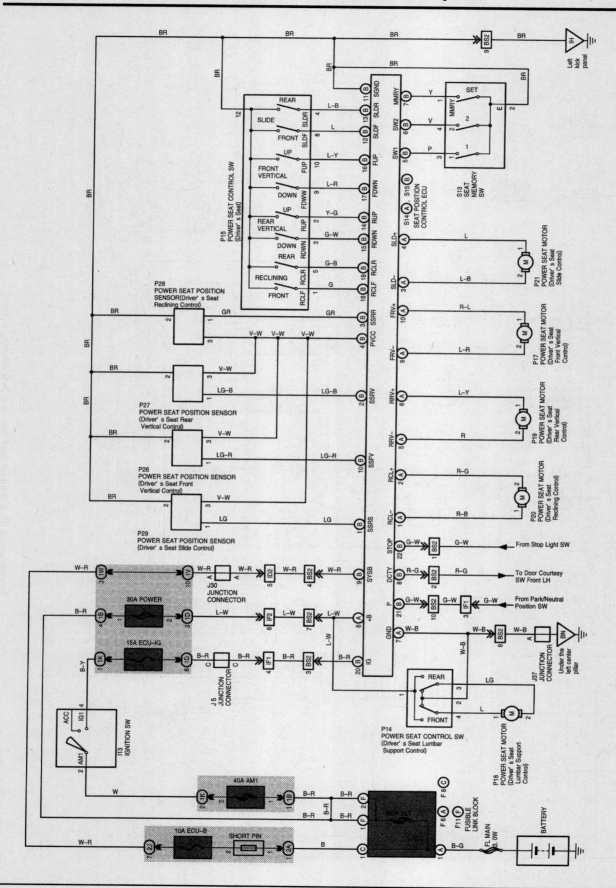

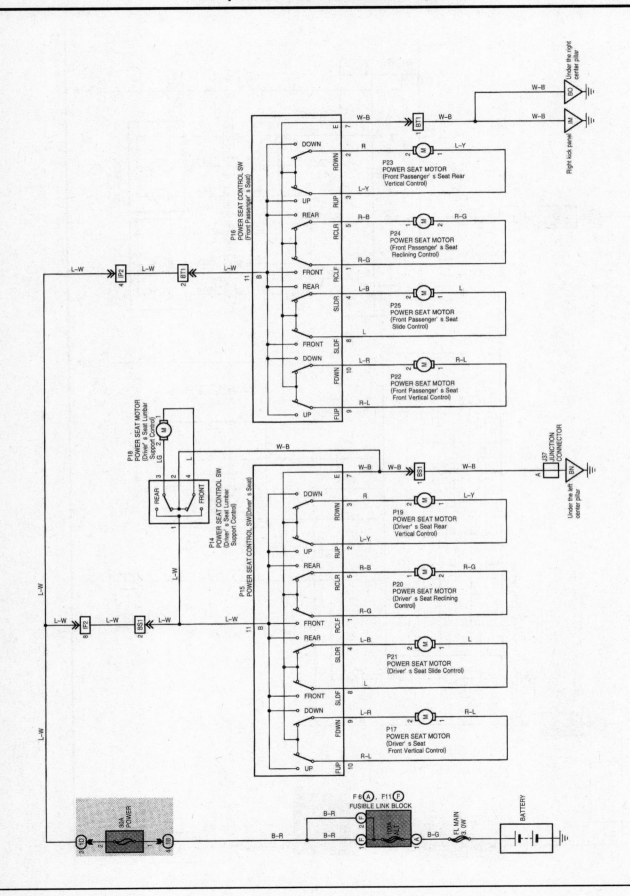

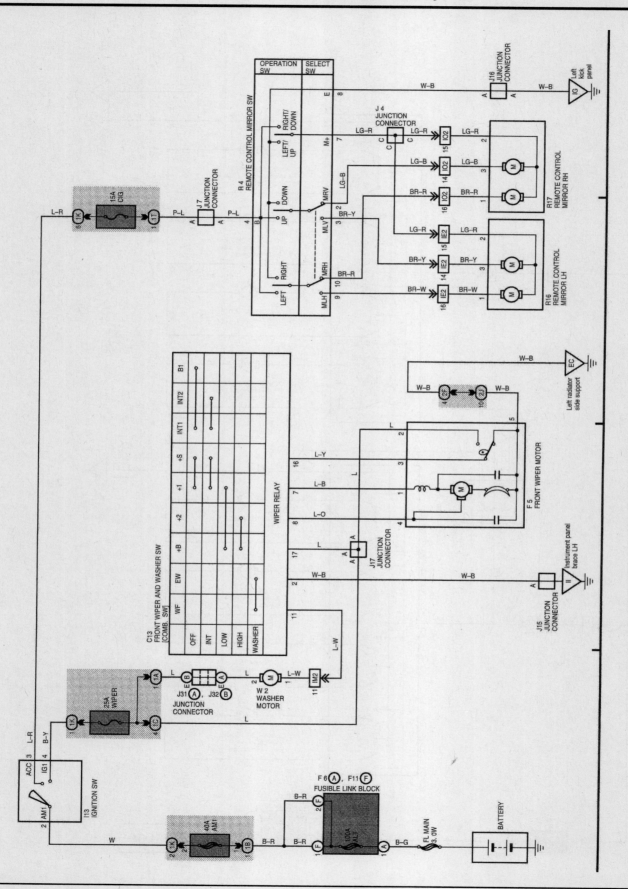

Typical Lexus ES 300 windshield wiper and washer system; power mirror system

Chapter 12 Chassis electrical system

12-73

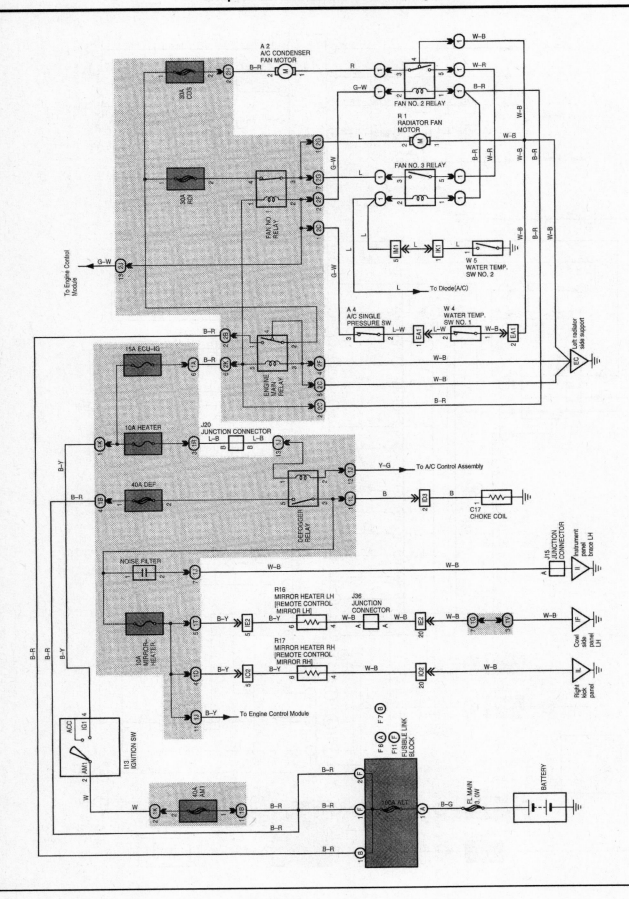

Typical Lexus ES 300 rear window defogger system

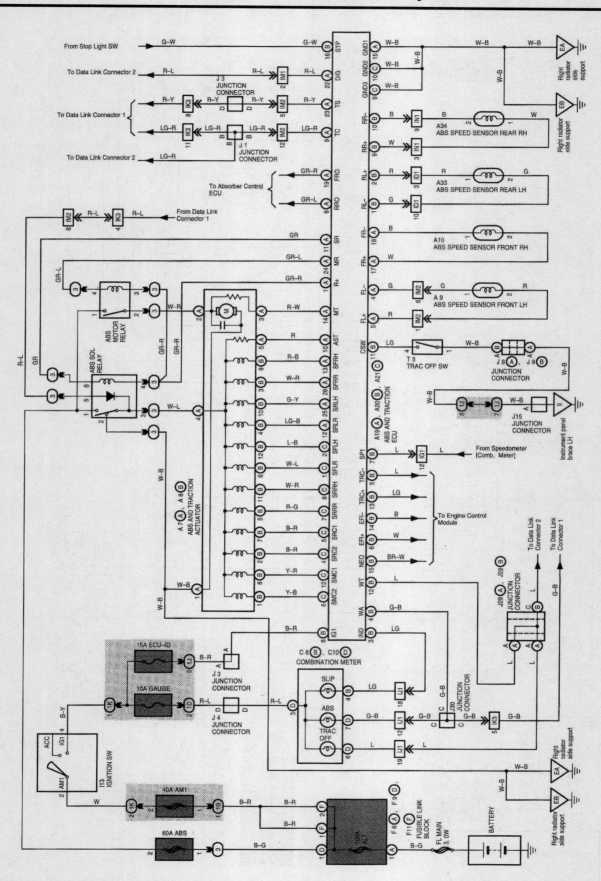

Typical Lexus ES 300 Anti-lock Brake and Traction Control system

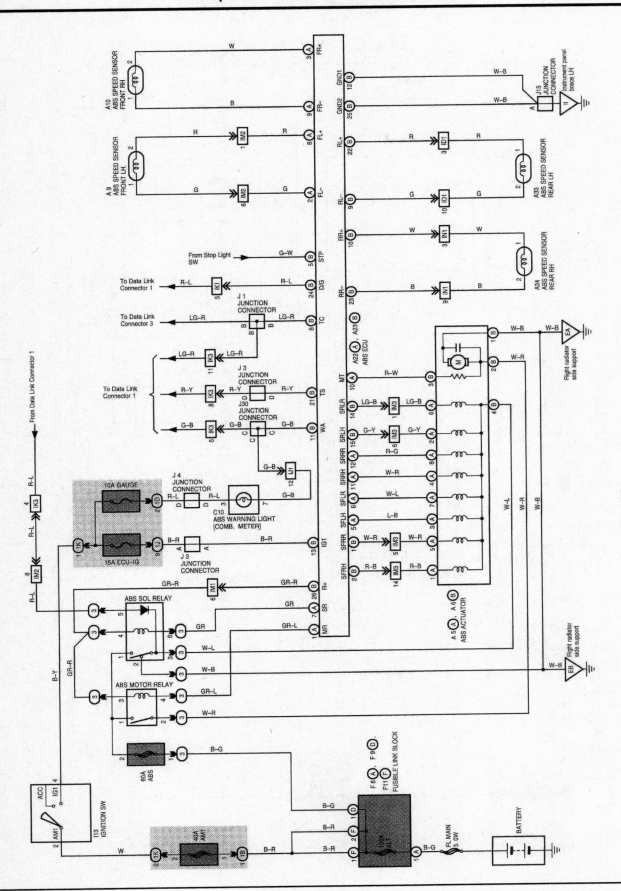

Typical Lexus ES 300 Anti-Lock Brake system

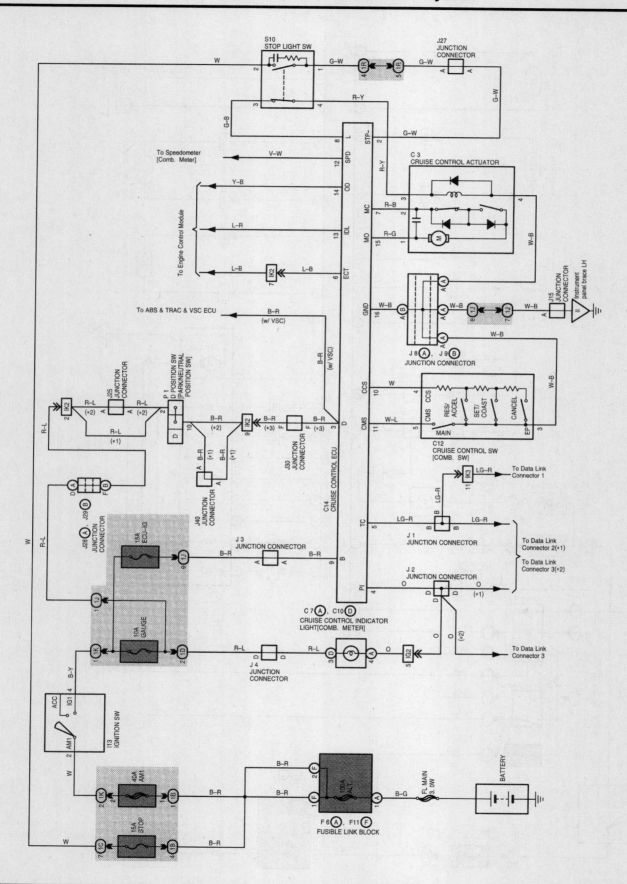

Typical Lexus ES 300 cruise control system

Chapter 12 Chassis electrical system

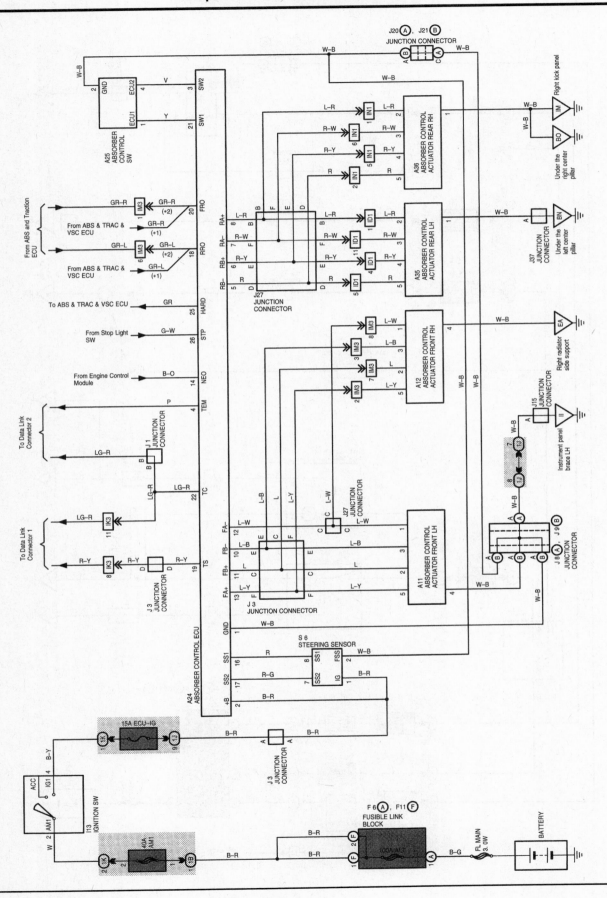

Typical Lexus ES 300 Electric Modulated Suspension system

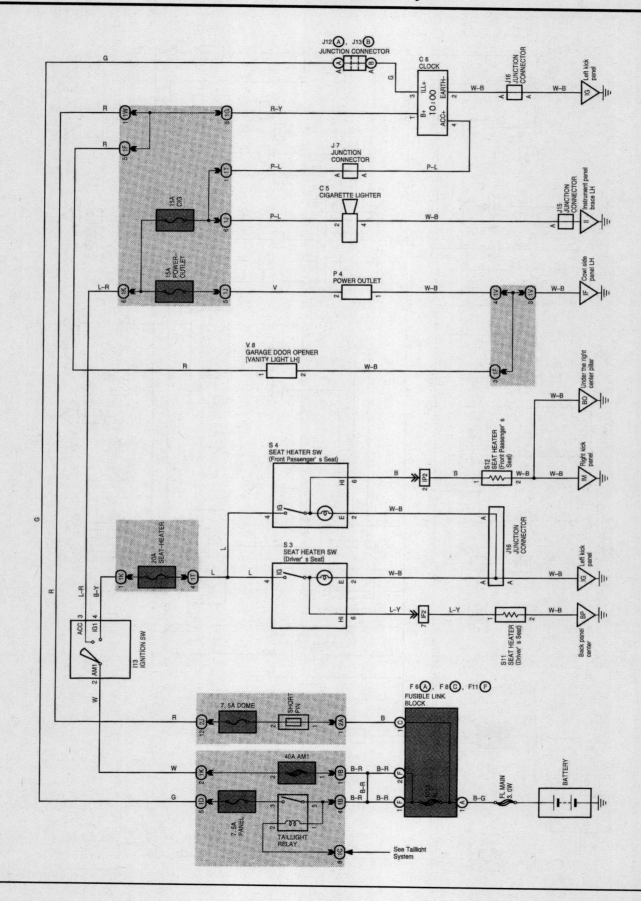

Typical Lexus ES 300 seat heater, power outlet and cigarette lighter system

Chapter 12 Chassis electrical system

12-79

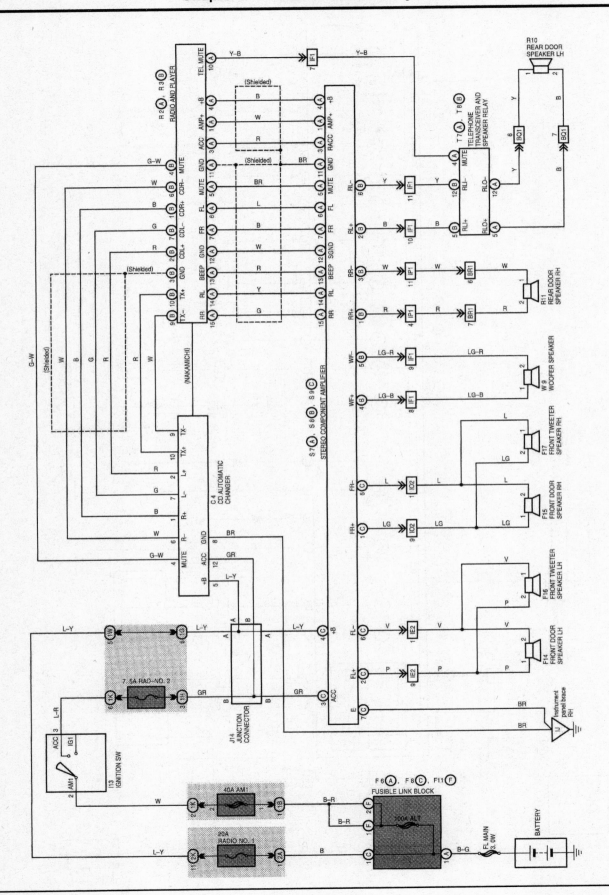

Typical Lexus ES 300 audio system

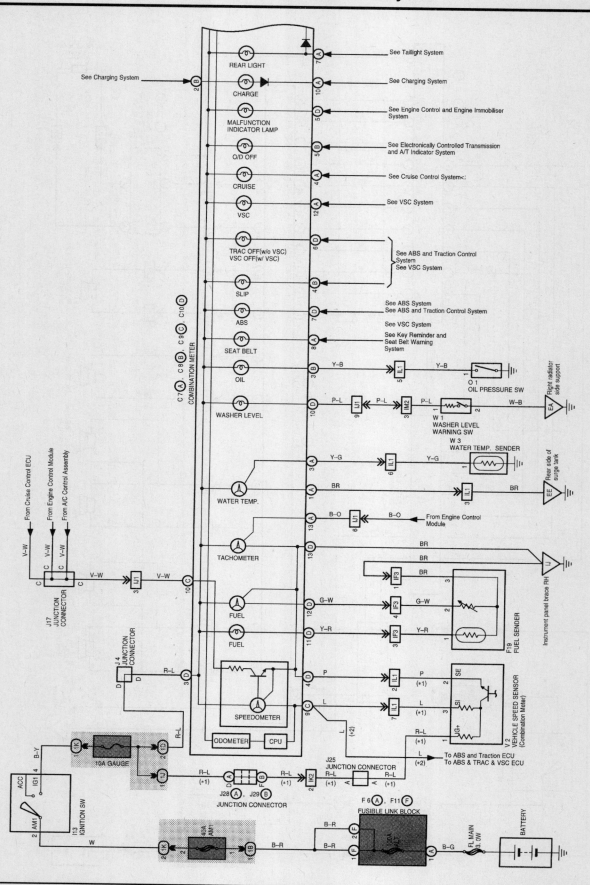

Chapter 12 Chassis electrical system

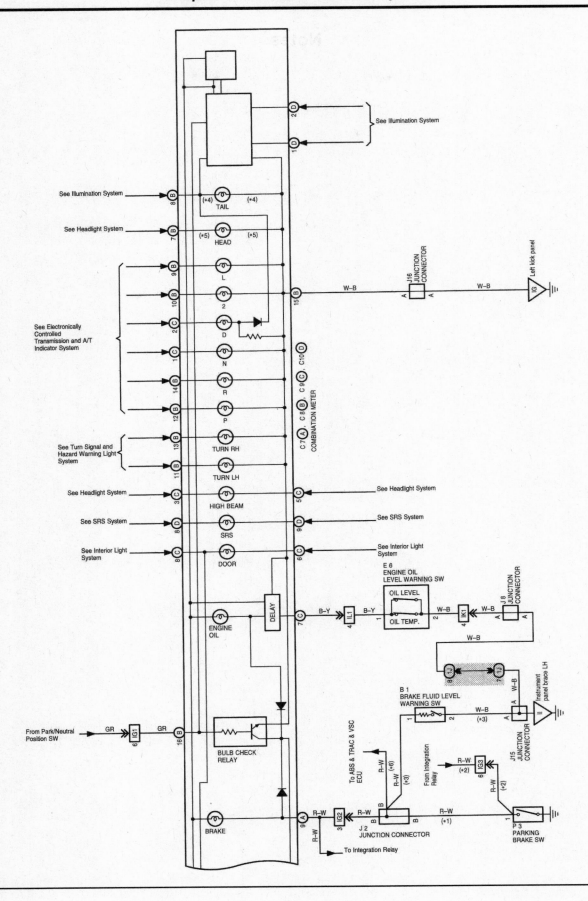

Typical Lexus ES 300 instrument cluster system (2 of 2)

Notes

Index

A

About this manual, 0-5
Accelerator cable, removal, installation and adjustment, 4-12
Acknowledgements, 0-2
Acoustic Control Induction System (ACIS) (V6 models only), 6-22
Airbag, general information, 12-22
Air cleaner assembly, removal and installation, 4-11
Air conditioning
   and heating system, check and maintenance, 3-13
   compressor clutch circuit, check, 3-16
   compressor, removal and installation, 3-17
   condenser, removal and installation, 3-17
   evaporator and expansion valve, removal and installation, 3-18
   receiver/drier, removal and installation, 3-17
Air filter replacement, 1-19
Alignment, general information, 10-19
Alternator, removal and installation, 5-10
Antenna, removal and installation, 12-12
Antifreeze, general information, 3-3
Anti-lock Brake System (ABS), general information, 9-3
Automatic transaxle, 7B-1 through 7B-14
   diagnosis, general, 7B-2
   differential lubricant change, 1-23
   differential lubricant level check, 1-19
   electronic control system, 7B-10
   fluid change, 1-23
   fluid level check, 1-10
   general information, 7B-2
   oil seal replacement, 7B-7
   Park/Neutral position switch, check, replacement and adjustment, 7B-4
   shift cable, adjustment and replacement, 7B-2
   shift lock system, description, check and component replacement, 7B-4
   Throttle Valve (TV) cable, check and adjustment, 7B-3
   removal and installation, 7B-8
Automotive chemicals and lubricants, 0-16

B

Back-up light switch, check and replacement, 7A-2
Balancer assembly (four-cylinder engine)
   backlash check, 2C-22
   installation, 2C-28
   removal, 2C-14
Balljoints, replacement, 10-9
Battery
   cables, check and replacement, 5-4
   check and replacement, 5-3
   check, maintenance and charging, 1-13
   emergency jump starting, 0-15
Blower motor
   circuit, check, 3-10
   removal and installation, 3-11
Body, 11-1 through 11-20
   bumper covers, removal and installation, 11-7
   center console, removal and installation, 11-15
   cowl cover, removal and installation, 11-19
   dashboard trim panels, 11-16
   door latch, lock cylinder and handle, removal and installation, 11-13
   door, removal, installation and adjustment, 11-12
   door trim panels, removal and installation, 11-11
   door window glass
      regulator, removal and installation, 11-14
      removal and installation, 11-14
   front fender, removal and installation, 11-9
   general information, 11-1
   hinges and locks, maintenance, 11-3
   hood latch and release cable, removal and installation, 11-6
   hood, removal, installation and adjustment, 11-6
   instrument panel, removal and installation, 11-17
   maintenance, 11-1
   mirrors, removal and installation, 11-15
   rear package shelf, removal and installation, 11-19
   repair
      major damage, 11-3
      minor damage, 11-2
   seats, removal and installation, 11-19

steering column covers, removal and installation, 11-17
trunk lid latch and lock cylinder, removal and installation, 11-10
trunk lid, removal, installation and adjustment, 11-10
trunk release and fuel door cable, removal and installation, 11-11
upholstery and carpets, maintenance, 11-2
vinyl trim, maintenance, 11-2
windshield and fixed glass, replacement, 11-3

Booster battery (jump) starting, 0-15
Brake check, 1-17
Brakes, 9-1 through 9-26
Anti-lock Brake System (ABS), general information, 9-3
brake disc, inspection, removal and installation, 9-12
brake light switch, removal, installation and adjustment, 9-26
disc brake
caliper, removal and installation, 9-9
pads, replacement, 9-5
drum brake shoes, replacement, 9-13
general information, 9-2
hoses and lines, inspection and replacement, 9-18
hydraulic system, bleeding, 9-19
master cylinder, removal and installation, 9-17
parking brake
adjustment, 9-22
cables, replacement, 9-24
shoes (rear disc brakes only), inspection and replacement, 9-20
power brake booster, check, removal and installation, 9-19
wheel cylinder, removal and installation, 9-16

Bulb replacement, 12-15
Bumper covers, removal and installation, 11-7
Buying parts, 0-7

C

Camshaft oil seal, replacement, four-cylinder engine, 2A-9
Camshaft Position sensor, check and replacement, 6-15
Camshafts and lifters, removal, inspection and installation
four-cylinder engine, 2A-10
V6 engine, 2B-12

Catalytic converter, 6-36
Center console, removal and installation, 11-15
Charging system
check, 5-9
general information and precautions, 5-9

Chassis and body fastener check, 1-25
Chassis electrical system, 12-1 through 12-82
Circuit breakers, general information, 12-3
Clutch
components, removal, inspection and installation, 8-2
description and check, 8-2
hydraulic system, bleeding, 8-6
master cylinder, removal and installation, 8-5
release bearing and lever, removal, inspection and installation, 8-4
release cylinder and accumulator, removal and installation, 8-5
start switch, check and adjustment, 8-6

Clutch and driveaxles, 8-1 through 8-12

Clutch/brake pedal height and freeplay, check and adjustment, 1-21
Control arm, removal, inspection and installation, 10-8
Conversion factors, 0-17
Coolant temperature sending unit, check and replacement, 3-9
Cooling system
check, 1-16
servicing (draining, flushing and refilling), 1-22

Cooling, heating and air conditioning systems, 3-1 through 3-20
air conditioning and heating system, check and maintenance, 3-13
antifreeze, general information, 3-3
blower motor
circuit, check, 3-10
removal and installation, 3-11
compressor clutch circuit, check, 3-16
compressor, removal and installation, 3-17
condenser, removal and installation, 3-17
coolant temperature sending unit, check and replacement, 3-9
engine cooling fans and circuit, check and replacement, 3-5
evaporator and expansion valve, removal and installation, 3-18
general information, 3-2
heater and air conditioning control assembly, removal and installation, cable adjustment and electrical checks, 3-11
heater core, removal and installation, 3-12
radiator and coolant reservoir, removal and installation, 3-6
receiver/drier, removal and installation, 3-17
thermostat, check and replacement, 3-3
water pump
check, 3-7
removal and installation, 3-7

Cowl cover, removal and installation, 11-19
Crankshaft
inspection, 2C-20
installation and main bearing oil clearance check, 2C-24
removal, 2C-15

Crankshaft front oil seal, replacement
four-cylinder engine, 2A-9
V6 engine, 2B-11

Crankshaft Position sensor, check and replacement, 6-14
Cruise control system, description and check, 12-16
Cylinder compression check, 2C-5
Cylinder head
cleaning and inspection, 2C-12
reassembly, 2C-14
disassembly, 2C-11
removal and installation
four-cylinder engine, 2A-13
V6 engine, 2B-15

Cylinder honing, 2C-18

D

Dashboard trim panels, 11-16
Daytime Running Lights (DRL), general information, 12-22
Disc brake
caliper, removal and installation, 9-9
pads, replacement, 9-5

Index

Door latch, lock cylinder and handle, removal and installation, 11-13
Door lock system, power, description and check, 12-20
Door, removal, installation and adjustment, 11-12
Door trim panels, removal and installation, 11-11
Door window glass
    regulator, removal and installation, 11-14
    removal and installation, 11-14
Driveaxle
    boot check, 1-21
    boot replacement, 8-9
    general information and inspection, 8-6
    removal and installation, 8-7
Drivebelt check, adjustment and replacement, 1-15
Drum brake shoes, replacement, 9-13

E

Electrical troubleshooting, general information, 12-1
Electric side view mirrors, description and check, 12-16
Electronic fuel injection (EFI) system
    check, 4-13
    general information, 4-12
Emissions and engine control systems, 6-1 through 6-36
    Acoustic Control Induction System (ACIS) (V6 models only), 6-22
    catalytic converter, 6-36
    Crankshaft Position sensor, check and replacement, 6-15
    Engine Coolant Temperature (ECT) sensor, check and replacement, 6-14
    Evaporative emissions control (EVAP) system, description, check and component replacement, 6-29
    Exhaust Gas Recirculation (EGR) system, 6-25
    general information, 6-2
    Idle Air Control (IAC) valve, check and replacement, 6-21
    Intake Air Temperature (IAT) sensor, check and replacement, 6-13
    knock sensor, check and replacement, 6-19
    Manifold Absolute Pressure (MAP) sensor (four-cylinder models), check and replacement, 6-11
    Mass Airflow (MAF) sensor (V6 models), check and replacement, 6-11
    On-Board Diagnosis (OBD) system and trouble codes, 6-4
    oxygen sensor and air/fuel sensor, check and replacement, 6-17
    Positive Crankcase Ventilation (PCV) system, 6-24
    Powertrain Control Module (PCM), removal and installation, 6-9
    Throttle Position Sensor (TPS), check and replacement, 6-10
    Vehicle Speed Sensor (VSS), check and replacement, 6-20
Engine Coolant Temperature (ECT) sensor, check and replacement, 6-14
Engine cooling fans and circuit, check and replacement, 3-5
Engine electrical systems, 5-1 through 5-14
    alternator, removal and installation, 5-10
    battery
        cables, check and replacement, 5-4
        check and replacement, 5-3
        check, maintenance and charging, 1-13
        emergency jump starting, 0-15
    charging system
        check, 5-9
        general information and precautions, 5-9
    igniter (V6 models), check and replacement, 5-8
    ignition coil(s), check and replacement, 5-7
    ignition system
        check, 5-5
        general information and precautions, 5-4
    ignition timing, check, 5-8
    starter motor
        and circuit check, 5-12
        removal and installation, 5-13
    starter solenoid, removal and installation, 5-13
    starting system, general information and precautions, 5-12
    voltage regulator and alternator brushes, replacement, 5-10
Engine oil and oil filter change, 1-11
Engine overhaul, reassembly sequence, 2C-22
Engines
    Four-cylinder engine, 2A-1 through 2A-18
        camshaft oil seal, replacement, 2A-9
        camshafts and valve lifters, removal, inspection and installation, 2A-10
        crankshaft front oil seal, replacement, 2A-9
        cylinder head, removal and installation, 2A-13
        exhaust manifold, removal and installation, 2A-5
        flywheel/driveplate, removal and installation, 2A-17
        general information, 2A-2
        intake manifold, removal and installation, 2A-4
        mounts, check and replacement, 2A-18
        oil pan, removal and installation, 2A-14
        oil pump, removal, inspection and installation, 2A-15
        rear main oil seal, replacement, 2A-17
        repair operations possible with the engine in the vehicle, 2A-2
        timing belt and sprockets, removal, inspection and installation, 2A-6
        Top Dead Center (TDC) for number one piston, locating, 2A-3
        valve cover, removal and installation, 2A-3
    General engine overhaul procedures, 2C-1 through 2C-28
        balancer assembly (four-cylinder engine)
            backlash check, 2C-22
            installation, 2C-28
            removal, 2C-14
        block, cleaning and inspection, 2C-16
        crankshaft
            inspection, 2C-20
            installation and main bearing oil clearance check, 2C-24
            removal, 2C-15
        cylinder compression check, 2C-5
        cylinder head
            cleaning and inspection, 2C-12
            disassembly, 2C-11
            reassembly, 2C-14
        cylinder honing, 2C-18
        engine overhaul
            disassembly sequence, 2C-9
            reassembly sequence, 2C-22
        engine rebuilding alternatives, 2C-9
        engine removal, methods and precautions, 2C-7
        engine, removal and installation, 2C-7

initial start-up and break-in after overhaul, 2C-28
main and connecting rod bearings, inspection and selection, 2C-20
oil pressure check, 2C-5
piston rings, installation, 2C-23
pistons/connecting rods
   inspection, 2C-19
   installation and rod bearing oil clearance check, 2C-26
   removal, 2C-14
rear main oil seal installation, 2C-25
vacuum gauge diagnostic checks, 2C-6
valves, servicing, 2C-14
V6 engine, 2B-1 through 2B-20
   camshafts and lifters, removal, inspection and installation, 2B-12
   cylinder heads, removal and installation, 2B-15
   exhaust manifolds, removal and installation, 2B-8
   flywheel/driveplate, removal and installation, 2B-20
   general information, 2B-3
   intake manifold, removal and installation, 2B-7
   mounts, check and replacement, 2B-20
   oil pan, removal and installation, 2B-16
   oil pump, removal, inspection and installation, 2B-17
   oil seals, replacement, 2B-11
   rear main oil seal, replacement, 2B-20
   repair operations possible with the engine in the vehicle, 2B-3
   timing belt and sprockets, removal, inspection and installation, 2B-8
   Top Dead Center (TDC) for number one piston, locating, 2B-3
   valve covers, removal and installation, 2B-6
   Variable Valve Timing (VVT) system, description, check and component replacement, 2B-4

Evaporative emissions control (EVAP) system, description, check and component replacement, 6-29
Evaporative emissions control system check, 1-23
Exhaust Gas Recirculation (EGR) system, description, check and component replacement, 6-25
Exhaust manifold, removal and installation
four-cylinder engine, 2A-5
V6 engine, 2B-8
Exhaust system
check, 1-21
servicing, general information, 4-19

F

Fender, removal and installation, 11-9
Fluid level checks, 1-7
brake and clutch fluid, 1-8
engine coolant, 1-8
engine oil, 1-7
windshield washer fluid, 1-8
Flywheel/driveplate, removal and installation
four-cylinder engine, 2A-17
V6 engine, 2B-20
Four-cylinder engine, 2A-1 through 2A-18
camshaft oil seal, replacement, 2A-9
camshafts and valve lifters, removal, inspection and installation, 2A-10
crankshaft front oil seal, replacement, 2A-9
cylinder head, removal and installation, 2A-13
exhaust manifold, removal and installation, 2A-5
flywheel/driveplate, removal and installation, 2A-17
general information, 2A-2
intake manifold, removal and installation, 2A-4
mounts, check and replacement, 2A-18
oil pan, removal and installation, 2A-14
oil pump, removal, inspection and installation, 2A-15
rear main oil seal, replacement, 2A-17
repair operations possible with the engine in the vehicle, 2A-2
timing belt and sprockets, removal, inspection and installation, 2A-6
Top Dead Center (TDC) for number one piston, locating, 2A-3
valve cover, removal and installation, 2A-3
Fraction/Decimal/Millimeter Equivalents, 0-18
Fuel
filter replacement, 1-22
system check, 1-19
tank cap gasket replacement, 1-30
Fuel and exhaust systems, 4-1 through 4-20
accelerator cable, removal, installation and adjustment, 4-12
air cleaner assembly, removal and installation, 4-11
electronic fuel injection (EFI) system
   check, 4-13
   general information, 4-12
exhaust system servicing, general information, 4-19
fuel level sending unit, check and replacement, 4-9
general information, 4-2
lines and fittings, repair and replacement, 4-5
pressure regulator, removal and installation, 4-9
pressure relief, 4-3
pump, removal and installation, 4-7
pump/fuel pressure, check, 4-3
rail and injectors, removal and installation, 4-18
tank
   cleaning and repair, general information, 4-11
   removal and installation, 4-11
throttle body, check, removal and installation, 4-13
Fuses, general information, 12-2
Fusible links, general information, 12-3

G

Gear boots, steering, replacement, 10-15
General engine overhaul procedures, 2C-1 through 2C-28
balancer assembly (four-cylinder engine)
   backlash check, 2C-22
   installation, 2C-28
   removal, 2C-14
block, cleaning and inspection, 2C-16
crankshaft
   inspection, 2C-20
   installation and main bearing oil clearance check, 2C-24
   removal, 2C-15
cylinder compression check, 2C-5

Index

cylinder head
  cleaning and inspection, 2C-12
  disassembly, 2C-11
  reassembly, 2C-14
cylinder honing, 2C-18
engine overhaul
  disassembly sequence, 2C-9
  reassembly sequence, 2C-22
engine rebuilding alternatives, 2C-9
engine removal, methods and precautions, 2C-7
engine, removal and installation, 2C-7
initial start-up and break-in after overhaul, 2C-28
main and connecting rod bearings, inspection and selection, 2C-20
oil pressure check, 2C-5
piston rings, installation, 2C-23
pistons/connecting rods
  inspection, 2C-19
  installation and rod bearing oil clearance check, 2C-26
  removal, 2C-14
rear main oil seal installation, 2C-25
vacuum gauge diagnostic checks, 2C-6
valves, servicing, 2C-14

H

Headlight bulbs, removal and installation, 12-13
Headlight housing, removal and installation, 12-14
Headlights
  adjustment, 12-13
  housing, removal and installation, 12-14
Heater and air conditioning control assembly, removal and installation, cable adjustment and electrical checks, 3-11
Heater core, removal and installation, 3-12
Hinges and locks, maintenance, 11-3
Hood latch and release cable, removal and installation, 11-6
Hood, removal, installation and adjustment, 11-6
Horn, check and replacement, 12-14
Hub and bearing assembly, removal and installation
  front, 10-9
  rear, 10-11

I

Idle Air Control (IAC) valve, check and replacement, 6-21
Igniter (V6 models), check and replacement, 5-8
Ignition coil(s), check and replacement, 5-7
Ignition switch and lock cylinder, check and replacement, 12-5
Ignition system
  check, 5-5
  general information and precautions, 5-4
Ignition timing, check, 5-8
Initial start-up and break-in after overhaul, 2C-28
Instrument cluster, removal and installation, 12-9
Instrument panel
  fuel and temperature gauges, check, 12-8
  removal and installation, 11-17
  switches, 12-6
Intake Air Temperature (IAT) sensor, check and replacement, 6-13

Intake manifold, removal and installation
  four-cylinder engine, 2A-4
  V6 engine, 2B-5
Introduction to the Toyota Camry, Avalon, Camry Solara and Lexus ES 300, 0-5

J

Jacking and towing, 0-15

K

Knock sensor, check and replacement, 6-19

M

Main and connecting rod bearings, inspection and selection, 2C-20
Maintenance
  schedule, 1-6
  techniques, tools and working facilities, 0-7
Manifold Absolute Pressure (MAP) sensor (four-cylinder models), check and replacement, 6-11
Manual transaxle, 7A-1 through 7A-6
  back-up light switch, check and replacement, 7A-2
  general information, 7A-1
  lubricant change, 1-25
  lubricant level check, 1-20
  overhaul, general information, 7A-5
  removal and installation, 7A-2
  shift and select cables, replacement, 7A-2
  shift lever assembly, removal and installation, 7A-2
Mass Airflow (MAF) sensor (V6 models), check and replacement, 6-11
Master cylinder, removal and installation, 9-17
Mirrors, removal and installation, 11-15

O

Oil pan, removal and installation
  four-cylinder engine, 2A-14
  V6 engine, 2B-16
Oil pressure check, 2C-5
Oil pump, removal, inspection and installation
  four-cylinder engine, 2A-15
  V6 engine, 2B-17
Oil seals, replacement, 2B-11
On-Board Diagnosis (OBD) system and trouble codes, 6-4
Oxygen sensor and air/fuel sensor, check and replacement, 6-17

P

Park/Neutral position switch, check, replacement and adjustment, 7B-4

Index

Parking brake
    adjustment, 9-22
    cables, replacement, 9-24
    shoes (rear disc brakes only), inspection and replacement, 9-20
Piston rings, installation, 2C-23
Pistons/connecting rods
    inspection, 2C-19
    installation and rod bearing oil clearance check, 2C-26
    removal, 2C-14
Positive Crankcase Ventilation (PCV) system, 6-24
Positive Crankcase Ventilation (PCV) valve check and replacement, 1-30
Power brake booster, check, removal and installation, 9-19
Power door lock system, description and check, 12-20
Power steering
    fluid level check, 1-10
    pump, removal and installation, 10-16
    system, bleeding, 10-18
Powertrain Control Module (PCM), removal and installation, 6-9

R

Radiator and coolant reservoir, removal and installation, 3-6
Radio and speakers, removal and installation, 12-11
Rear axle carrier, removal and installation, 10-12
Rear main oil seal
    installation (during overhaul), 2C-25
    replacement (in-vehicle)
        four-cylinder engine, 2A-17
        V6 engine, 2B-20
Rear package shelf, removal and installation, 11-19
Relays, general information and testing, 12-3
Repair operations possible with the engine in the vehicle
    four-cylinder engine, 2A-2
    V6 engine, 2B-3

S

Safety first!, 0-19
Seats, removal and installation, 11-19
Shift and select cables, replacement, 7A-2
Shift cable, adjustment and replacement, 7B-2
Shift lever assembly, removal and installation, 7A-2
Shift lock system, description, check and component replacement, 7B-4
Spark plug
    check and replacement, 1-25
    wire, check and replacement, 1-28
Stabilizer bar and bushings, removal and installation
    front, 10-4
    rear, 10-10
Starter motor
    and circuit check, 5-12
    removal and installation, 5-13
Starter solenoid, removal and installation, 5-13
Starting system, general information and precautions, 5-12
Steering and suspension check, 1-20

Steering column
    covers, removal and installation, 11-17
    switches, check and replacement, 12-4
Steering gear, removal and installation, 10-15
Steering knuckle and hub, removal and installation, 10-9
Steering system
    alignment, general information, 10-19
    general information, 10-13
    power steering
        pump, removal and installation, 10-16
        system, bleeding, 10-18
    steering gear boots, replacement, 10-15
    steering gear, removal and installation, 10-15
    steering wheel, removal and installation, 10-13
    tie-rod ends, removal and installation, 10-14
    wheels and tires, general information, 10-18
Steering wheel, removal and installation, 10-13
Stereo anti-theft system precaution, 0-6
Strut assembly, removal, inspection and installation
    front, 10-4
    rear, 10-10
Strut rod, removal and installation, 10-11
Strut/coil spring assembly, replacement, 10-5
Suspension and steering systems, 10-1 through 10-20
Suspension arms, removal and installation, 10-11
Suspension system
    balljoints, replacement, 10-9
    control arm, removal, inspection and installation, 10-8
    hub and bearing assembly, removal and installation
        front, 10-9
        rear, 10-11
    rear axle carrier, removal and installation, 10-12
    stabilizer bar and bushings, removal and installation
        front, 10-4
        rear, 10-10
    steering knuckle and hub, removal and installation, 10-9
    strut assembly, removal, inspection and installation
        front, 10-4
        rear, 10-10
    strut/coil spring assembly, replacement, 10-5
    strut rod, removal and installation, 10-11
    suspension arms, removal and installation, 10-11

T

Thermostat, check and replacement, 3-3
Throttle body, check, removal and installation, 4-13
Throttle Position Sensor (TPS), check and replacement, 6-10
Throttle Valve (TV) cable, check and adjustment, 7B-3
Tie-rod ends, removal and installation, 10-14
Timing belt and sprockets, removal, inspection and installation
    four-cylinder engine, 2A-6
    V6 engine, 2B-8
Tire and tire pressure checks, 1-9
Tire rotation, 1-17
Tools, 0-10
Top Dead Center (TDC) for number one piston, locating
    four-cylinder engine, 2A-3
    V6 engine, 2B-3

Index

Transaxle, automatic, 7B-1 through 7B-14
    diagnosis, general, 7B-2
    differential lubricant change, 1-23
    differential lubricant level check, 1-19
    electronic control system, 7B-10
    fluid change, 1-23
    fluid level check, 1-10
    general information, 7B-2
    oil seal replacement, 7B-7
    Park/Neutral position switch, check, replacement and adjustment, 7B-4
    shift cable, adjustment and replacement, 7B-2
    shift lock system, description, check and component replacement, 7B-4
    Throttle Valve (TV) cable, check and adjustment, 7B-3
    removal and installation, 7B-8

Transaxle, manual, 7A-1 through 7A-6
    back-up light switch, check and replacement, 7A-2
    general information, 7A-1
    lubricant change, 1-25
    lubricant level check, 1-20
    overhaul, general information, 7A-5
    removal and installation, 7A-2
    shift and select cables, replacement, 7A-2
    shift lever assembly, removal and installation, 7A-2

Troubleshooting, 0-20
Trunk lid latch and lock cylinder, removal and installation, 11-10
Trunk lid, removal, installation and adjustment, 11-10
Trunk release and fuel door cable, removal and installation, 11-11
Tune-up and routine maintenance, 1-1 through 1-30
Tune-up general information, 1-7
Turn signal and hazard flasher, check and replacement, 12-4

U

Underhood hose check and replacement, 1-16
Upholstery and carpets, maintenance, 11-2

V

V6 engine, 2B-1 through 2B-20
Vacuum gauge diagnostic checks, 2C-6
Valve clearance check and adjustment, 1-28
Valve cover, removal and installation
    four-cylinder engine, 2A-3
    V6 engine, 2B-6
Valves, servicing, 2C-14
Variable Valve Timing (VVT) system, description, check and component replacement, 2B-4
Vehicle identification numbers, 0-6
Vehicle Speed Sensor (VSS), check and replacement, 6-20
Vinyl trim, maintenance, 11-2
Voltage regulator and alternator brushes, replacement, 5-10

W

Water pump
    check, 3-7
    removal and installation, 3-7
Wheel cylinder, removal and installation, 9-16
Wheels and tires, general information, 10-18
Window defogger, rear, check and repair, 12-12
Window system, power, description and check, 12-17
Windshield and fixed glass, replacement, 11-3
Windshield wiper blade inspection and replacement, 1-12
Wiper motor, check and replacement, 12-10
Wiring diagrams, general information, 12-28
Working facilities, 0-14

Haynes Automotive Manuals

NOTE: If you do not see a listing for your vehicle, consult your local Haynes dealer for the latest product information.

ACURA
- 12020 Integra '86 thru '89 & Legend '86 thru '90
- 12021 Integra '90 thru '93 & Legend '91 thru '95
- Integra '94 thru '00 - see HONDA Civic (42025)
- MDX '01 thru '07 - see HONDA Pilot (42037)
- 12050 Acura TL all models '99 thru '08

AMC
- Jeep CJ - see JEEP (50020)
- 14020 Concord/Hornet/Gremlin/Spirit '70 thru '83
- 14025 (Renault) Alliance & Encore '83 thru '87

AUDI
- 15020 4000 all models '80 thru '87
- 15025 5000 all models '77 thru '83
- 15026 5000 all models '84 thru '88
- Audi A4 '96 thru '01 - see VW Passat (96023)
- 15030 Audi A4 '02 thru '08

AUSTIN
- Healey Sprite - see MG Midget (66015)

BMW
- 18020 3/5 Series '82 thru '92
- 18021 3 Series including Z3 models '92 thru '98
- 18022 3-Series incl. Z4 models '99 thru '05
- 18023 3-Series '06 thru '10
- 18025 320i all 4 cyl models '75 thru '83
- 18050 1500 thru 2002 except Turbo '59 thru '77

BUICK
- 19010 Buick Century '97 thru '05
- Century (front-wheel drive) - see GM (38005)
- 19020 Buick, Oldsmobile & Pontiac Full-size (Front wheel drive) '85 thru '05
- 19025 Buick, Oldsmobile & Pontiac Full-size (Rear wheel drive) '70 thru '90
- 19030 Mid-size Regal & Century '74 thru '87
- Regal - see GENERAL MOTORS (38010)
- Skyhawk - see GM (38030)
- Skylark - see GM (38020, 38025)
- Somerset - see GENERAL MOTORS (38025)

CADILLAC
- 21015 CTS & CTS-V '03 thru '12
- 21030 Cadillac Rear Wheel Drive '70 thru '93
- Cimarron, Eldorado & Seville - see GM (38015, 38030, 38031)

CHEVROLET
- 10305 Chevrolet Engine Overhaul Manual
- 24010 Astro & GMC Safari Mini-vans '85 thru '05
- 24015 Camaro V8 all models '70 thru '81
- 24016 Camaro all models '82 thru '92
- Cavalier - see GM (38015)
- Celebrity - see GM (38005)
- 24017 Camaro & Firebird '93 thru '02
- 24020 Chevelle, Malibu, El Camino '69 thru '87
- 24024 Chevette & Pontiac T1000 '76 thru '87
- Citation - see GENERAL MOTORS (38020)
- 24027 Colorado & GMC Canyon '04 thru '10
- 24032 Corsica/Beretta all models '87 thru '96
- 24040 Corvette all V8 models '68 thru '82
- 24041 Corvette all models '84 thru '96
- 24045 Full-size Sedans Caprice, Impala, Biscayne, Bel Air & Wagons '69 thru '90
- 24046 Impala SS & Caprice and Buick Roadmaster '91 thru '96
- Impala '00 thru '05 - see LUMINA (24048)
- 24047 Impala & Monte Carlo all models '06 thru '11
- Lumina '90 thru '94 - see GM (38010)
- 24048 Lumina & Monte Carlo '95 thru '05
- Lumina APV - see GM (38035)
- 24050 Luv Pick-up all 2WD & 4WD '72 thru '82
- Malibu '00 '05 - see GM (38026)
- 24055 Monte Carlo all models '70 thru '88
- Monte Carlo '95 thru '01 - see LUMINA
- 24059 Nova all models '69 thru '79
- 24060 Nova/Geo Prizm '85 thru '92
- 24064 Pick-ups '67 thru '87 - Chevrolet & GMC
- 24065 Pick-ups '88 thru '98 - Chevrolet & GMC
- 24066 Pick-ups '99 thru '06 - Chevrolet & GMC
- 24067 Chevy Silverado & GMC Sierra '07 thru '12
- 24070 S-10 & GMC S-15 Pick-ups '82 thru '93
- 24071 S-10, Sonoma & Jimmy '94 thru '04
- 24072 Chevy TrailBlazer, GMC Envoy & Oldsmobile Bravada '02 thru '09
- 24075 Sprint '85 thru '88, Geo Metro '89 thru '01
- 24080 Vans - Chevrolet & GMC '68 thru '96
- 24081 Full-size Vans '96 thru '10

CHRYSLER
- 10310 Chrysler Engine Overhaul Manual
- 25015 Chrysler Cirrus, Dodge Stratus, Plymouth Breeze, '95 thru '00
- 25020 Full-size Front-Wheel Drive '88 thru '93
- K-Cars - see DODGE Aries (30008)
- Laser - see DODGE Daytona (30030)
- 25025 Chrysler LHS, Concorde & New Yorker, Dodge Intrepid, Eagle Vision, '93 thru '97
- 25026 Chrysler LHS, Concorde, 300M, Dodge Intrepid '98 thru '04
- 25027 Chrysler 300, Dodge Charger & Magnum '05 thru '10
- 25030 Chrysler/Plym. Mid-size '82 thru '95
- Rear-wheel Drive - see DODGE (30050)
- 25035 PT Cruiser all models '01 thru '10
- 25040 Chrysler Sebring '95 thru '06, Dodge Stratus '01 thru '06, Dodge Avenger

DATSUN
- 28005 200SX all models '80 thru '83
- 28007 B-210 all models '73 thru '78
- 28009 210 all models '78 thru '82
- 28012 240Z, 260Z & 280Z Coupe '70 thru '78
- 28014 280ZX Coupe & 2+2 '79 thru '83
- 300ZX - see NISSAN (72010)
- 28018 510 & PL521 Pick-up '68 thru '73
- 28020 510 all models '78 thru '81
- 28022 620 Series Pick-up all models '73 thru '79
- 720 Series Pick-up - see NISSAN (72030)
- 28025 810/Maxima all gas models '77 thru '84

DODGE
- 400 & 600 - see CHRYSLER (25030)
- 30008 Aries & Plymouth Reliant '81 thru '89
- 30010 Caravan & Ply. Voyager '84 thru '95
- 30011 Caravan & Ply. Voyager '96 thru '02
- 30012 Challenger/Plymouth Saporro '78 thru '83
- Challenger '67-'76 - see DART (30025)
- 30013 Caravan, Chrysler Voyager, Town & Country '03 thru '07
- 30016 Colt/Plymouth Champ '78 thru '87
- 30020 Dakota Pick-ups all models '87 thru '96
- 30021 Durango '98 & '99, Dakota '97 thru '99
- 30022 Durango '00 thru '03, Dakota '00 thru '04
- 30023 Durango '04 thru '09, Dakota '05 thru '11
- 30025 Dart, Challenger/Plymouth Barracuda & Valiant 6 cyl models '67 thru '76
- 30030 Daytona & Chrysler Laser '84 thru '89
- Intrepid - see Chrysler (25025, 25026)
- 30034 Dodge & Plymouth Neon '95 thru '99
- 30035 Omni & Plymouth Horizon '78 thru '90
- 30036 Dodge & Plymouth Neon '00 thru '05
- 30040 Pick-ups all full-size models '74 thru '93
- 30041 Pick-ups all full-size models '94 thru '01
- 30042 Pick-ups full-size models '02 thru '08
- 30045 Ram 50/D50 Pick-ups & Raider and Plymouth Arrow Pick-ups '79 thru '93
- 30050 Dodge/Ply./Chrysler RWD '71 thru '89
- 30055 Shadow/Plymouth Sundance '87 thru '94
- 30060 Spirit & Plymouth Acclaim '89 thru '95
- 30065 Vans - Dodge & Plymouth '71 thru '03

EAGLE
- Talon - see MITSUBISHI (68030, 68031)
- Vision - see CHRYSLER (25025)

FIAT
- 34010 124 Sport Coupe & Spider '68 thru '78
- 34025 X1/9 all models '74 thru '80

FORD
- 10320 Ford Engine Overhaul Manual
- 10355 Ford Automatic Transmission Overhaul
- 11500 Mustang '64-1/2 thru '70 Restoration Guide
- 36004 Aerostar Mini-vans '86 thru '97
- Aspire - see FORD Festiva (36030)
- 36006 Contour/Mercury Mystique '95 thru '00
- 36008 Courier Pick-up all models '72 thru '82
- 36012 Crown Victoria & Mercury Grand Marquis '88 thru '10
- 36016 Escort/Mercury Lynx '81 thru '90
- 36020 Escort/Mercury Tracer '91 thru '02
- Expedition - see FORD Pick-up (36059)
- 36022 Escape & Mazda Tribute '01 thru '11
- 36024 Explorer & Mazda Navajo '91 thru '01
- 36025 Explorer/Mercury Mountaineer '02 thru '09
- 36028 Fairmont & Mercury Zephyr '78 thru '83
- 36030 Festiva & Aspire '88 thru '97
- 36032 Fiesta all models '77 thru '80
- 36034 Focus all models '00 thru '11
- 36036 Ford & Mercury Full-size '75 thru '87
- 36044 Ford & Mercury Mid-size '75 thru '86
- 36045 Ford Fusion & Mercury Milan '06 thru '10
- 36048 Mustang V8 all models '64-1/2 thru '73
- 36049 Mustang II 4 cyl, V6 & V8 '74 thru '78
- 36050 Mustang & Mercury Capri '79 thru '93
- 36051 Mustang all models '94 thru '04
- 36052 Mustang '05 thru '10
- 36054 Pick-ups and Bronco '73 thru '79
- 36058 Pick-ups and Bronco '80 thru '96
- 36059 Pick-ups & Expedition '97 thru '09
- 36060 Super Duty Pick-up, Excursion '99 thru '10
- 36061 F-150 full-size '04 thru '10
- 36062 Pinto & Mercury Bobcat '75 thru '80
- 36066 Probe all models '89 thru '92
- Probe '93 thru '97 - see MAZDA 626 (61042)
- 36070 Ranger/Bronco II gas models '83 thru '92
- 36071 Ford Ranger '93 thru '10 &
- 36074 Mazda Pick-ups '94 thru '09
- 36075 Taurus & Mercury Sable '86 thru '95
- 36078 Taurus & Mercury Sable '96 thru '01
- 36082 Tempo & Mercury Topaz '84 thru '94
- 36086 Thunderbird/Mercury Cougar '83 thru '88
- 36090 Thunderbird/Mercury Cougar '89 thru '97
- 36094 Vans all V8 Econoline models '69 thru '91
- 36097 Vans full size '92 thru '10
- Windstar Mini-van '95 thru '07

GENERAL MOTORS
- 10360 GM Automatic Transmission Overhaul
- 38005 Buick Century, Chevrolet Celebrity, Olds Cutlass Ciera & Pontiac 6000 '82 thru '96
- 38010 Buick Regal, Chevrolet Lumina, Oldsmobile Cutlass Supreme & Pontiac Grand Prix front wheel drive '88 thru '07
- 38015 Buick Skyhawk, Cadillac Cimarron, Chevrolet Cavalier, Oldsmobile Firenza Pontiac J-2000 & Sunbird '82 thru '94
- 38016 Chevrolet Cavalier/Pontiac Sunfire '95 thru '05
- 38017 Chevrolet Cobalt & Pontiac G5 '05 thru '11
- 38020 Buick Skylark, Chevrolet Citation, Olds Omega, Pontiac Phoenix '80 thru '85
- 38025 Buick Skylark & Somerset, Olds Achieva, Calais & Pontiac Grand Am '85 thru '98
- 38026 Chevrolet Malibu, Olds Alero & Cutlass, Pontiac Grand Am '97 thru '03
- 38027 Chevrolet Malibu '04 thru '10
- 38030 Cadillac Eldorado & Oldsmobile Toronado '71 thru '85, Seville '80 thru '85, Buick Riviera '79 thru '85
- 38031 Cadillac Eldorado & Seville '86 thru '91, DeVille & Buick Riviera '86 thru '93, Fleetwood & Olds Toronado '86 thru '92
- 38032 DeVille '94 thru '05, Seville '92 thru '04, Cadillac DTS '06 thru '10
- 38035 Chevrolet Lumina APV, Olds Silhouette & Pontiac Trans Sport '90 thru '96
- 38036 Chevrolet Venture, Olds Silhouette, Pontiac Trans Sport & Montana '97 thru '05
- GM Full-size RWD - see BUICK (19025)
- 38040 Chevrolet Equinox '05 thru '09
- Pontiac Torrent '06 thru '09
- 38070 Chevrolet HHR '06 thru '11

GEO
- Metro - see CHEVROLET Sprint (24075)
- Prizm - see CHEVROLET (24060) or TOYOTA (92036)
- 40030 Storm all models '90 thru '93
- Tracker - see SUZUKI Samurai (90010)

GMC
- Vans & Pick-ups - see CHEVROLET

HONDA
- 42010 Accord CVCC all models '76 thru '83
- 42011 Accord all models '84 thru '89
- 42012 Accord all models '90 thru '93
- 42013 Accord all models '94 thru '97
- 42014 Accord all models '98 thru '02
- 42015 Accord '03 thru '07
- 42020 Civic 1200 all models '73 thru '79
- 42021 Civic 1300 & 1500 CVCC '80 thru '83
- 42022 Civic 1500 CVCC all models '75 thru '79
- 42023 Civic all models '84 thru '91
- 42024 Civic & del Sol '92 thru '95
- 42025 Civic '96 thru '00, CR-V '97 thru '01, Acura Integra '94 thru '00
- 42026 Civic '01 thru '10, CR-V '02 thru '09
- 42035 Odyssey all models '99 thru '10
- Passport - see ISUZU Rodeo (47017)
- 42037 Honda Pilot '03 thru '07, Acura MDX '01 thru '07
- 42040 Prelude CVCC all models '79 thru '89

HYUNDAI
- 43010 Elantra all models '96 thru '10
- 43015 Excel & Accent all models '86 thru '09
- 43050 Santa Fe all models '01 thru '06
- 43055 Sonata all models '99 thru '08

INFINITI
- G35 '03 thru '08 - see NISSAN 350Z (72011)

ISUZU
- Hombre - see CHEVROLET S-10 (24071)
- 47017 Rodeo, Amigo & Honda Passport '89 thru '02
- 47020 Trooper '84 thru '91, Pick-up '81 thru '93

JAGUAR
- 49010 XJ6 all 6 cyl models '68 thru '86
- 49011 XJ6 all models '88 thru '94
- 49015 XJ12 & XJS all 12 cyl models '72 thru '85

JEEP
- 50010 Cherokee, Comanche & Wagoneer Limited all models '84 thru '01
- 50020 CJ all models '49 thru '86
- 50025 Grand Cherokee all models '93 thru '04
- 50026 Grand Cherokee '05 thru '09
- 50029 Grand Wagoneer & Pick-up '72 thru '91
- 50030 Wrangler all models '87 thru '11
- 50035 Liberty '02 thru '07

KIA
- 54050 Optima '01 thru '10
- 54070 Sephia '94 thru '01, Spectra '00 thru '09, Sportage '05 thru '10

LEXUS
- ES 300/330 - see TOYOTA Camry (92007) (92008)
- RX 330 - see TOYOTA Highlander (92095)

LINCOLN
- Navigator - see FORD Pick-up (36059)
- 59010 Rear Wheel Drive all models '70 thru '10

MAZDA
- 61010 GLC (rear wheel drive) '77 thru '83
- 61011 GLC (front wheel drive) '81 thru '85
- 61012 Mazda3 '04 thru '11
- 61015 323 & Protegé '90 thru '03
- 61016 MX-5 Miata '90 thru '09
- 61020 MPV all models '89 thru '98
- Navajo - see FORD Explorer (36024)
- 61030 Pick-ups '72 thru '93
- Pick-ups '94 on - see Ford (36071)
- 61035 RX-7 all models '79 thru '85
- 61036 RX-7 all models '86 thru '91
- 61040 626 (rear wheel drive) '79 thru '82
- 61041 626 & MX-6 (front wheel drive) '83 thru '92
- 61042 626 '93 thru '01, & MX-6/Ford Probe '93 thru '02
- 61043 Mazda6 '03 thru '11

MERCEDES-BENZ
- 63012 123 Series Diesel '76 thru '85
- 63015 190 Series 4-cyl gas models, '84 thru '88
- 63020 230, 250 & 280 6 cyl sohc '68 thru '72
- 63025 280 123 Series gasoline models '77 thru '81
- 63030 350 & 450 all models '71 thru '80
- 63040 C-Class: C230/C240/C280/C320/C350 '01 thru '07

MERCURY
- 64200 Villager & Nissan Quest '93 thru '01
- All other titles, see FORD listing.

MG
- 66010 MGB Roadster & GT Coupe '62 thru '80
- 66015 MG Midget & Austin Healey Sprite Roadster '58 thru '80

MINI
- 67020 Mini '02 thru '11

MITSUBISHI
- 68020 Cordia, Tredia, Galant, Precis & Mirage '83 thru '93
- 68030 Eclipse, Eagle Talon & Plymouth Laser '90 thru '94
- 68031 Eclipse '95 thru '05, Eagle Talon '95 thru '98
- 68035 Galant '94 thru '10
- 68040 Pick-up '83 thru '96, Montero '83 thru '93

NISSAN
- 72010 300ZX all models incl. Turbo '84 thru '89
- 72011 350Z & Infiniti G35 all models '03 thru '08
- 72015 Altima all models '93 thru '06
- 72016 Altima '07 thru '10
- 72020 Maxima all models '85 thru '92
- 72021 Maxima all models '93 thru '11
- 72025 Murano '03 thru '10
- 72030 Pick-ups '80 thru '97, Pathfinder '87 thru '95
- 72031 Frontier Pick-up, Xterra, Pathfinder '96 thru '04
- 72032 Frontier & Xterra '05 thru '11
- 72040 Pulsar all models '83 thru '86
- 72050 Sentra all models '82 thru '94
- 72051 Sentra & 200SX all models '95 thru '06
- 72060 Stanza all models '82 thru '90
- 72070 Titan pick-ups '04 thru '10, Armada '05 thru '10

OLDSMOBILE
- 73015 Cutlass '74 thru '88
- For other OLDSMOBILE titles, see BUICK, CHEVROLET or GM listings.

PLYMOUTH
- For PLYMOUTH titles, see DODGE.

PONTIAC
- 79008 Fiero all models '84 thru '88
- 79018 Firebird V8 models except Turbo '70 thru '81
- 79019 Firebird all models '82 thru '92
- 79025 G6 all models '05 thru '09
- 79040 Mid-size Rear-wheel Drive '70 thru '87
- Vibe '03 thru '11 - see TOYOTA Matrix (92060)
- For other PONTIAC titles, see BUICK, CHEVROLET or GM listings.

PORSCHE
- 80020 911 Coupe & Targa models '65 thru '89
- 80025 914 all 4 cyl models '69 thru '76
- 80030 924 all models incl. Turbo '76 thru '82
- 80035 944 all models '83 thru '89

RENAULT
- Alliance, Encore - see AMC (14020)

SAAB
- 84010 900 including Turbo '79 thru '88

SATURN
- 87010 Saturn all S-series models '91 thru '02
- 87011 Saturn Ion '03 thru '07
- 87020 Saturn all L-series models '00 thru '04
- 87040 Saturn VUE '02 thru '07

SUBARU
- 89002 1100, 1300, 1400 & 1600 '71 thru '79
- 89003 1600 & 1800 2WD & 4WD '80 thru '94
- 89100 Legacy models '90 thru '99
- 89101 Legacy & Forester '00 thru '06

SUZUKI
- 90010 Samurai/Sidekick/Geo Tracker '86 thru '01

TOYOTA
- 92005 Camry all models '83 thru '91
- 92006 Camry all models '92 thru '96
- 92007 Camry/Avalon/Solara/Lexus ES 300 '97 thru '01
- 92008 Toyota Camry, Avalon and Solara & Lexus ES 300/330 all models '02 thru '06
- 92009 Camry '07 thru '11
- 92015 Celica Rear Wheel Drive '71 thru '85
- 92020 Celica Front Wheel Drive '86 thru '99
- 92025 Celica Supra all models '79 thru '92
- 92030 Corolla all models '75 thru '79
- 92032 Corolla rear wheel drive models '80 thru '87
- 92035 Corolla front wheel drive models '84 thru '92
- 92036 Corolla & Geo Prizm '93 thru '02
- 92037 Corolla models '03 thru '11
- 92040 Corolla Tercel all models '80 thru '82
- 92045 Corona all models '74 thru '82
- 92050 Cressida all models '78 thru '82
- 92055 Land Cruiser FJ40/43/45/55 '68 thru '82
- 92056 Land Cruiser FJ60/62/80/FZJ80 '80 thru '96
- 92060 Matrix & Pontiac Vibe '03 thru '11
- 92065 MR2 all models '85 thru '87
- 92070 Pick-up all models '69 thru '78
- 92075 Pick-up all models '79 thru '95
- 92076 Tacoma, 4Runner & T100 '93 thru '04
- 92077 Tacoma all models '05 thru '09
- 92078 Tundra '00 thru '06, Sequoia '01 thru '07
- 92079 4Runner all models '03 thru '09
- 92080 Previa all models '91 thru '95
- 92081 Prius '01 thru '08
- 92082 RAV4 all models '96 thru '10
- 92085 Tercel all models '87 thru '94
- 92090 Sienna all models '98 thru '09
- 92095 Highlander & Lexus RX-330 '99 thru '07

TRIUMPH
- 94007 Spitfire all models '62 thru '81
- 94010 TR7 all models '75 thru '81

VW
- 96008 Beetle & Karmann Ghia '54 thru '79
- 96009 New Beetle '98 thru '11
- 96016 Rabbit, Jetta, Scirocco, & Pick-up gas models '75 thru '92 & Convertible '80 thru '92
- 96017 Golf, GTI & Jetta '93 thru '98, Cabrio '95 thru 02
- 96018 Golf, GTI & Jetta '98 thru '05
- 96019 Jetta, Rabbit, GTI & Golf '05 thru '11
- 96020 Rabbit, Jetta, Pick-up diesel '77 thru '84
- 96023 Passat '98 thru '05, Audi A4 '96 thru '01
- 96030 Transporter 1600 all models '68 thru '79
- 96035 Transporter 1700, 1800, 2000 '72 thru '79
- 96040 Type 3 1500 & 1600 '63 thru '73
- 96045 Vanagon air-cooled models '80 thru '83

VOLVO
- 97010 120, 130 Series & 1800 Sports '61 thru '73
- 97015 140 Series all models '66 thru '74
- 97020 240 Series all models '76 thru '93
- 97040 740 & 760 Series all models '82 thru '88

TECHBOOK MANUALS
- 10205 Automotive Computer Codes
- 10206 OBD-II & Electronic Engine Management
- 10210 Automotive Emissions Control Manual
- 10215 Fuel Injection Manual, '78 thru '85
- 10220 Fuel Injection Manual, '86 thru '99
- 10225 Holley Carburetor Manual
- 10230 Rochester Carburetor Manual
- 10240 Weber/Zenith/Stromberg/SU Carburetor
- 10305 Chevrolet Engine Overhaul Manual
- 10310 Chrysler Engine Overhaul Manual
- 10320 Ford Engine Overhaul Manual
- 10330 GM and Ford Diesel Engine Repair
- 10333 Engine Performance Manual
- 10340 Small Engine Repair Manual
- 10345 Suspension, Steering & Driveline
- 10355 Ford Automatic Transmission Overhaul
- 10360 GM Automatic Transmission Overhaul
- 10405 Automotive Body Repair & Painting
- 10410 Automotive Brake Manual
- 10415 Automotive Detailing Manual
- 10420 Automotive Electrical Manual
- 10425 Automotive Heating & Air Conditioning
- 10430 Automotive Reference Dictionary
- 10435 Automotive Tools Manual
- 10440 Used Car Buying Guide
- 10445 Welding Manual
- 10450 ATV Basics
- 10452 Scooters 50cc to 250cc

SPANISH MANUALS
- 98903 Reparación de Carrocería & Pintura
- 98904 Manual de Carburador Spanish Holley & Rochester
- 98905 Códigos Automotrices de la Computadora
- 98906 OBD-II & Sistemas de Control Electrónico del Motor
- 98910 Frenos Automotriz
- 98913 Electricidad Automotriz
- 98915 Inyección de Combustible '86 al '99
- 99040 Chevrolet & GMC Camionetas '67 al '87
- 99041 Chevrolet & GMC Camionetas '88 al '98
- 99042 Chevrolet Camionetas Cerradas '68 al '95
- 99043 Chevrolet/GMC Camionetas '94 al '04
- 99048 Chevrolet/GMC Camionetas '99 al '06
- 99055 Dodge Caravan/Ply. Voyager '84 al '95
- 99075 Ford Camionetas y Bronco '80 al '94
- 99076 Ford F-150 '97 al '09
- 99077 Ford Camionetas Cerradas '69 al '91
- 99088 Ford Modelos de Tamaño Mediano '75 al '86
- 99089 Ford Ranger '93 al '10
- 99091 Ford Taurus & Mercury Sable '86 al '95
- 99095 GM Modelos de Tamaño Grande '70 al '90
- 99100 GM Modelos de Tamaño Mediano '70 al '88
- 99106 Jeep Cherokee, Wagoneer & Comanche '84 al '00
- 99110 Nissan Camionetas & Pathfinder '80 al '96
- 99118 Nissan Sentra '82 al '94
- 99125 Toyota Camionetas y 4-Runner '79 al '95

Over 100 Haynes motorcycle manuals also available

Haynes North America, Inc., 859 Lawrence Drive, Newbury Park, CA 91320 • (805) 498-6703 • http://www.haynes.com